Praise for *Profile of Western North America*

"*Profile of Western North America* paints a comprehensive picture of the people, places, and economies of the western United States, Canada, and Mexico, and how our nations are growing closer. A useful, forward-looking reference book."

—Senator Pete Domenici, New Mexico

"Dr. Philip M. Burgess and the Center for the New West have come up with an outstanding book. *Profile of Western North America* is a 'must.' Easy to read, with a wealth of sociological and statistical facts enlightening to policymakers, businesses, the educational world, and the people of our three nations, it not only portrays a 'continent ascendant,' but the common destiny of Canada, Mexico, and the United States. ... a guiding light."

—Fausto C. Miranda,
Founder and Senior Partner,
Miranda, Estavillo y Hernandez, Mexico City

"With extensive data and a unique way of focusing on a continental economic powerhouse, *Profile of Western North America* brings to life the opportunities and role that this region should play in the upcoming decades. The ever-accelerating modernization and growth of the Mexican and South American economies ... and western North America's position as the geographic and cultural gateway to Asia make this book a must read for domestic and foreign investors, company strategic planners, and state and provincial economic development leaders."

—Jeremy Davies,
Managing Partner,
Price Waterhouse, Los Angeles

"An enormous amount of important data that will be invaluable to those of us working on economic and international trade issues in the western states."

—Morgan Smith,
Director,
Colorado International Trade Office, Denver

"I love books with lots of understandable data. This is a great one."

—Ben Wattenberg,
American Enterprise Institute, Washington, D.C.,
and host of *Think Tank* on PBS

"A treat to use. *Profile of Western North America* is packed with information about the geography, people, and economies of the western United States, Canada, and Mexico, presented in easy-to-follow graphs, charts, and tables, with text that is interesting as well as informative."

—Ken Cole,
Vice President, Allied Signal Corp.,
and President, USA–NAFTA, Washington, D.C.

"*Profile of Western North America* is the most complete data compilation on North America. ... In determining where to establish operations, information is the biggest factor. This information is ideal for developing a business strategy. ... *Profile* provides the ideal format for a user-friendly document on demographics, market trends, and other pertinent data. It is a must for those who are just awakening to the prospects and opportunities that exist as a result of the new North American 'common market' called NAFTA."

—William V. Stephenson,
Executive, Arizona Public Service,
and Chair, Border Trade Alliance, Phoenix

PROFILE OF WESTERN NORTH AMERICA

Indicators of an Emerging Continental Market

Philip M. Burgess and Michael Kelly
Center for the New West

North American Press
A Division of Fulcrum Publishing
1995

Cover design by Robert Spooner Design

Library of Congress Cataloging-in-Publication Data

Burgess, Philip M.

Profile of western North America: indicators of an emerging continental market / Philip M. Burgess and Michael Kelly.

p. cm.

Includes bibliographical references and index.

ISBN 1-55591-907-3

1. West (U.S.)—Economic conditions—Statistics. 2. Canada—Economic conditions—1991—Statistics. 3. Mexico—Economic conditions—1982—Statistics. I. Kelly, Michael. II. Title.

HC95.B87 1994 94-31787

330.978'0021—dc20 CIP

Printed in the United States of America

0 9 8 7 6 5 4 3 2 1

North American Press
A Division of Fulcrum Publishing
350 Indiana Street, Suite 350
Golden, Colorado 80401-5093

TABLE OF CONTENTS

FOREWORD

BEYOND BORDERS

Co-chair, Western U.S. Senate Coalition, Dennis DeConcini (D-Ariz.)

Profile of Western North America is a snapshot of a continent vast in size, rich in natural resources and exceptional in the energy and diversity of its peoples.

Our task now is to bring the snapshot to life, to make it real for people and communities throughout our continent.

The historic ratification of the North American Free Trade Agreement (NAFTA) in November of 1993 gave political recognition to an important new economic reality: North America has a large, growing, and increasingly integrated continental economy.

Canada is the number one trading partner for the United States. Mexico currently is the United States' number three trading partner. But if trade between the United States and Mexico continues to grow at the rate it has grown since 1986, Mexico will move past Japan into second place before the end of the year. Both Canada and Mexico trade more with the United States than with any other country. Trade between Canada and Mexico is small but growing very rapidly.

The United States, Canada, and Mexico are more like partners than competitors in the integrated world economy of the 1990s. An example is the automobile industry. The "American" automobile industry is really a North American industry. GM, Ford, and Chrysler may be headquartered in Detroit, but value is added from the suburbs of Toronto to the suburbs of Chihuahua. Canada's principal exports and imports in 1993 were automobiles and automotive parts. Mexico's principal imports that year were automobiles and automotive parts, and, after oil, Mexico's principal exports also were automobiles and automotive parts.

The economic health of each North American nation is, in large part, dependent on the health of its continental partners. North American neighbors are the largest customers for each others' products and partners in building products sold in other parts of the world. As Benjamin Franklin said: "We must all hang together, or surely, we shall all hang separately."

Politics tend to trail behind social and economic reality. There was evidence of a continent knitting itself together long before there was political recognition of the merger. This is especially true along the U.S.–Mexico border. San Diego, California, and Tijuana, Mexico, have long been one metropolitan area, separated by artificial lines drawn on a map. The same is true for Nogales, Arizona, and Nogales, Mexico; El Paso, Texas, and Ciudad Juárez, Mexico; and for Laredo, Texas, and Nuevo Laredo, Mexico, and other border towns. Residents on both sides of the artificial lines share the same problems and have a stake in their common destiny.

In a world of trading blocs where people are concerned about protectionism and trade wars, NAFTA is a bold statement for free trade and international cooperation.

NAFTA is a major step toward a "win-win-win" situation for North America and for North America's role in expanded hemispheric cooperation. With one of the most liberal accession clauses, NAFTA is a model for "open" regionalism and a vehicle for the easy accession of the other nations of the Western Hemisphere.

NAFTA has removed some of the political obstacles to expanded continental trade, and with expanded trade will come greater understanding, respect, and mutual trust. But there is much more to be done. We need to lay down new infrastructure and new management regimes through which expanded trade can flow.

We must begin at the border.

We have an archaic understanding of borders. Traditionally, borders have been barriers, places to keep out unwanted people, ideas, and things. In the new world of the 1990s, *we must see borders more as gateways, avenues for the orderly interchange of people, ideas, and things*. In our mind's eye, the welcome mat must replace the barbed-wire fence when we think of borders.

Unfortunately, the spirit of NAFTA has not yet caught up with how we administer border crossings. Operations at the border still tend to delay rather than to facilitate trade and commerce. This deplorable fact is caricatured in this joke among truckers: *How long is it from Monterrey, Mexico, to the border? Three hours. How long is it from the border to San Antonio? Three hours. How long is it from Monterrey to San Antonio? Three days.*

Our problems at the border clearly require investments in new infrastructure, and solutions also require serious attention to problems of administration and management. A study conducted by the Center for the New West for the U.S. Department of Transportation found there was precious little coordination among the U.S. agencies that have border inspection responsibilities, much less between U.S. border officials and their Mexican and Canadian counterparts. No single government agency has overall responsibility for border management. Among those involved are the Customs Service, the Immigration and Naturalization Service, the Border Patrol, the General Services Administration, and the Animal and Plant Health Inspection Service of the Department of Agriculture. One important California crossing is, in effect, open for commercial traffic only a few hours a day because the U.S. and Mexican border officials work on different schedules.

But there are serious infrastructure problems as well, especially along the U.S.–Mexico border. With one major exception, border crossings between the United States and Canada in the West are in *rural* areas. The most important crossings between the United States and Mexico are in *urban* areas that have special problems urban congestion brings, compounding those caused by inadequate physical facilities and fragmented administration and management. Transport movements associated with international trade, commerce, and tourism must compete for access within a system that is already congested by local demands for urban transportation. More public and private investments are required if we are to maximize the potential for continental trade.

Beyond the border gateways, transportation networks in the United States and Canada are well developed. This is less true in Mexico. Moreover, while the flow of road and rail lines in the United States and Canada is predominantly east-west, the flow in Mexico is predominantly north-south, and there are relatively few connections between the U.S. and Mexican systems—thirty-seven crossings on a two-thousand-mile border. Western United States trade with Mexico is growing very rapidly. But existing transportation routes favor the shipment of goods to and from the U.S. Northeast and Midwest. Investment in Mexican infrastructure—especially in the new Pacific port of Topolobampo and the land routes leading from it to the U.S. border—may be among the most productive that U.S. entrepreneurs could make.

Another major challenge is pollution. Contaminants in the air and water are even less respectful of artificial lines drawn on a map than are movements of trade and people.

Increasing economic activity along the border means increasing environmental challenges. Because residents on both sides of the U.S.–Mexico border have the same stake in preserving air and water quality in the border regions, there must be intergovernmental cooperation for the health and safety of all.

Border policy will be among the most important challenges for all three North American nations—but especially for the United States and Mexico—in the 1990s. Transnational solutions must be found to transnational problems.

We are marching in the right direction. The Free Trade Agreement between the United States and Canada was an important first step. Ratification of NAFTA was a major advance.

The publication of *Profile of Western North America* also moves us forward. A solid source of pertinent information about Mexico and the western United States and Canada, presented in a form businessmen and women and government officials will find easy to use, *Profile of Western North America* helps us see our continent as an integrated whole rather than as three separate parts. This fresh perspective is especially helpful at this time, when the economic futures of all three countries so clearly depend on "hanging together." Knowledge is the key to understanding, and understanding is the key to trust and cooperation. *Profile of Western North America* is a cogent source of the knowledge we need to build our continent together.

INTRODUCTION

Three nations.

Two borders.

One market on one continent.

A shared destiny among many peoples.

North America is a core element of the New West—an area that includes the expanding new industrial economies of Asia; the newly emerging economies of China and the Russian Far East; and the rapidly expanding economies of Chile and other market-oriented Latin American nations that border the Pacific. North America is an ascendant continent in an exciting new world bound together by electrons. And western North America is leading the ascent.

Vast in size, rich in natural resources, spectacular in beauty, the North American West has long been the focus of the hopes and dreams of peoples around the world, and as the twentieth century draws to a close, those dreams are coming true.

The North American West is poised to lead the world into a new century of global economic interdependence and cooperation. Its geography and ethnic variety place it in an ideal position to capitalize on expanding world trade—especially with the Asian nations on the Pacific Rim, the fastest-growing economies in the world.

Tourism is the world's largest industry, and the West's natural beauty provides a magnet for tourists from all over the world. The West contains more than 90 percent of the national parkland in the United States. The Canadian West contains nearly 97 percent of Canada's national parkland. In the western United States and Canada together, there are 137,817 square miles (356,946 sq. km) of parkland, an area nearly as large as Montana. Mexico boasts world-class resorts on both coasts, in particular Los Cabos, Mazatlán, Puerto Vallarta, Ixtapa, Acapulco, Huatulco, Cancún, and Cozumel. In 1990, Mexico ranked ninth in the world in hotel supply and eleventh in tourist revenues.

The mineral and timber resources of the West are as sought after by firms across the Pacific as by firms east of the Mississippi and the Great Lakes. Japan's construction industry depends chiefly upon timber from British Columbia, Washington, and Oregon.

The West has a higher proportion of the high-technology industries (computers, software, semiconductors, aerospace, biotechnology) likely to be the chief source of jobs and profits in the Information Age, and a lower proportion of the older manufacturing concerns (steel, automobiles, textiles) that are increasingly buffeted by lower-cost competition from the emerging nations of the Pacific Rim. Of twenty-two computer manufacturers in the *Business Week* 1000 for 1993, fourteen are located in the West. Of thirty-one computer software firms in the *Business Week* 1000, eighteen are in the West. Of fourteen semiconductor manufacturers, eleven are in the West.

But the greatest contribution the North American West can make to the world isn't a product of its natural beauty, natural resources, or the imagination and industry of its peoples. The greatest contribution is an example of how peoples divided by race, religion, language, ethnic origin, and national boundaries can work together in peace and harmony for the betterment of all. For the rest of the world, the North American West is as much a state of mind as a physical place. The North American West has become synonymous with individual liberty and economic opportunity.

The Challenge

The West is at a crossroads. Fundamental changes in world markets—including the changing structures of supply (e.g., new technologies of discovery, production, and processing) and the changing structure of demand (e.g., the impact of conservation and materials substitution) for energy, minerals, agriculture, and timber—indicate the boom-and-bust economy that has characterized much of the U.S. and Canadian West will continue unless there is more aggressive economic diversification.

Other mainstays of the West's economy—including its expanding high-technology, manufacturing, and service sectors—are facing serious and growing competition from abroad. Global trends such as the growth in international trade, the free flow of capital, the formation of regional trading blocs, state-directed competition, and targeted marketing by foreign competitors mean that businesses must change the way they operate. Governments and businesses must also change the way they deal with each other and with the outside world.

Mexico, as it prepares its leap into the developed world and joins the world's leading economies, faces special challenges. The greatest of these are how to provide jobs, food, education, medical care, and adequate shelter to one of the world's youngest populations, a population that in the 1980s was growing faster than the economy.

Other challenges face Mexico as it strives to become an equal partner in forging North America's destiny. Mexico must come up with ways to strengthen its democratic institutions and deal with urban problems—including air and water pollution and problems of governance in densely populated metropolitan areas.

Economic development in the West is hampered by the fact that many critical decisions affecting the region are made elsewhere. In some ways, the situation in the West today is analogous to that of the thirteen original colonies when they were still ruled by Great Britain.

In both the United States and Canada, the bulk of the natural resources and the growth industries are in the West, but the bulk of the political power is in the East. And Mexicans have long complained that important decisions affecting them have been made not in another part of their own country, but in another country. The looming shadow of the United States has darkened much of Mexican history. U.S. leaders must become aware that domestic decisions have continental implications.

The vast distances in the West make the region much more dependent on federal spending for infrastructure (roads, bridges, highways, airports) to knit markets together than is the case for states in the East. In both the United States and Canada, federal agencies, with control over environmental policies, water development, native peoples, borders, ocean and coastal management, and public lands, make decisions that have far-reaching consequences for the political stability, economic competitiveness, and growth potential of the western economy.

In the private sector, the influence of outside forces looms large. Most major money centers and corporate headquarters are based outside the region. Only 147 companies of the *Fortune* 500 for 1993, and only 181 companies of the *Business Week* 1000 are headquartered in the West. Most of these (41 of the *Fortune* 500, 117 of the *Business Week* 1000) are in California, the world's seventh largest "nation." Although Calgary has the second largest grouping of Canadian corporate headquarters (after Toronto), and Vancouver is fourth (after Montréal), none of the top ten firms in the Canada 1000 listing compiled by the *Toronto Globe and Mail,* and only two of the top twenty-five, are headquartered in the West. Too much of the West is largely a "branch plant–subsidiary" region—highly vulnerable to restructuring decisions made elsewhere; in too many cases, the West is powerless to influence the negative impacts of mergers and acquisition decisions. Perhaps most importantly, much of the West lacks access to the talent and leadership resources that a headquarters operation injects into a community and its surrounding region.

Government and business are more centralized in Mexico than in either the United States or Canada. If Toronto is to Canada what New York, Chicago, and Los Angeles combined are to the United States, then Mexico City is to Mexico what New York, Chicago, Los Angeles, *and* Boston, Atlanta, Houston, Dallas, St. Louis, Minneapolis, Milwaukee, Detroit, Cleveland, and Washington, D.C., are to the United States. Mexico City is the seat of a federal government vastly more powerful in relation to state and local governments than are the federal governments of Canada or the United States, and the city is headquarters to a federal bureaucracy proportionately larger than those of Mexico's North American neighbors. A majority of major business corporations in Mexico are also headquartered there. It was said in the ancient world that "all roads lead to Rome." In contemporary Mexico, virtually all roads lead to Mexico City. There is little direct highway and rail communication among other regions of the country. However, a major consequence of the Information Age is to push important decision-making downward and outward.

If the West wants to realize its promise, it must reorganize itself institutionally. Its great coastal cities must forge stronger ties to the hinterland—to the states and cities that lie along the Rocky Mountain spine and to those further east. Western leaders must develop the institutional capacity that will permit them to identify and mobilize the region's resources (natural and human), anticipate change, act strategically, and unite the region's civic leadership behind sound strategies and effective action. They must coordinate these activities within the region, in Washington and in Ottawa, in national and offshore capital markets, and with major trading partners overseas, particularly those along the Pacific Rim.

The western United States and Canada need to look east less, and north and south more. For often, westerners have more in common with each other than with their countrymen in the East.

Our destinies have long been joined. Roughly one-sixth of all Canadians are descended from Americans who moved north from the thirteen original colonies after the American Revolution. Nearly half of what is now the Western United States was once part of Mexico.

The visionaries who formed thirteen separate British colonies into what was to become a single independent nation understood that the disparate societies stretching from Boston to South Carolina had more in common than they had separating them, and that each could realize its destiny only if all worked together. "We must all hang together," Benjamin Franklin said, "or surely we shall all hang separately." In the frequently cutthroat global economic competition of today, what Benjamin Franklin said of the thirteen colonies is increasingly true for all of North America.

The economic integration of North America is well advanced. Witness the network among most major privately owned corporations in Canada, Mexico, and the United States. It is time for social and political institutions to catch up.

The most significant development in international trade is the formation of huge regional trading blocs and the consequent diminution of the economic influence of the nation-state and the international economic institutions established in the wake of World War II (a development discussed in detail in Appendix One). Artificial barriers to trade and commerce like those currently existing between the United States, Canada, and Mexico (although much diminished by NAFTA) once existed among the thirteen original U.S. states. One of the smartest things the Founding Fathers ever did was to abolish them. The results were an explosion of trade and greater prosperity for all. We need leaders with the vision to realize that trade barriers between North American neighbors make no more sense than trade barriers between New York and South Carolina.

Many of those leaders can be found in the West, where "sub-continental regionalism" is on the rise. The Pacific Northwest Economic Region, also called Cascadia, links the states of Washington, Oregon, Idaho, and Montana with the Canadian provinces of Alberta and British Columbia. The Red River Trade Corridor unites Minnesota and North Dakota with the Canadian province of Manitoba. The Border Trade Alliance links business, government, and civic associations on both sides of the Mexican border.

Increasingly, regions of countries are discovering their economic futures are more closely tied to similar regions in neighboring countries than to other parts of their own country. And many industries are finding that their operations are increasingly integrated with subsidiaries or joint-venture partners across a national border.

This is especially true in the West, where people in Seattle and Portland are learning that their economic well-being is more dependent on what is happening in Vancouver than on what is going on in Philadelphia or Miami. Texas, New Mexico, Arizona, and California are learning there is profit to be made by expanding and deepening historic relationships with the northern states of Mexico. The West can lead all of North America away from an obsolescent Eurocentric view of trade and commerce to the enormous possibilities within this continent and hemisphere, and across the Pacific.

The West is young—geologically, historically, and demographically. We must build the leadership and institutions that can expand choices, increase confidence, and give westerners the means to take charge and make things work for our region and our country.

Profile of Western North America is designed to be a key reference tool for civic, business, government, education and community leaders who are directing the New West toward bold horizons in a changing global marketplace. Designed as a serial publication of the region's key social, economic, environmental, and demographic indicators, *Profile of Western North America* strives to impart accurate, comprehensive, timely, and accessible information.

This volume contains nearly 350 indicators of demographics, economics, finance, business, natural resources, trade, commerce, technology, health, education, and welfare for the North American West—the twenty-three U.S. states west of the Mississippi (minus Louisiana); the ten provinces and two territories of Canada (including highlights of the four western provinces—British Columbia, Alberta, Saskatchewan, and Manitoba—and the two territories, both in the West—The Yukon and Northwest Territories); and the thirty-one states of Mexico plus the Distrito Federal. The data are presented in a fashion that highlights the interdependence of western North America. It supplements conventional analytical data (e.g., population, income, education) with useful institutional data (e.g., *Fortune* 500 companies, trade organizations, chambers of commerce).

Many people contributed to the design, review, and production of this volume. First and foremost, we are indebted to foundation and corporate sponsors of the Center for the New West, whose financial support makes possible this kind of research and publication—Allied Signal, America West Airlines, Arizona Public Service, Banc One, Coopers and Lybrand, Cyprus Climax Metals, El Pomar Foundation, Goldman Sachs, GTE Telephone Operations, MAGMA, M. A. Mortenson (construction), Motorola, Phelps Dodge, The Dial Corporation, Tucson Electric Power Company, US West, Inc. (communications), Western Fuels Assocation, and others.

Major credit goes to Center associate Doug Linkhart for assembling and verifying the data presented in this book from sources in the U.S., Mexico, and Canada, and to Sam Scinta, who assisted in data gathering and conversion, verified the data, and turned the data into graphics.

This volume also reflects many of the lessons learned from the Center's 1993 study of North American transportation corridors, including the border gateways with Canada and Mexico. Entitled *Making Things Work: Transportation and Trade Expansion in Western North America*, the project's final report includes seven volumes submitted to the Federal Highway Administration and available through the National Technical Information Service (NTIS).

We are particularly grateful to project codirectors and Center Senior Fellows Lou Higgs and Paulette Hansen and to the project study team, especially Rick Donnelly, Barton-Aschman in San Francisco; Boris Kozolchyk, University of Arizona's National Law Center for Inter-American Free Trade; John McCray, McCray Research in San Antonio; Gerry Schwebel, Border Trade Alliance; Lee McPheters, College of Business at Arizona State University; Joedy Cambridge, Cambridge and Campbell, Inc., in Alexandria, Virginia; Larry Swanson, University of Montana; Ted Chambers, University of Alberta; Jerry Nagel, Red River Trade Corridor in Crookston, Minnesota; David Albright, Alliance for Transportation Research in Albuquerque, New Mexico; John Hunt, University of Idaho; William Beyers, University of Washington; and many others.

We are also grateful for the many comments and suggestions of other colleagues in Canada and Mexico who reviewed draft copies as the book was being prepared—and for the comments and suggestions of our Center colleagues, especially Andy Bane, Roberta Bhasin, Kent Briggs, Colleen Murphy, and Bob Wurmstedt—and to Senior Fellows Joel Kotkin, author and commentator on California, and Sergio Muñoz, former editor of *La Opinion*, who have helped us learn about and appreciate the growing interdependence of people and business in North America.

Finally, we appreciate the continuing interest of U.S. Senator Dennis DeConcini in our efforts to advance North American cooperation and his own leadership of initiatives to improve both the management systems and infrastructure conditions on both the Canadian and Mexican borders of the United States. We are also appreciative of the support of Fulcrum publisher Robert Baron, who shares and encourages our desire to heighten awareness of the potential of North America among the continent's civic leaders. We are especially grateful to Fulcrum editor Shirley Lambert for her professional assistance in the design, research, and production stages of this effort. Fulcrum Publishing, an international publishing house headquartered in the Rocky Mountain West, is a rapidly growing institution that gives life to the growing diversity defining the New West of North America.

Despite our best efforts, this volume undoubtedly contains errors of fact and judgement as well as other deficiencies. Accordingly, we invite readers and users to help us identify problems in the current volume, and we invite suggestions for ways we can improve the utility of future editions. Although the view is still fuzzy owing to problems of availability and comparability of data—especially the lack of common metrics and common definitional standards among the three countries—we have learned from this exercise and hope this volume will be a useful contribution to others who are trying to bring the picture of Western North America gradually into focus.

Profile of Western North America will be successful if it strengthens the capacity of all of us in the North American West to better understand ourselves and each other, the continent we share, and its relationship to the larger world. Out of that understanding will come the vision and the will to strengthen our institutions, make our case in national and international forums, and achieve higher levels of cooperation between the public and private sectors and among our diverse peoples.

Philip M. Burgess
Michael Kelly
Denver, Colorado

CONVERSION TABLE

	U.S.	Metric	Metric	U.S.
Linear Measure	1 foot	0.31 meters	1 meter	3.28 feet
	1 mile	1,609.3 meters	1 kilometer	0.62 miles
		1.61 kilometers		
Area	1 sq. foot	0.09 sq. meters	1 sq. meter	10.76 sq. feet
	1 sq. mile	2.59 sq. kilometers	1 sq. kilometer	0.39 sq. miles
		259 hectares		247.11 acres
	1 acre	0.41 hectares	1 hectare	2.47 acres
		4,046.86 square meters		.004 sq. miles
Volume and Capacity	1 gallon	3.79 liters	1 liter	0.26 gallons
			1 cubic meter	6.29 barrels (oil)
Weight	1 pound	0.45 kilograms	1 kilogram	2.21 pounds
	1 short ton	907.18 kilograms	1 metric ton	1.10 short tons
		0.91 metric tons		0.98 long tons
	1 long ton	1,016 kilograms		2,204.6 pounds
		1.02 metric tons		
Temperature	Fahrenheit	(Celsius*9/5)+32	Celsius	(Fahrenheit-32)*5/9
		Canadian Dollars Per U.S. Dollar		**Mexican New Per U.S. Dollar**
Currency Exchange	1990	1.17	1990	2.95
	1991	1.15	1991	3.08
	1992	1.21	1992	3.12
	1993	1.29	1993	3.11
	6/30/94	1.38	6/30/94	3.39

Chapter One
Basic Resources

Land and Climate

North America is vast. Canada is the world's second largest nation in area, the United States is fourth, and Mexico is thirteenth. Canada and the United States together are about the size of Russia. The United States is slightly smaller than China.

Geographically, the West dominates North America. The four western provinces and two territories of Canada, the twenty-three U.S. states west of the Mississippi, and the thirty-one states of Mexico comprise 5.8 million square miles (15 million sq. km)—73 percent of the total land area of North America.

Within the West, Mexico has the smallest land area—761,604 square miles (approximately 1.97 million sq. km), 13 percent of the North American West, 9 percent of North America.

The western United States, with 2.6 million square miles (6.7 million sq. km), is the largest part of North America, slightly larger than Western Canada (2.5 million square miles [6.3 million sq. km]). The western United States comprises almost three quarters of the land area of the United States. Western Canada accounts for 70 percent of the land area of Canada.

Figure 1.1 **North America***

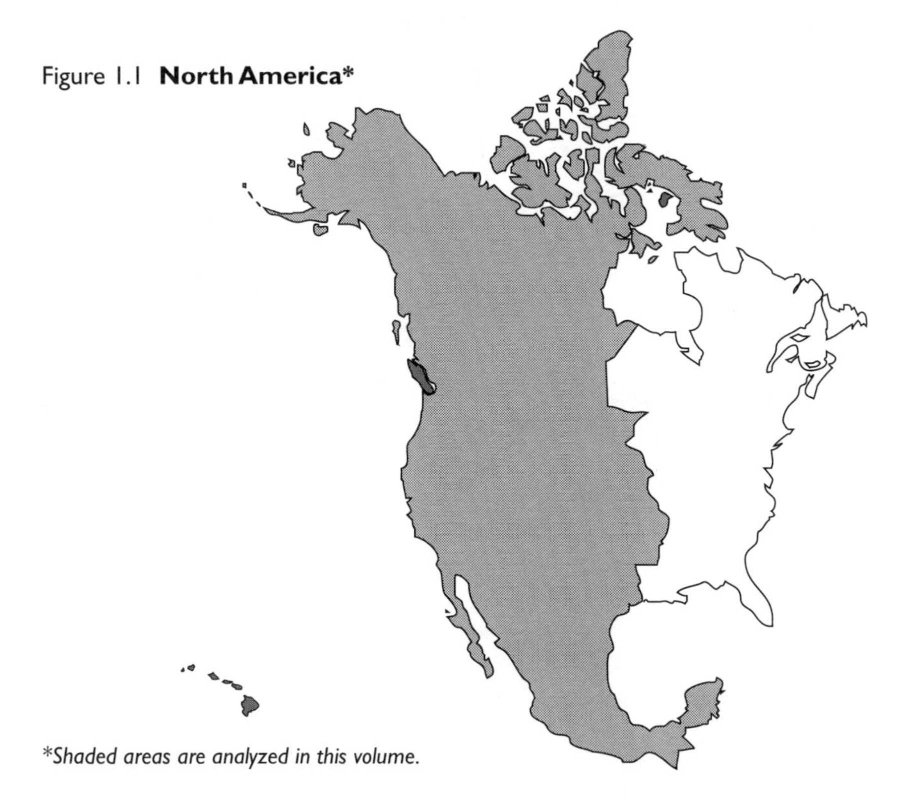

**Shaded areas are analyzed in this volume.*

Figure 1.2a **United States: The Western States**

Figure 1.2b **Canada**

Figure 1.2c **Mexico**

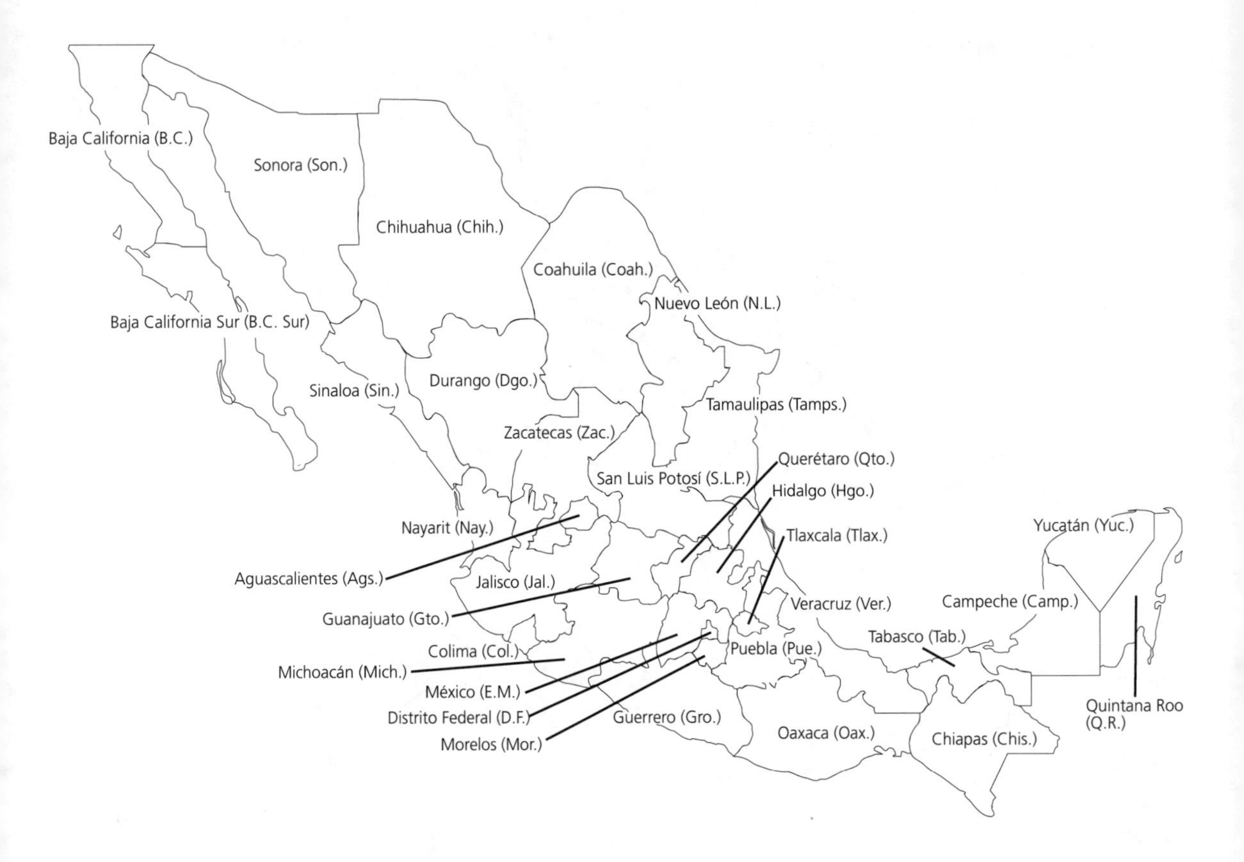

Each western Canadian province is larger than any of the U.S. States except Alaska and Texas. British Columbia is larger than Texas and Oklahoma combined. The largest Mexican State, Chihuahua, is about the size of Oregon or Wyoming.

Figure 1.3 **Western United States Area of States, 1990**

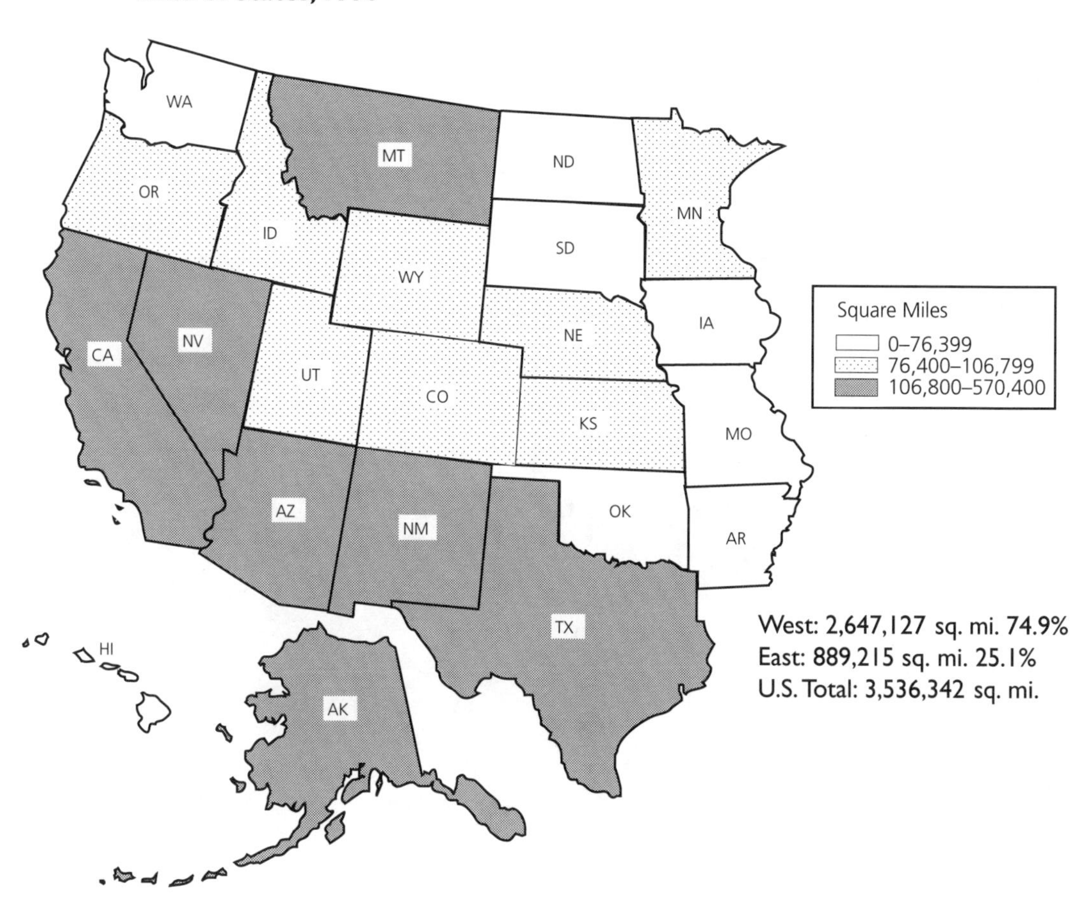

Source: U.S. Census Bureau

	Square Miles	Percent of Total U.S. Land Area		Square Miles	Percent of Total U.S. Land Area
Alaska	570,374	16.1%	**Kansas**	81,823	2.3%
Texas	261,914	7.4%	**Minnesota**	79,617	2.3%
California	155,973	4.4%	**Nebraska**	76,878	2.2%
Montana	145,556	4.1%	**South Dakota**	75,898	2.2%
New Mexico	121,365	3.4%	**North Dakota**	68,994	2.0%
Arizona	113,642	3.2%	**Missouri**	68,898	2.0%
Nevada	109,806	3.1%	**Oklahoma**	68,679	1.9%
Colorado	103,730	2.9%	**Washington**	66,582	1.9%
Wyoming	97,105	2.8%	**Arkansas**	57,075	1.6%
Oregon	96,003	2.7%	**Iowa**	55,875	1.6%
Idaho	82,751	2.3%	**Hawaii**	6,423	0.2%
Utah	82,168	2.3%			

Figure 1.4 **Canada**
Area of Provinces, 1990

Source: Statistics Canada, Canada Yearbook

	Square Miles	Percent of Total Canada Land Area
Northwest Territories	1,322,903	34.4%
Québec	594,857	15.5%
Ontario	412,579	10.7%
British Columbia	365,946	9.5%
Alberta	255,286	6.6%
Saskatchewan	251,865	6.5%
Manitoba	250,946	6.5%
Yukon	186,660	4.8%
Newfoundland	156,649	4.1%
New Brunswick	28,355	0.7%
Nova Scotia	21,425	0.6%
Prince Edward Island	2,185	0.1%

Figure 1.5 **Mexico**
Area of States, 1990

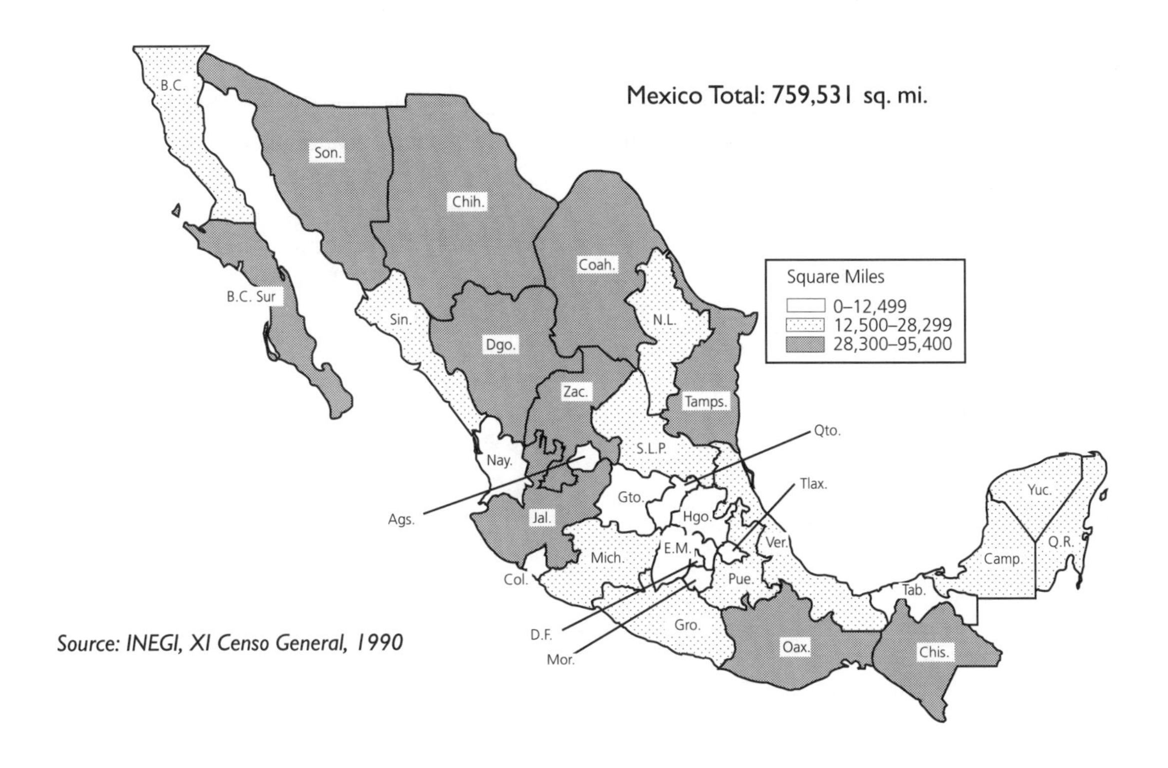

Source: INEGI, XI Censo General, 1990

	Square Miles	Percent of Total Mexico Land Area
Chihuahua	95,400	12.6%
Sonora	71,403	9.4%
Coahuila	58,522	7.7%
Durango	46,196	6.1%
Oaxaca	36,820	4.8%
Jalisco	30,941	4.1%
Tamaulipas	30,822	4.1%
Zacatecas	28,973	3.8%
Chiapas	28,528	3.8%
Baja California Sur	28,447	3.7%
Veracruz	28,114	3.7%
Baja California	27,071	3.6%
Nuevo León	24,925	3.3%
Guerrero	24,631	3.2%
San Luis Potosí	24,266	3.2%
Michoacán	23,114	3.0%
Sinaloa	22,429	3.0%
Campeche	20,013	2.6%
Quintana Roo	19,440	2.6%
Yucatán	15,189	2.0%
Puebla	13,096	1.7%
Guanajuato	11,810	1.6%
Nayarit	10,664	1.4%
Tabasco	9,522	1.3%
México	8,286	1.1%
Hidalgo	8,103	1.1%
Querétaro	4,544	0.6%
Aguascalientes	2,158	0.3%
Colima	2,106	0.3%
Morelos	1,908	0.3%
Tlaxcala	1,511	0.2%
Distrito Federal	579	0.1%

North America stretches from the Arctic to the tropics, but most of the continent is located in the temperate zone.

Because of its considerable north-south extension and variations in altitude, Mexico experiences great variation in climate. The Tropic of Cancer, which demarcates the northern edge of the tropics, bisects the country just north of Mazatlán.

Climatically, Mexico is divided vertically: The *tierra caliente* (hot zone) runs from sea level to approximately 2,000 feet (610 m) in altitude. It includes all of the coastal plains, the Yucatán peninsula, Baja California, the Isthmus of Tehuantepec, and the north-central portion of the Mesa del Norte. The *tierra templada* (temperate zone) includes the highland valleys that range from about 3,000 to 6,000 feet (915–1,830 m) in altitude. The *tierra fria* (cold zone) begins above 6,000 feet (1,830 m). Summers are rarely hot here, and frosts during winter nights are not uncommon.

The warmest places in North America are in Death Valley in southern California and in the deserts of Baja California, Sonora, and Chihuahua, where summer temperatures frequently exceed 110° F (43° C). The coldest portions of the continent are the Yukon, the Arctic islands of Alaska, and the Northwest Territories and along Hudson Bay in Canada, where winter temperatures often fall below -50° F (-45° C).

Figure 1.6 **Western United States**
Mean Daily Temperatures, January and July, 1961–1990

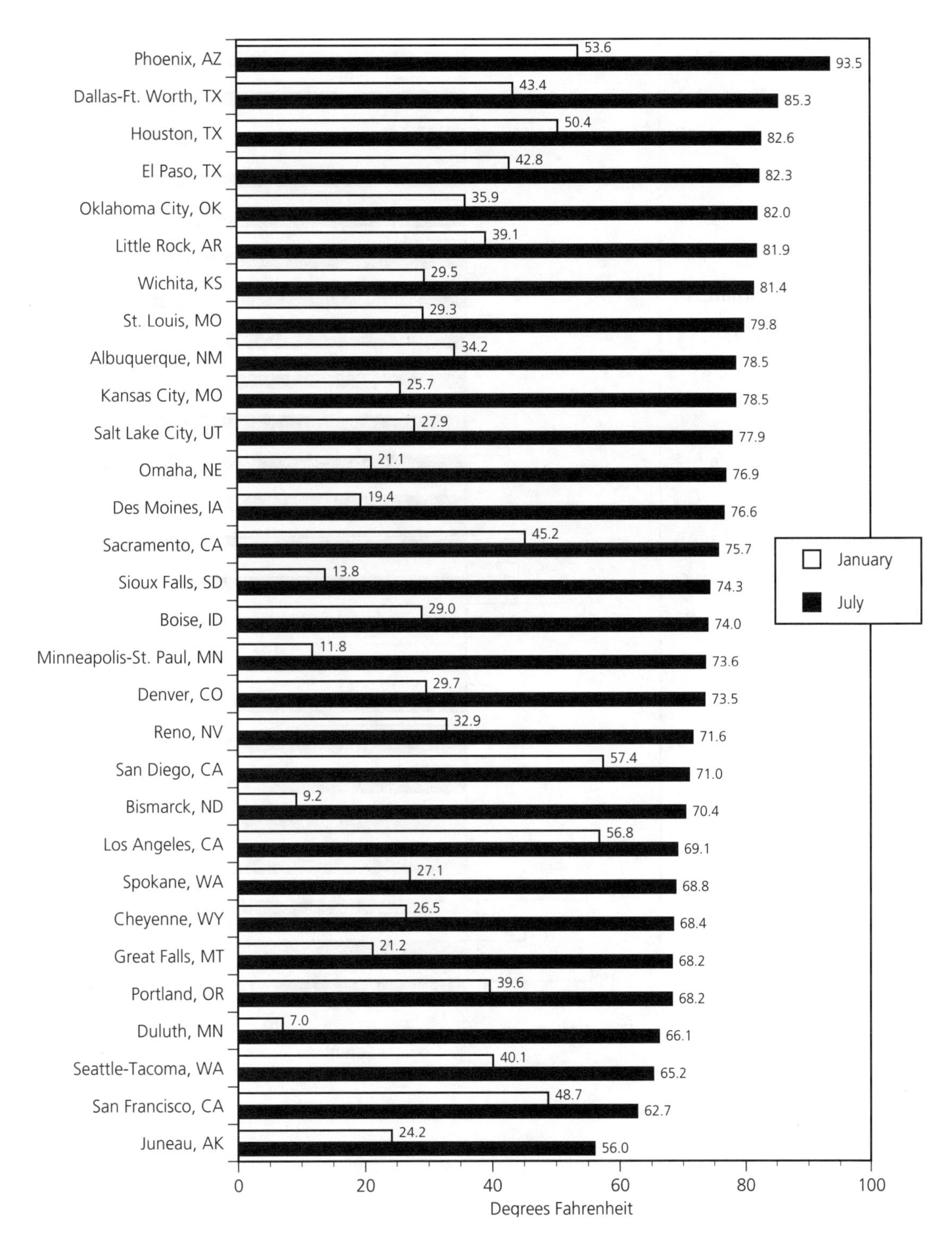

Source: U.S. National Oceanic and Atmospheric Administration

Figure 1.7 **Canada**
Mean Daily Temperatures, January and July

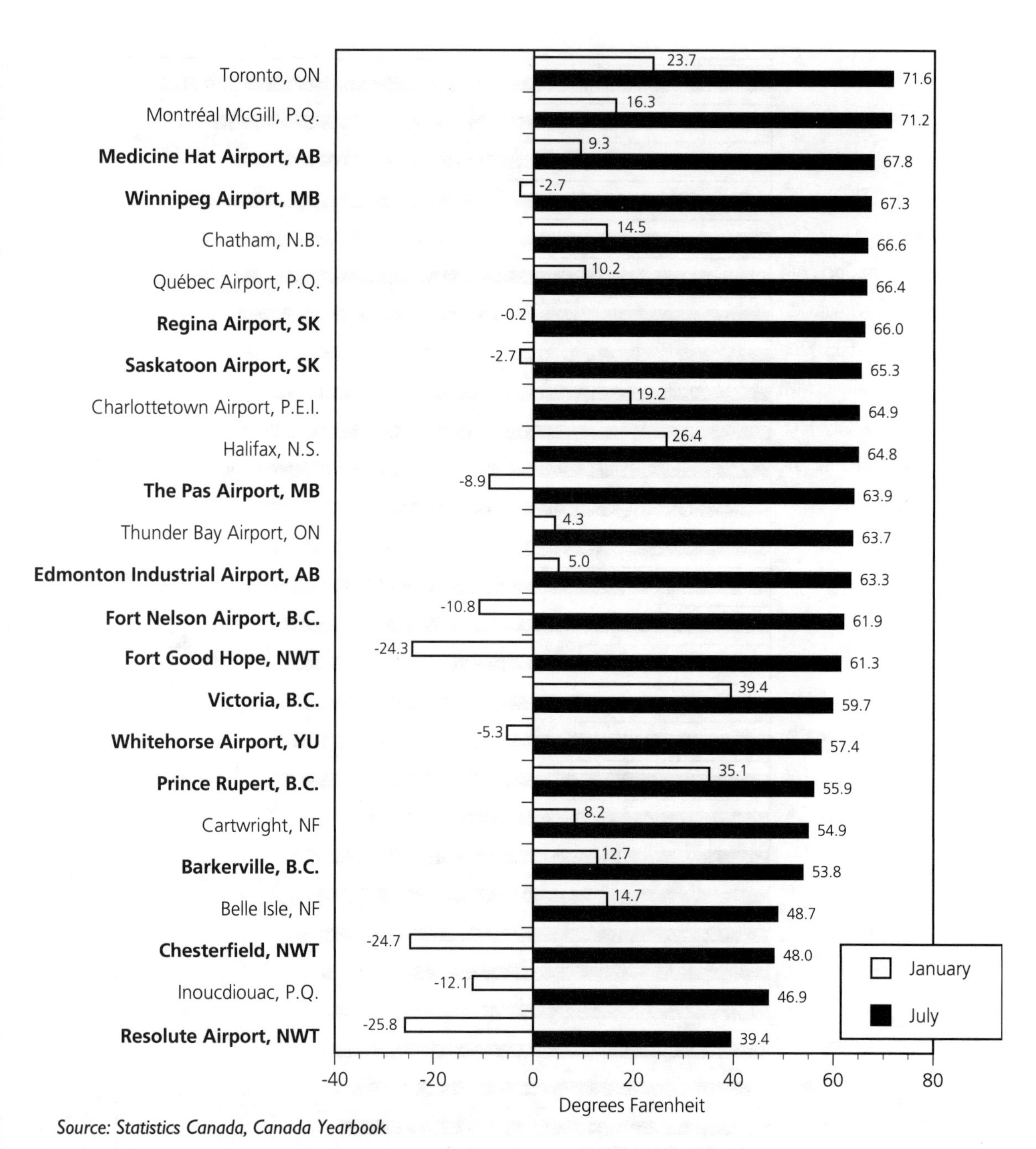

Source: Statistics Canada, Canada Yearbook

Figure 1.8 **Mexico**
Mean Annual Temperatures, 1982–1983

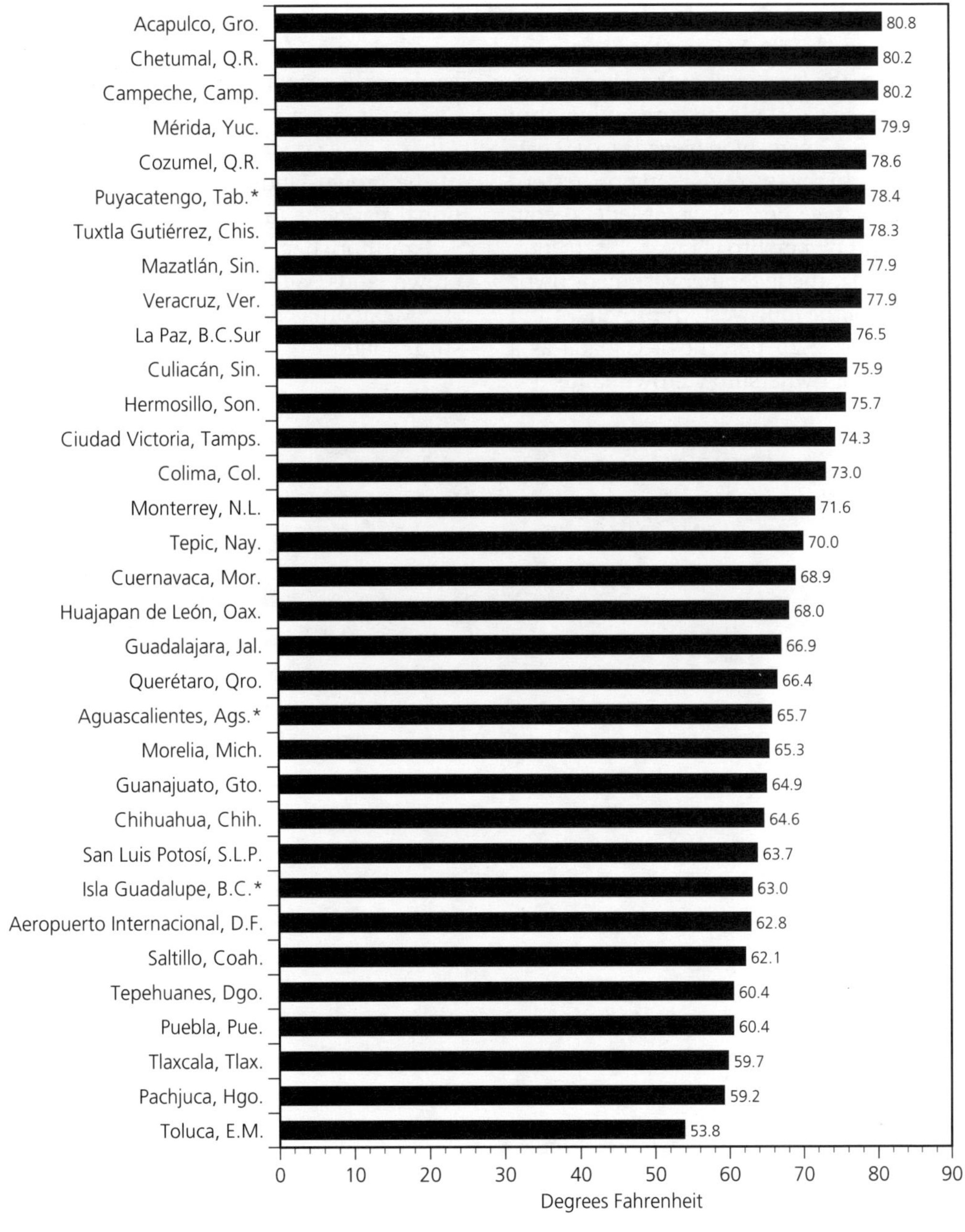

Source: INEGI, Anuario Estadistico

Government Ownership of Land

The biggest landowners in North America are the federal governments of the United States and Canada. The U.S. government owns 37 percent of the land west of the Mississippi. The percentage of federal land ownership ranges from 82.7 percent in Nevada to 0.9 percent in Iowa. In the twenty-seven nonwestern states, the average percentage of federal government land ownership is just 5.1 percent.

In Canada, the federal government owns 3.1 percent of the four western provinces and two territories, 2.25 percent of Canada as a whole. Federal ownership ("crown lands") is higher in the West because most of Canada's national parkland is in the West. Of 22.5 million hectares (55.5 million acres) of crown lands in the provinces, 11.45 million hectares (28 million acres) are in British Columbia, Alberta, Saskatchewan, and Manitoba. Most of the crown lands in the provinces are national parks or military bases. Alberta, with 7.3 million hectares (18 million acres) of crown lands, is the leader.

Mexico has discontinued federal ownership of land.

Figure 1.9 **Western United States**
Land Owned by the Federal Government, 1989
(by Percent of State and Acreage)*

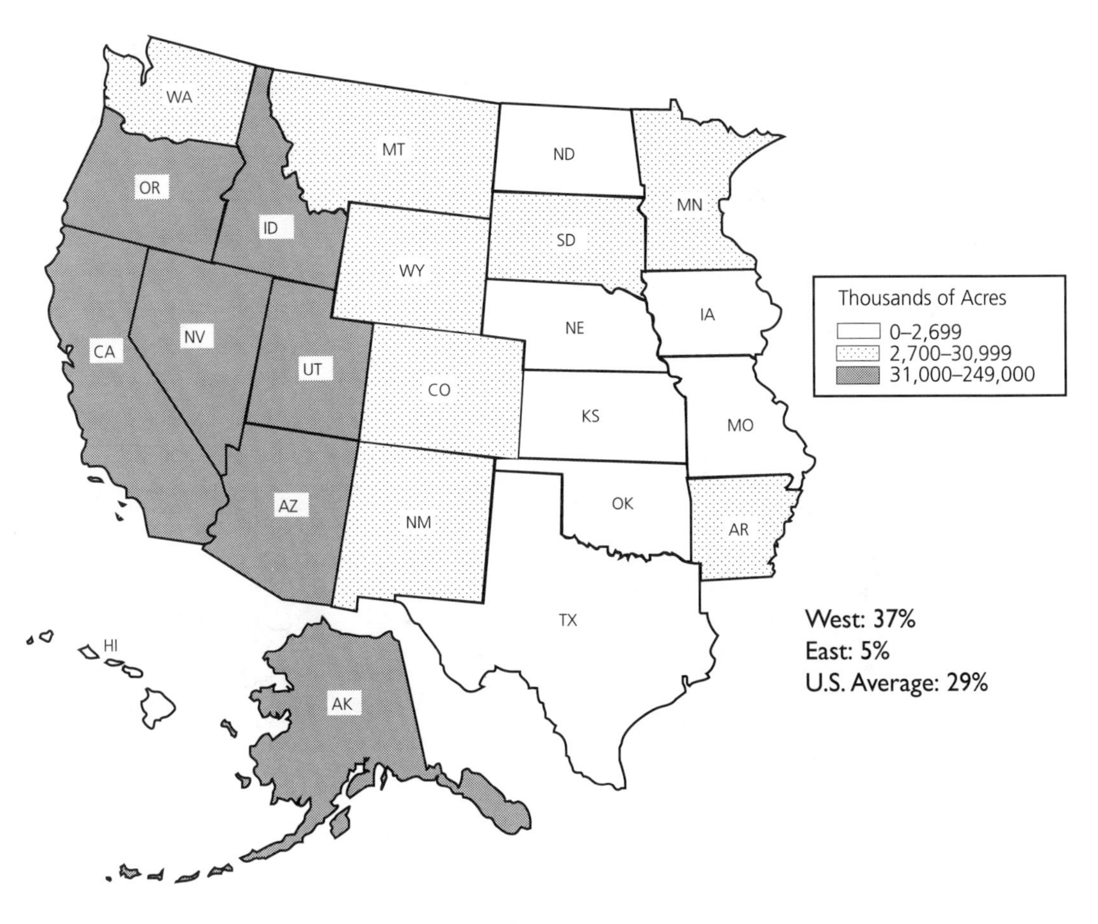

(* Does not include freshwater area.)

Source: U.S. General Services Administration

	Thousands of Acres	Percent of Total Land Area		Thousands of Acres	Percent of Total Land Area
Alaska	248,040	67.9%	**Minnesota**	5,365	10.5%
Nevada	58,135	82.7%	**South Dakota**	2,807	5.7%
California	44,541	44.4%	**Arkansas**	2,762	8.2%
Arizona	34,236	47.1%	**Texas**	2,247	1.3%
Utah	33,620	63.8%	**Missouri**	2,093	4.7%
Idaho	32,672	61.7%	**North Dakota**	1,879	4.2%
Oregon	32,289	52.4%	**Nebraska**	710	1.4%
Wyoming	30,416	48.8%	**Oklahoma**	704	1.6%
Montana	26,143	28.0%	**Hawaii**	637	15.5%
New Mexico	25,729	33.1%	**Kansas**	422	0.8%
Colorado	24,069	36.2%	**Iowa**	337	0.9%
Washington	12,374	29.0%			

Figure 1.10 **Canada**
Land Owned and Leased by the Federal Government, June 1992

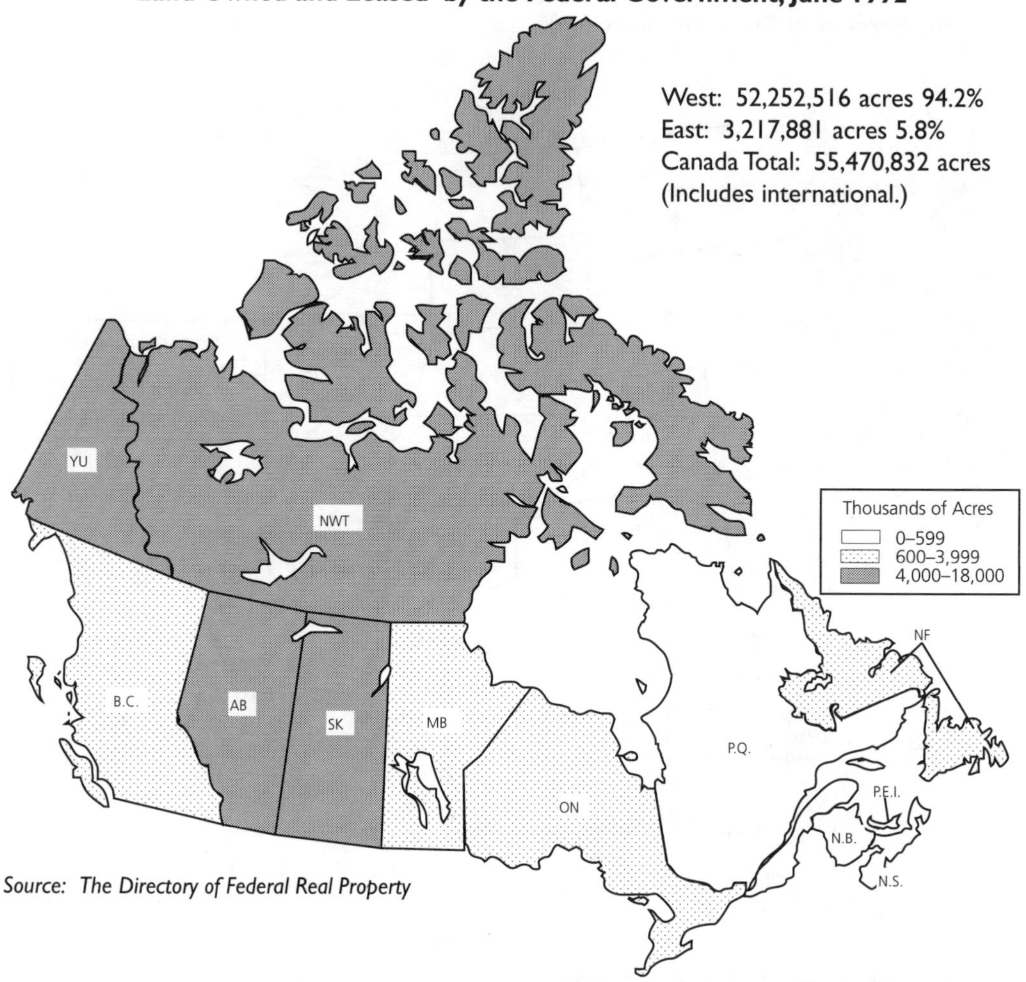

Source: The Directory of Federal Real Property

	Thousands of Acres	Percent of Total Land Area
Alberta	17,980	11.3%
Northwest Territories	16,003	2.0%
Yukon	7,974	6.7%
Saskatchewan	4,649	3.3%
Manitoba	3,774	2.8%
British Columbia	1,894	0.8%
Ontario	1,022	0.5%
Newfoundland	700	0.8%
New Brunswick	561	3.1%
Québec	518	0.2%
Nova Scotia	410	3.1%
Prince Edward Island	10	0.7%

Forests

Forest land in North America is concentrated in the Northwest, from Alaska, the Yukon, and the Northwest Territories through Washington, Oregon, Idaho, western Montana, and northern California. Most of the forest land in Mexico is in the far south, in the rainforest of the Yucatán Peninsula, and in the Tarahumara Sierra in the northern state of Chihuahua.

Figure 1.11 **Western United States Forest Land by State, 1987**

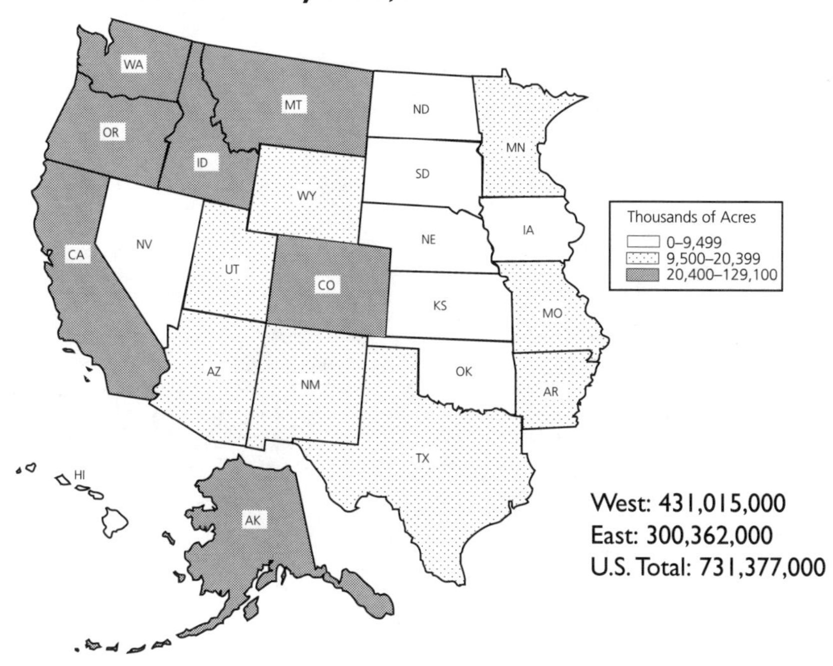

Source: U.S. Forest Service

Figure 1.12 **Canada**
Forest Land by Province, 1991

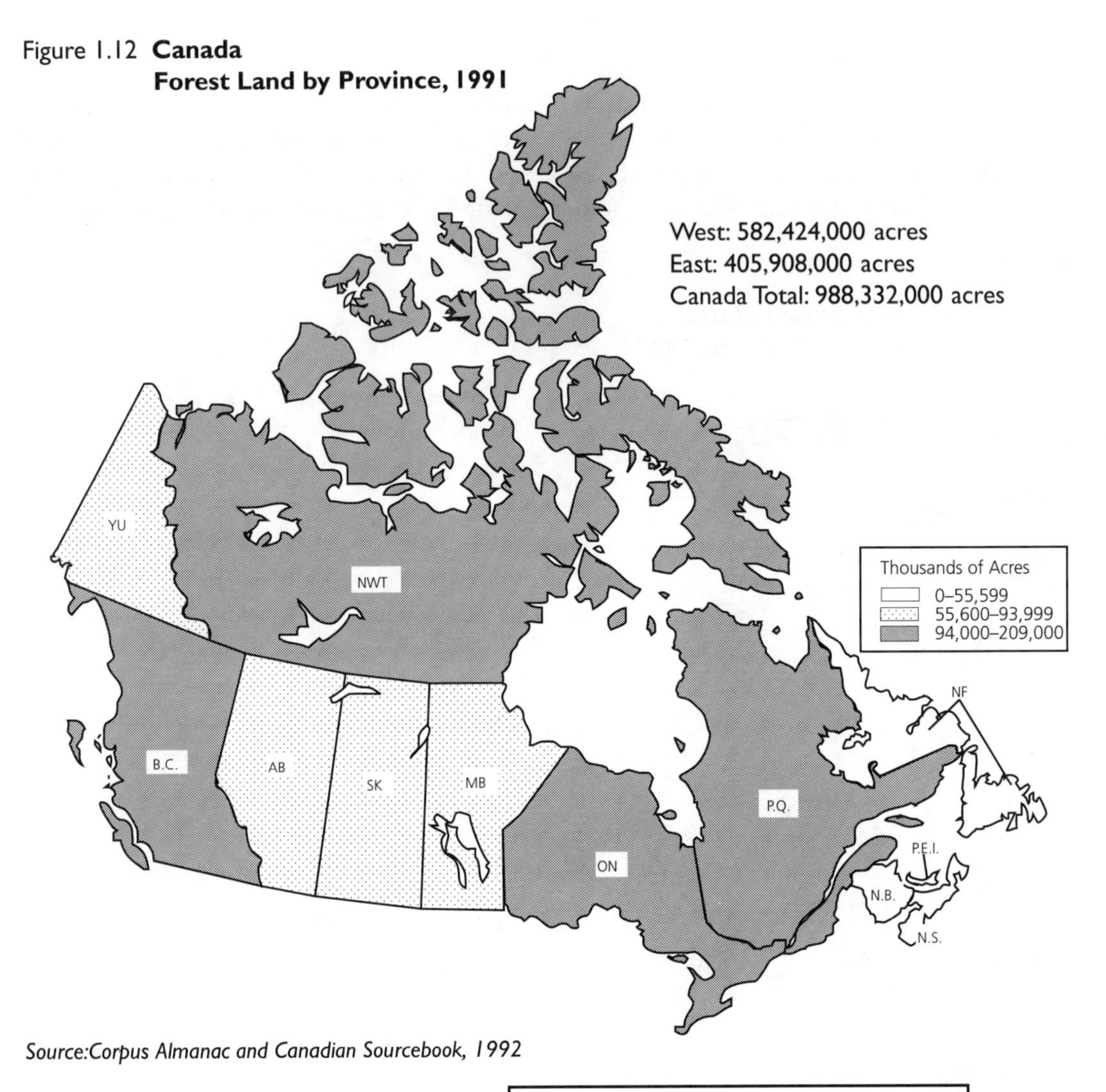

Source:Corpus Almanac and Canadian Sourcebook, 1992

	Thousands of Acres
Québec	208,836
Northwest Territories	151,821
British Columbia	149,028
Ontario	115,151
Alberta	93,282
Yukon	67,633
Manitoba	62,072
Saskatchewan	58,588
Newfoundland	55,549
New Brunswick	15,667
Nova Scotia	9,988
Prince Edward Island	717

Figure 1.13 **Canada, Mexico, United States Distribution of Land by Use, 1986–1988**

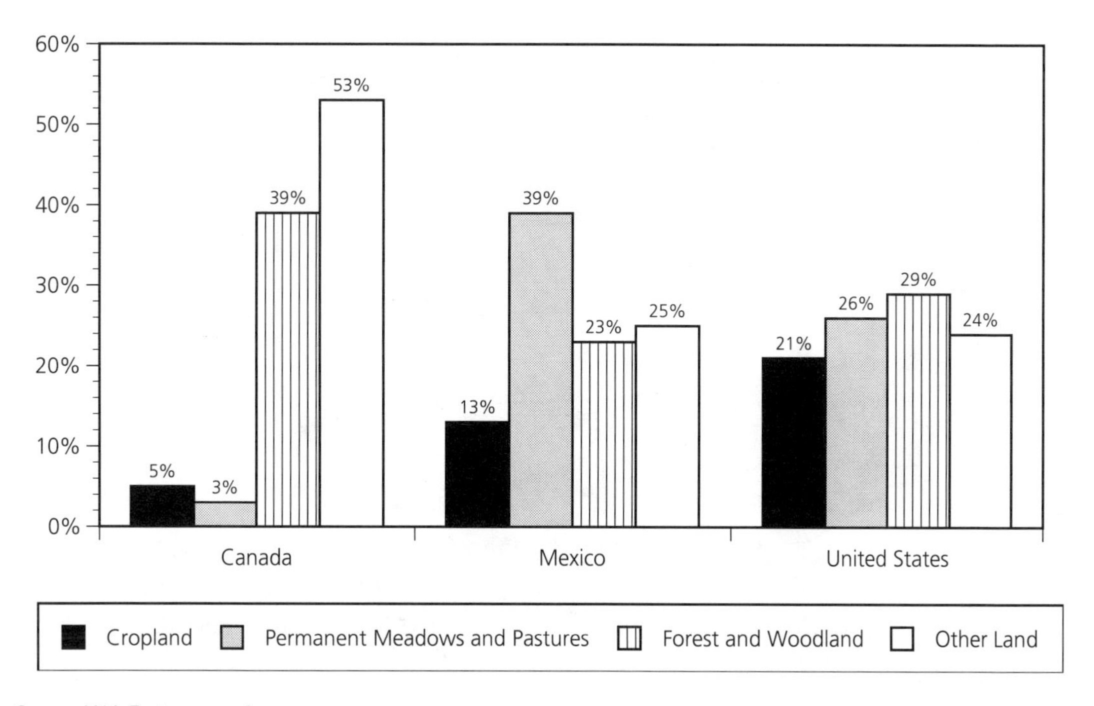

Source: U.N. Environment Programme

Timber

North America is the world's foremost source of timber products. The United States produced nearly 43 billion board feet of timber in 1990. Of that, more than 25 billion board feet (58 percent) was produced in the twenty-three states west of the Mississippi. Canada produced over 80 billion board feet in 1991. Of that, more than 44 billion board feet (55 percent) was produced in the Canadian West. Mexico produced more than 3 billion board feet of timber in 1990, which ranked thirty-seventh in the world that year.

British Columbia, with 36 billion board feet of production, is the largest timber producer in the North American West, more than four times as much as Oregon (more than 7 billion board feet), the region's number two timber producer, and larger than all of the western United States combined. California is third, with production of more than 5 billion board feet in 1990.

Figure 1.14 **Western United States**
Timber Production by State, 1990 (Millions of Board Feet)

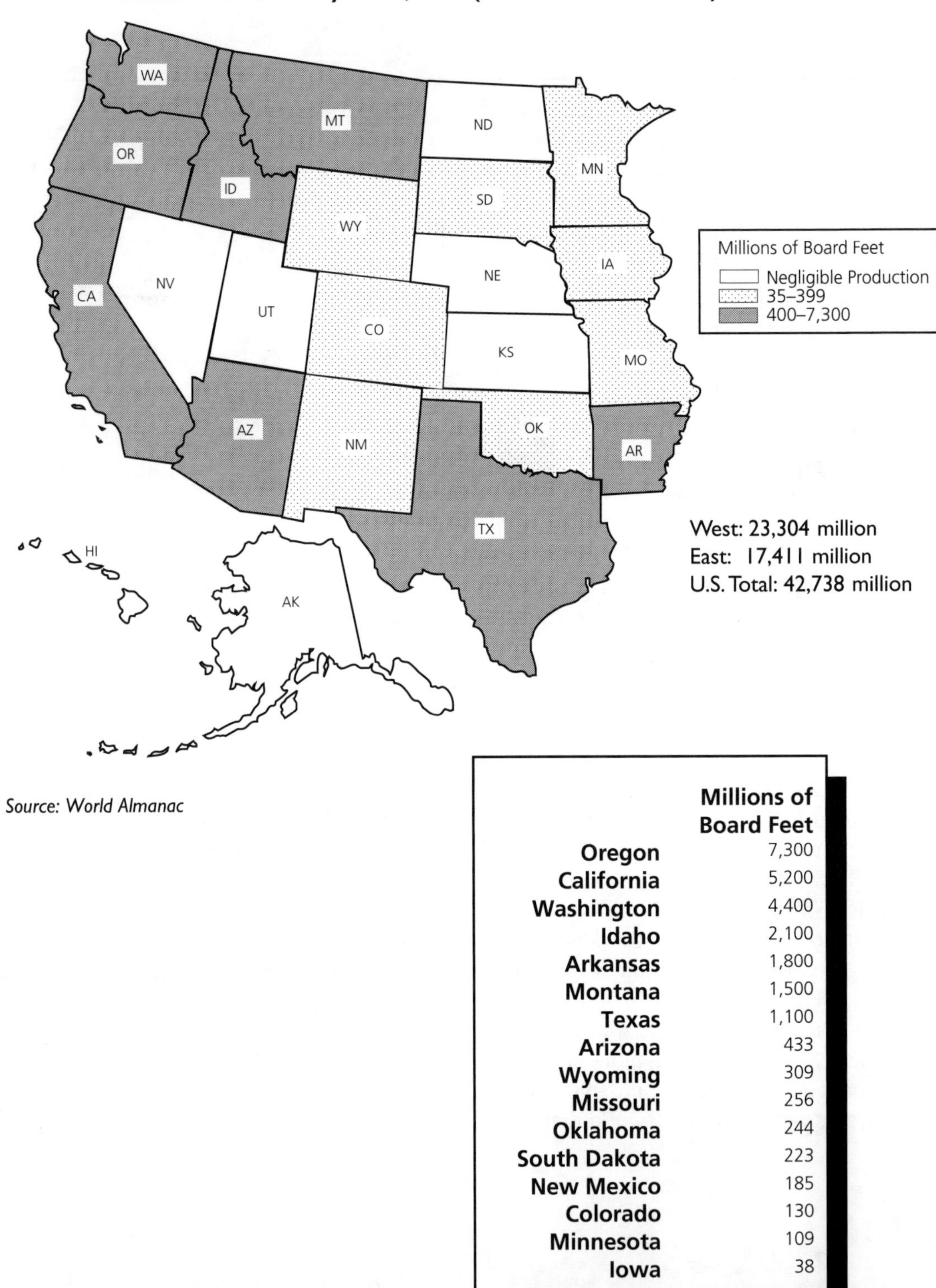

Source: World Almanac

	Millions of Board Feet
Oregon	7,300
California	5,200
Washington	4,400
Idaho	2,100
Arkansas	1,800
Montana	1,500
Texas	1,100
Arizona	433
Wyoming	309
Missouri	256
Oklahoma	244
South Dakota	223
New Mexico	185
Colorado	130
Minnesota	109
Iowa	38

Figure 1.15 **Canada**
Timber Production by Province, 1988
(Millions of Board Feet)

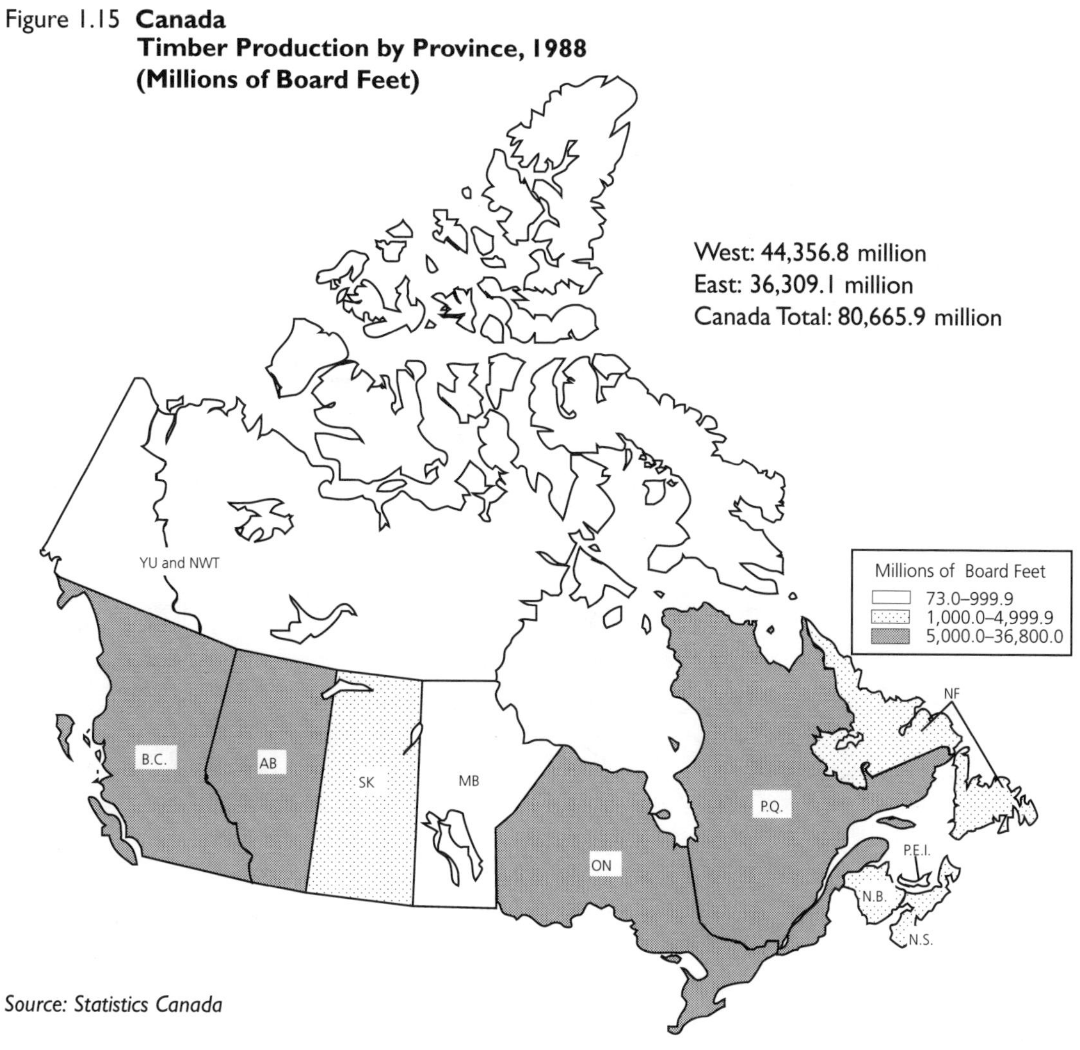

Source: Statistics Canada

	Millions of Board Feet
British Columbia	36,787.2
Québec	16,688.4
Ontario	12,432.7
Alberta	5,080.8
New Brunswick	3,940.8
Nova Scotia	1,976.4
Saskatchewan	1,617.6
Newfoundland	1,064.4
Manitoba	798.0
Prince Edward Island	206.4
NWT and Yukon	73.2

Figure 1.16 **Mexico**
Timber Production by State, 1990 (Millions of Board Feet)

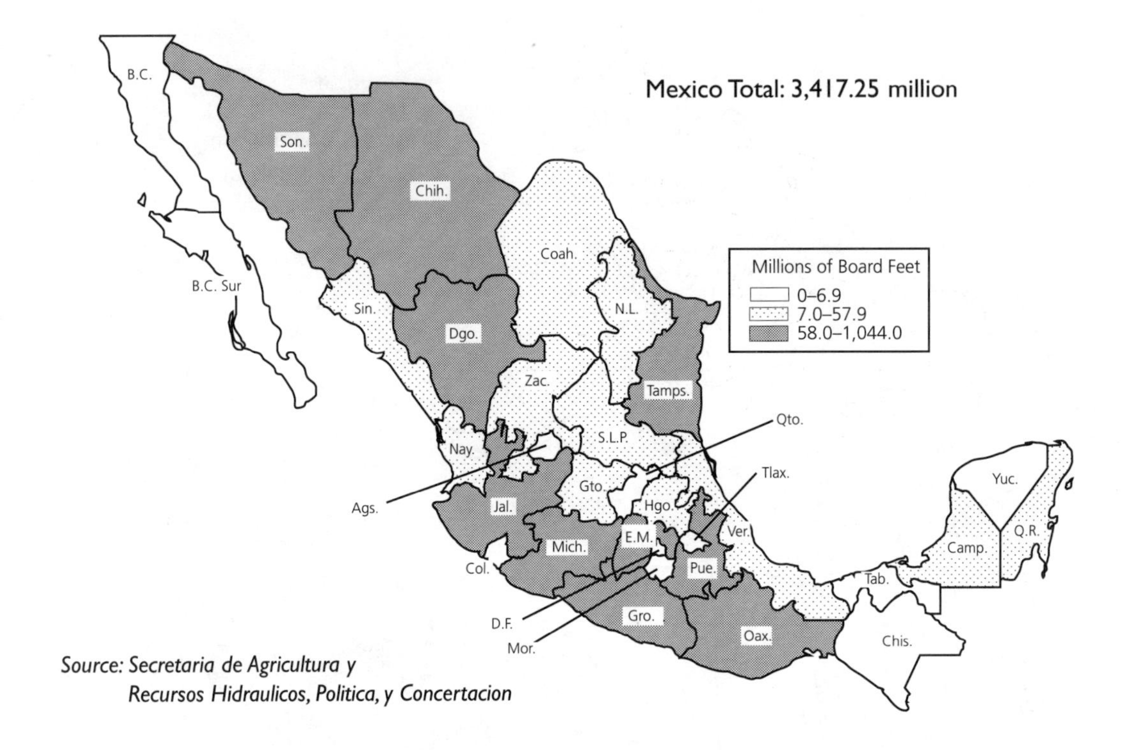

Source: Secretaria de Agricultura y Recursos Hidraulicos, Politica, y Concertacion

Fresh Water and Precipitation

Mexico is more arid than its northern neighbors, with 47.5 square miles (123 sq. km) of fresh water. The western United States and western Canada each have about 135,000 square miles (349.650 sq. km) of fresh water areas (lakes, streams, rivers, etc.).

Figure 1.17 **Western United States Fresh Water Area by State, 1990**

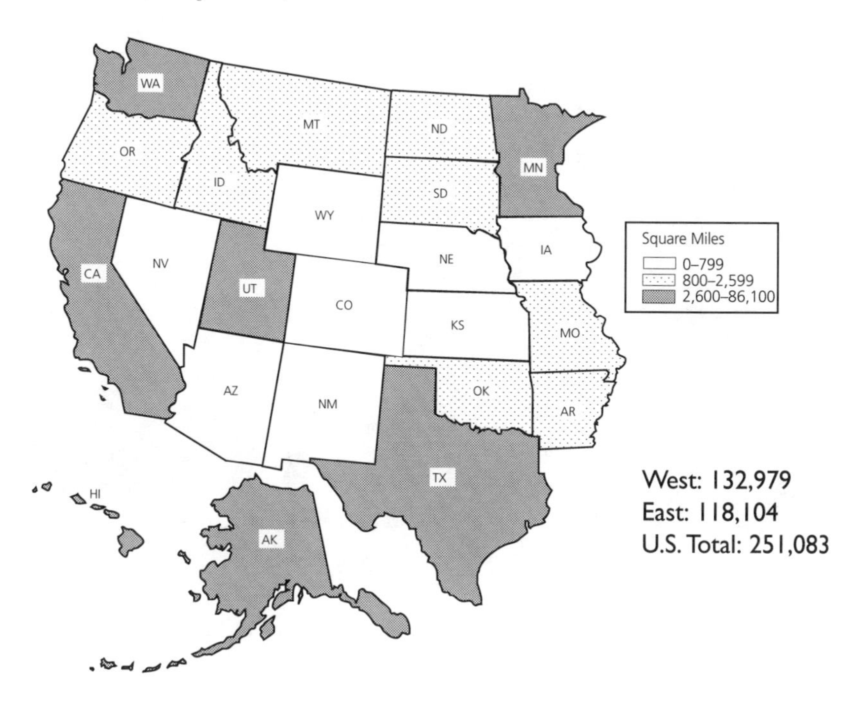

Source: U.S. Bureau of the Census, 1990

	Square Miles		Square Miles
Alaska	86,051	**Idaho**	823
California	7,734	**Missouri**	811
Minnesota	7,326	**Nevada**	761
Texas	6,687	**Wyoming**	714
Washington	4,721	**Nebraska**	481
Hawaii	4,508	**Kansas**	459
Utah	2,736	**Iowa**	401
Oregon	2,383	**Colorado**	371
North Dakota	1,710	**Arizona**	364
Oklahoma	1,224	**New Mexico**	234
South Dakota	1,224	**Montana**	149
Arkansas	1,107		

Figure 1.18 **Canada**
Fresh Water Area by Province, 1992

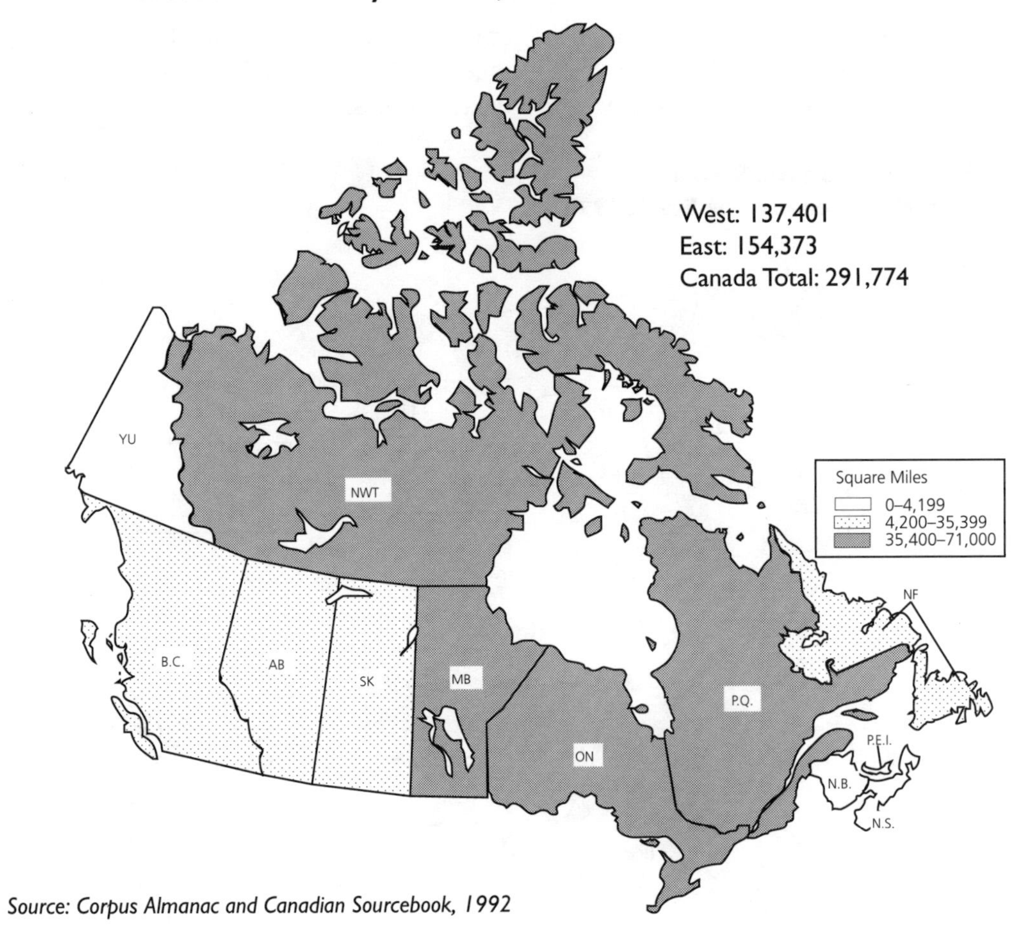

Source: Corpus Almanac and Canadian Sourcebook, 1992

	Square Miles
Québec	71,000
Ontario	68,490
Northwest Territories	51,467
Manitoba	39,224
Saskatchewan	31,517
Newfoundland	13,139
British Columbia	6,977
Alberta	6,486
Yukon	1,730
Nova Scotia	1,023
New Brunswick	521
Prince Edward Island	0

Figure 1.19 **North America**
Renewable Water Resources (Cubic Kilometers)

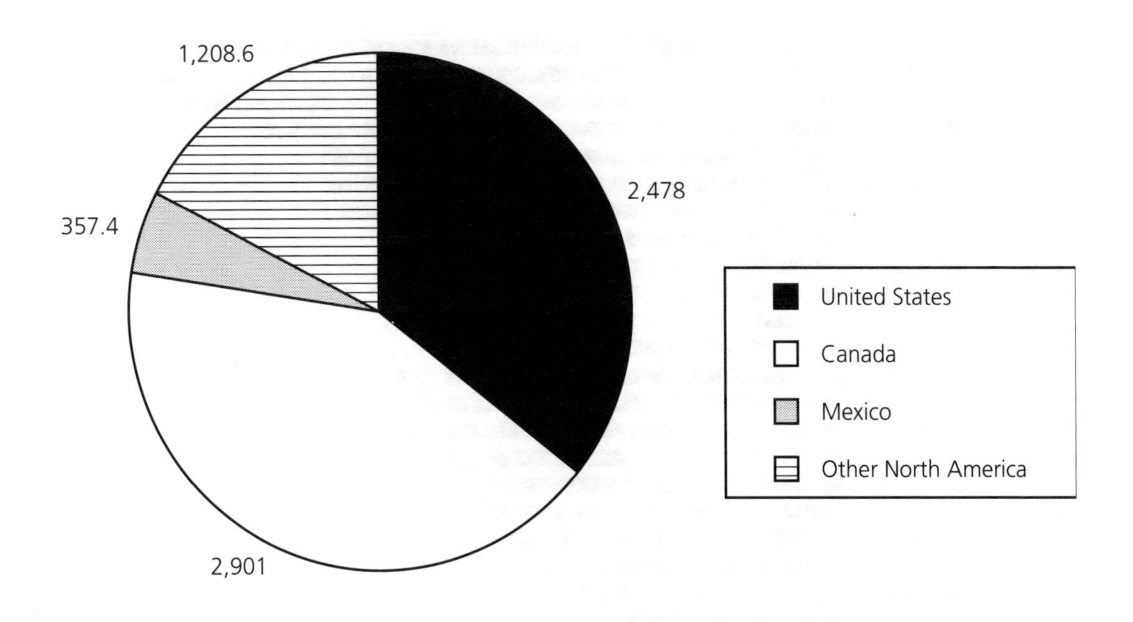

Source: U.N. Environment Programme

Annual rainfall increases south to north as well. In the urbanized areas of the North American West, precipitation ranges from 4.2 inches (10.8 cm) a year in Las Vegas to 53.2 inches (135.13 cm) per year in Juneau, Alaska, based on a thirty-year average for the years 1951–1980. In 1980, North America's wettest locale was Yakutat, Alaska, with 142 inches (360.7 cm) of precipitation a year, making it the twelfth wettest city in the world.

Half of the Canadian cities whose rainfall exceeds 5.1 inches (12.95 cm) per year are in the West. The average is 22 inches (55.9 cm) annually.

Figure 1.20 **Western United States**
Average Precipitation, Selected Cities (Based on 30-Year Mean, 1951–1980)

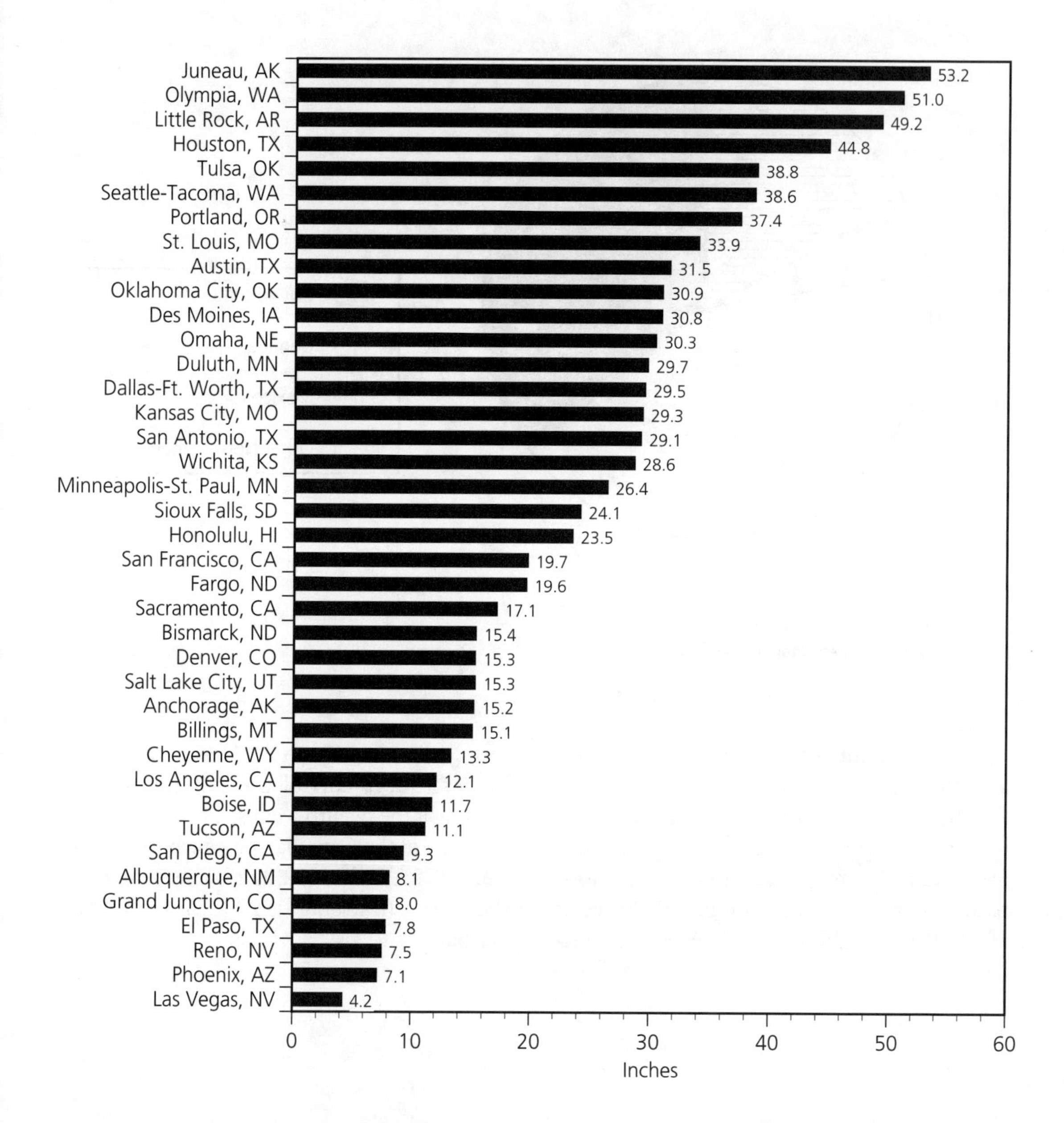

Source: National Oceanic and Atmospheric Administration

Figure 1.21 **Canada**
Average Precipitation, Selected Cities (Based on 30-Year Mean, 1951–1980)

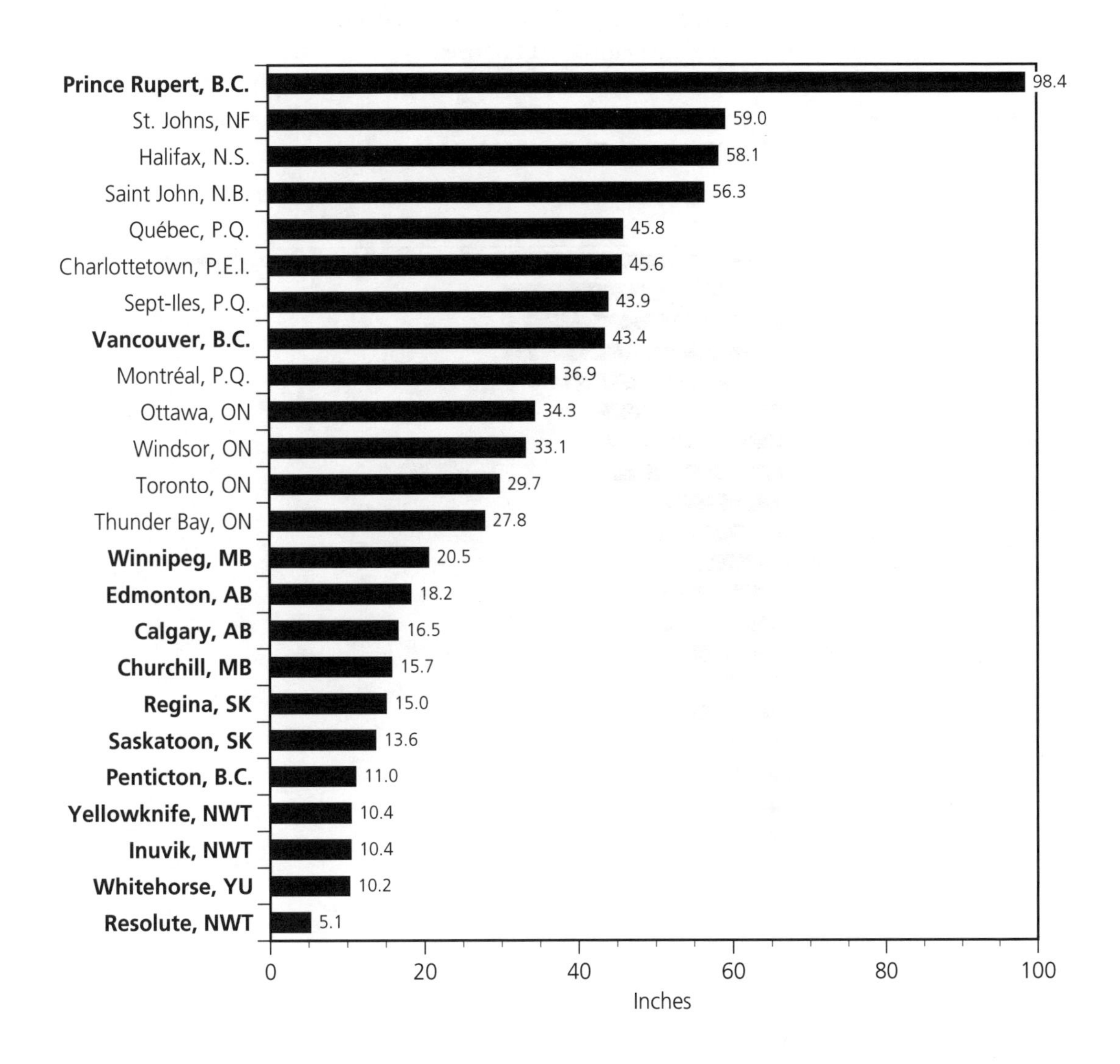

Source: Environment Canada

Precipitation patterns vary more dramatically in Mexico than in either the western United States or Canada. But throughout most of the country, rainfall is inadequate. Only about one-eighth of Mexico gets adequate rainfall year-round. The driest part of Mexico is the Baja California peninsula. The wettest states are Veracruz, Tabasco, and Chiapas, which get an average of 65.7 inches (166.88 cm) a year.

Figure 1.22 **Mexico**
Total Precipitation, Selected Cities, 1982–1983

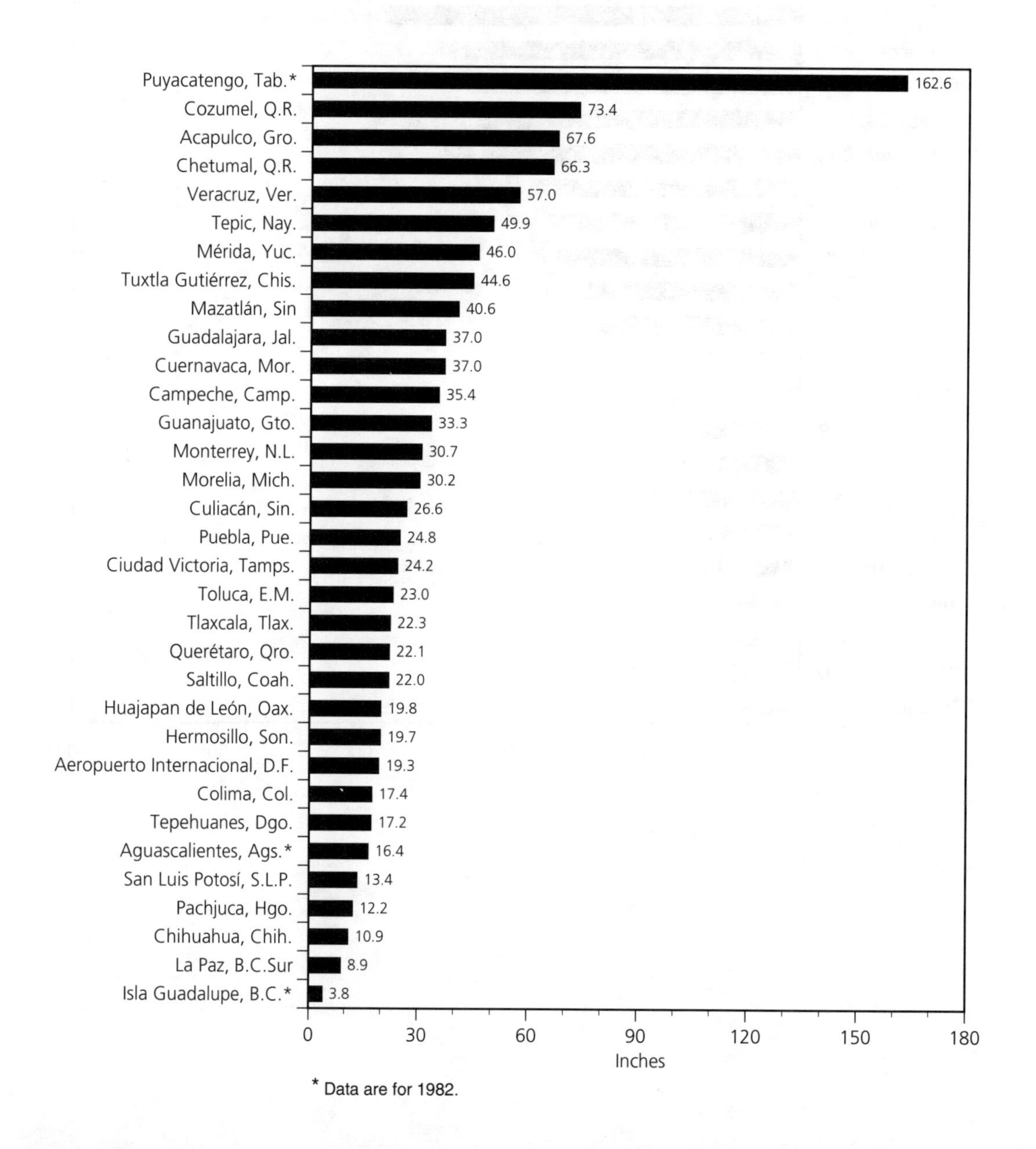

Source: INEGI

Petroleum

Western North America is a major player in global petroleum and is a net exporter of petroleum products to other regions in North America or to world markets. North America has approximately 80 billion barrels of crude oil reserves, 8.4 percent of the estimated world total. The size of these reserves pales compared to those of the Middle East (662 billion barrels, 66 percent), but are substantially larger than the oil reserves in any other part of the world. Nearly all these reserves are in the U.S. and Canadian West, and along Mexico's Caribbean coast.

Mexico possesses most of the continent's oil, an estimated 44.4 billion barrels as of January 1993. That's the eighth largest reserve in the world, more than the United States (26.2 billion barrels) and Canada (7.3 billion barrels) combined.

In the United States, Texas, Alaska, and California have the most reserves. In Canada, the province of Alberta accounts for 69.8 percent of that country's proven reserves. Nearly 57 percent of Mexican oil reserves are offshore in the Gulf of Mexico.

Crude oil production, however, is much higher in the western United States than it is in Mexico. In 1990, the United States produced 2.7 billion barrels of oil, compared to 976.4 million barrels for Mexico, and 634.5 million barrels of oil for Canada. The United States accounted for 62 percent of North America's oil production. Western North America accounts for 51 percent of the continental total.

Figure 1.23 Canada, Mexico, and the United States Crude Oil Production and Reserves, 1990 and 1992*

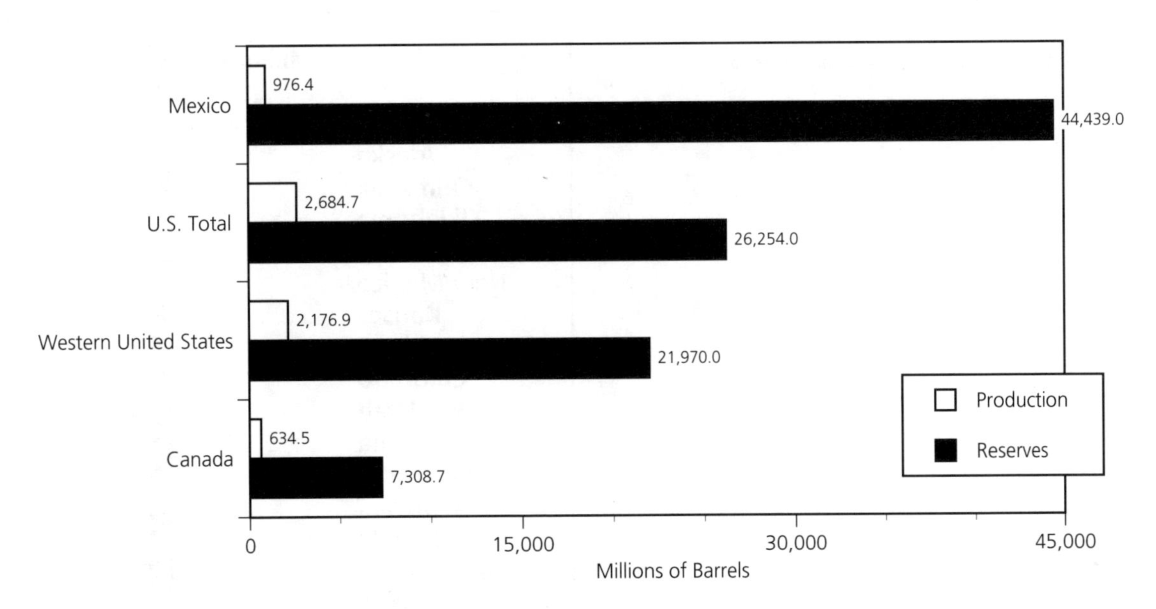

(U.S. data are for 1990. Canada and Mexico data are for 1992.)

Source: U.S. Energy Information Administration, Statistics Canada, INEGI

Figure 1.24 **Western United States Crude Oil Production by State, 1990**

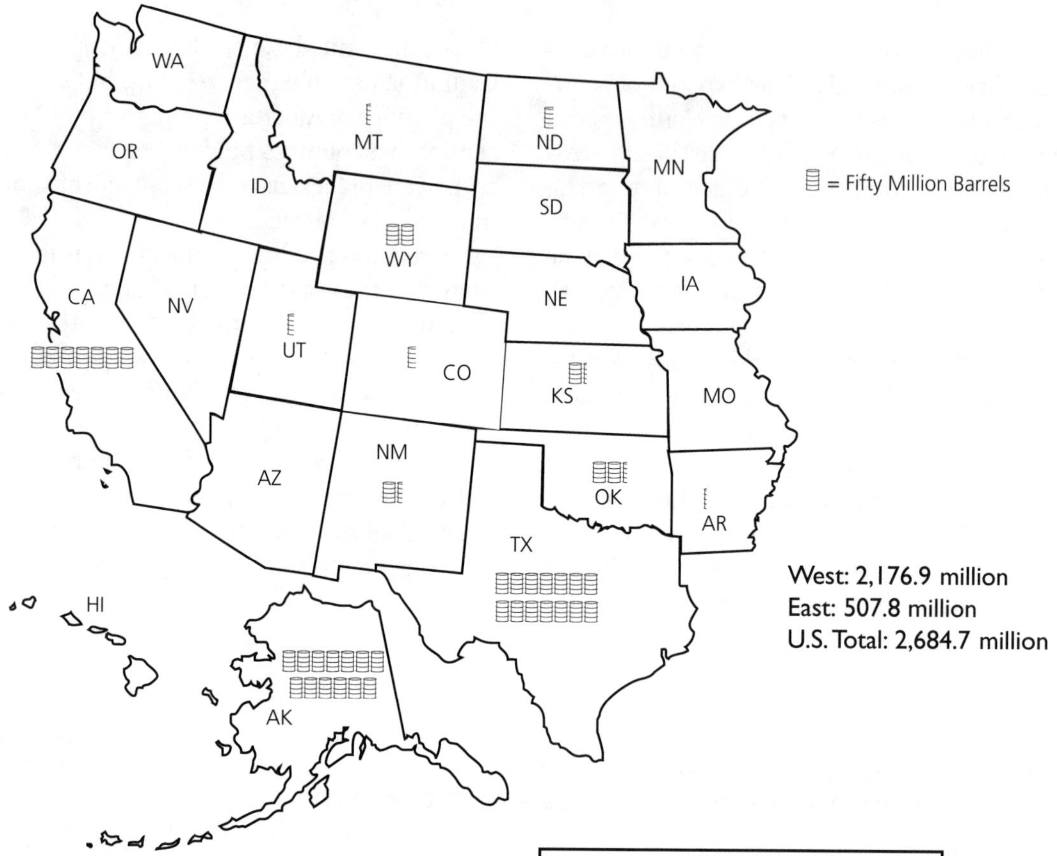

= Fifty Million Barrels

West: 2,176.9 million
East: 507.8 million
U.S. Total: 2,684.7 million

Source: U.S. Energy Information Administration

	Millions of Barrels
Texas	703.1
Alaska	647.3
California	350.9
Oklahoma	112.3
Wyoming	103.9
New Mexico	67.2
Kansas	55.4
North Dakota	36.7
Colorado	30.5
Utah	27.6
Montana	19.8
Arkansas	10.4
Nebraska	5.9
Nevada	4.0
South Dakota	1.6
Missouri	0.1
Arizona	0.1

Figure 1.25 **Canada Crude Oil Production by Province, 1992**

Source: Statistics Canada

	Millions of Barrels
Alberta	513.6
Saskatchewan	84.9
British Columbia	14.7
Northwest Territories	12
Manitoba	4.1
Nova Scotia	3.7
Ontario	1.4

Figure 1.26 **Mexico Crude Oil Production by Geographic Region, 1992**

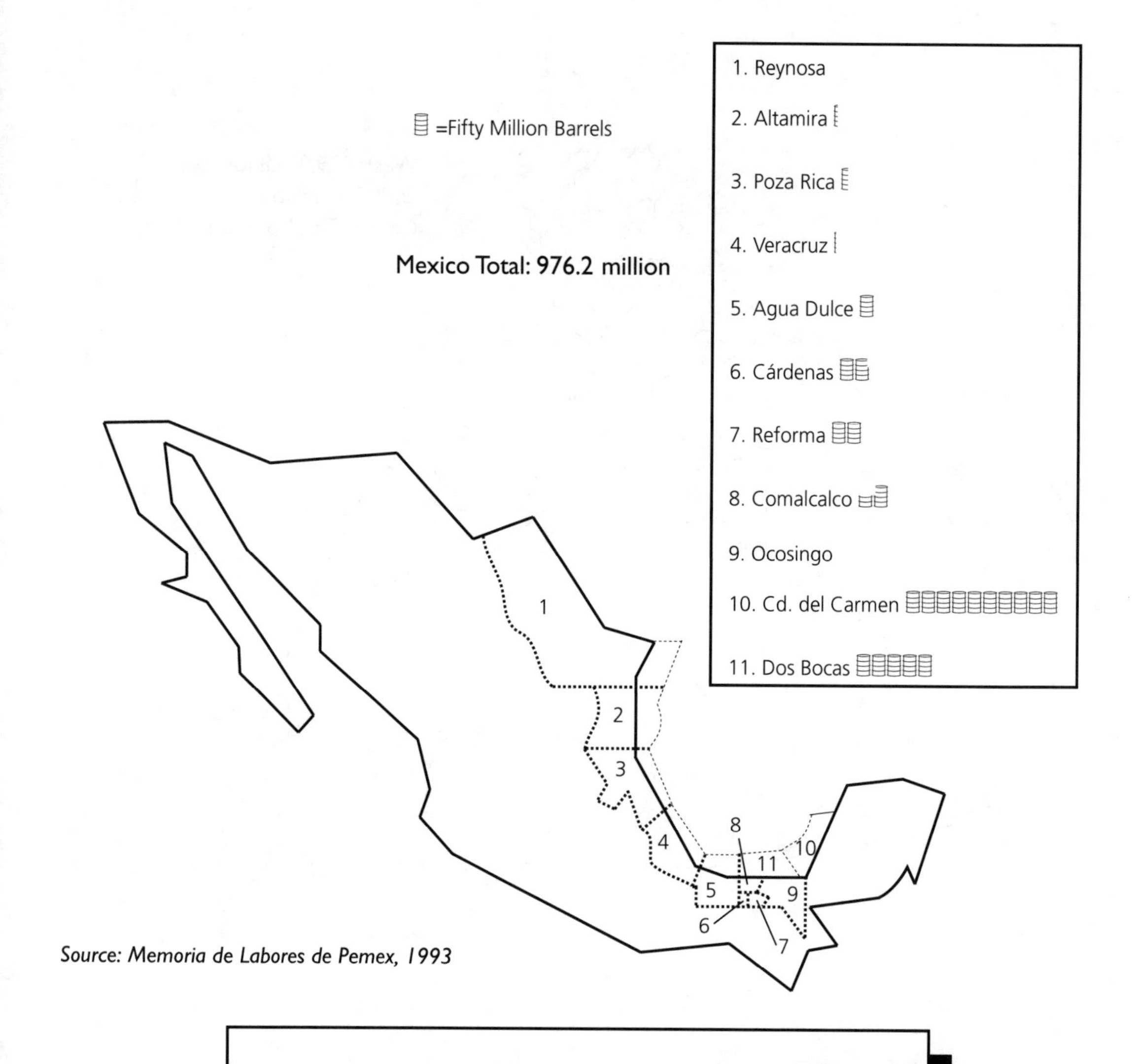

Source: Memoria de Labores de Pemex, 1993

	Millions of Barrels
Sea Region: Distrito Ciudad del Carmen	474.2
Sea Region: Distrito Dos Bocas	226.4
South Region: Distrito Cárdenas	88.1
South Region: Distrito Reforma	68.4
South Region: Distrito Comalcalco	60.7
North Region: Distrito Poza Rica	24.1
South Region: Distrito Agua Dulce	22.0
North Region: Distrito Altamira	9.9
North Region: Distrito Veracruz	2.2
South Region: Distrito Ocosingo	0.2
North Region: Distrito Reynosa	0.0

Figure 1.27 **World's Largest Oil Producers, 1990**

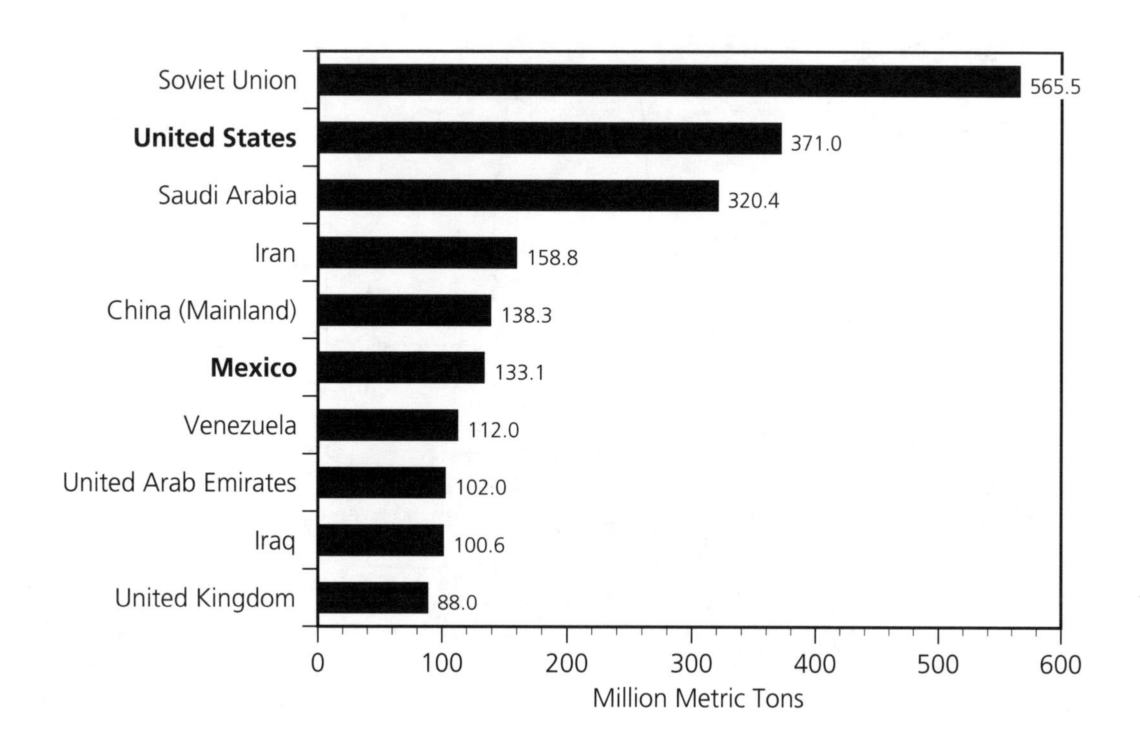

Source: Statistical Office of the U.N., Energy Statistics

Natural Gas

All three North American nations rank among the top ten producers of natural gas in the world. The United States ranks second, Canada ranks third, and Mexico ranks sixth, according to the *United Nations Energy Statistics Yearbook*. As with crude oil, virtually all the natural gas production in the United States and Canada is in the West.

The number one natural gas producer on the continent is Texas, which produced 6.3 trillion cubic feet of natural gas in 1990. Second is the Canadian province of Alberta, with production in 1992 of 3.3 trillion.

Figure 1.28 **Western United States Natural Gas Production by State, 1990**

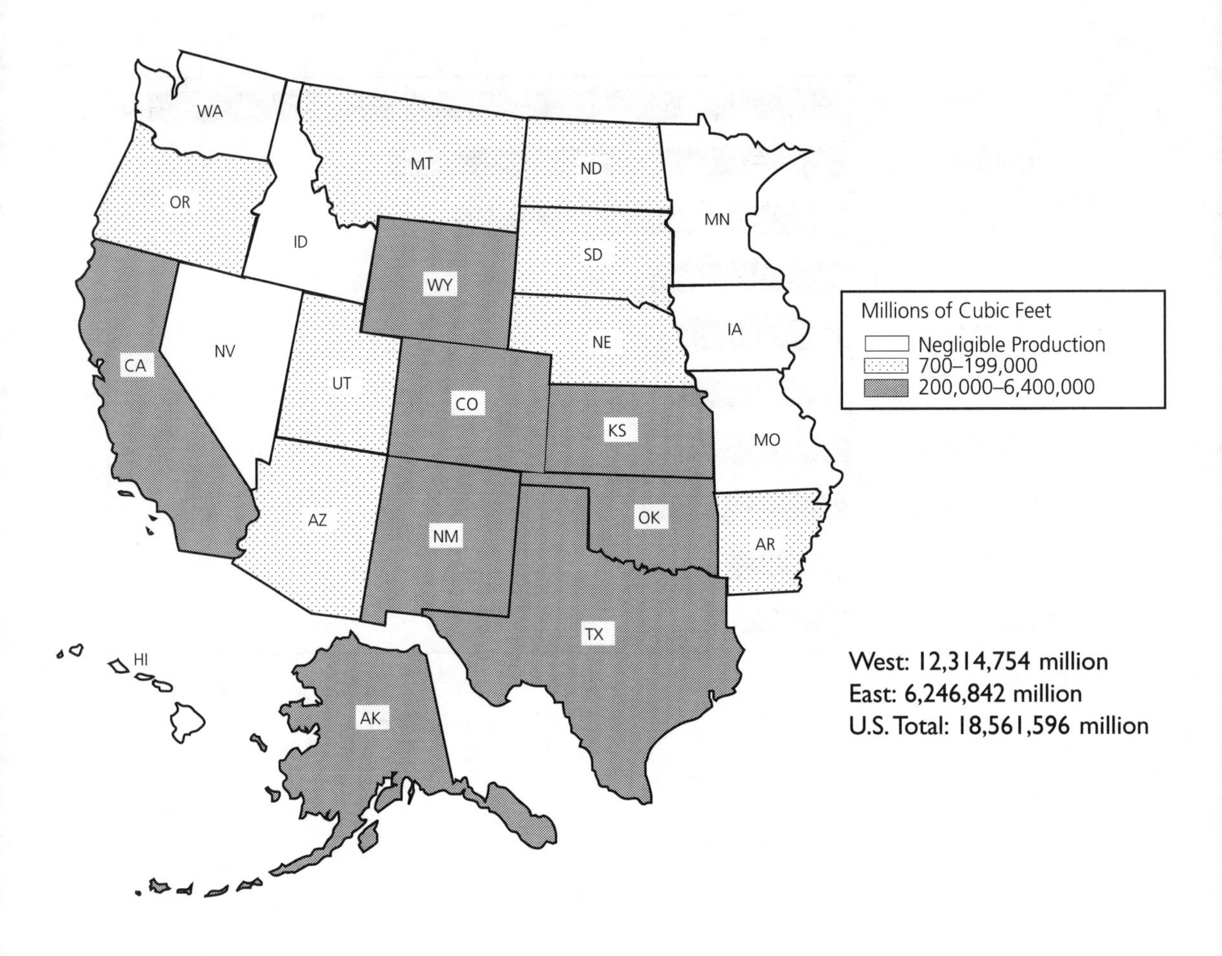

Source: U.S. Energy Information Administration

	Millions of Cubic Feet		Millions of Cubic Feet
Texas	6,343,146	**Utah**	145,875
Oklahoma	2,258,471	**North Dakota**	52,169
New Mexico	965,104	**Montana**	50,429
Wyoming	735,728	**Oregon**	2,815
Kansas	573,603	**Arizona**	2,125
Alaska	402,907	**South Dakota**	881
California	362,748	**Nebraska**	793
Colorado	242,997	**Missouri**	7
Arkansas	174,956		

Figure 1.29 **Canada**
Natural Gas Production, 1992

Source: Statistics Canada, Energy, Mines, and Resources

	Millions of Cubic Feet
Alberta	3,340,799
British Columbia	516,700
Saskatchewan	218,320
Ontario	23,208
Yukon	4,287
Québec	85

Figure 1.30 **Mexico**
Natural Gas Production by Region, 1992

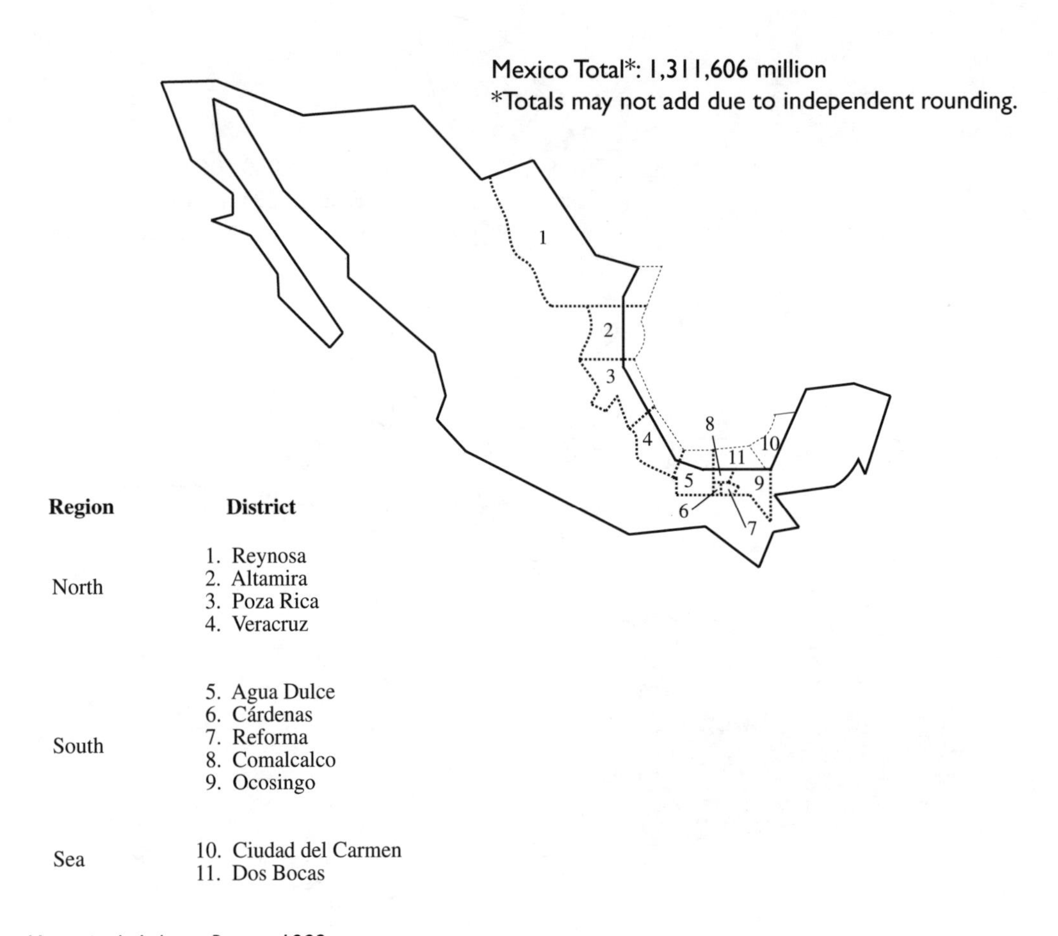

Source: Memoria de Labores Pemex, 1992

	Millions of Cubic Feet
South Region: Distrito Reforma	300,856
Sea Region: Distrito Dos Bocas	223,862
Sea Region: Distrito Ciudad del Carmen	205,798
South Region: Distrito Comalcalco	157,007
South Region: Distrito Cárdenas	143,297
North Region: Distrito Reynosa	87,289
South Region: Distrito Ocosingo	70,115
South Region: Distrito Agua Dulce	40,862
North Region: Distrito Veracruz	40,802
North Region: Distrito Poza Rica	25,401
North Region: Distrito Altamira	16,317

The Western United States also leads the continent in proven reserves of natural gas: 112.9 trillion cubic feet. Canada has reserves of 95.7 trillion, and Mexico has proven reserves of 70 trillion. The three North American countries together possess 8 percent of the world's proven reserves of natural gas.

Figure 1.31 **Western United States Natural Gas Reserves by State, 1990**

West: 112,927 billion
East: 24,916 billion
U.S. Total: 169,346* billion

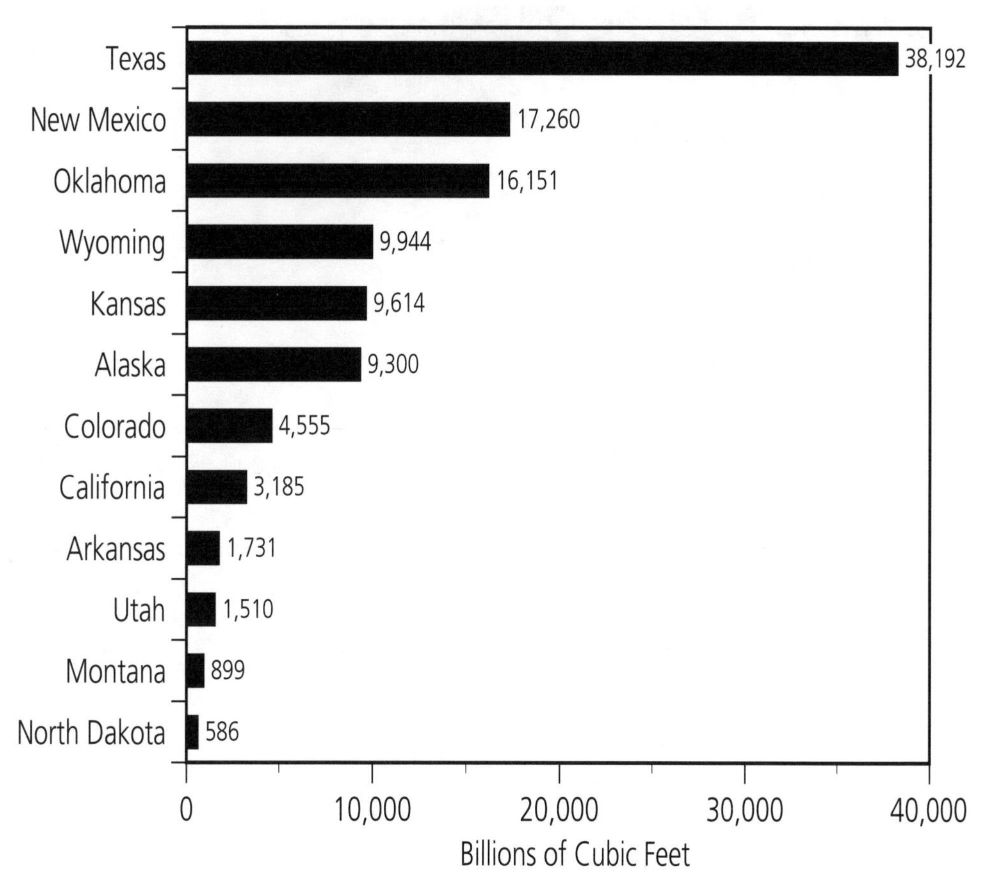

(*Includes miscellaneous and federal offshore.
Arizona, Iowa, Minnesota, Missouri, Nebraska, Oregon, South Dakota, and Washington are included in the national miscellaneous total.)

Source: U.S. Energy Administration

Figure 1.32 **Canada**
Natural Gas Reserves by Province, 1992

West: 95,141.3 billion
East: 589.8 billion
Canada Total: 95,734.6* billion

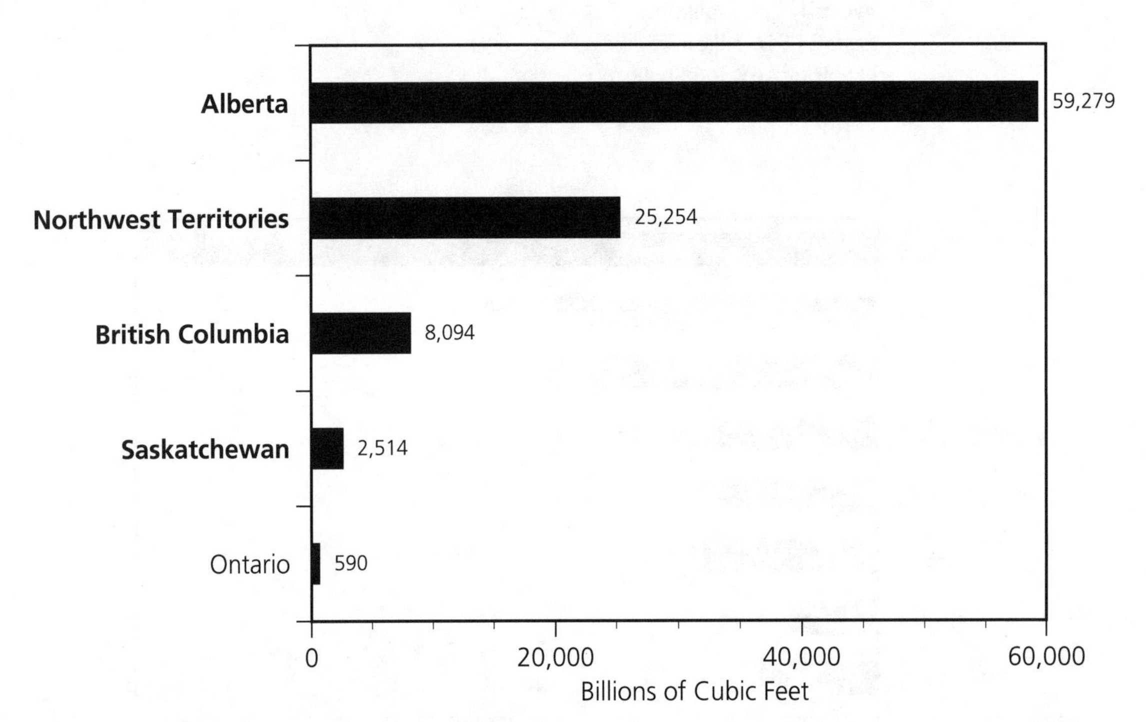

(*Totals may not add due to independent rounding and inclusion of other eastern province reserves too small to be included independently.)

Source: Statistics Canada, Energy, Mines, and Resources

Figure 1.33 **Mexico**
Natural Gas Reserves by Region, 1992

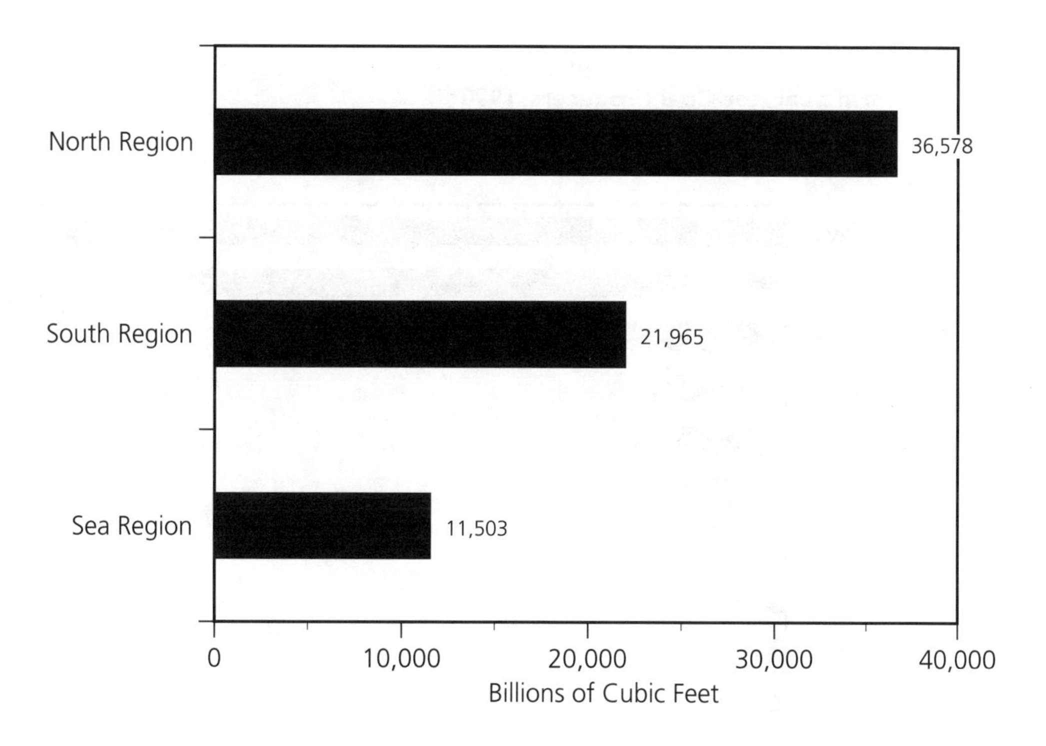

Source: Memoria de Labores de Pemex, 1992

COAL

In 1990, the United States ranked second in the world in coal production. Canada was tenth, Mexico sixteenth. The United States was the world's second largest exporter of coal in 1989, just behind Australia. Canada was the world's sixth largest coal exporter in the same year. The western United States produced approximately 40 percent of their total.

Figure 1.34 **World's Largest Coal Producers, 1990***

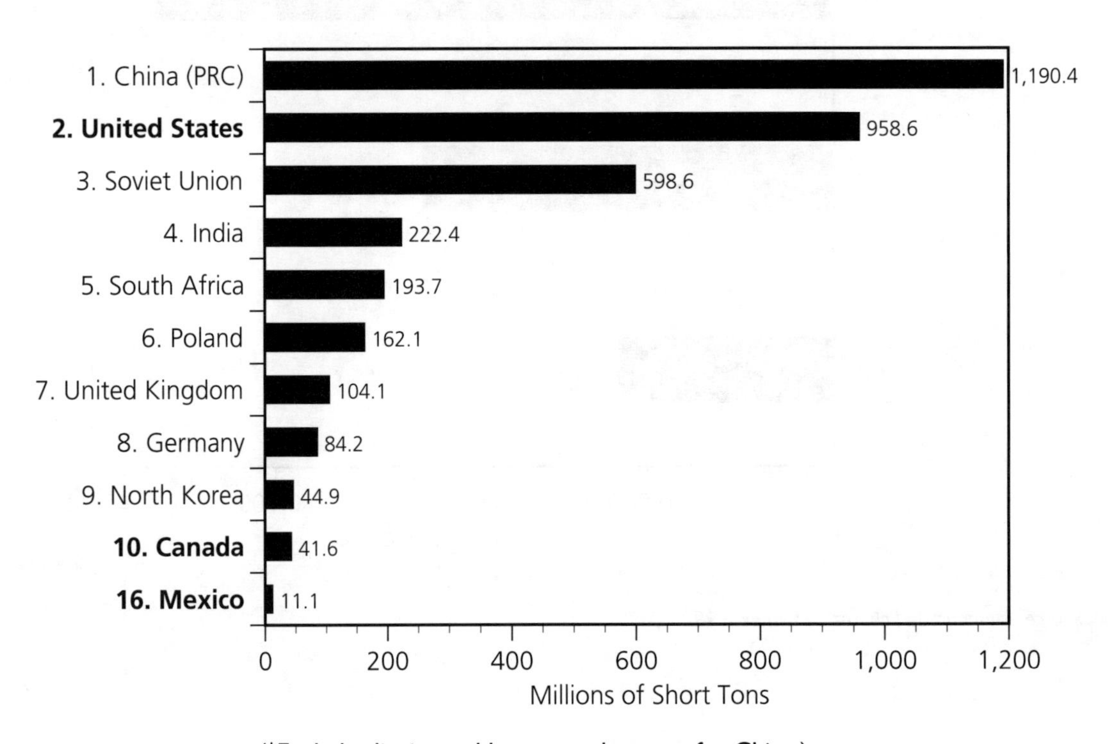

(*Excludes lignite and brown coal, except for China.)

Source: Statistical Office of the U.N., Energy Statistics

Wyoming, the leading coal-producing state in the United States, accounts for nearly half of all western coal production. The next four largest producers—Kentucky, West Virginia, Pennsylvania, and Illinois—are located in the East. Alberta is the leading coal producer in Canada, although its margin over British Columbia is slight. Mexican coal production is almost exclusively in the state of Coahuila.

Coal reserves in the western United States were estimated at 256 billion short tons (232.2 billion mt) in 1990 (second in the world). Canadian reserves were estimated in 1985 to be 7.3 billion short tons (6.7 billion mt) (twelfth in the world). Estimated Mexican reserves of 1.9 billion short tons (1.7 billion mt) in 1985 ranked twenty-seventh.

Figure 1.35 **Largest Coal-Producing States in the United States, 1990**

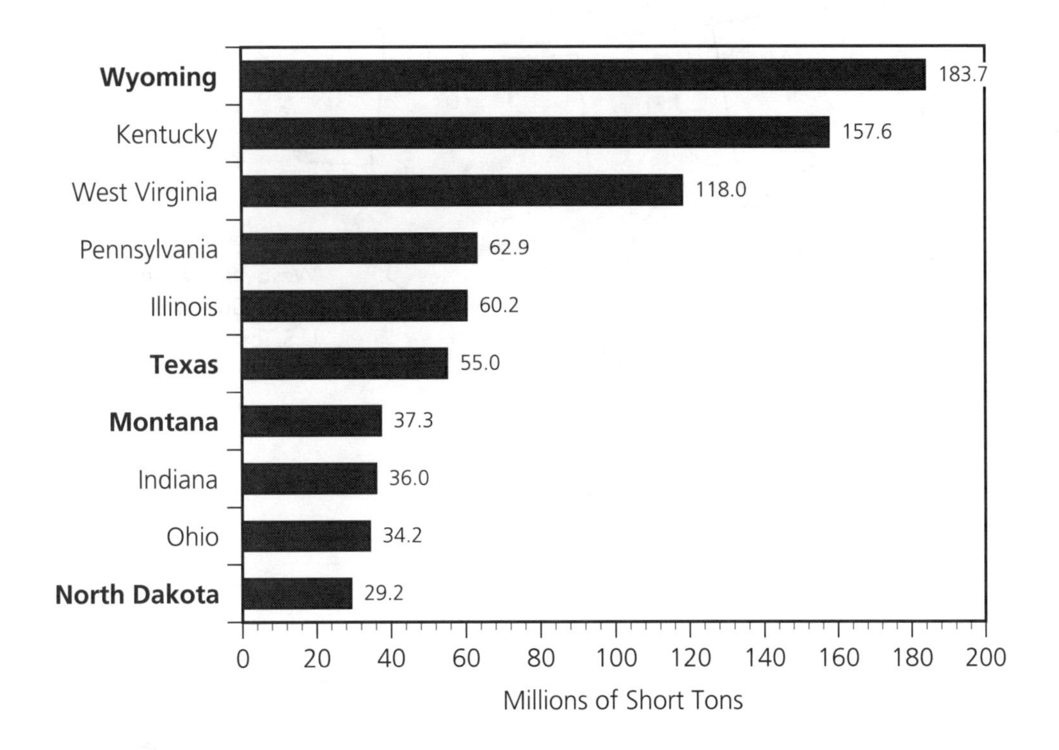

Source: U.S. Energy Information Administration

Figure 1.36 **Western United States Coal Production by State, 1990**

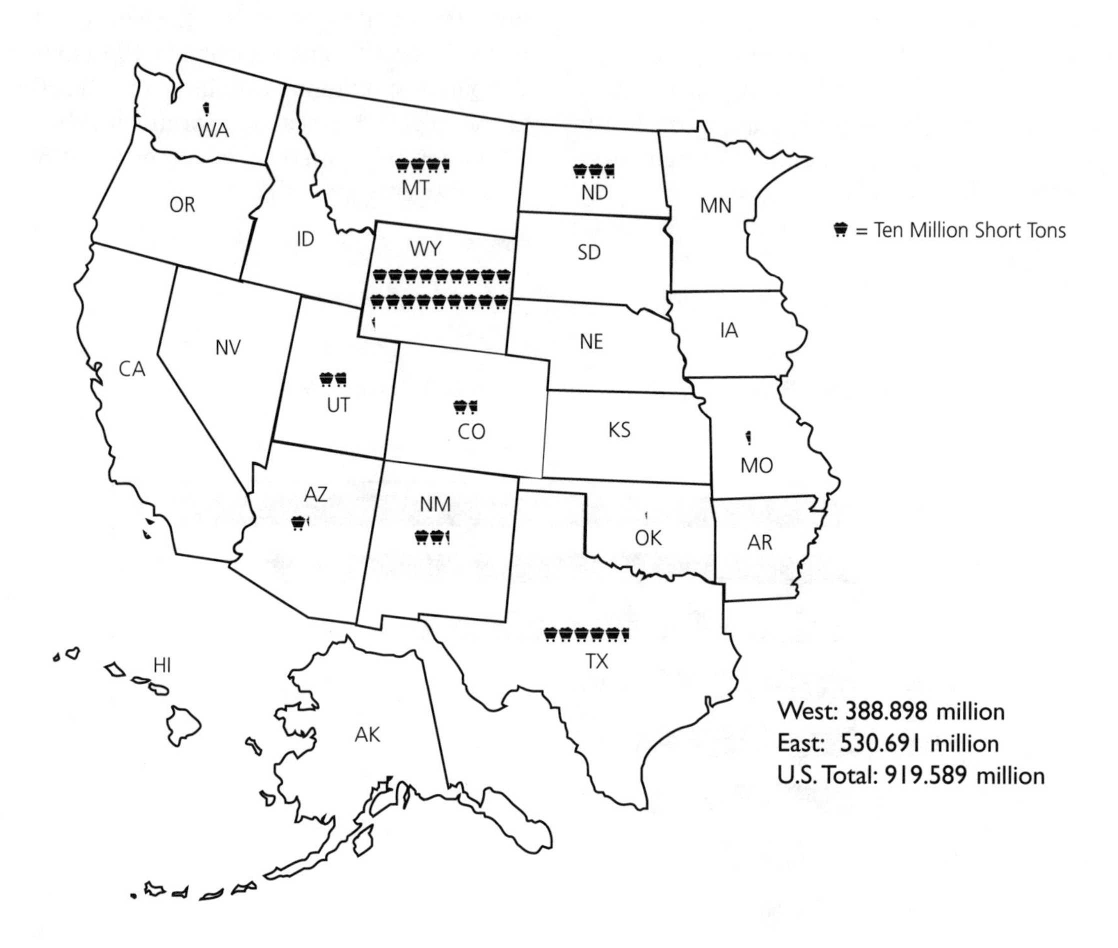

Source: U.S. Energy Information Administration

	Millions of Short Tons		Millions of Short Tons
Wyoming	183.71	**Washington**	4.89
Texas	55.02	**Missouri**	2.49
Montana	37.29	**Oklahoma**	1.73
North Dakota	29.21	**Alaska**	0.78
New Mexico	23.38	**Kansas**	0.73
Utah	19.76	**Iowa**	0.38
Colorado	18.02	**California**	0.06
Arizona	11.45	**Arkansas**	0.01

Figure 1.37 **Canada**
Coal Production by Province, 1991

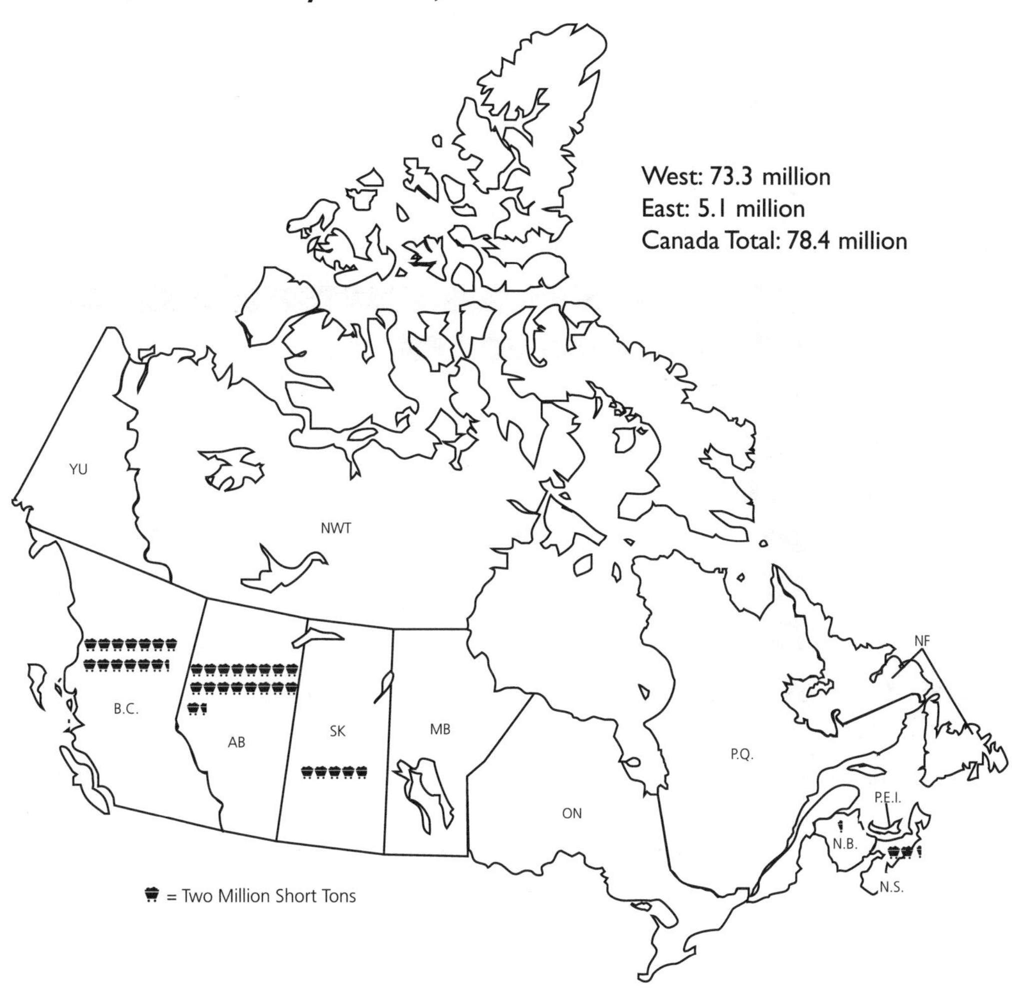

Source: Statistics Canada

	Millions of Short Tons
Alberta	32.53
British Columbia	27.52
Saskatchewan	9.90
Nova Scotia	4.56
New Brunswick	0.55

Figure 1.38 **Western United States Demonstrated Coal Reserve Base by State, 1990**

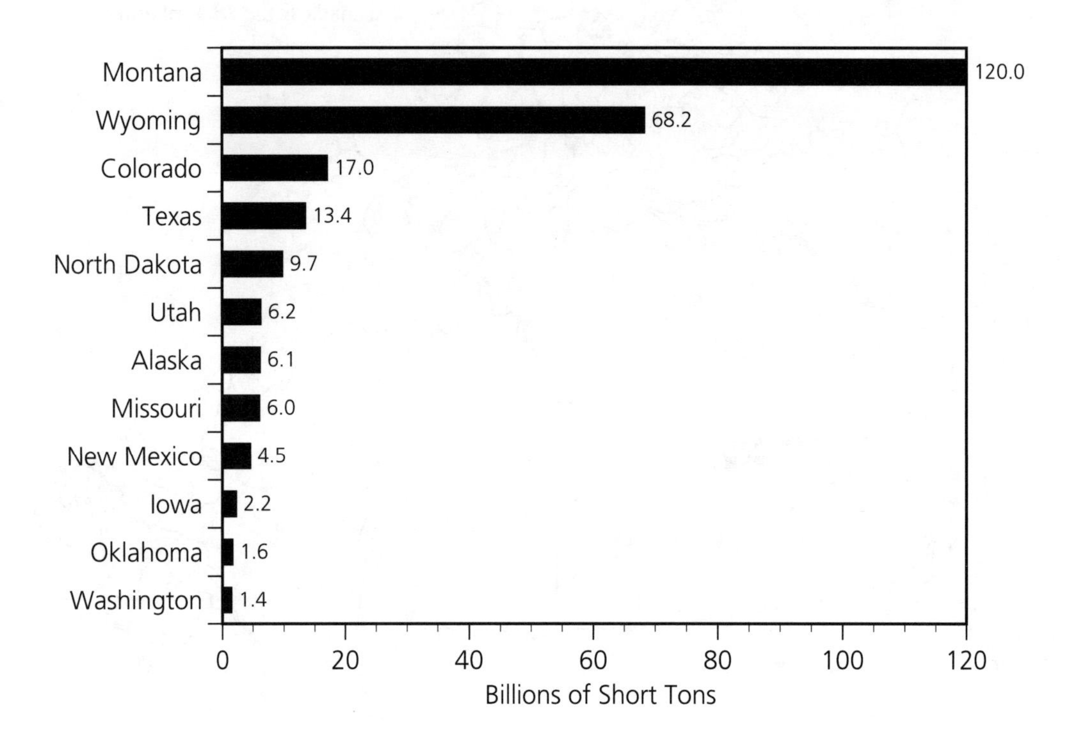

(*Totals may differ due to separate rounding. Arkansas and Kansas comprise the interior region miscellaneous total of two billion short tons. Arizona, Idaho, Oregon, and South Dakota comprise the western region miscellaneous total of 0.7 billion short tons.)

Source: U.S. Energy Information Administration

Figure 1.39 **Canada**
Recoverable Coal Reserves, 1985

West: 6.8 billion
East: 0.5 billion
Canada Total: 7.3 billion

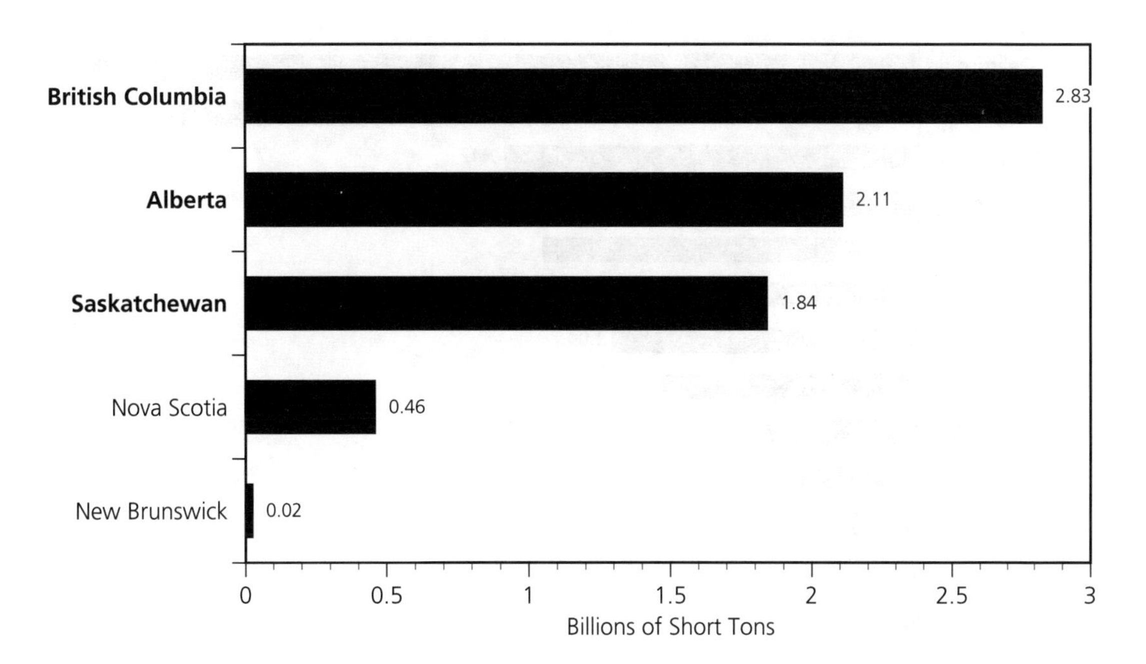

Source: Geological Survey of Canada

Figure 1.40 **Western United States**
Value of Non-Fuel Mineral Production by State, 1990

West: $16,564,870
East: $16,754,130
U.S. Total: $33,319,000

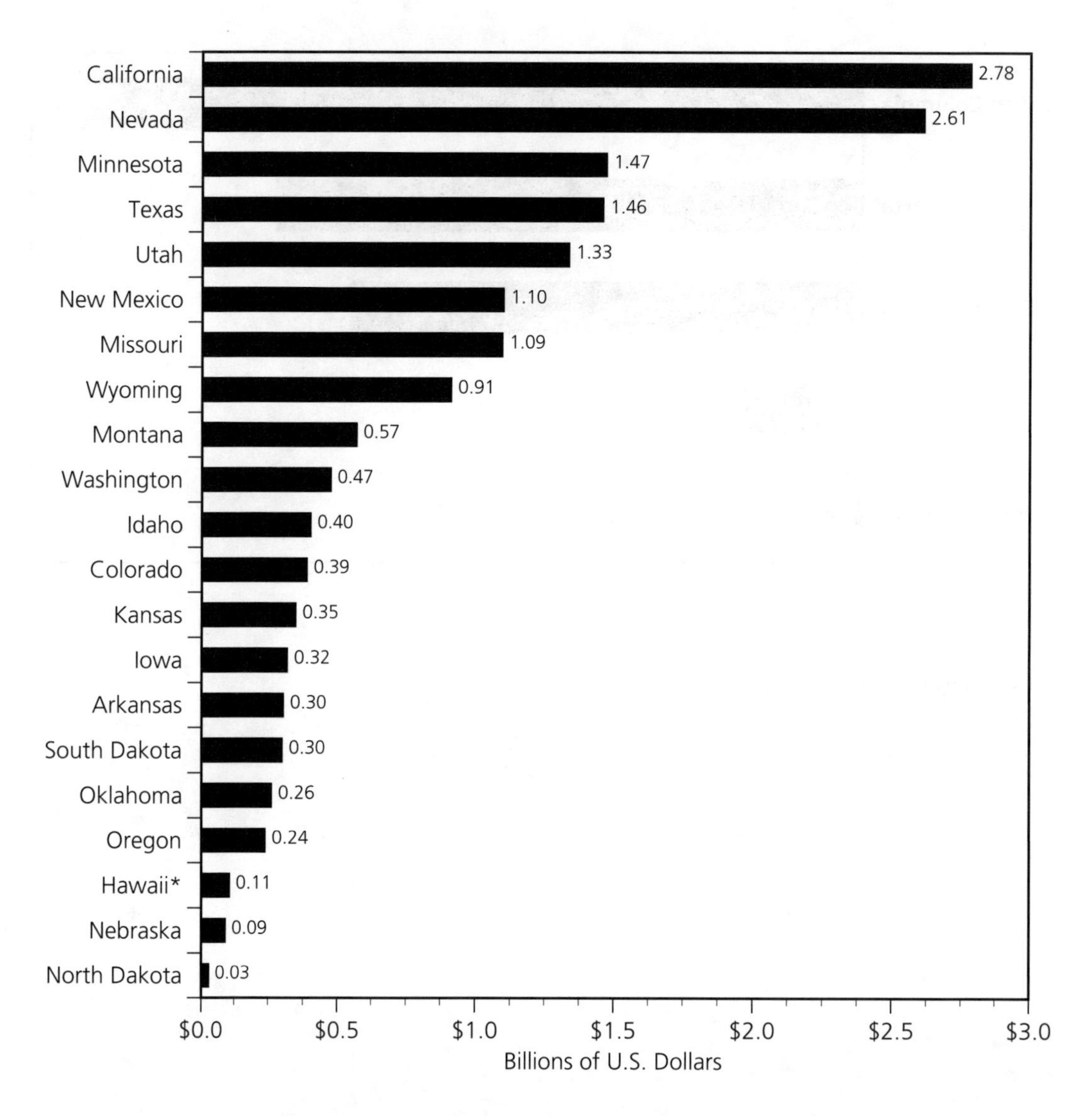

* Partial total (does not include values that are concealed to avoid revealing proprietary data).

Source: U.S. Department of the Interior, Bureau of Mines

NON-FUEL MINERALS

The United States produced $33.3 billion worth of non-fuel minerals—chiefly copper, iron ore, lead, zinc, and gold—in 1990, of which $16.6 billion was produced in the West. The leading producers were California, with $2.8 billion, and Nevada, with $2.6 billion.

Canada is among the world's top producers of zinc, uranium, nickel, potash, asbestos, gypsum, sulfur, and titanium. Most non-fuel minerals in Canada are mined in the East, which produced $9.1 billion of the Canadian total of $14.6 billion in 1992. The largest producers in the Canadian West are British Columbia and Saskatchewan, but their output combined is barely half that of Ontario, the leading Canadian producer.

Mexico mined $2.3 billion worth of non-fuel minerals in 1988.

Figure 1.41 **Canada**
Value of Non-Fuel Mineral Production by Province, 1992

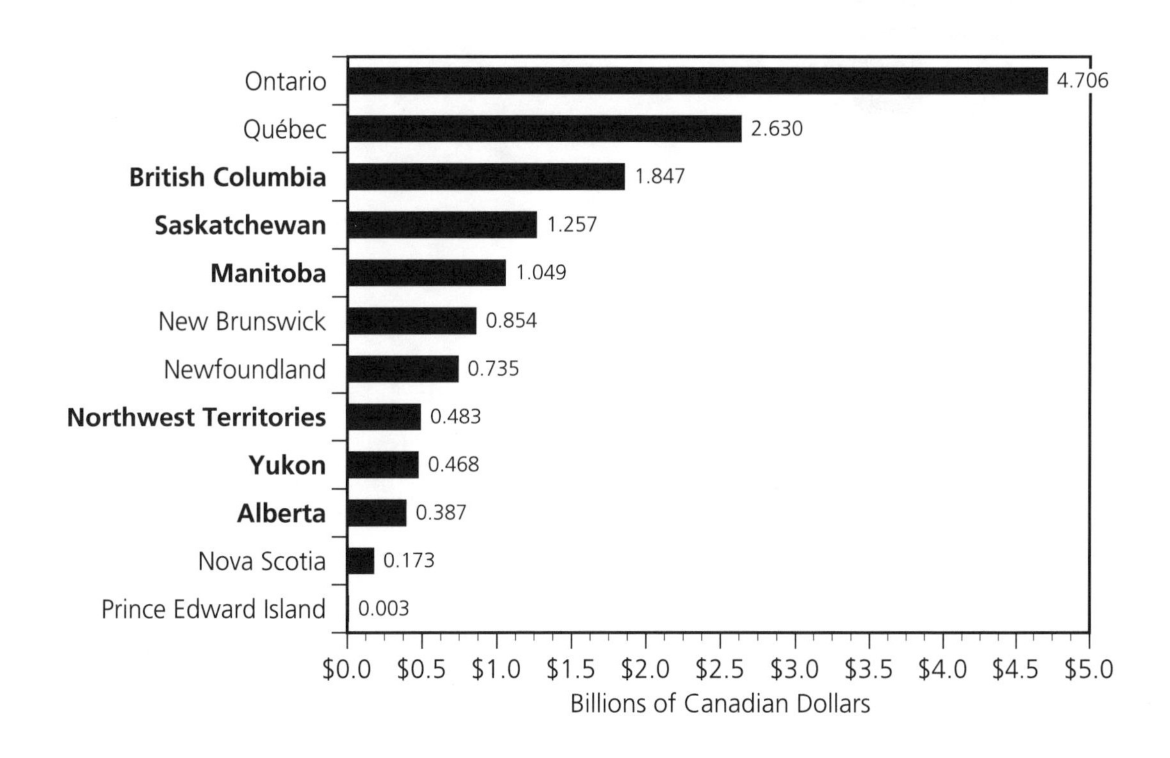

Source: Statistics Canada

Figure 1.42 **Mexico**
Value of Non-Fuel Mineral Production by State, 1988

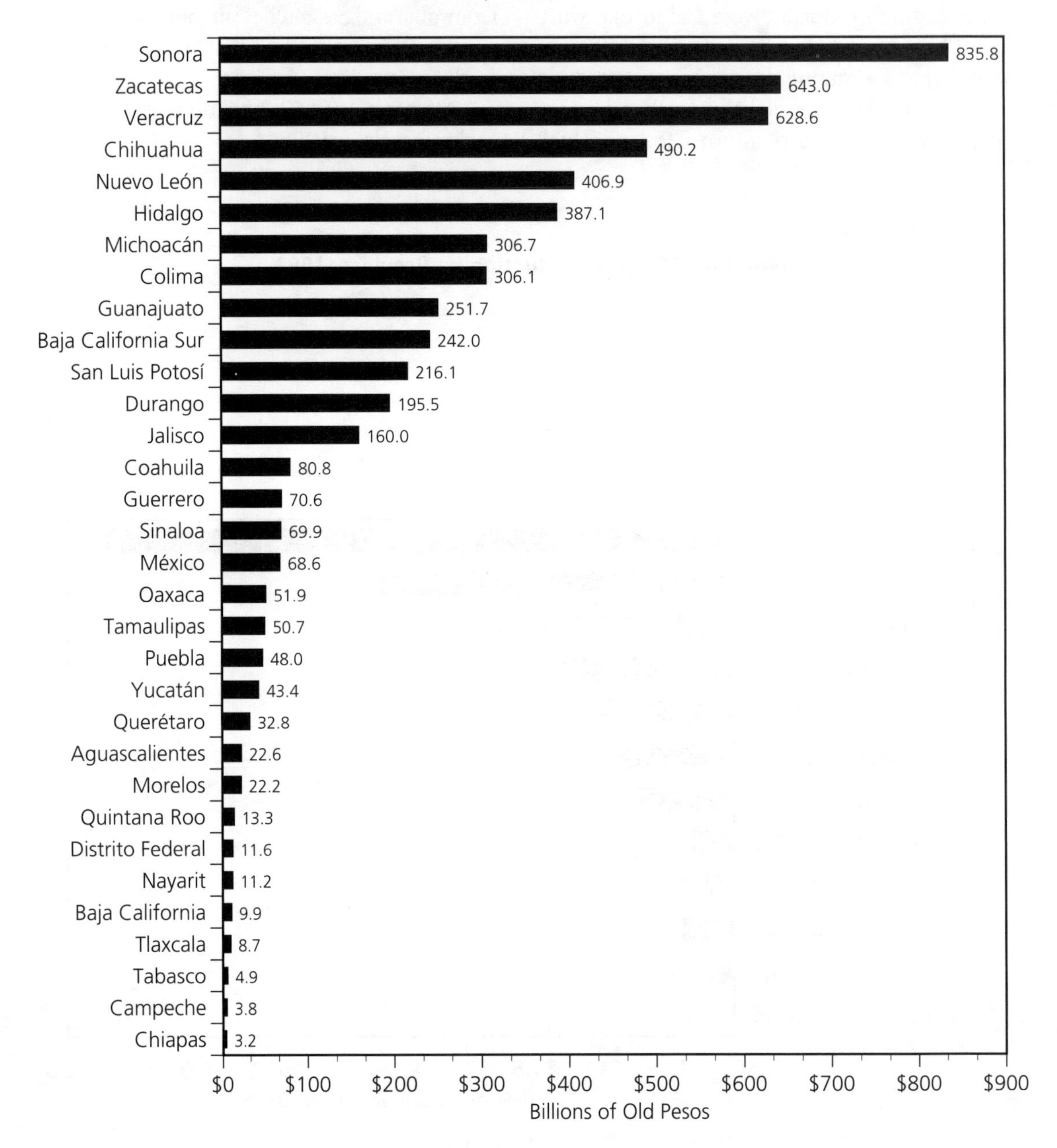

Source: INEGI, XIII Censo Industrial

Figure 1.43 **Largest Exporters of Non-Fuel Minerals and Metals, 1990**

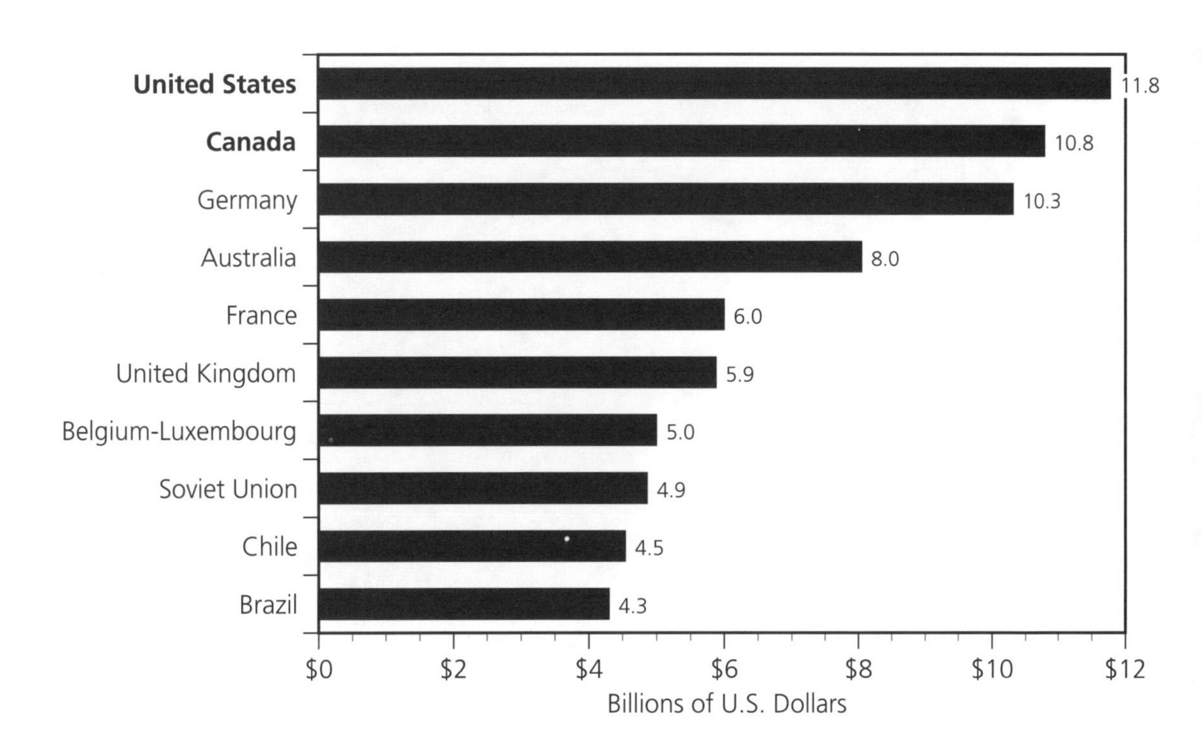

Source: United Nations Conference on Trade and Development (UNCTAD)

Chapter Two
Farms and Fisheries

FARMS

Most U.S. farm produce is grown on western farms and ranches. In 1991 the region accounted for 65 percent of the harvested acreage of principal crops. Only about 8 percent of Canada's land is suitable for farming, but 80 percent of that land is in the West.

Figure 2.1 **Western United States Acreage Harvested by State, 1992 preliminary**

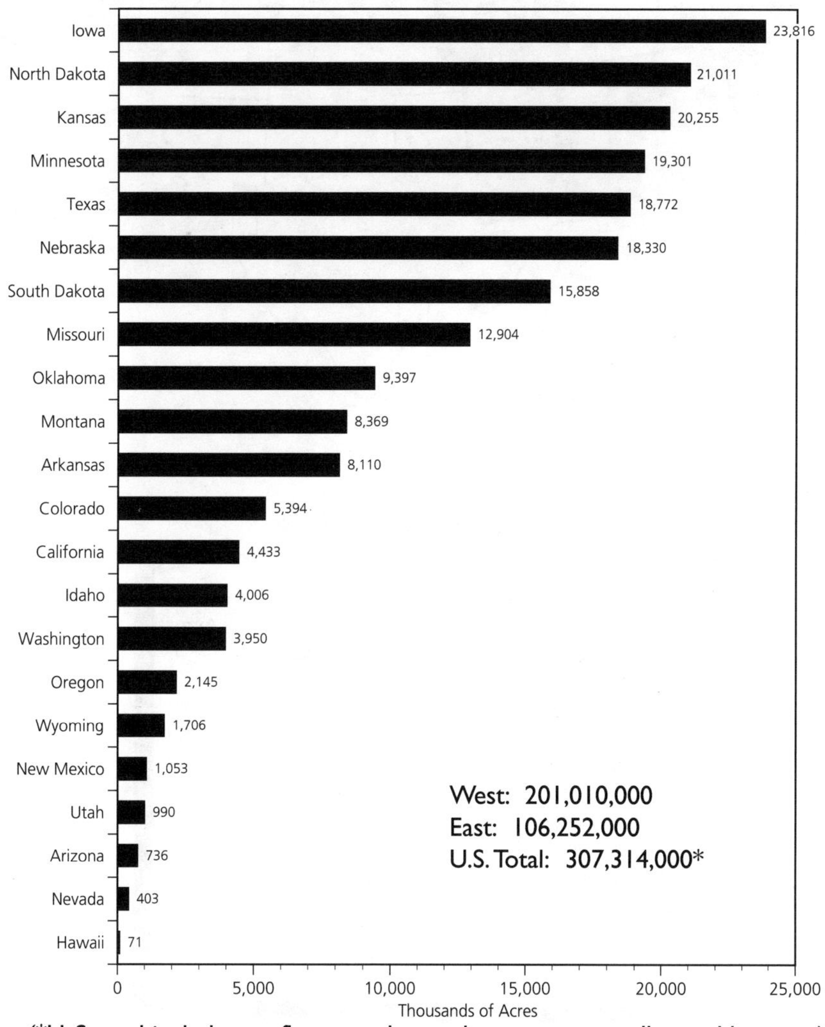

(*U. S. total includes sunflower and sugar beet acreage unallocated by state.)

Source: U.S. Department of Agriculture, National Agricultural Statistics Service

Of 2.1 million farms in the United States in 1991, more than 1 million are in the West. The size of the average farm in the United States is 467 acres (189 hectares). In the West, Texas has the most farms, but California has the most farm income.

Figure 2.2 **Western United States Average Acreage per Farm by State, 1991 preliminary**

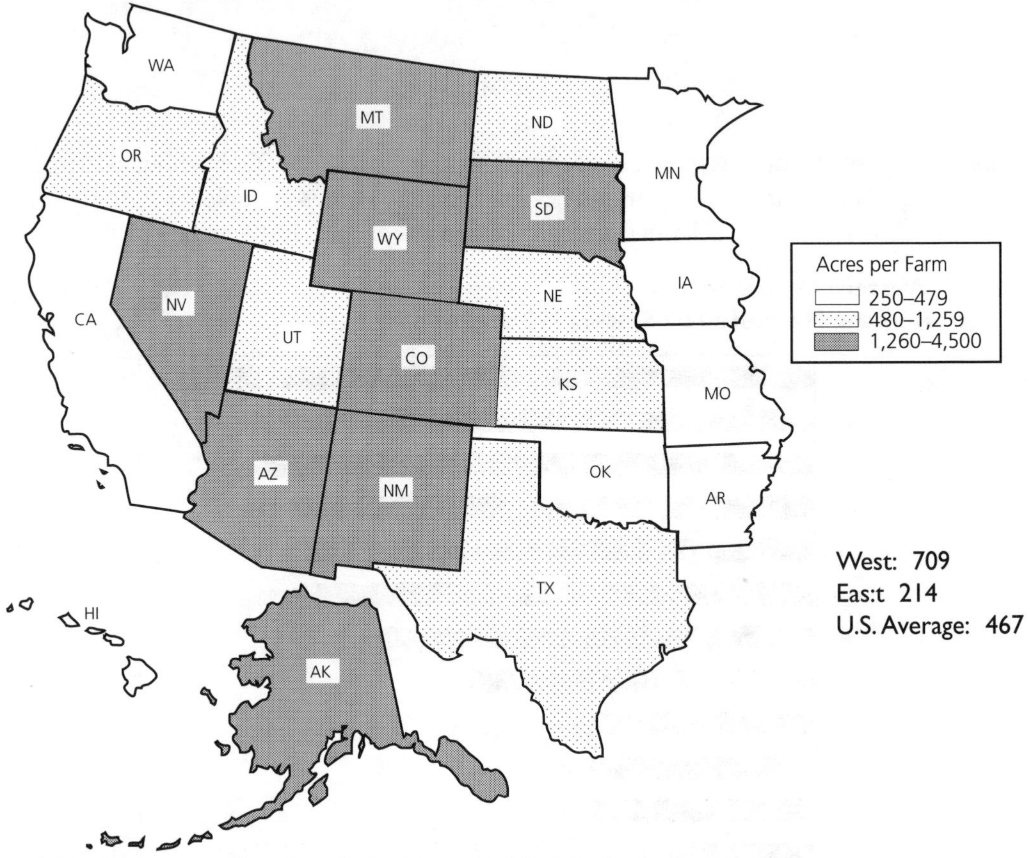

Source: U.S. Department of Agriculture, National Agricultural Statistics Service

	Average Acres		Average Acres
Arizona	4,500	**Kansas**	694
Wyoming	3,867	**Idaho**	631
Nevada	3,560	**Oregon**	481
New Mexico	3,281	**Oklahoma**	471
Montana	2,431	**Washingto**	432
Alaska	1,768	**Hawaii**	372
South Dakota	1,263	**California**	361
Colorado	1,262	**Minnesota**	341
North Dakota	1,224	**Arkansas**	337
Utah	850	**Iowa**	328
Nebraska	841	**Missouri**	284
Texas	708		

Figure 2.3 **Western United States**
Number of Farms per State, 1991 preliminary

West: 1,074,000
East: 1,031,000
U.S. Total: 2,105,000

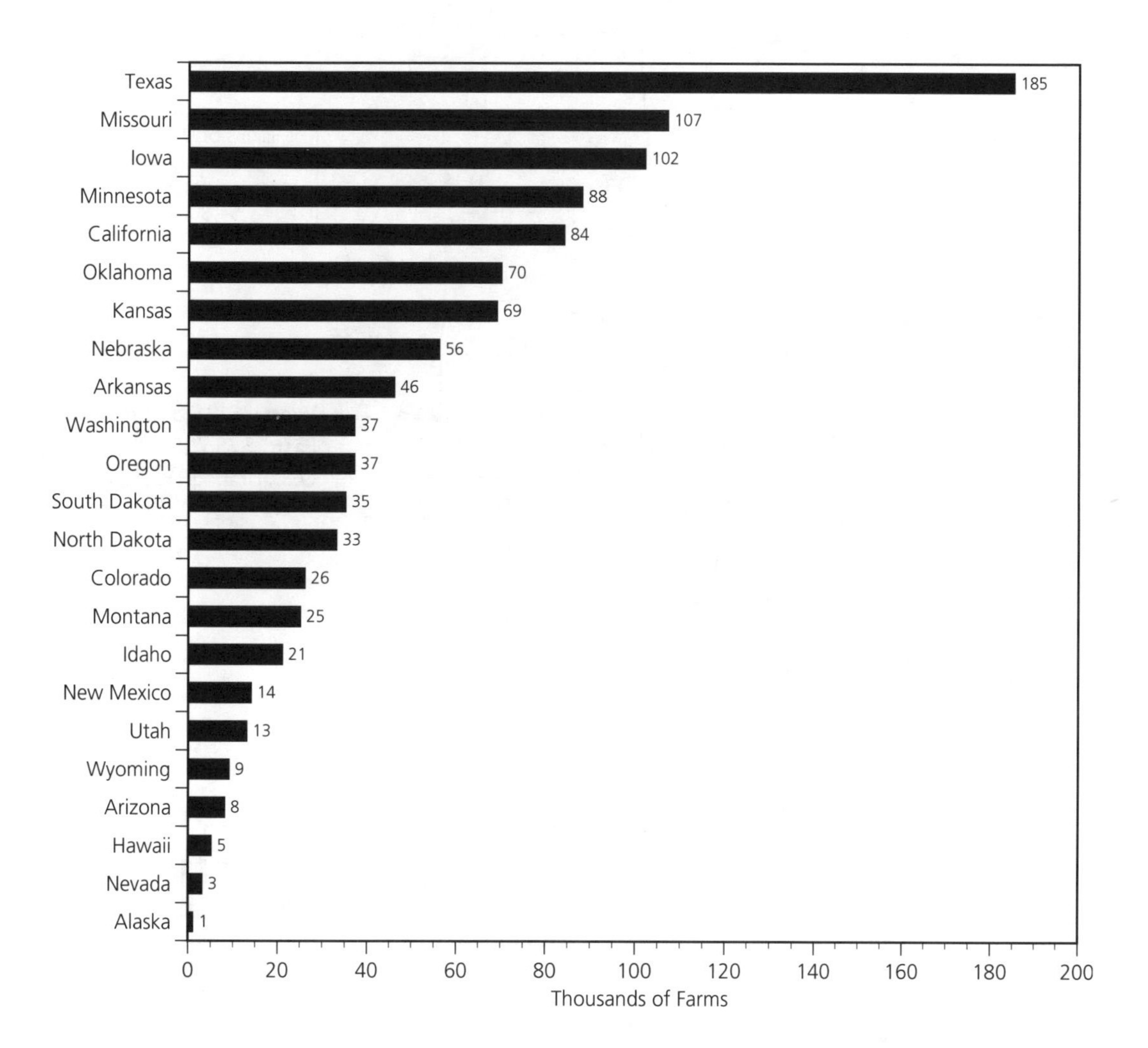

Source: U.S. Department of Agriculture, National Agricultural Statistics Service

Figure 2.4 **Western United States Gross Farm Income by State, 1990**

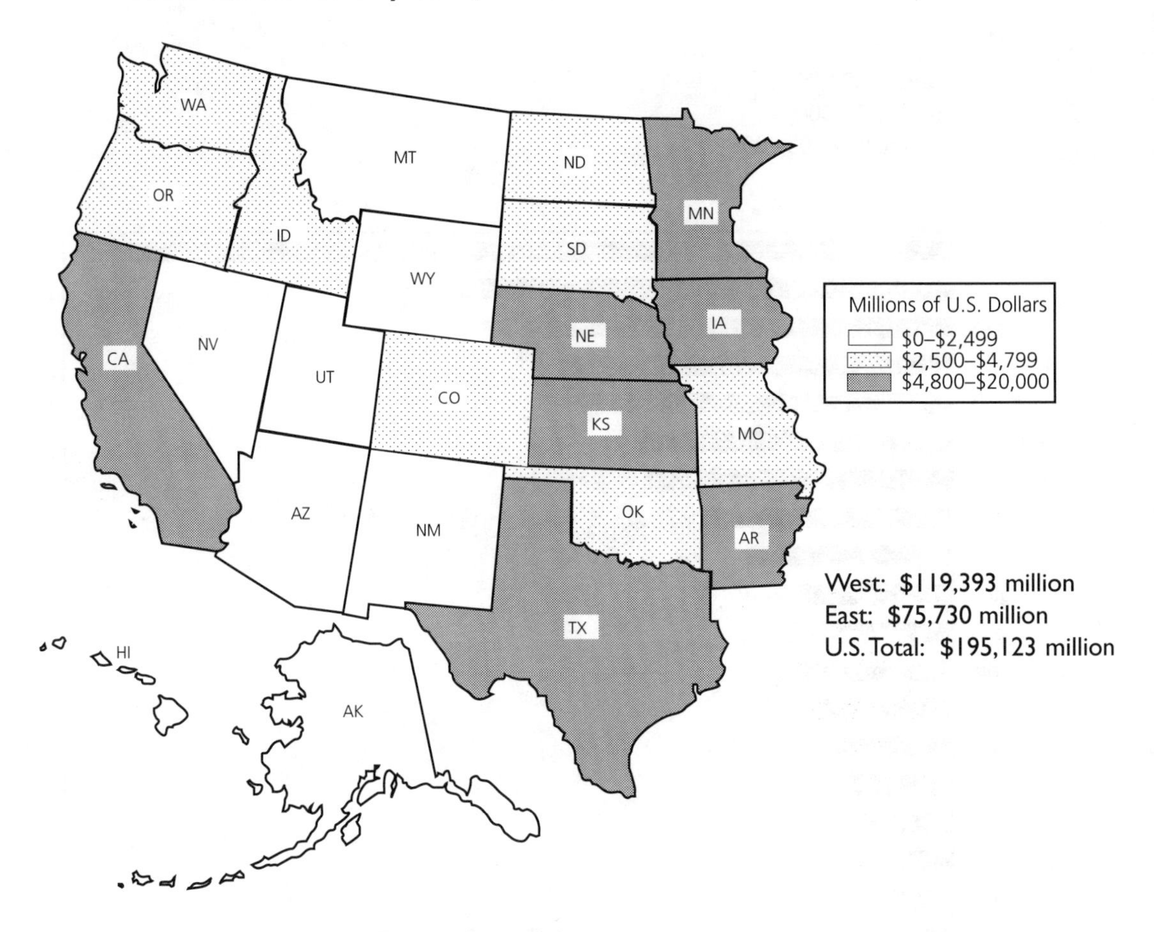

Source: U.S. Department of Agriculture, Economic Research Service

	Millions of U.S. Dollars		Millions of U.S. Dollars
California	$19,962	**North Dakota**	$3,463
Texas	$14,489	**Idaho**	$3,332
Iowa	$12,045	**Oregon**	$2,898
Nebraska	$10,408	**Montana**	$2,098
Kansas	$8,561	**Arizona**	$2,039
Minnesota	$8,398	**New Mexico**	$1,641
Arkansas	$4,810	**Utah**	$883
Colorado	$4,785	**Wyoming**	$857
Missouri	$4,705	**Hawaii**	$614
Washington	$4,525	**Nevada**	$341
Oklahoma	$4,429	**Alaska**	$32
South Dakota	$4,078		

Canada had 280,000 farms in 1991, of which 163,000 were in the Canadian West. Ontario in the East, followed by Saskatchewan and Alberta in the West, had the most farms. Farms in Ontario are relatively small, however. Saskatchewan had the most farm acreage—followed by Alberta and Manitoba. Farm income was greatest in Ontario, followed by Alberta and Saskatchewan.

Figure 2.5 **Canada**
Average Acreage per Farm by Province, 1991

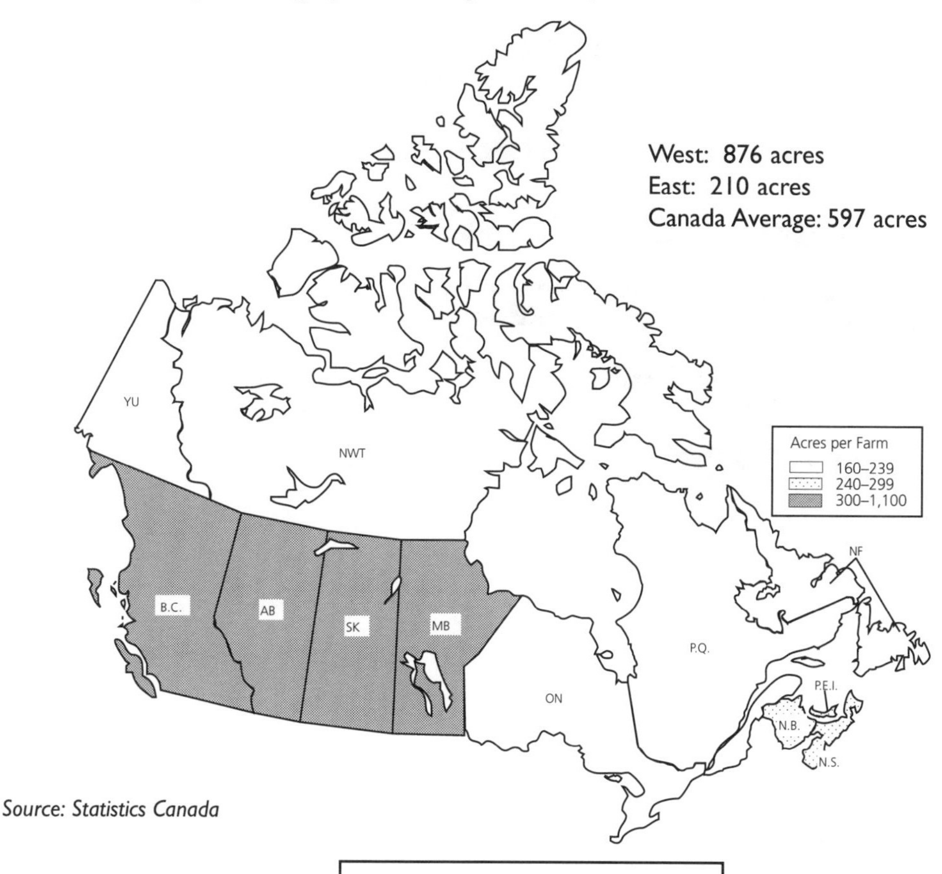

Source: Statistics Canada

	Average Acres
Saskatchewan	1,091
Alberta	898
Manitoba	743
British Columbia	307
New Brunswick	285
Prince Edward Island	271
Nova Scotia	247
Québec	223
Ontario	196
Newfoundland	161

Figure 2.6 **Canada**
Number of Farms by Province, 1991

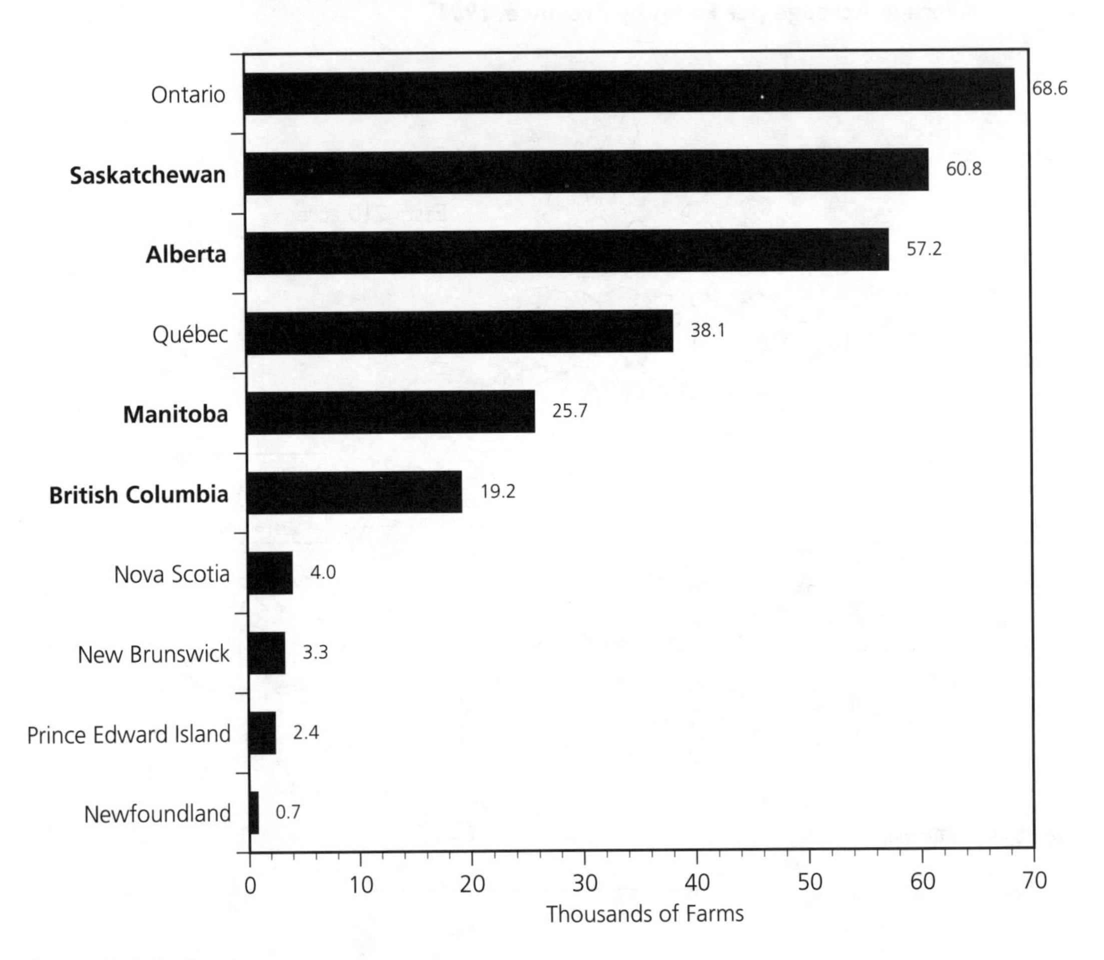

Source: Statistics Canada

Figure 2.7 **Canada**
Gross Farm Income by Province, 1991

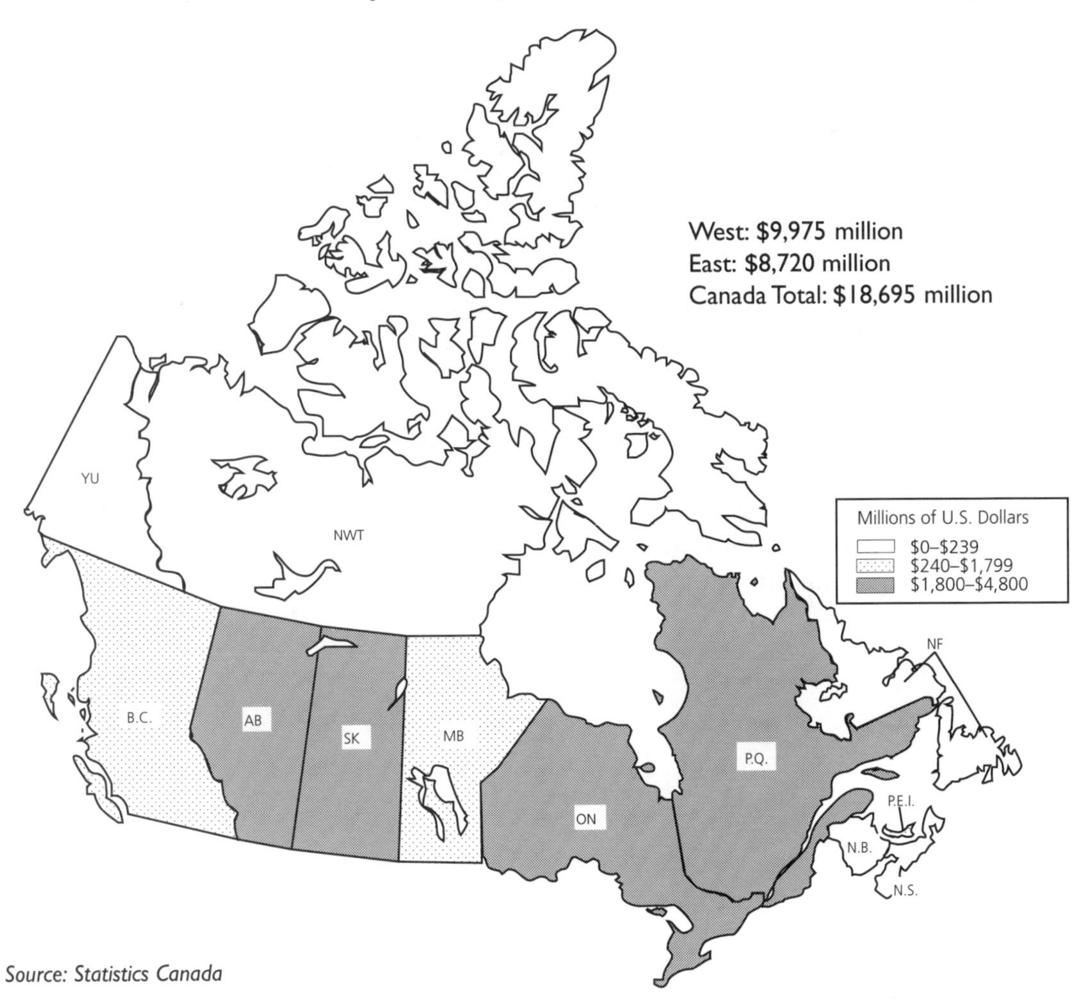

Source: Statistics Canada

	Millions of U.S. Dollars
Ontario	$4,716
Alberta	$3,675
Saskatchewan	$3,500
Québec	$3,250
Manitoba	$1,728
British Columbia	$1,073
Nova Scotia	$269
New Brunswick	$222
Prince Edward Island	$209
Newfoundland	$54

Most of Mexico is unsuitable for agriculture. Much of the northern region is desert. Two mountain ranges run the length of the country. Topsoil in the Yucatán Peninsula is so thin that little can grow. Even where the soil is good enough to plant and flat enough to plow, water is often insufficient for most crops. Fifty-two percent of Mexican territory is arid, 32.5 percent is semi-arid, 10.5 percent is semi-humid, and only 7 percent is humid. About 50 percent is considered too steep for cultivation. Only about 15 percent of Mexico is considered ideal arable land. That's about double the proportion of Canada that is considered suitable for agriculture. But Canada is much larger than Mexico, and has a much smaller population to support.

Because of a history of collective farming dating from pre-Columbian times (the *ejido* system), Mexico had 4.3 million farms in 1991. Averaging 32 acres (12.95 hectares) or less, most of these were too small to be farmed economically, and most of the peasant farmers who farmed them didn't have the capital to farm efficiently. Consequently, most farming in Mexico is at a near subsistence level. However, in 1992, President Salinas proposed a comprehensive agrarian reform bill to the Mexican Congress, which approved it. The bill allows *ejidatarios* and *comuneros* to privatize their parcels with a qualified vote of more than two-thirds within their own assemblies.

Mexico is a major importer of food. The chief Mexican farm products are corn and beans—the staples of the Mexican diet. The leading foodstuffs imported are meat, sorghum, and soybean seed. The principal supplier is the United States.

Figure 2.8 **Mexico**
Average Acreage per Farm by State, 1991

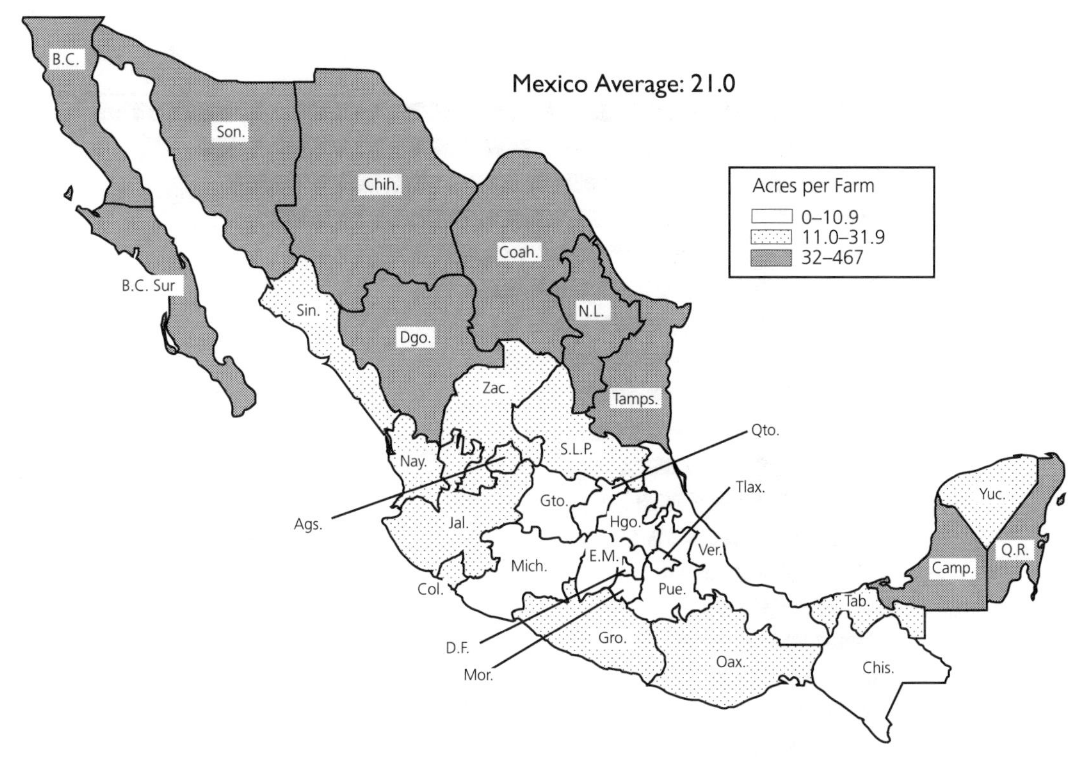

Source: INEGI, VII Censo Agropecuario, 1991

	Average Acres		Average Acres
Baja California Sur	466.5	**Yucatán**	17.1
Baja California	284.1	**San Luis Potosí**	16.6
Sonora	145.4	**Oaxaca**	12.9
Coahuila	113.0	**Tabasco**	11.9
Chihuahua	109.2	**Aguascalientes**	11.2
Durango	64.3	**Michoacán**	10.8
Campeche	55.2	**Chiapas**	10.5
Nuevo León	49.9	**Querétaro**	9.0
Quintana Roo	45.7	**Guanajuato**	8.6
Tamaulipas	34.3	**Veracruz**	8.0
Sinaloa	28.8	**Morelos**	4.1
Guerrero	27.6	**Hidalgo**	3.8
Zacatecas	24.2	**Puebla**	3.4
Nayarit	20.2	**México**	2.5
Jalisco	18.6	**Tlaxcala**	2.3
Colima	18.1	**Distrito Federal**	1.7

Figure 2.9 **Mexico**
Number of Farms by State, 1991

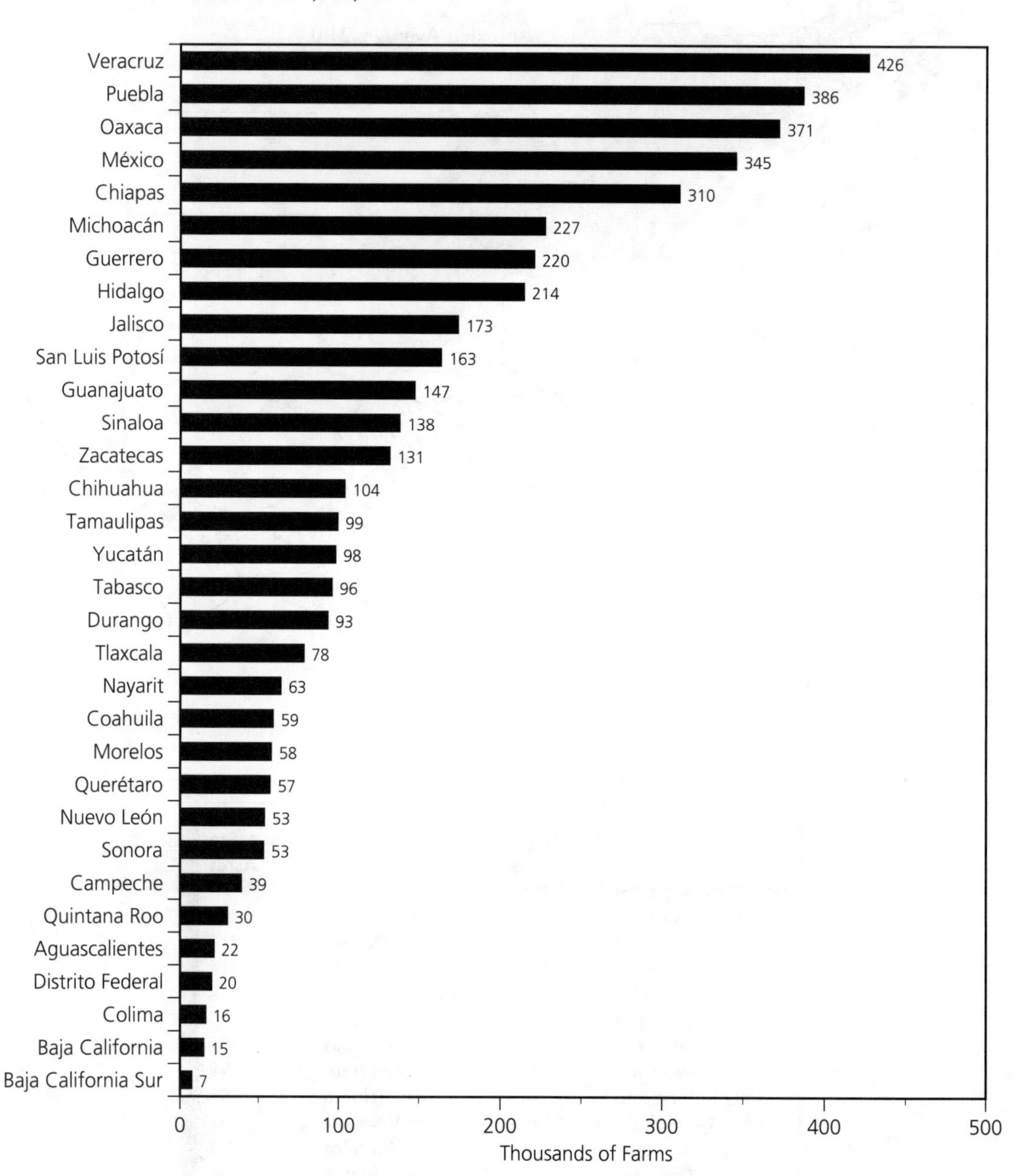

Source: INEGI, VII Censo Agropecuario, 1991

Livestock—chiefly cattle—is the principal product of farms and ranches in the western United States. Wheat, canola, and vegetables are the chief products from Canadian farms. Corn and wheat are the principal products from Mexican farms.

Figure 2.10 **Western United States Principal Farm Products by State, 1990**

Alaska	Greenhouse	Dairy	Potatoes	Hay
Arizona	Cattle	Cotton	Dairy	Hay
Arkansas	Broilers	Cattle	Soybeans	Rice
California	Dairy	Greenhouse	Cattle	Grapes
Colorado	Cattle	Corn	Wheat	Dairy
Hawaii	Sugar	Pineapples	Greenhouse	Nuts
Idaho	Cattle	Potatoes	Dairy	Wheat
Iowa	Hogs	Corn	Cattle	Soybeans
Kansas	Cattle	Wheat	Corn	Hogs
Minnesota	Dairy	Corn	Soybeans	Hogs
Missouri	Cattle	Soybeans	Hogs	Dairy
Montana	Cattle	Wheat	Barley	Hay
Nebraska	Cattle	Corn	Hogs	Soybeans
Nevada	Cattle	Hay	Dairy	Potatoes
New Mexico	Cattle	Dairy	Hay	Chili
North Dakota	Wheat	Cattle	Barley	Sunflower
Oklahoma	Cattle	Wheat	Greenhouse	Broilers
Oregon	Cattle	Greenhouse	Dairy	Wheat
South Dakota	Cattle	Hogs	Wheat	Soybeans
Texas	Cattle	Cotton	Dairy	Greenhouse
Utah	Cattle	Dairy	Hay	Turkeys
Washington	Dairy	Cattle	Apples	Wheat
Wyoming	Cattle	Sugar Beets	Hay	Sheep

Source: U.S. Department of Agriculture, Economic Research Service

Figure 2.11 **Canada**
Principal Farm Products by Province, 1990

Alberta	Barley	Hay	Canola	Wheat	Oats
British Columbia	Hay	Corn	Wheat	Oats	Barley
Manitoba	Hay	Oats	Flaxseed	Wheat	Barley
New Brunswick	Potatoes	Hay	Oats	Barley	Wheat
Newfoundland	Potatoes	Hay			
Nova Scotia	Potatoes	Hay	Corn	Oats	Barley
Ontario	Hay	Corn	Mixed Grains	Wheat	Barley
Prince Edward Island	Potatoes	Hay	Mixed Grains	Oats	Barley
Québec	Hay	Potatoes	Corn	Oats	Barley
Saskatchewan	Hay	Barley	Canola	Wheat	Oats

Source: Statistics Canada, Canada Yearbook

Figure 2.12 **Mexico**
Principal Farm Products by State, 1990

Aguascalientes	Grapes	**Morelos**	Sugar Cane
Baja California	Wheat	**Nayarit**	Tobacco
Baja California Sur	Wheat	**Nuevo León**	Sorghum
Campeche	Rice	**Oaxaca**	Coffee
Coahuila	Alfalfa	**Puebla**	Corn
Colima	Lemons	**Querétaro**	Corn
Chiapas	Corn	**Quintana Roo**	Sugar Cane
Chihuahua	Corn	**San Luis Potosí**	Corn
Distrito Federal	Pears	**Sinaloa**	Wheat
Durango	Beans	**Sonora**	Wheat
Guanajuato	Sorghum	**Tabasco**	Cocoa
Guerrero	Corn	**Tamaulipas**	Sorghum
Hidalgo	Corn	**Tlaxcala**	Corn
Jalisco	Corn	**Veracruz**	Sugar Cane
México	Corn	**Yucatán**	Oranges
Michoacán	Corn	**Zacatecas**	Beans

Source: INEGI

North America is the breadbasket of the world. The world's three great cereal crops are wheat, corn, and rice. The United States is the leading exporter of wheat and corn, and the world's number two exporter of rice. Canada is the world's second largest exporter of wheat, and sixth largest exporter of corn. Together, the United States and Canada export nearly as much wheat as the rest of the world combined.

Figure 2.13 **Largest Exporters of Wheat, 1990**

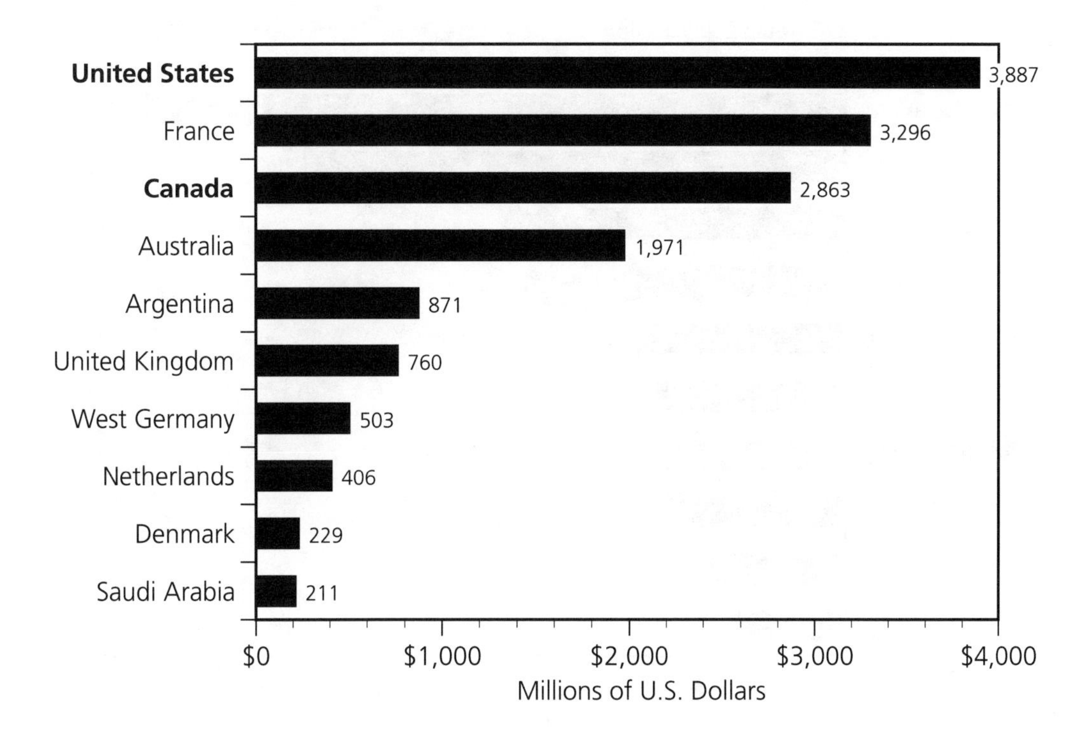

Source: U.S. Department of Agriculture

The largest wheat-producing states are Kansas and North Dakota. Saskatchewan is by far the leading wheat producer in Canada, producing roughly as much as the rest of Canada combined.

Figure 2.14 **United States**
Ten Largest Producers of Wheat, 1992

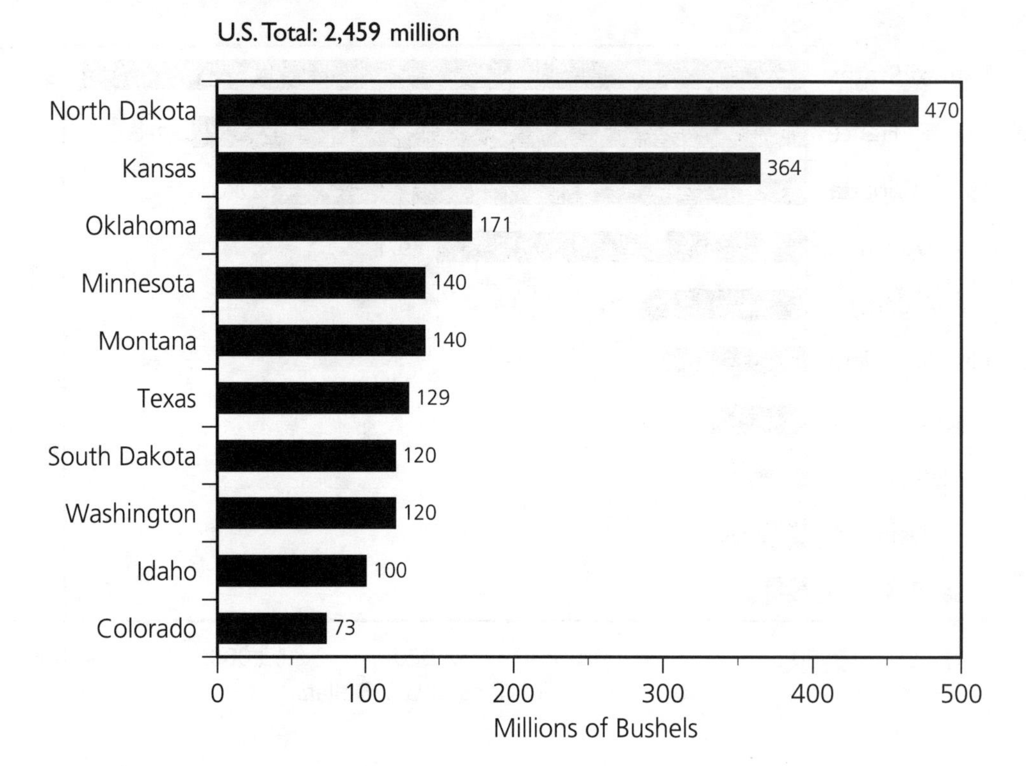

Source: U.S. Department of Agriculture, Agricultural Statistics Service

Figure 2.15 **Canada**
Wheat Production by Province, 1990

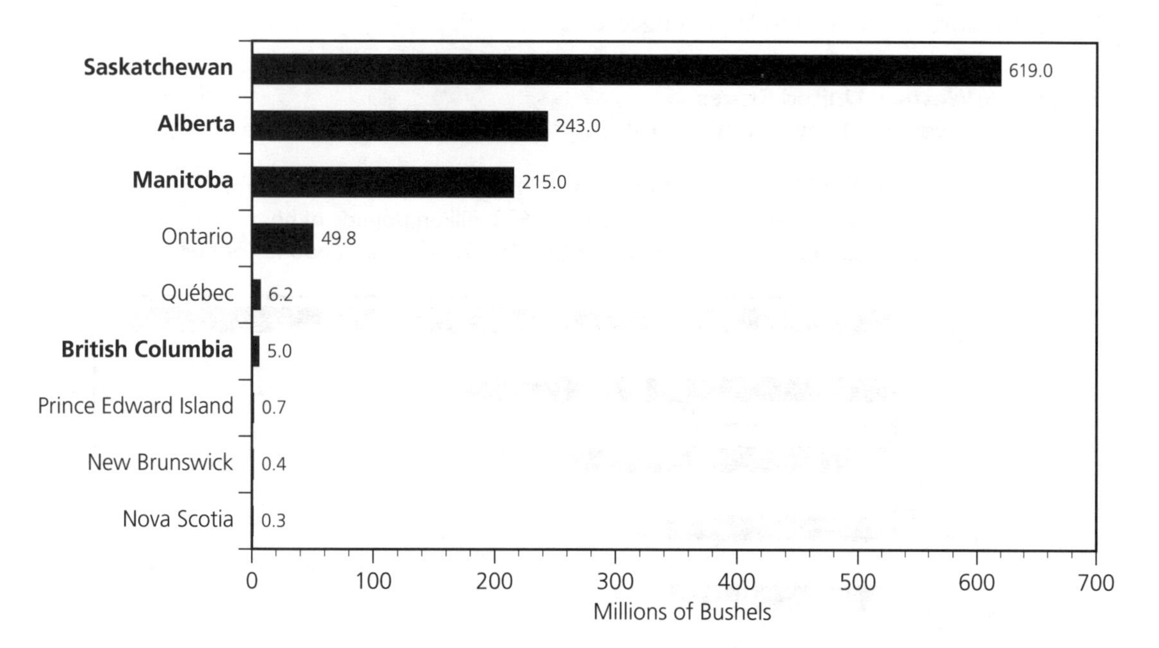

Source: Statistics Canada, Canada Yearbook

The West accounted for 69 percent of the beef produced in the United States in 1992, 92 percent of the sheep and lambs, and 58 percent of the hogs. The largest beef-producing states in the western United States are Texas, Nebraska, and Kansas. Texas, California, Idaho, and Wyoming lead in production of sheep and lambs. Iowa, Minnesota, and Nebraska are the leading hog producers in the western United States.

In Canada, beef is the principal livestock produced. Production is greatest in Alberta.

Figure 2.16 **Western United States Livestock Production by State, 1991**

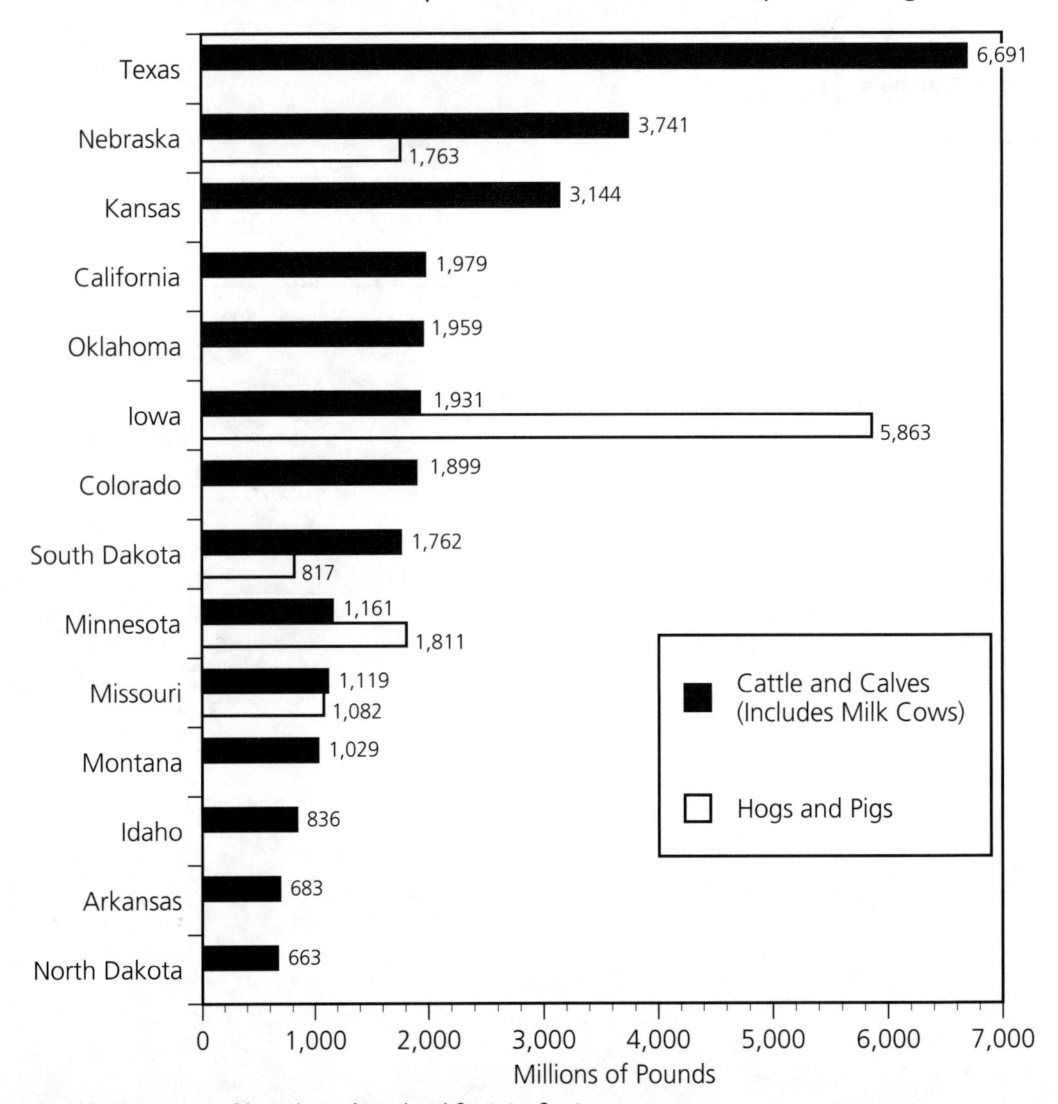

Source: U.S. Department of Agriculture, Agricultural Statistics Service

Figure 2.17 **Canada**
Number of Livestock by Province, 1991

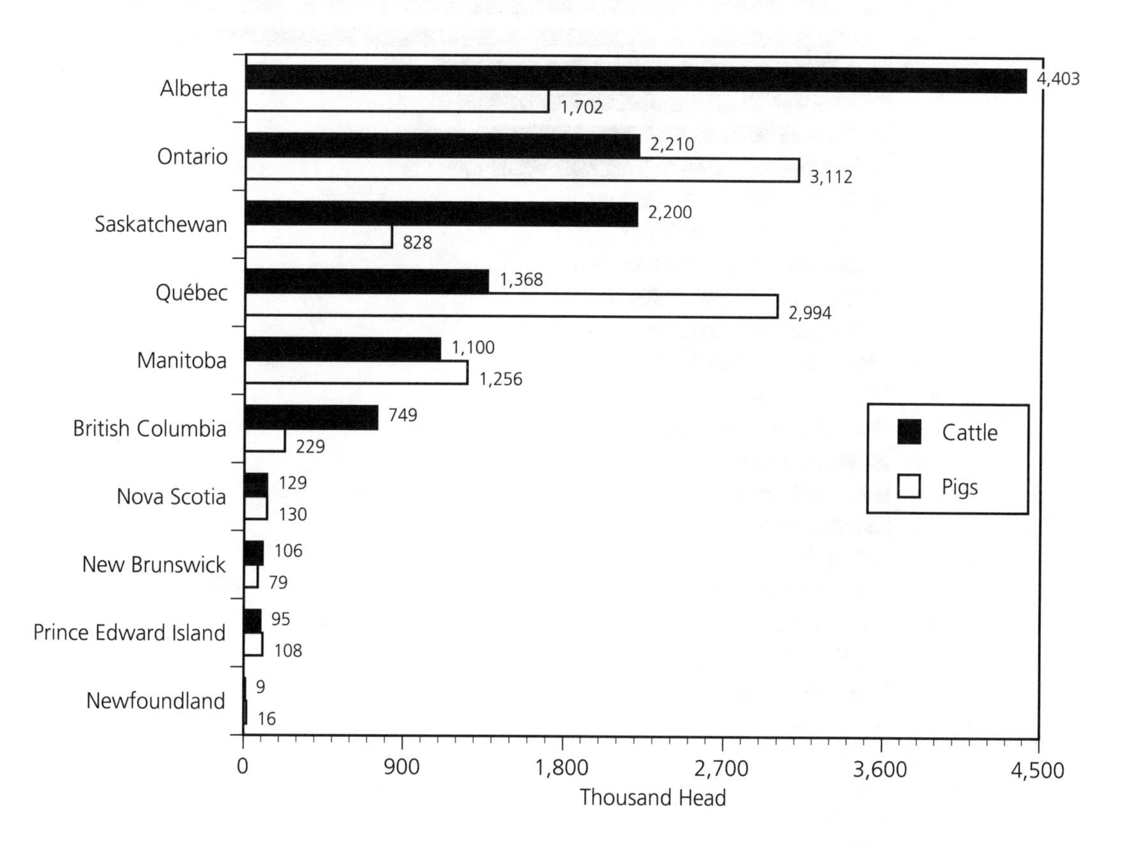

Source: Statistics Canada, Canada Yearbook

Figure 2.18 **Mexico**
Beef Production by State, 1990

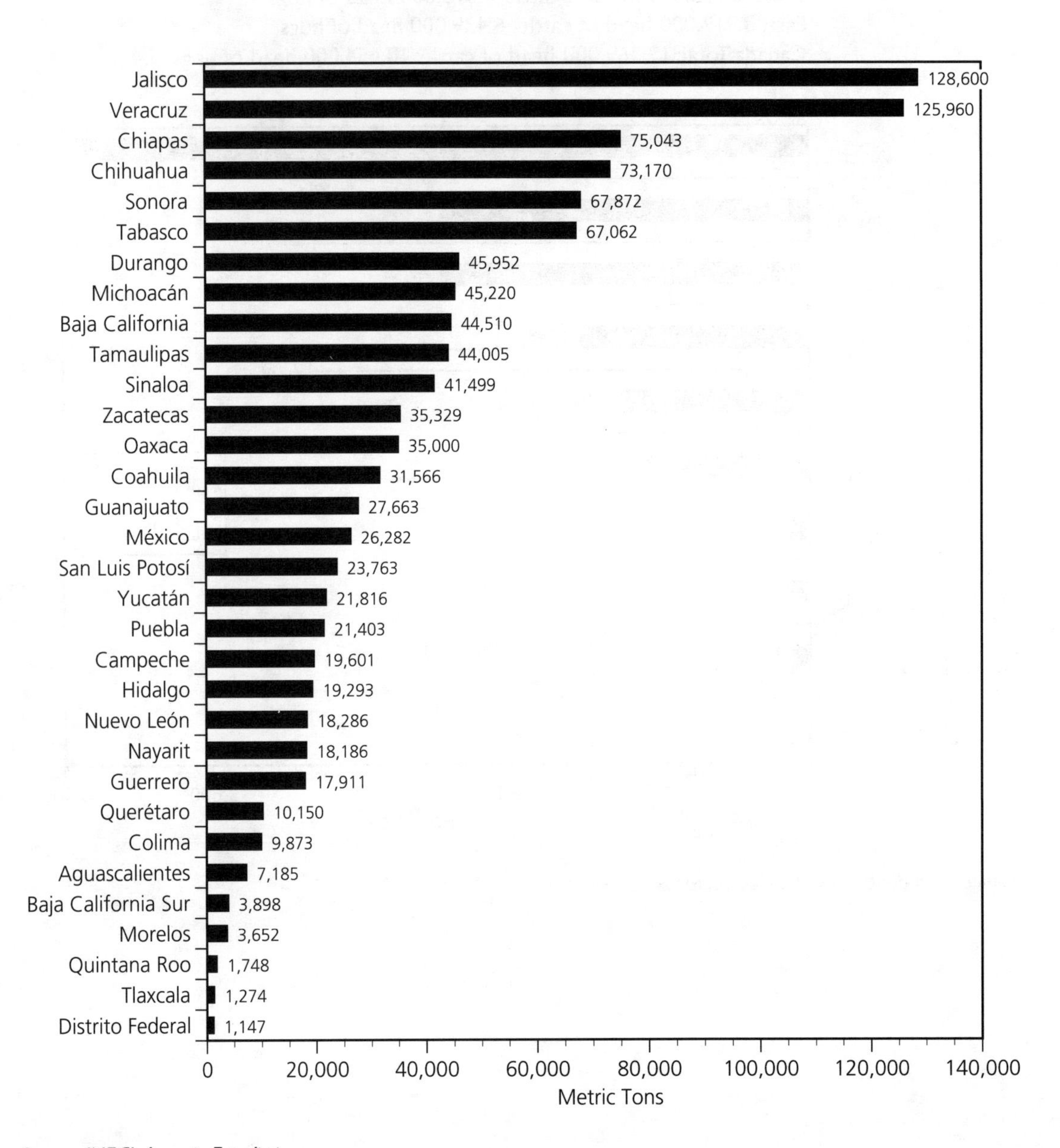

Source: INEGI, Anuario Estadistico

FISHERIES

Total U.S. income from commercial fishing in 1991 was $3.3 billion. More than half came from eight western states. Alaska, with $1.2 billion, had more than twice the commercial fishing income of the other seven states combined, and more than twice the commercial fishing income of western Canada.

Figure 2.19 **Western United States Fishery Income by State, 1991**

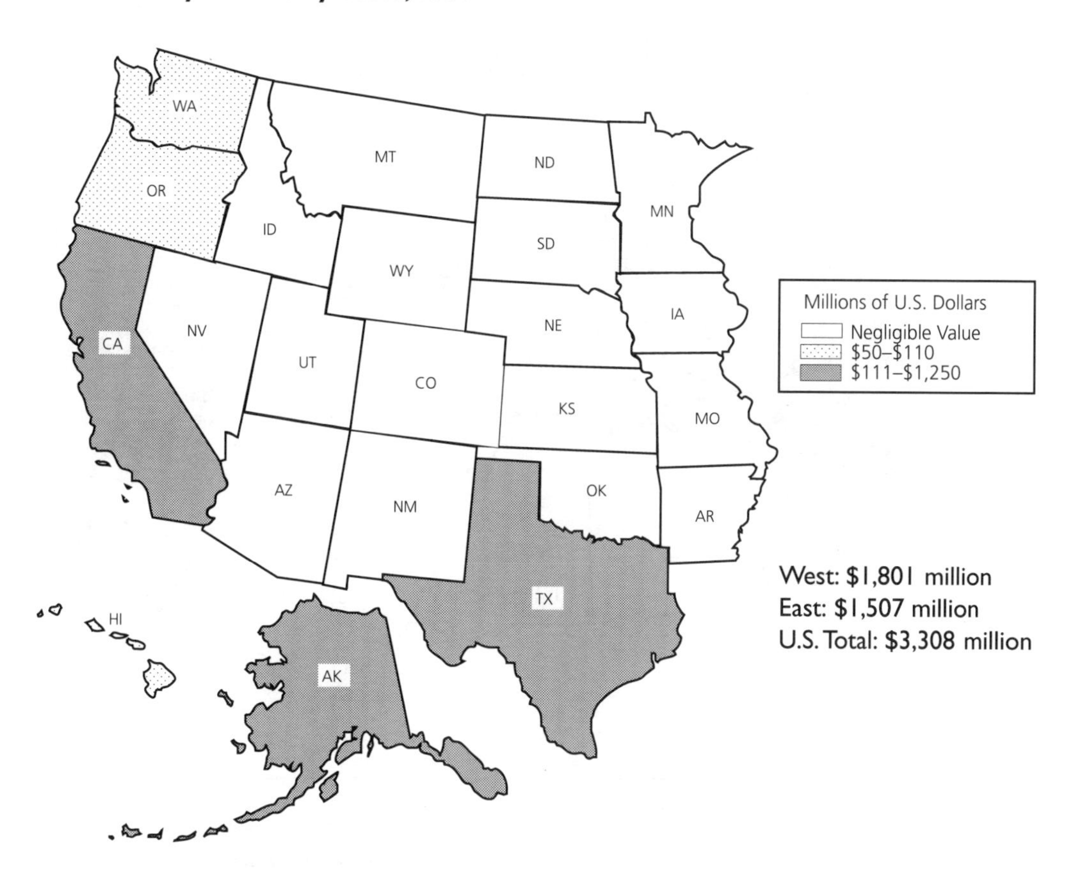

Source: U.S. National Oceanic and Atmospheric Administration, National Marine Fisheries Service

	Millions of U.S. Dollars
Alaska	$1,216
Texas	$214
California	$140
Washington	$110
Oregon	$63
Hawaii	$58

British Columbia, the only western Canadian province with significant income from commercial fishing, ranks second in fishery income among Canada's provinces, accounting for nearly 30 percent of the national total in 1989.

Figure 2.20 **Canada Fishery Income by Province, 1989**

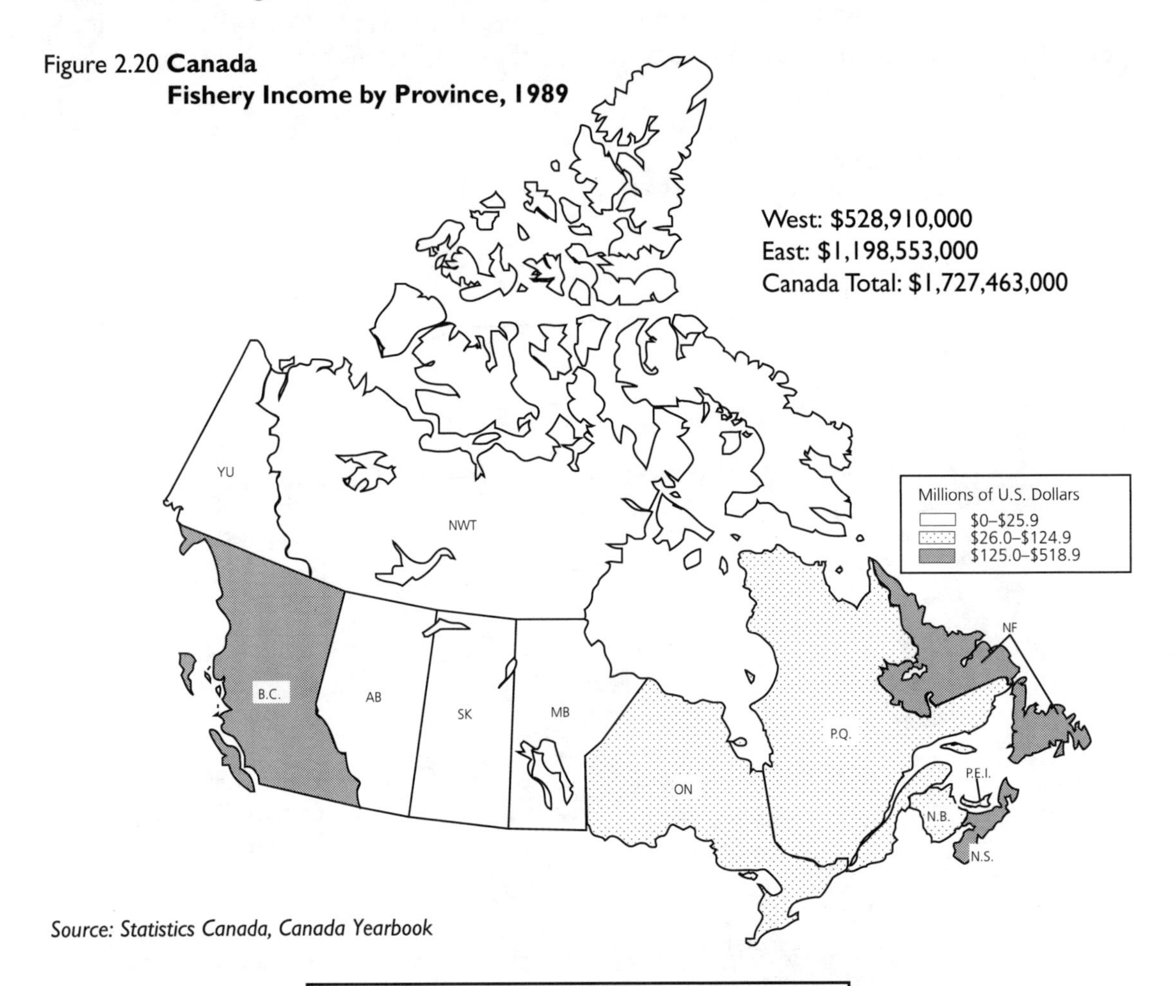

Source: Statistics Canada, Canada Yearbook

	Millions of U.S. Dollars
Nova Scotia	$518.9
British Columbia	$493.0
Newfoundland	$315.4
New Brunswick	$124.3
Québec	$100.3
Prince Edward Island	$82.7
Ontario	$57.0
Manitoba	$25.5
Saskatchewan	$4.9
NWT and Yukon	$3.2
Alberta	$2.3

Although commercial fishing income is relatively small, Canada ranks first in the value of fish exports, chiefly because domestic consumption is so small.

All but two of Mexico's states earn money from commercial fishing, but the earnings are small compared to those of the United States and Canada. Baja California is the leader, earning $129.1 million of the total $647 million of Mexican fishing earnings in 1988.

Figure 2.21 **Mexico Fishery Income by State, 1988**

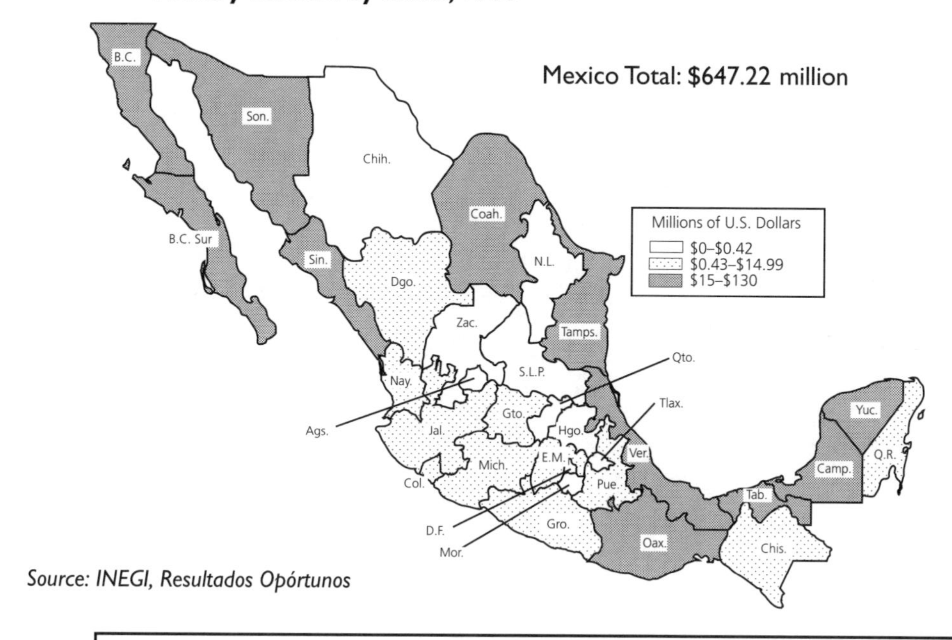

Source: INEGI, Resultados Opórtunos

	Millions of U.S. Dollars		Millions of U.S. Dollars
Baja California	$129.1	**Chihuahua**	$2.2
Sinaloa	$123.4	**México**	$1.7
Sonora	$90.2	**Durango**	$1.1
Campeche	$53.4	**Guanajuato**	$0.7
Veracruz	$48.4	**Puebla**	$0.4
Tamaulipas	$36.0	**Chiapas**	$0.4
Baja California Sur	$32.6	**Hidalgo**	$0.4
Yucatán	$21.3	**Colima**	$0.3
Oaxaca	$21.2	**San Luis Potosí**	$0.2
Tabasco	$18.6	**Morelos**	$0.2
Coahuila	$18.1	**Zacatecas**	$0.1
Nayarit	$14.4	**Tlaxcala**	$0.0
Michoacán	$11.7	**Aguascalientes**	$0.0
Quintana Roo	$10.8	**Querétaro**	$0.0
Guerrero	$6.6	**Distrito Federal**	$0.0
Jalisco	$4.8	**Nuevo León**	$0.0

Chapter Three
Population

Population and Population Growth

In 1991, the United States had 252 million people, the most of the three North American nations. Mexico is second with 81.1 million. Canada is the smallest with 27.4 million. The combined population of North America is 360 million and growing rapidly, primarily because of high birth rates in Mexico. Even though the United States has the largest population of the three North American countries, its population is growing the slowest. The median age for Americans is 32.9 years old. The median age for Canadians is 34.1 years old. For Mexicans, the median age is 20.4 years old. In the United States, westerners are younger than easterners, 31.8 years versus 34.2 years. Only 4.1 percent of the population is age 65 or older in Mexico, compared to 11.5 percent in the western United States and 11.6 percent in Canada. More than 60 percent of the population is age 24 or younger in Mexico, compared to 38 percent in the western United States, and 35 percent in Canada.

Figure 3.1 **Canada**
Population by Age Group, 1991

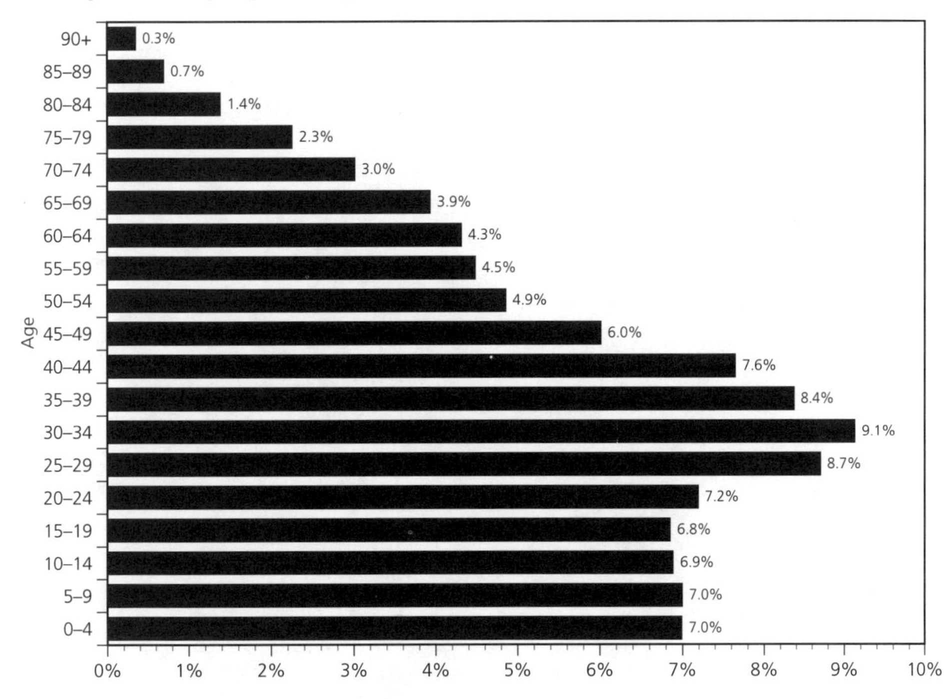

Source: Statistics Canada: Age, Sex, and Marital Status

Figure 3.2 **Canada**
Population Age Groups by Province, 1991

	0–9 Years	10–19 Years	20–39 Years	40–64 Years	65 and Older
Alberta	16.3%	14.3%	35.8%	24.6%	9.1%
British Columbia	13.7%	13.0%	32.5%	28.0%	12.9%
Manitoba	14.9%	14.4%	31.9%	25.4%	13.4%
New Brunswick	13.6%	15.3%	32.6%	26.4%	12.2%
Newfoundland	14.1%	17.9%	33.2%	25.0%	9.7%
Northwest Territories	23.5%	17.5%	37.6%	18.5%	2.8%
Nova Scotia	13.6%	14.3%	32.9%	26.6%	12.6%
Ontario	13.9%	13.3%	33.6%	27.6%	11.7%
Prince Edward Island	15.0%	15.4%	30.7%	25.8%	13.2%
Québec	13.0%	13.6%	33.5%	28.8%	11.2%
Saskatchewan	16.1%	15.3%	30.2%	24.3%	14.1%
Yukon	17.4%	13.9%	38.5%	26.2%	4.0%
Canada Average	**14.0%**	**13.7%**	**33.4%**	**27.3%**	**11.6%**

Source: Statistics Canada: Age, Sex, and Marital Status

Figure 3.3 **Western United States Population by Age Group, 1990**

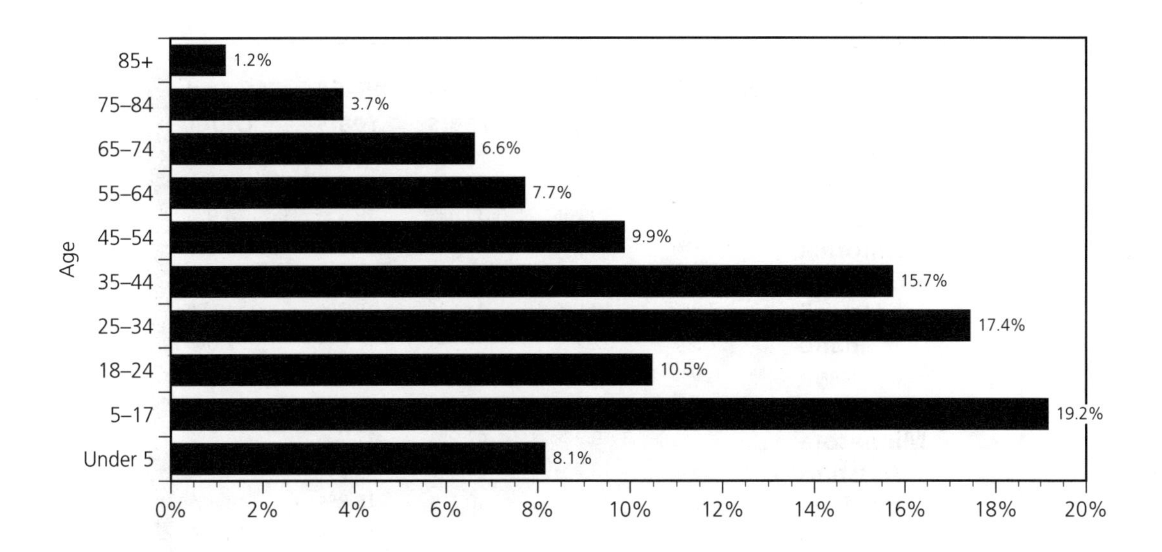

Source: U.S. Bureau of the Census

Figure 3.4 **Western United States Population Age Groups by State, 1990**

	Under 5	5–17 Years	18–44 Years	45–64 Years	65 and Older
Alaska	10.0%	21.6%	48.9%	15.5%	4.0%
Arizona	8.3%	18.7%	42.2%	17.6%	13.3%
Arkansas	7.2%	19.2%	39.4%	19.3%	14.9%
California	8.7%	18.1%	45.6%	17.0%	10.5%
Colorado	7.6%	18.5%	45.8%	18.0%	10.1%
Hawaii	7.8%	17.5%	44.8%	18.3%	11.5%
Idaho	7.9%	22.7%	39.9%	17.4%	12.0%
Iowa	6.9%	19.0%	39.9%	18.7%	15.4%
Kansas	7.6%	19.3%	41.4%	17.8%	13.9%
Minnesota	7.6%	19.2%	42.9%	17.7%	12.5%
Missouri	7.3%	18.7%	41.0%	19.0%	14.1%
Montana	7.3%	20.4%	40.1%	18.8%	13.4%
Nebraska	7.5%	19.8%	40.6%	17.9%	14.1%
Nevada	7.9%	17.1%	44.2%	19.9%	10.8%
New Mexico	8.4%	21.2%	41.9%	17.6%	10.9%
North Dakota	7.3%	20.0%	41.2%	17.2%	14.4%
Oklahoma	7.2%	19.4%	40.9%	18.9%	13.5%
Oregon	7.2%	18.5%	42.0%	18.7%	13.7%
South Dakota	7.7%	20.8%	39.4%	17.4%	14.8%
Texas	8.4%	20.2%	44.2%	17.1%	10.1%
Utah	9.8%	26.4%	40.9%	14.1%	8.8%
Washington	7.6%	18.6%	43.9%	18.1%	11.7%
Wyoming	7.4%	22.2%	42.3%	17.6%	10.5%
West	**8.1%**	**19.2%**	**43.6%**	**17.6%**	**11.5%**
East	**7.3%**	**17.6%**	**42.7%**	**19.1%**	**13.2%**
U.S. Average	**7.6%**	**18.2%**	**43.1%**	**18.5%**	**12.6%**

Source: U.S. Bureau of the Census

Figure 3.5 **Mexico**
Population by Age Group, 1990

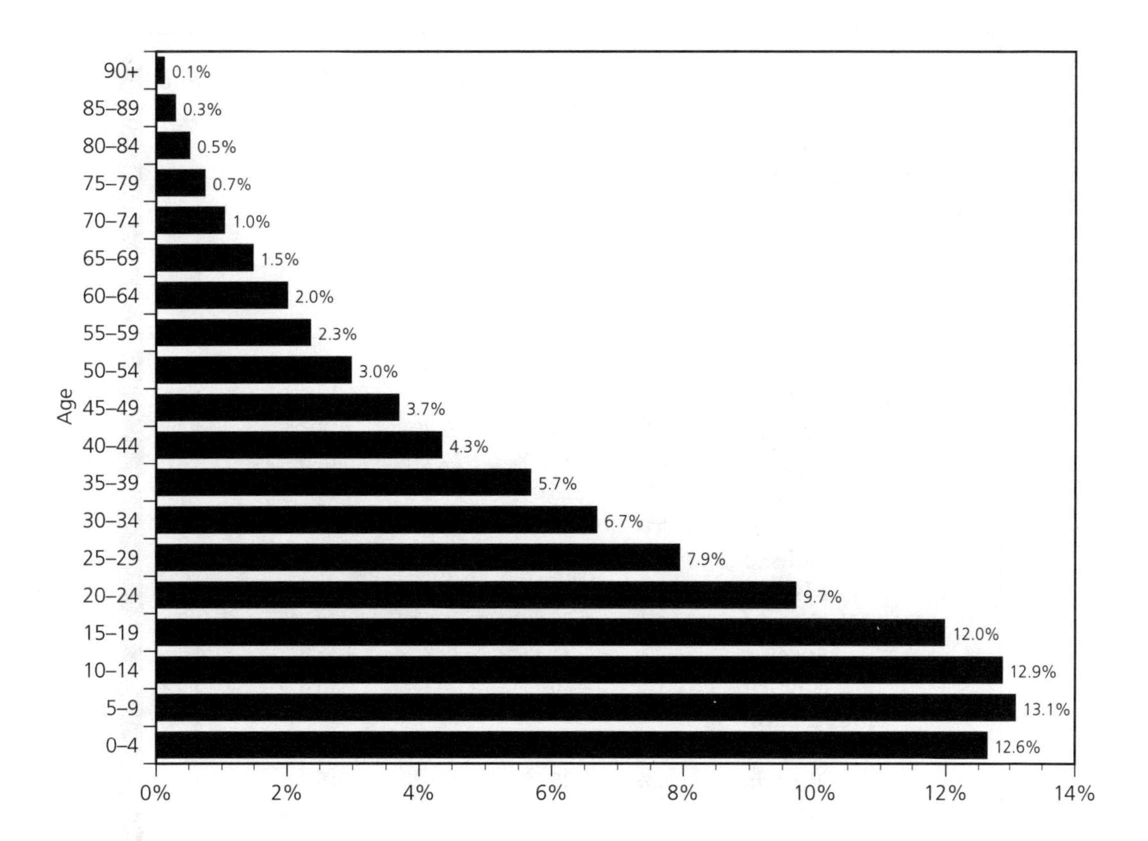

Source: INEGI, Anuario Estadistico

Figure 3.6 **Mexico**
Population Age Groups by State, 1990

	0–9 Years	10–19 Years	20–39 Years	40–64 Years	65 and Older
Aguascalientes	27.3%	25.5%	29.3%	13.8%	4.1%
Baja California	23.8%	23.5%	34.1%	15.1%	3.5%
Baja California Sur	24.5%	24.2%	33.1%	14.8%	3.4%
Campeche	27.2%	24.5%	29.8%	14.7%	3.9%
Coahuila	23.8%	25.2%	31.0%	15.9%	4.1%
Colima	25.2%	25.6%	29.8%	14.9%	4.6%
Chiapas	30.7%	25.1%	28.0%	13.1%	3.1%
Chihuahua	23.9%	24.1%	31.6%	16.3%	4.1%
Distrito Federal	20.4%	22.1%	35.1%	17.6%	4.8%
Durango	26.7%	26.4%	27.2%	15.3%	4.3%
Guanajuato	27.8%	26.1%	27.7%	14.0%	4.4%
Guerrero	28.9%	26.0%	26.4%	14.6%	4.1%
Hidalgo	27.5%	25.1%	27.8%	15.1%	4.4%
Jalisco	26.0%	25.2%	28.7%	15.2%	4.9%
México	25.7%	25.0%	32.1%	14.3%	3.0%
Michoacán	27.4%	26.5%	26.6%	14.7%	4.8%
Morelos	24.7%	25.1%	29.8%	15.9%	4.4%
Nayarit	26.1%	26.1%	27.1%	15.8%	5.0%
Nuevo León	22.1%	24.6%	32.8%	16.4%	4.1%
Oaxaca	28.6%	24.8%	26.0%	15.8%	4.8%
Puebla	27.6%	25.2%	27.6%	15.0%	4.5%
Querétaro	28.3%	25.5%	29.2%	13.3%	3.7%
Quintana Roo	28.3%	23.7%	34.0%	11.8%	2.1%
San Luis Potosí	27.5%	25.2%	27.1%	15.5%	4.7%
Sinaloa	25.1%	26.4%	29.2%	15.4%	4.0%
Sonora	23.8%	24.3%	31.7%	16.2%	4.0%
Tabasco	28.4%	25.6%	29.8%	13.1%	3.1%
Tamaulipas	23.2%	24.3%	31.5%	16.6%	4.4%
Tlaxcala	27.0%	25.6%	28.5%	14.3%	4.6%
Veracruz	25.8%	24.4%	29.7%	16.0%	4.2%
Yucatán	25.3%	23.9%	28.6%	16.5%	5.6%
Zacatecas	27.4%	27.0%	25.8%	14.9%	5.0%
Mexico Average	**25.7%**	**24.8%**	**30.0%**	**15.3%**	**4.2%**

Source: INEGI, Anuario Estadistico

In 1991, the U.S. population was 252.2 million. More than one out of three people—37.5 percent or 94.7 million—lived in the West. California, the region's and the nation's most populous state (30 million), was home to nearly one out of three people in the western region.

Figure 3.7 **Western United States Population by State, 1991**

West: 94,767,000
East: 157,410,000
U.S. Total: 252,177,000

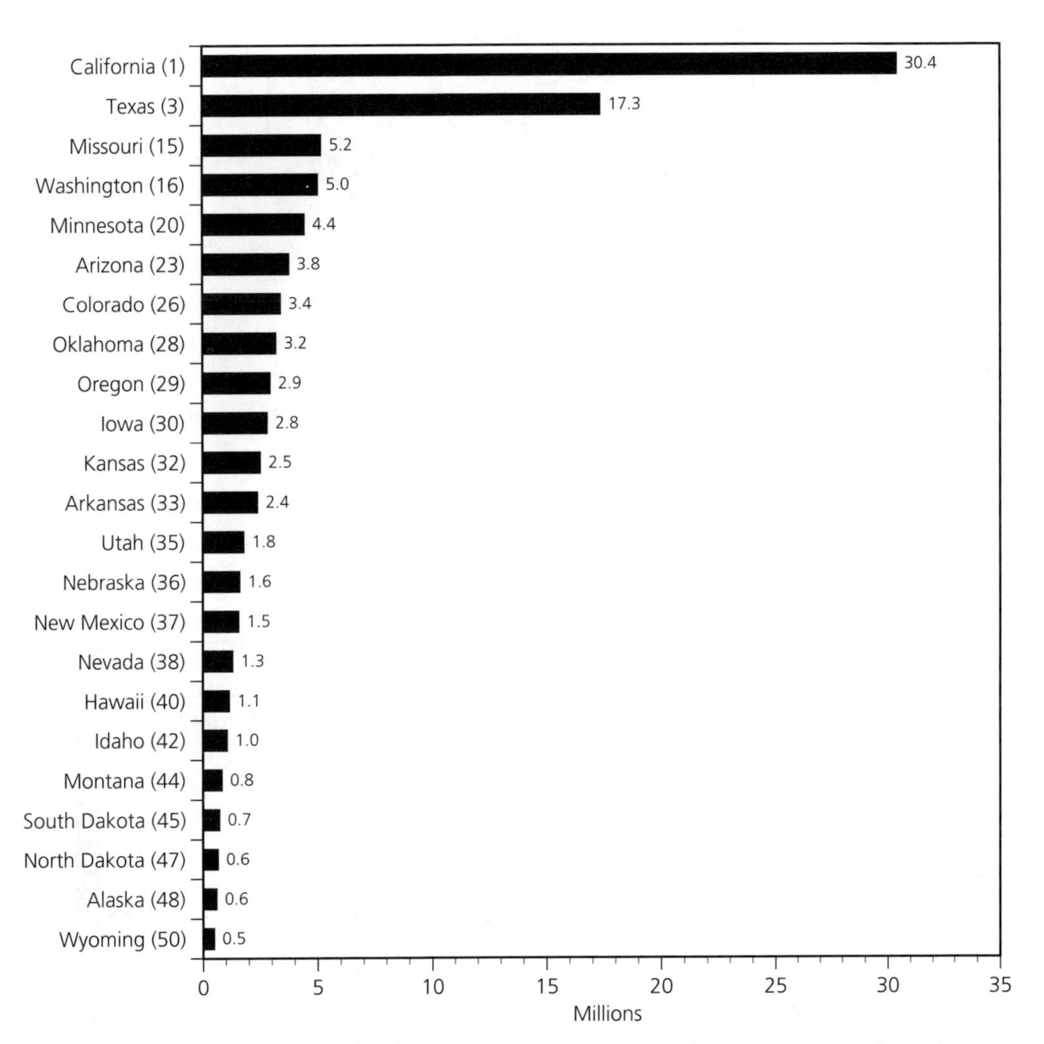

(Numbers in parentheses indicate state's rank in national population.)

Source: U.S. Bureau of the Census

California, with 194.8 persons per square mile (75.2 persons per sq. km), is the most densely populated state in the West. Alaska, with only one person per square mile (2.4 persons per sq. km), is the least densely populated.

Figure 3.8 **Western United States Population Density by State, 1991**

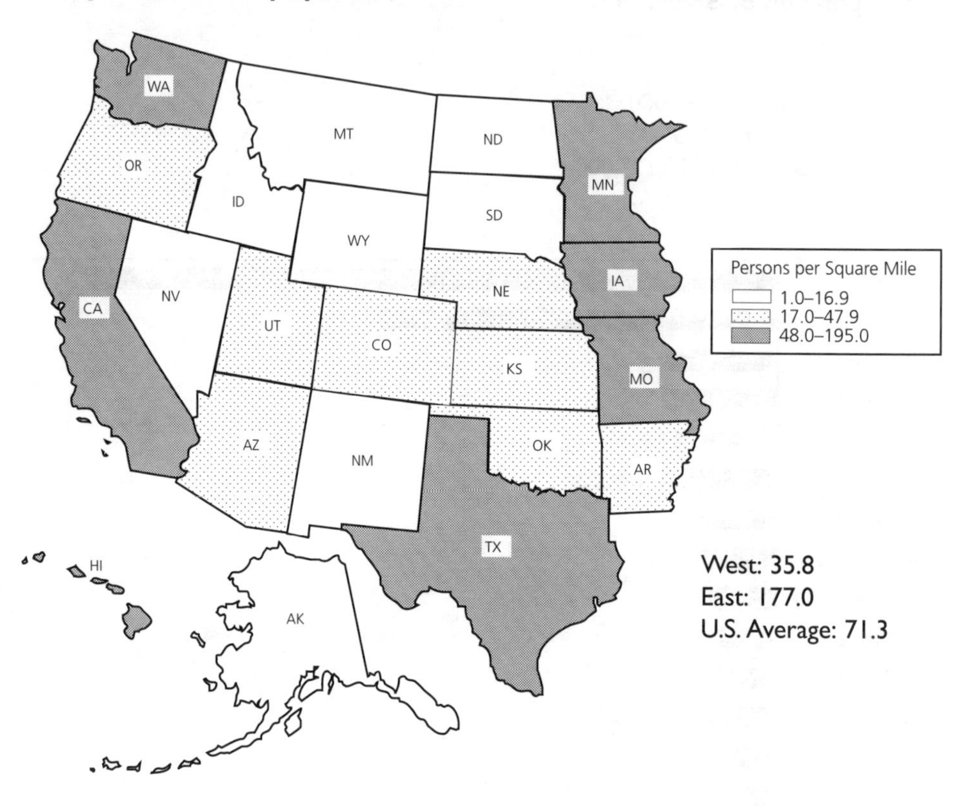

Source: U.S. Bureau of the Census

	Persons per Square Mile		Persons per Square Mile
California	194.8	**Oregon**	30.4
Hawaii	176.7	**Utah**	21.5
Washington	75.4	**Nebraska**	20.7
Missouri	74.9	**New Mexico**	12.8
Texas	66.2	**Idaho**	12.6
Minnesota	55.7	**Nevada**	11.7
Iowa	50.0	**South Dakota**	9.3
Oklahoma	46.2	**North Dakota**	9.2
Arkansas	45.5	**Montana**	5.6
Arizona	33.0	**Wyoming**	4.7
Colorado	32.6	**Alaska**	1.0
Kansas	30.5		

In 1991 Canada had a population of 26.8 million, of whom nearly 30 percent (7.9 million), lived in the Canadian West. The most populous western Canadian province is British Columbia, with 3.2 million people, 40 percent of the regional total.

Figure 3.9 **Canada**
Population by Province, 1991 (western provinces in bold)

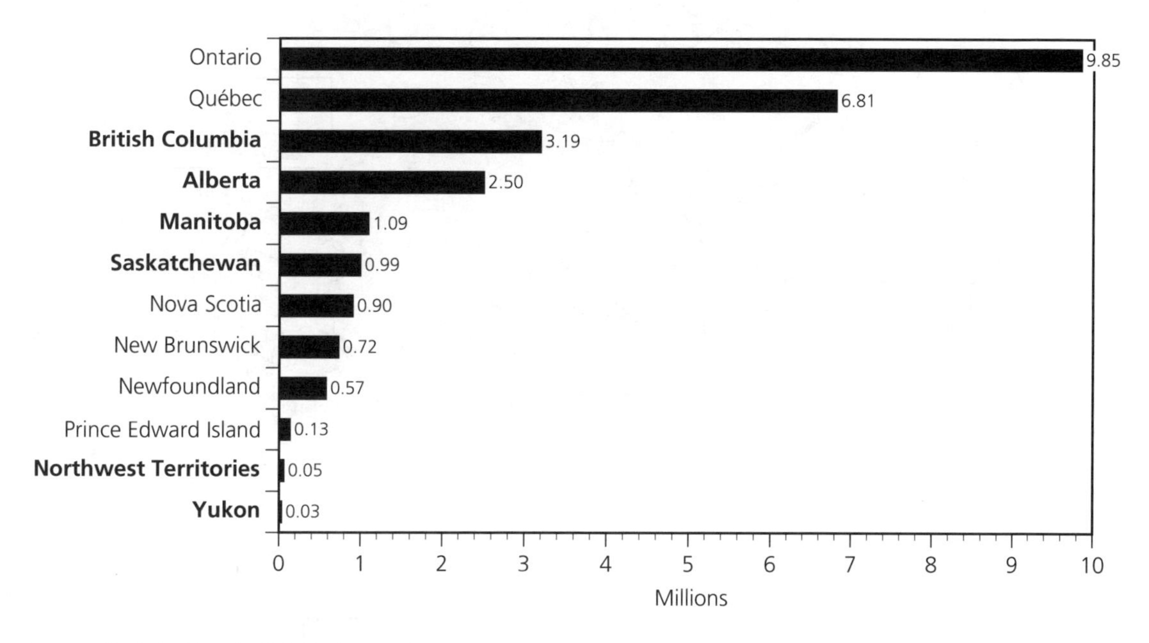

Source: Statistics Canada, Canada Yearbook

Figure 3.10 **Canada**
Population Density by Province, 1992

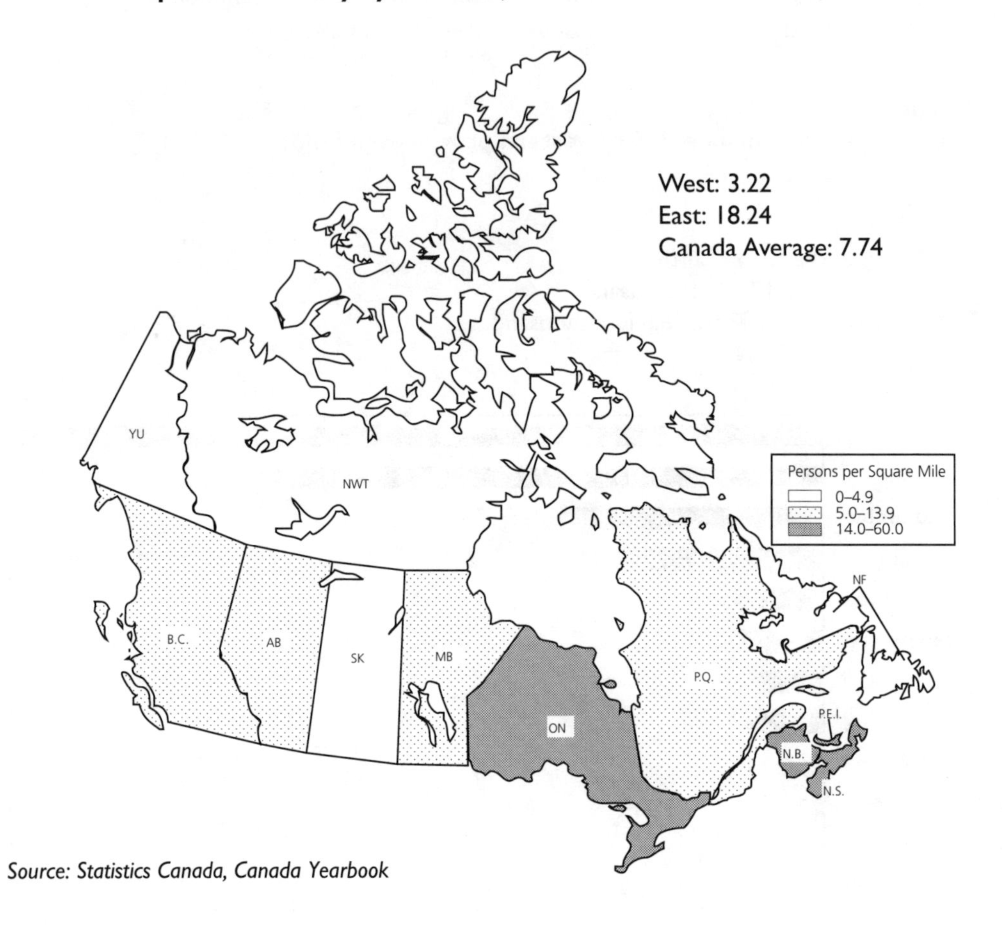

Source: Statistics Canada, Canada Yearbook

	Persons per Square Mile
Prince Edward Island	59.63
Nova Scotia	44.05
Ontario	28.61
New Brunswick	26.04
Québec	13.01
Alberta	10.30
British Columbia	8.89
Manitoba	5.15
Saskatchewan	4.50
Newfoundland	3.98
Yukon	0.14
Northwest Territories	0.04

The most densely populated part of the Canadian West is Alberta, with 10.3 persons per square mile (4 persons per sq. km). The least densely populated part of Canada is the Northwest Territories, with only .04 persons per square mile (.015 person per sq. km). For Canada as a whole, population density averages 7.7 persons per square mile (3 persons per sq. km).

Canada's population is concentrated in the south, much of it close to the border. In fact, most Canadians live within an hour's drive of the U.S. border. There are only two sizeable Canadian communities—Calgary and Edmonton in Alberta—that are more than 150 miles from the U.S. border.

Mexico, with a 1990 population of 81.3 million, is the most densely populated portion of western North America. Population densities range from a low of 11.2 persons per square mile (4.3 persons per sq. km) in Baja California Sur to 14,422 persons per square mile (5,568 persons per sq. km) in the Distrito Federal, Mexico City's metropolitan area.

The national average for Mexico is 107.7 persons per square mile (41.6 persons per sq. km), compared to 71.3 persons per square mile (27.5 per sq. km.) for the United States, 35.8 per square mile (13.8 per sq. km.) for the western United States, 7.7 per square mile (3 per sq. km.) for Canada, and 3.2 per square mile (1.2 per sq. km.) for western Canada. The population density for Baja California del Sur, the least densely populated Mexican state, is more than twice as great as the average population density for all of western Canada, and is greater than the population density of the U.S. states of Alaska, Wyoming, Montana, North Dakota, and South Dakota. The average population density of Mexico is greater than any western U.S. states except California and Hawaii.

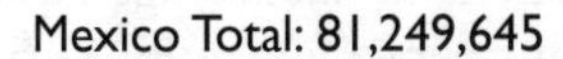
Figure 3.11 **Mexico Population by State, 1990**

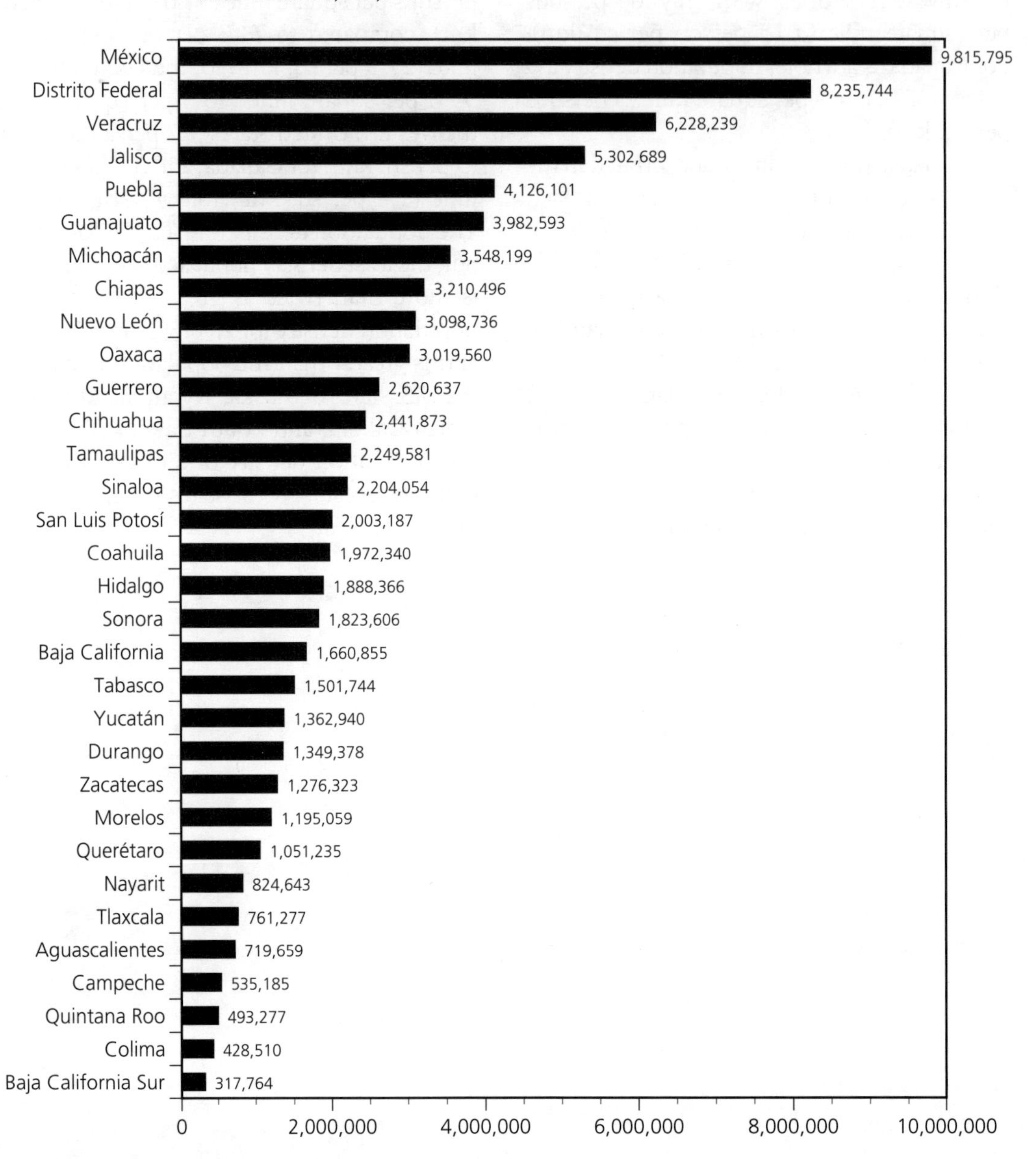

Source: INEGI, Anuario Estadistico

Figure 3.12 **Mexico**
Population Density by State, 1990

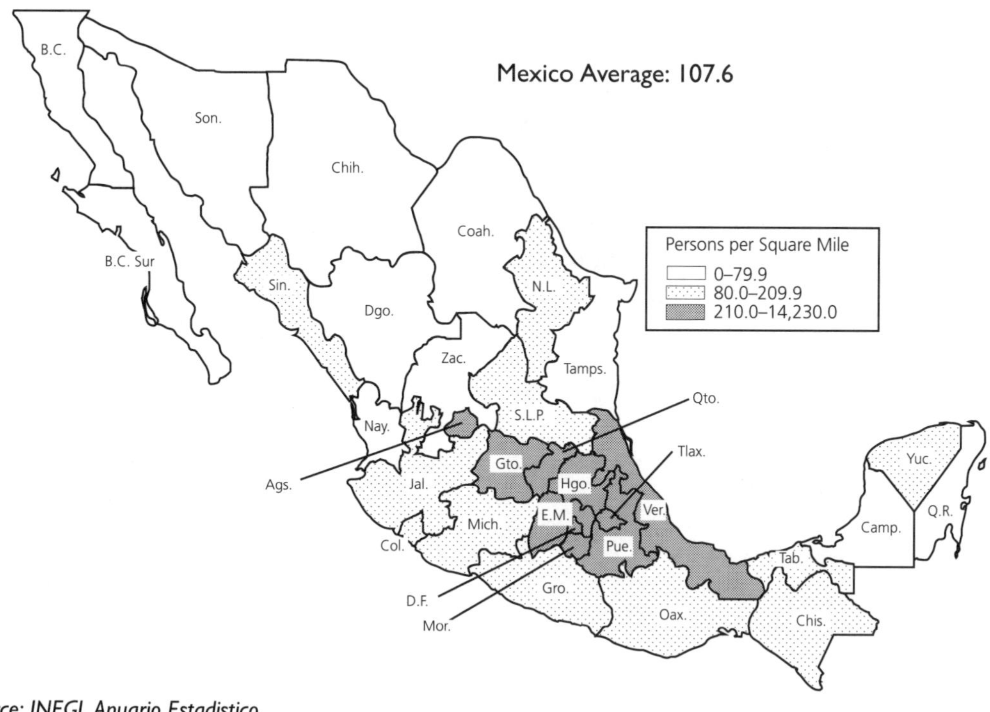

Source: INEGI, Anuario Estadistico

	Persons per Square Mile		Persons per Square Mile
Distrito Federal	14,229.8	**Guerrero**	106.4
México	1,184.6	**Sinaloa**	98.3
Morelos	626.4	**Yucatán**	89.7
Tlaxcala	503.8	**San Luis Potosí**	82.6
Guanajuato	337.2	**Oaxaca**	82.0
Aguascalientes	333.5	**Nayarit**	77.3
Puebla	315.1	**Tamaulipas**	73.0
Hidalgo	233.0	**Baja California**	61.4
Querétaro	231.3	**Zacatecas**	44.1
Veracruz	221.5	**Coahuila**	33.7
Colima	203.5	**Durango**	29.2
Jalisco	171.4	**Campeche**	26.7
Tabasco	157.7	**Chihuahua**	25.6
Michoacán	153.5	**Sonora**	25.5
Nuevo León	124.3	**Quintana Roo**	25.4
Chiapas	112.5	**Baja California Sur**	11.2

Urbanization

All three North American countries are urbanized, but none to the extent that Japan or many European companies are. Mexico is the least urbanized, with 66.3 percent of the population living in urban areas in 1988. Canada is the most, with 76.4 percent. The United States ranks in the middle with 73.7 percent. Worldwide, Canada ranks thirtieth in degree of urbanization, the United States is thirty-third, and Mexico is forty-fourth.

Contrary to popular belief, in both the United States and Canada, the West is more urbanized than the East as people settle in urban clusters, creating urban archipelagos dotting the region's vast land area. Of the various regions of western North America (excluding the Distrito Federal), the most urbanized state in all three countries is Nuevo León. The least is Oaxaca. In all of North America, the least urbanized state/province is Québec. The least urbanized state in the western United States is South Dakota. The most, California.

Figure 3.13 **Canada**
Urban Population by Province, 1992

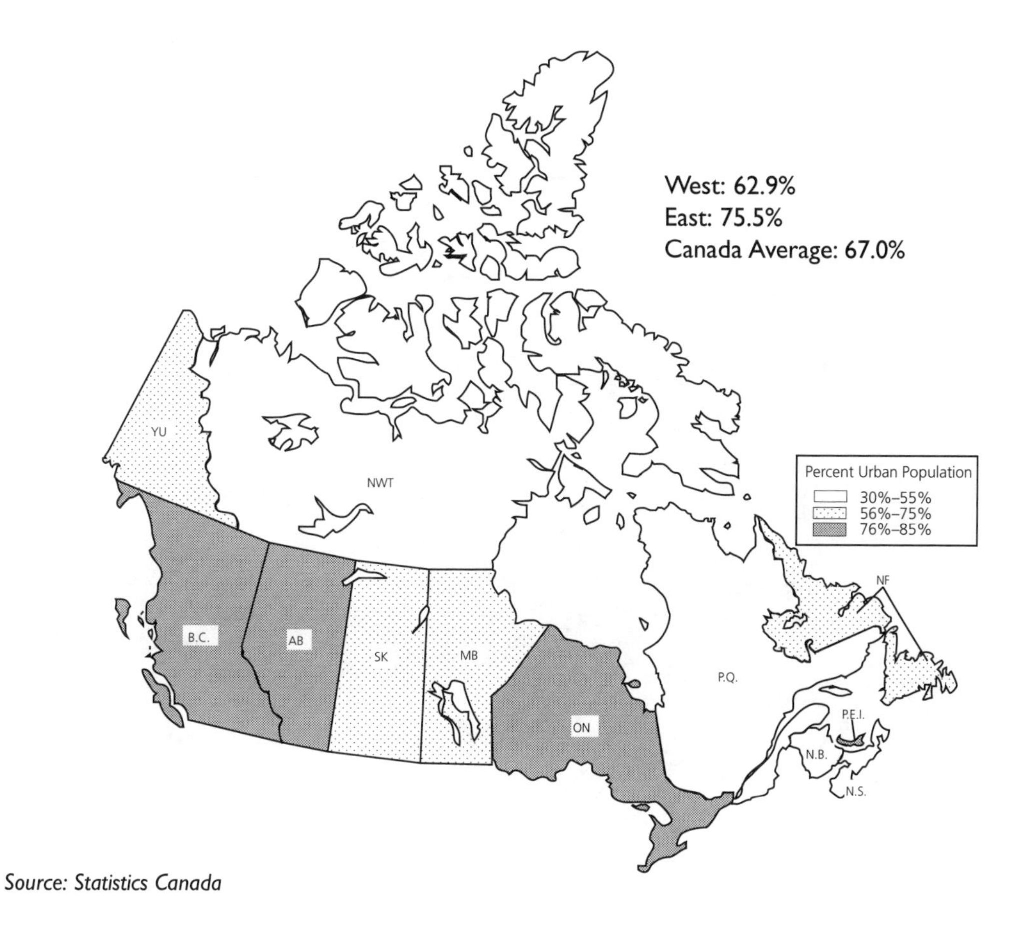

Source: Statistics Canada

	Percent Urban Population
Ontario	82%
Alberta	79%
British Columbia	79%
Prince Edward Island	78%
Manitoba	72%
Yukon	65%
Saskatchewan	61%
Newfoundland	57%
Nova Scotia	54%
New Brunswick	49%
Northwest Territories	46%
Québec	38%

Figure 3.14 **Western United States Urban Population by State, 1990**

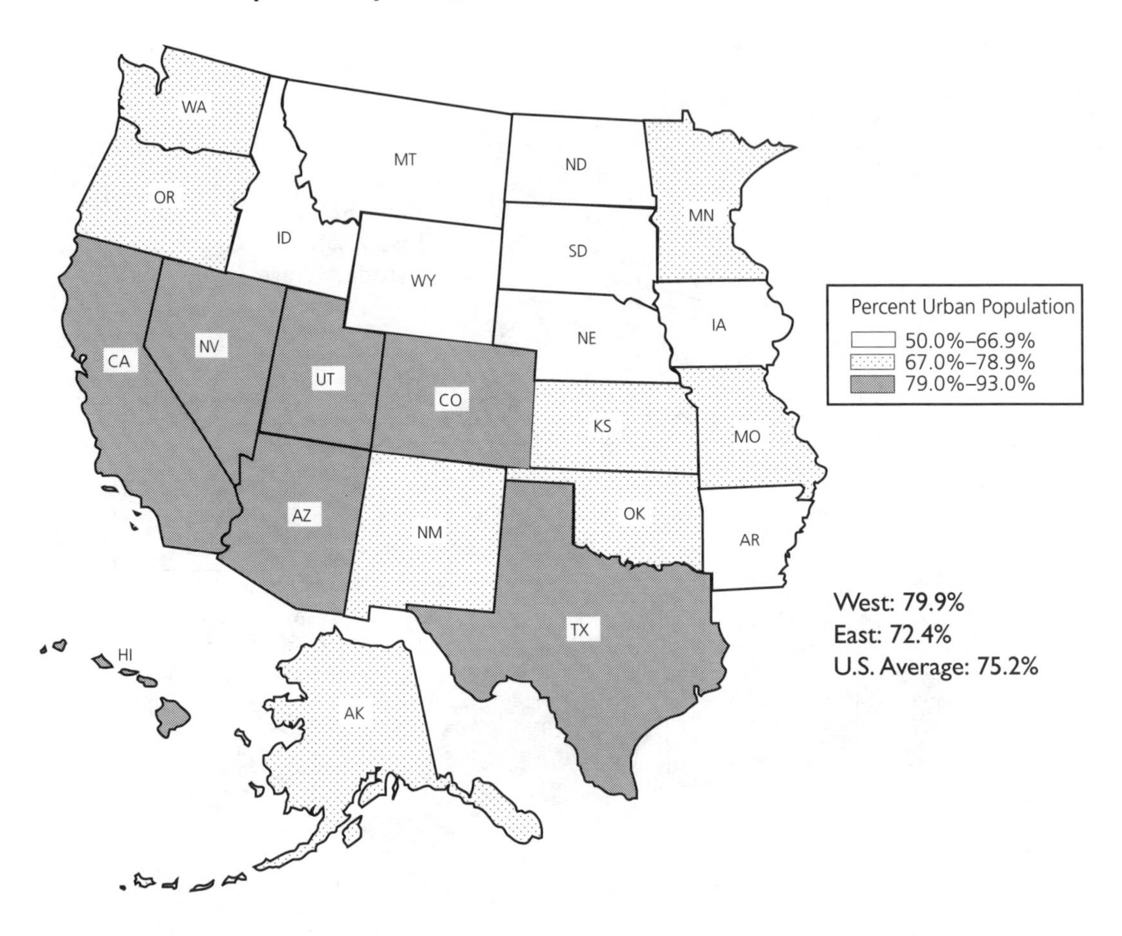

Source: U.S. Bureau of the Census

	Percent Urban Population		Percent Urban Population
California	92.6%	**Missouri**	68.7%
Hawaii	89.0%	**Oklahoma**	67.7%
Nevada	88.3%	**Alaska**	67.5%
Arizona	87.5%	**Nebraska**	66.1%
Utah	87.0%	**Wyoming**	65.0%
Colorado	82.4%	**Iowa**	60.6%
Texas	80.3%	**Idaho**	57.4%
Washington	76.4%	**Arkansas**	53.5%
New Mexico	73.0%	**North Dakota**	53.3%
Oregon	70.5%	**Montana**	52.5%
Minnesota	69.9%	**South Dakota**	50.0%
Kansas	69.1%		

Figure 3.15 **Mexico**
Urban Population by State, 1990

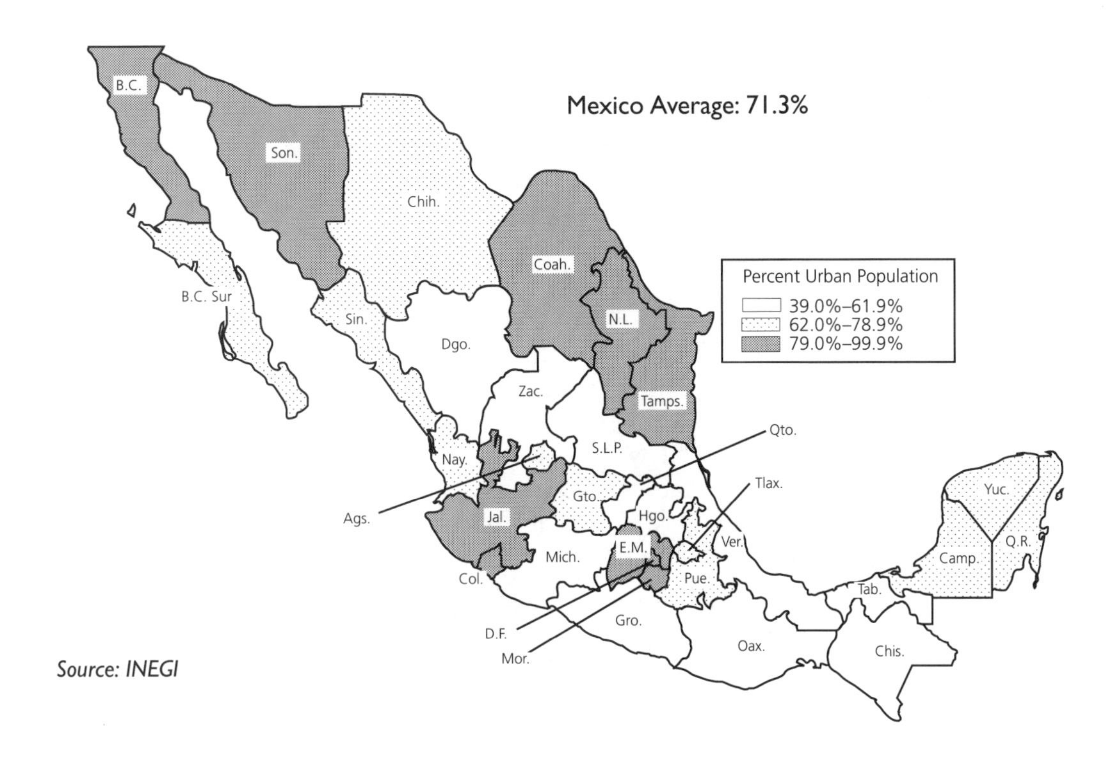

Source: INEGI

	Percent Urban Population		Percent Urban Population
Distrito Federal	99.7%	**Campeche**	70.0%
Nuevo León	92.0%	**Puebla**	64.3%
Baja California	90.9%	**Sinaloa**	64.1%
Coahuila	86.1%	**Guanajuato**	63.4%
Morelos	85.6%	**Nayarit**	62.1%
México	84.4%	**Michoacán**	61.6%
Colima	83.3%	**Querétaro**	59.7%
Jalisco	81.9%	**Durango**	57.4%
Tamaulipas	81.1%	**Veracruz**	56.2%
Sonora	79.1%	**San Luis Potosí**	55.2%
Yucatán	78.6%	**Guerrero**	52.3%
Baja California Sur	78.3%	**Tabasco**	49.7%
Chihuahua	77.4%	**Zacatecas**	45.9%
Aguascalientes	76.5%	**Hidalgo**	44.8%
Tlaxcala	76.5%	**Chiapas**	40.4%
Quintana Roo	73.9%	**Oaxaca**	39.5%

The largest city in western North America, Mexico City, has more people (20 million) than the largest cities in each of the twenty-three western U.S. states and four western Canadian provinces combined.

Figure 3.16 **Canada**
Twenty Largest Metropolitan Areas and Cities, June 1991

Source: Statistics Canada

Figure 3.17 **Western United States Twenty Largest Cities, 1990**

Source: U.S. Bureau of the Census

Figure 3.18 **Mexico**
Twenty Largest Cities, 1989

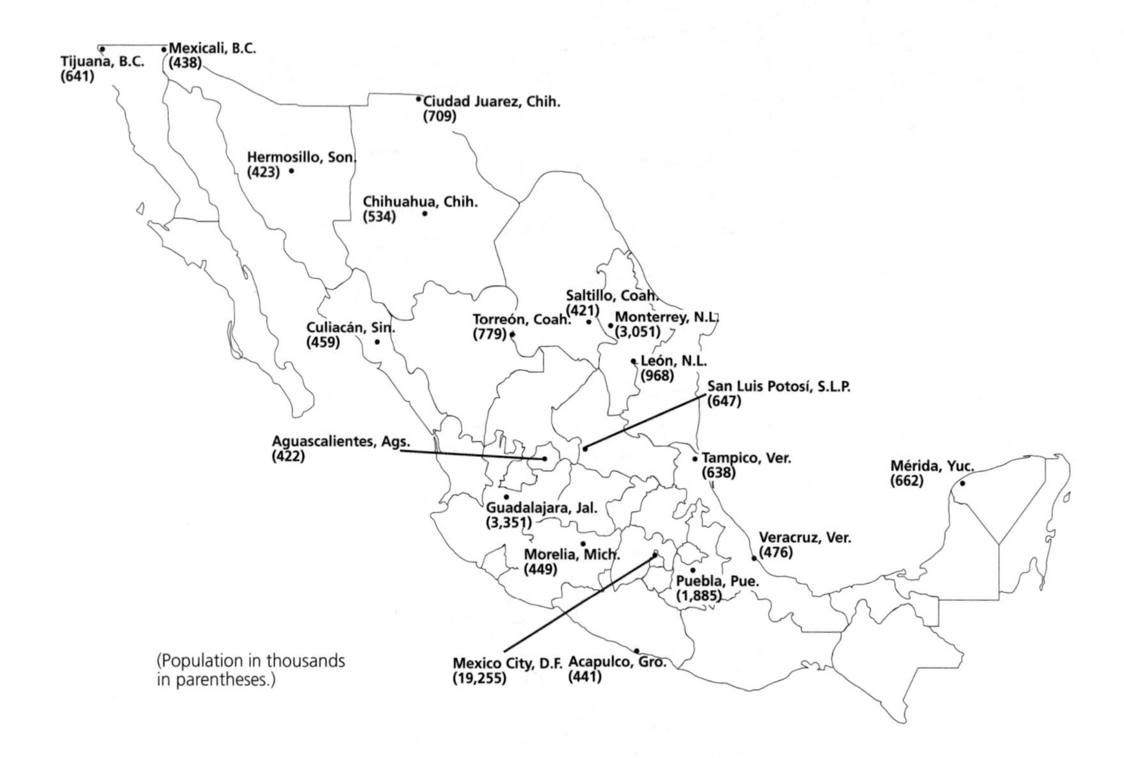

Source: INEGI

The proliferation of suburbs and the existence of "twin cities" throughout the western U.S. (e.g., Minneapolis/St. Paul, Kansas City Missouri/Kansas, Seattle/Tacoma, San Francisco/Oakland) make comparisons of the populations of U.S. cities with those in Canada and Mexico somewhat misleading. Examined from the perpective of larger "metropolitan areas" Mexico City is about the same size as the Los Angeles and San Francisco metro areas combined.

Figure 3.19 Western United States Twenty Largest Metropolitan Areas, 1990

Source: U.S. Bureau of the Census

Population Growth

Population is growing much more rapidly in Mexico than in either the United States or Canada, although the rate of population growth in Mexico is slowing. Between 1980 and 1990, Mexico's population grew 21 percent, ranging from a low of 12 percent in Zacatecas to a whopping 118 percent in Quintana Roo. An exception was the Distrito Federal, which lost 7 percent in population.

Canada's population grew 11.5 percent between 1982 and 1992, 12 percent in the western provinces and territories. Most of that growth was in British Columbia, the fastest-growing Canadian province, where population grew 18.6 percent between 1982 and 1992.

In the United States, population grew 9.8 percent in the 1980s. The fastest-growing western state was Nevada. Population in Iowa, Wyoming, and North Dakota declined.

Figure 3.20 **Western United States**
Percent Change in Population by State, 1980–1990

West: 16.3%
East: 6.2%
U.S. Average: 9.8%

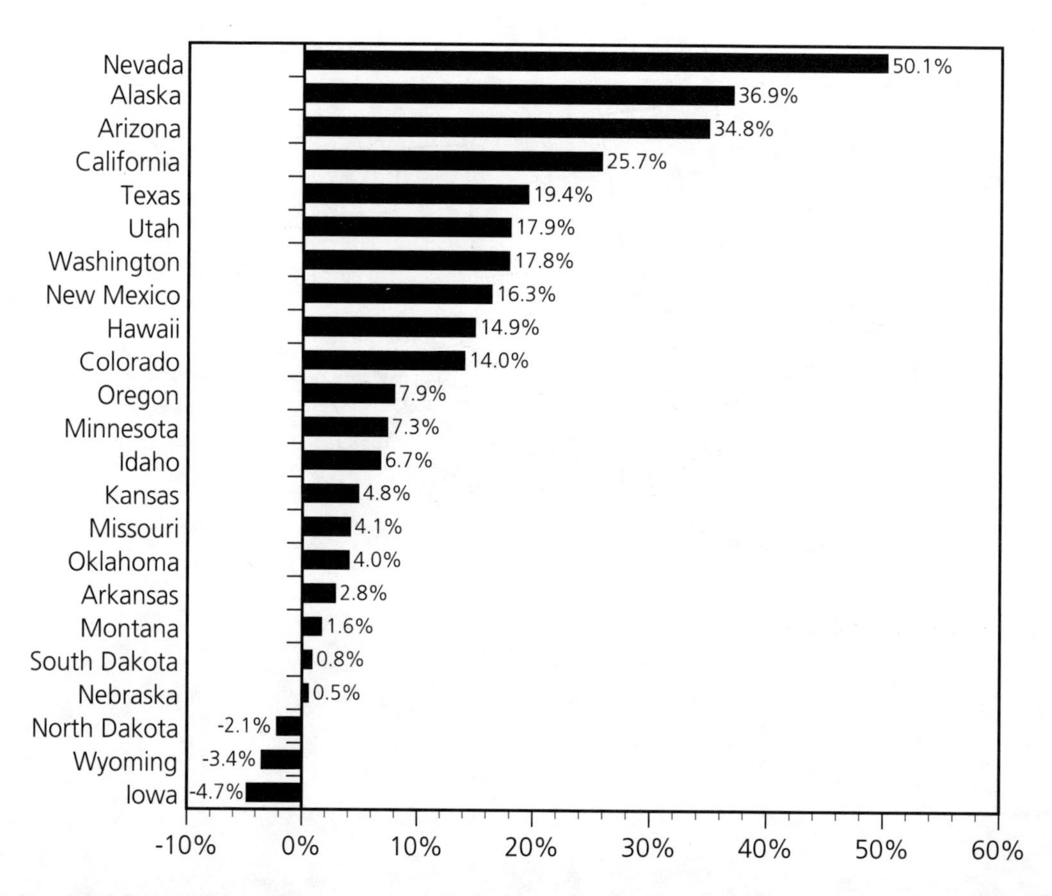

Source: U.S. Bureau of the Census

Figure 3.21 **Canada**
Percent Change in Population by Province, 1982–1992

West: 12.0%
East: 11.3%
Canada Average: 11.5%

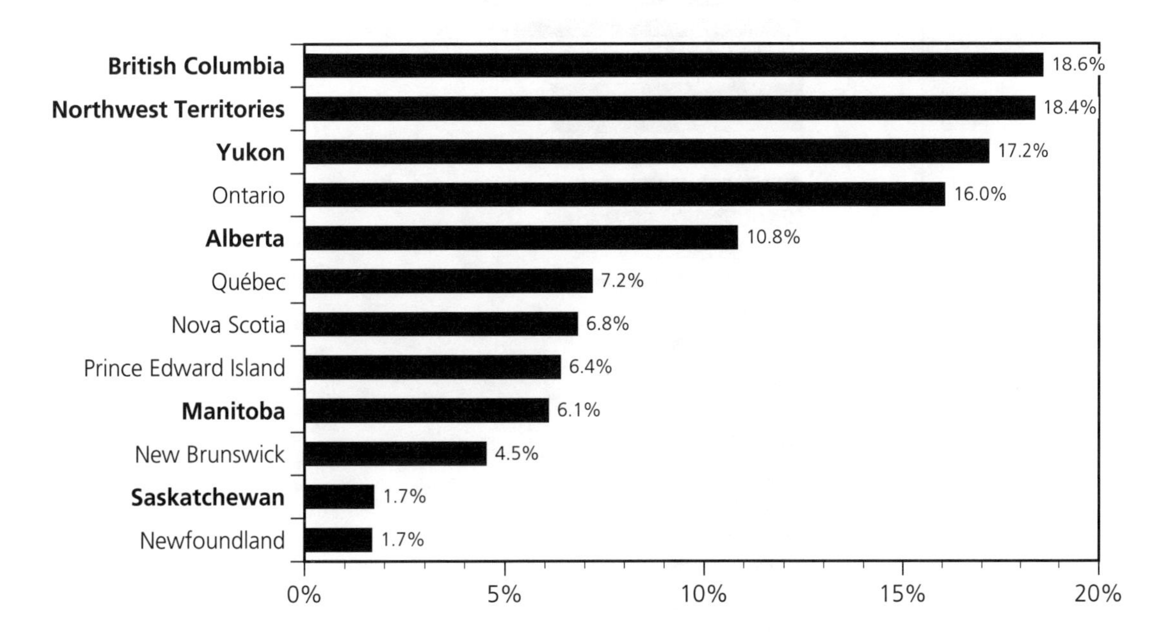

Source: Statistics Canada

Figure 3.22 **Mexico**
Percent Change in Population by State, 1980–1990

Mexico Average: 21%

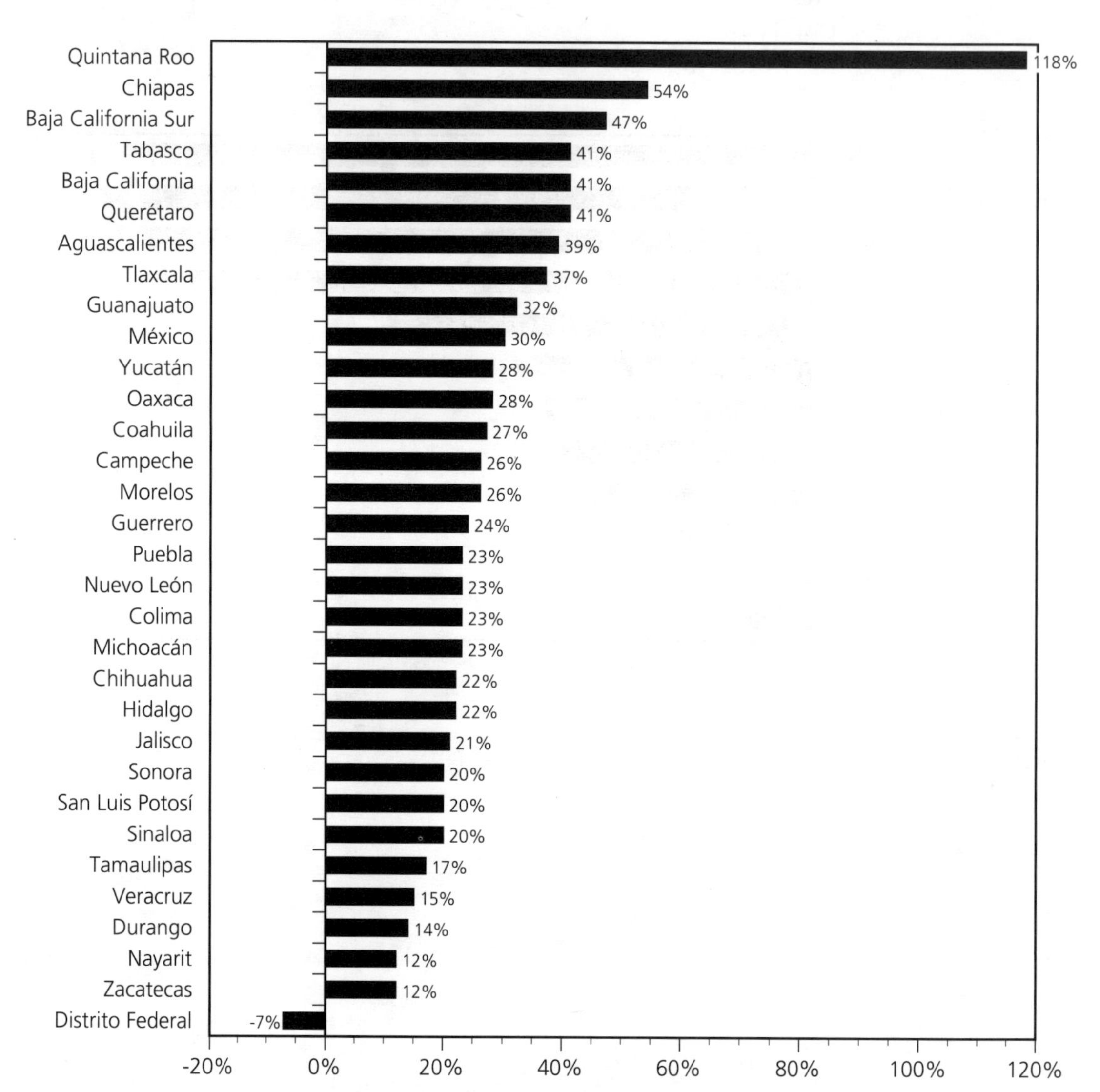

(High prices and increasing commercialization in Mexico City proper are said to account for the 7 percent decline in population as more people move to the suburbs and edge cities surrounding the Distrito Federal.)

Source: INEGI

Population Distribution

There are more women than men in western North America, chiefly because women live longer. Men most outnumber women in the Yukon and Northwest territories of Canada, the U.S. state of Alaska, and the Mexican state of Quintana Roo. Women most outnumber men in the Canadian province of Québec, the western U.S. states of Missouri and Arkansas, and the Distrito Federal in Mexico.

Figure 3.23 **Canada**
Population Distribution by Sex, by Province, 1991

West: 49.8% Male/ 50.2% Female
East: 49.1% Male/ 50.9% Female
Canada Average: 49.3% Male/ 50.7% Female

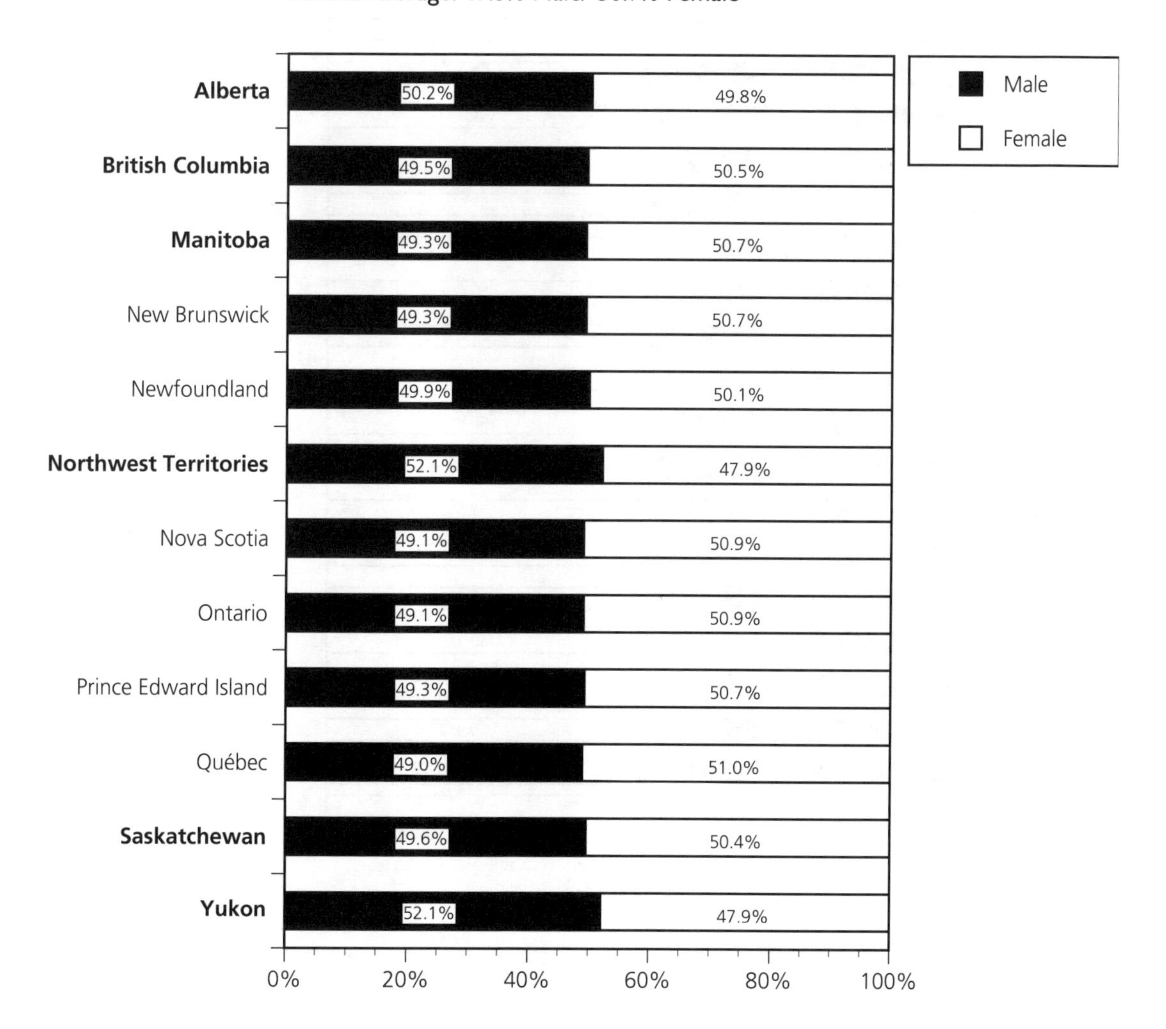

Source: Statistics Canada: Age, Sex and Marital Status

Figure 3.24 **Western United States**
Population Distribution by Sex, by State, 1990

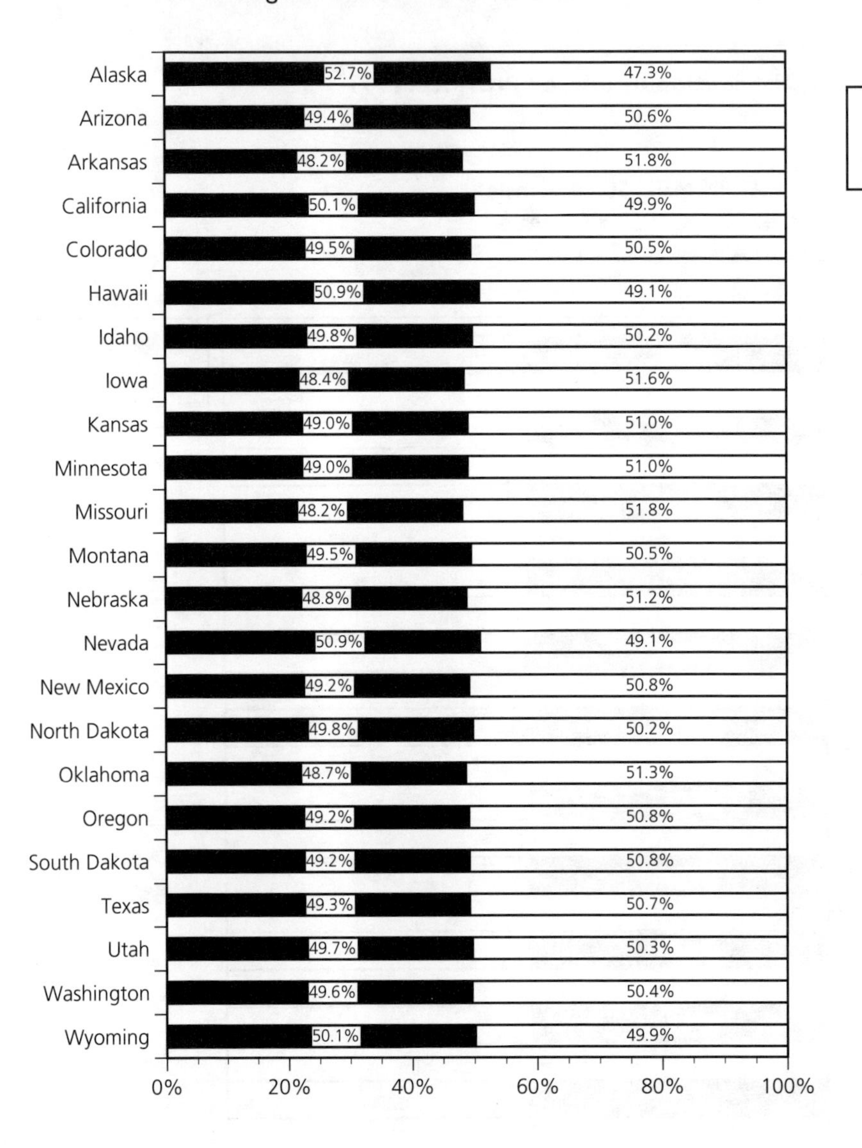

Source: U.S. Bureau of the Census

Figure 3.25 **Mexico**
Population Distribution by Sex, by State, 1990

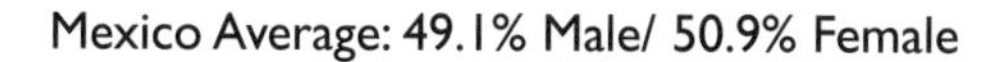

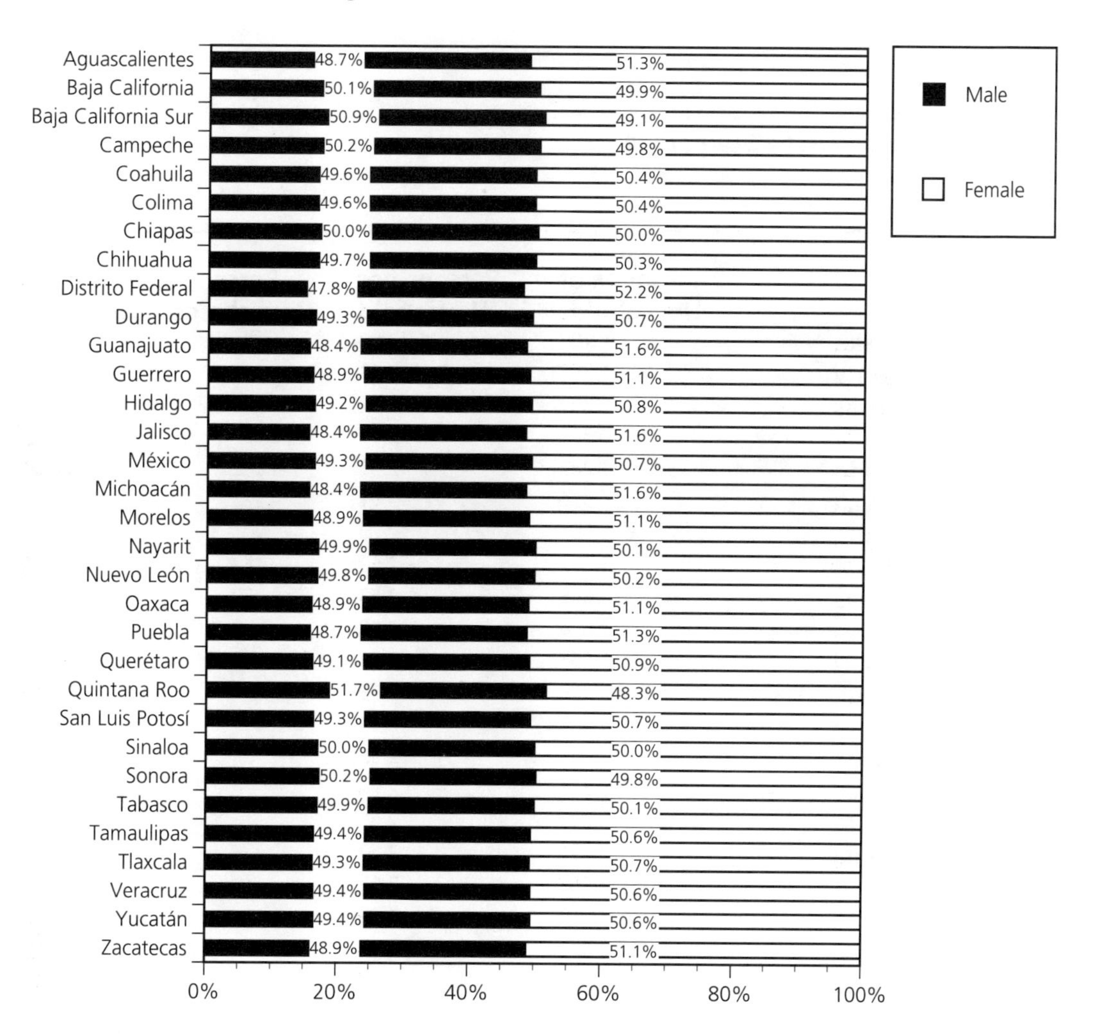

Source: INEGI, XI Censo General

Households

A "household" is defined as "individuals or groups with common housekeeping arrangements." The United States had 91.9 million households in 1990, of which 24.5 million, or 27 percent, were in the West. The average household size was 2.63 persons. The number of households in the West grew 17 percent during the decade, compared to a U.S. average of 14 percent. Household growth was greatest in Nevada, Alaska, and Arizona and least in Iowa and Wyoming.

Figure 3.26 **Western United States Households by State, 1990**

	Thousands of Households	Persons per Household	Percent Change in Households, 1980–90
Alaska	189	2.80	43.7%
Arizona	1,369	2.62	43.0%
Arkansas	891	2.57	9.2%
California	10,381	2.79	20.3%
Colorado	1,282	2.51	20.8%
Hawaii	356	3.01	21.2%
Idaho	361	2.73	11.3%
Iowa	1,064	2.52	1.1%
Kansas	945	2.53	8.3%
Minnesota	1,648	2.58	14.0%
Missouri	1,961	2.54	9.4%
Montana	306	2.53	7.9%
Nebraska	602	2.54	5.4%
Nevada	466	2.53	53.2%
New Mexico	543	2.74	22.9%
North Dakota	241	2.55	5.8%
Oklahoma	1,206	2.53	7.8%
Oregon	1,103	2.52	11.3%
South Dakota	259	2.59	6.8%
Texas	6,071	2.73	23.2%
Utah	537	3.15	19.8%
Washington	1,872	2.53	21.5%
Wyoming	169	2.63	1.9%
U.S. Total	**91,947**	**2.63**	**14.4%**

Source: U.S. Bureau of the Census

Canada had 10 million households in 1991, of which 2.9 million households, or 29 percent of the national total, were in the West. Overall, the number of households in Canada grew 21 percent between 1981 and 1991, and 23 percent in the West. The average number of persons in a Canadian household was 2.7.

Figure 3.27 **Canada**
Households by Province, 1991

	Thousands of Households	Persons per Household	Percent Change in Households, 1981–91
Alberta	910	2.7	20.1%
British Columbia	1,244	2.6	24.8%
Manitoba	405	2.6	13.2%
New Brunswick	254	2.8	18.0%
Newfoundland	174	3.2	17.6%
Northwest Territories	16	3.5	39.6%
Nova Scotia	324	2.7	18.7%
Ontario	3,638	2.7	22.5%
Prince Edward Island	44	2.8	18.1%
Québec	2,634	2.6	21.2%
Saskatchewan	363	2.7	9.1%
Yukon	10	2.7	30.5%
Canada Total*	**10,018**	**2.7**	**21.0%**

(*National total difference is due to rounding.)

Source: Statistics Canada

Mexico had 16 million households in 1990, an increase of 33 percent over 1980. Mexican households were nearly twice the size of American and Canadian households, at an average of 5.1 persons.

Figure 3.28 **Mexico**
Households by State, 1990

	Thousands of Households	Persons per Household	Percent Change in Households, 1980–90
Aguascalientes	130	5.5	55.0%
Baja California	363	4.6	52.0%
Baja California Sur	67	4.7	69.7%
Campeche	108	5.0	42.2%
Coahuila	405	4.9	43.1%
Colima	89	4.8	37.9%
Chiapas	594	5.4	60.4%
Chihuahua	530	4.6	35.3%
Distrito Federal	1,789	4.6	2.4%
Durango	262	5.1	32.2%
Guanajuato	687	5.8	44.7%
Guerrero	502	5.2	32.8%
Hidalgo	363	5.2	33.4%
Jalisco	1,029	5.2	32.5%
México	1,877	5.2	46.5%
Michoacán	663	5.3	34.1%
Morelos	245	4.9	39.7%
Nayarit	168	4.9	27.2%
Nuevo León	642	4.8	39.3%
Oaxaca	587	5.1	30.9%
Puebla	772	5.3	31.0%
Querétaro	193	5.4	60.5%
Quintana Roo	103	4.8	131.4%
San Luis Potosí	379	5.3	34.0%
Sinaloa	422	5.2	32.0%
Sonora	379	4.8	36.7%
Tabasco	285	5.3	57.7%
Tamaulipas	489	4.6	28.7%
Tlaxcala	137	5.6	48.5%
Veracruz	1,263	4.9	24.3%
Yucatán	274	5.0	36.3%
Zacatecas	239	5.3	29.7%
Mexico Total	**16,035**	**5.1**	**32.8%**

Source: INEGI

Birth and Death Rates

Birth rates are highest, and death rates lowest, in Mexico. The birthrate per thousand in Mexico in 1990 was 33.7, ranging from a low of 26.4 in Nuevo León to a high of 61.9 in Chiapas. The death rate per thousand in Mexico in 1990 was 5.2, ranging from a low of 3.3 in Quintana Roo to 6.7 in Puebla.

Figure 3.29 **Mexico**
Birth Rates and Death Rates by State, 1990

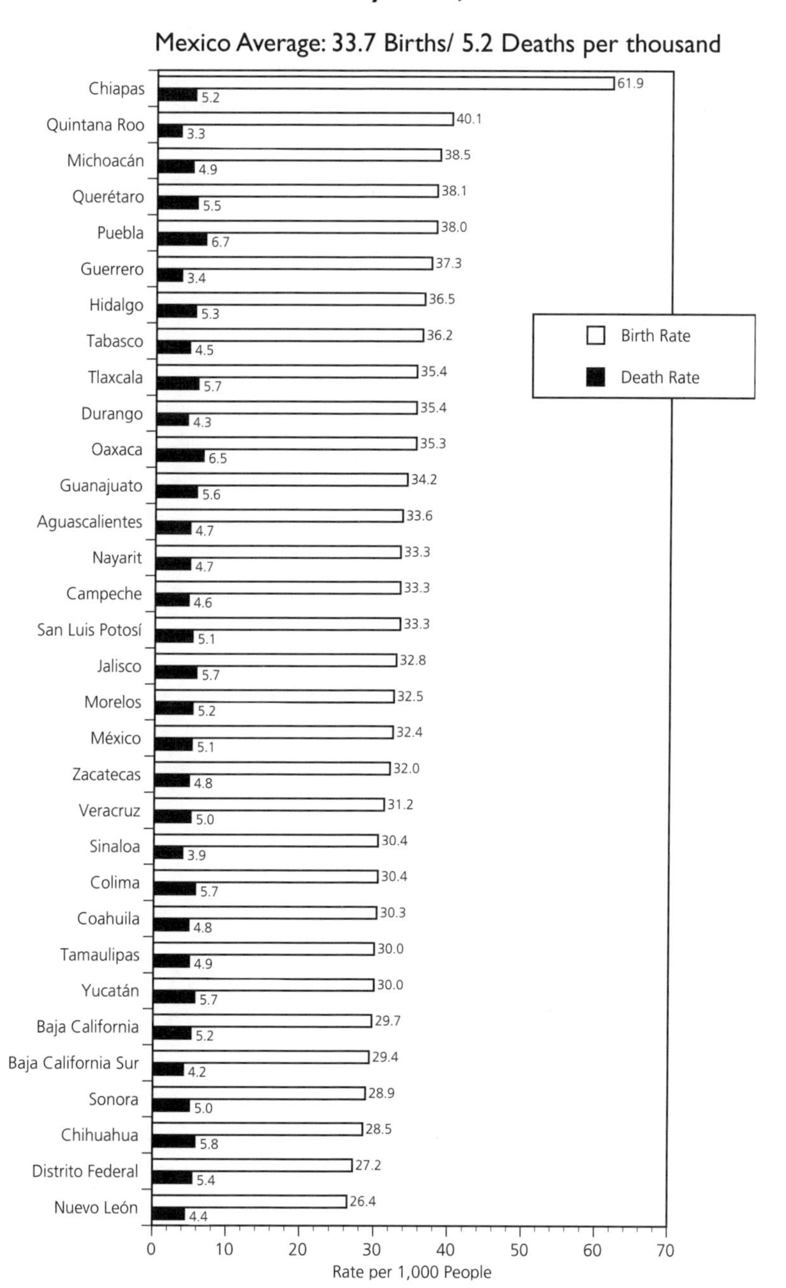

Source: INEGI

Birth rates are lower in Canada than they are in the United States, and so are death rates. Birth rates are higher, and death rates lower, in western Canada than they are in the East. The highest birth rates, and lowest death rates, were in the Yukon, Northwest Territories, and Alberta.

Figure 3.30 **Canada**
Birth Rates and Death Rates by Province, 1990

West: 16.1 Births/ 7.1 Deaths per thousand
East: 15.0 Births/ 7.3 Deaths per thousand
Canada Average: 15.3 Births/ 7.2 Deaths per thousand

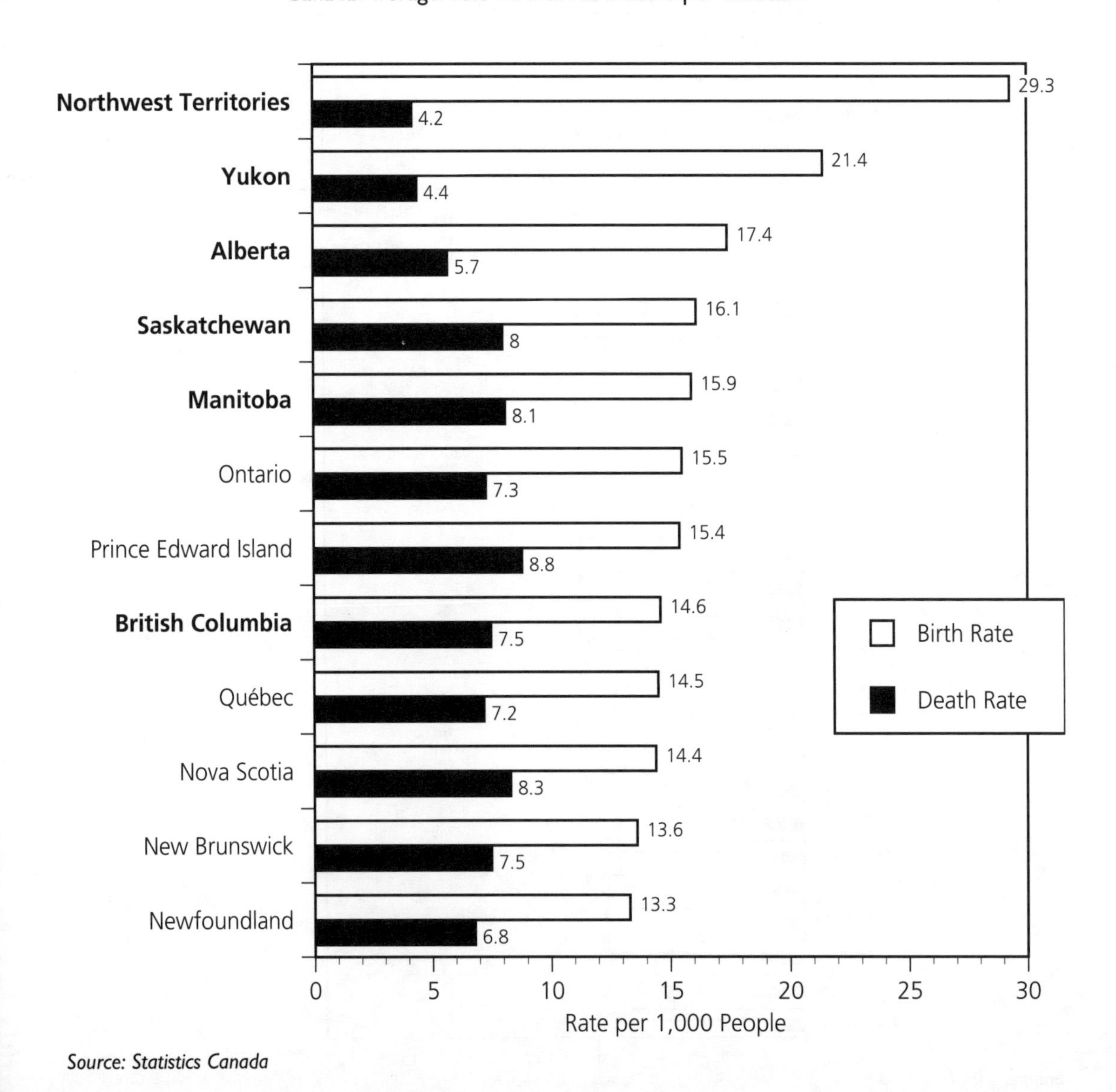

Source: Statistics Canada

The western United States has North America's highest death rate and a birth rate similar to that of Canada, but one-half that of Mexico. In the western United States, birth rates were highest, and death rates lowest, in Alaska and Utah. Death rates were highest in Missouri and Arkansas. Eight western states had birth rates higher than the U.S. average. Fourteen had death rates lower than the U.S. average. Generally, high birth rates and low death rates coincide with the age of the population, and states with higher birth rates have lower death rates. California is something of an anomaly, with both a high birth rate (third in the West), and a relatively high death rate (fifteenth in the West).

Figure 3.31 **Western United States Birth Rates and Death Rates by State, 1990**

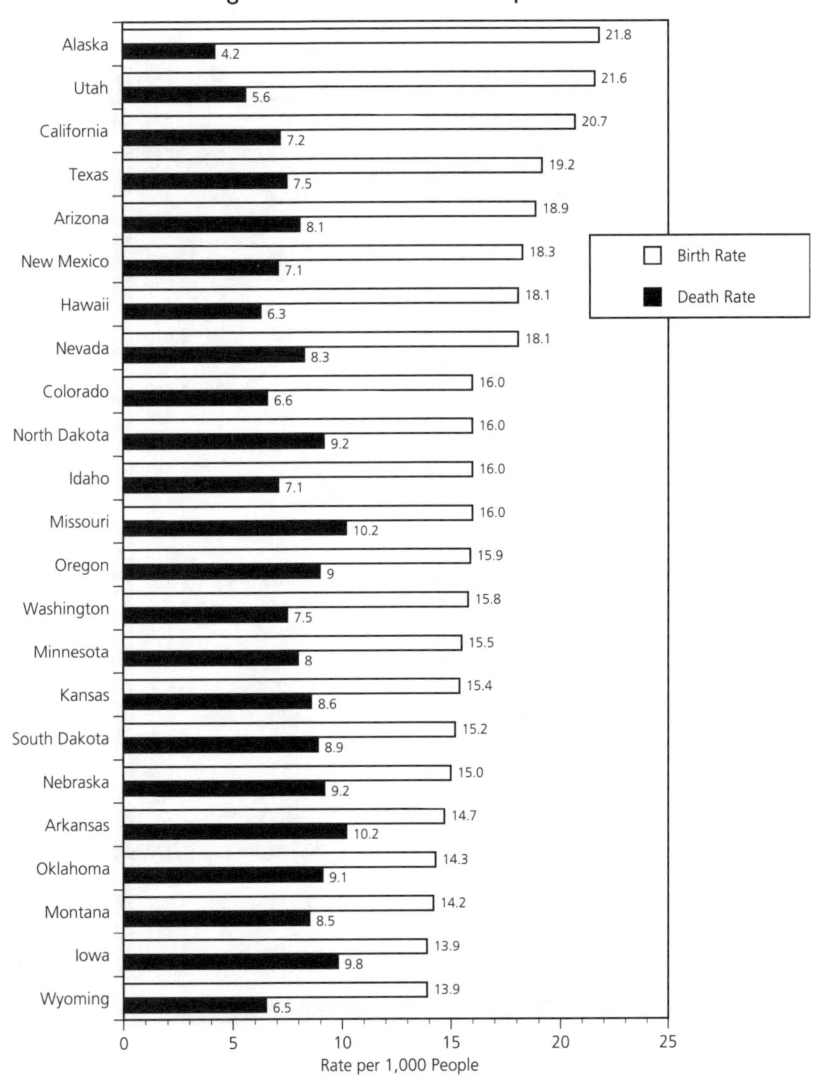

Source: U.S. National Center for Health Statistics

Marriages and Divorces

Mexicans are much more likely to get married and stay married than are Canadians or U.S. citizens, and Canadians are more likely to get married and stay married than are U.S. citizens.

Divorces in Mexico in 1990 averaged 7.2 percent of marriages for that year, compared to 48 percent for the United States, and 42 percent for Canada. Mexican divorce rates ranged from a low of 1.5 percent in Guerrero to a high of 19 percent in Chihuahua.

Figure 3.32 **Mexico**
Marriages and Divorces by State, 1990

Mexico Total: 642,201 Marriages/46,254 Divorces

	Marriages	Divorces
Aguascalientes	5,997	330
Baja California	15,650	1,610
Baja California Sur	2,423	269
Campeche	4,922	240
Coahuila	19,334	2,009
Colima	3,384	504
Chiapas	17,785	1,648
Chihuahua	21,124	4,062
Distrito Federal	59,582	5,610
Durango	11,425	608
Guanajuato	35,017	1,600
Guerrero	21,967	346
Hidalgo	13,325	608
Jalisco	48,953	2,885
México	70,616	3,723
Michoacán	34,050	1,586
Morelos	9,592	914
Nayarit	6,187	459
Nuevo León	30,047	2,522
Oaxaca	21,496	508
Puebla	25,361	1,407
Querétaro	8,966	440
Quintana Roo	4,636	435
San Luis Potosí	15,894	936
Sinaloa	17,453	1,798
Sonora	14,160	1,138
Tabasco	11,999	1,251
Tamaulipas	20,388	1,278
Tlaxcala	7,616	323
Veracruz	37,976	3,050
Yucatán	12,473	1,435
Zacatecas	12,403	722

Source: INEGI, Anuario Estadistico

Divorces in Canada averaged 42 percent of marriages in 1989, ranging from a low of 24 percent in Prince Edward Island and one in four in Newfoundland to 59 percent in Québec. Divorce rates in the western provinces mirrored the national average, ranging from a low of 37 percent in Saskatchewan and Manitoba to a high of 42 percent in British Columbia.

Figure 3.33 **Canada**
Marriages and Divorces by Province, 1989

West: 59,932 Marriages/ 24,329 Divorces
East: 130,708 Marriages/ 56,387 Divorces
Canada Total: 190,640 Marriages/ 80,716 Divorces

	Marriages	Divorces
Alberta	19,888	8,227
British Columbia	25,170	10,630
Manitoba	7,800	2,847
New Brunswick	5,254	1,647
Newfoundland	3,905	981
Northwest Territories	223	92
Nova Scotia	6,828	2,524
Ontario	80,377	31,202
Prince Edward Island	1,019	243
Québec	33,325	19,790
Saskatchewan	6,637	2,451
Yukon	214	82

Source: Statistics Canada, Canada Yearbook

Figure 3.34 **Western United States**
Marriages and Divorces by State, 1990 preliminary

U.S. Total: 2,448,000 Marriages/ 1,175,000 Divorces

	Thousands of Marriages	Thousands of Divorces
Alaska	5.7	2.9
Arizona	37.0	25.1
Arkansas	35.7	16.8
California*	236.7	137.5
Colorado	31.5	18.4
Hawaii	18.1	5.2
Idaho	15.0	6.6
Iowa	24.8	11.1
Kansas	23.4	12.6
Minnesota	33.7	15.4
Missouri	49.3	26.4
Montana	7.0	4.1
Nebraska	12.5	6.5
Nevada*	106.5	13.3
New Mexico	13.2	7.7
North Dakota	4.8	2.3
Oklahoma	33.2	24.9
Oregon	25.2	15.9
South Dakota	7.7	2.6
Texas	182.8	95.1
Utah	19.0	8.8
Washington	48.6	28.8
Wyoming	4.8	3.1

*Divorce data for California and marriage data for Nevada are from 1985. California divorce data include legal separations.

Source: U.S. National Center for Health Statistics

RACE AND ETHNICITY

The United States is the most racially diverse of the North American nations. Although blacks comprise the largest minority group in the United States, Hispanics are the largest minority in the twenty-three western states. In Hawaii, whites are the largest minority group. In Alaska, Montana, North Dakota, and South Dakota, Native Americans are the largest minority group. In Washington and Oregon, Asians are the largest minority group.

Canada is less racially diverse than is the United States, but is the most ethnically and linguistically diverse of the North American nations. The population is overwhelmingly Caucasian. Roughly 25 percent of Canadians are descended from the original French settlers, another 25 percent from the original English settlers. The remainder of the Caucasians are from later immigrant groups (chiefly Irish, German, Italian, and Ukrainian), and mixtures of various ethnicities. Asians are the largest racial minority (16 percent in 1986). Blacks comprise about 3 percent of the Canadian population. Indians and Inuits (Eskimos) account for 3 percent. Canada is officially a bilingual nation. Approximately 24 percent of Canadians are francophones. French speakers are concentrated overwhelmingly in the province of Québec.

All three North American nations are ethnically diverse compared to the rest of the world. The Academy of Sciences in Moscow developed the *Ethnic and Linguistic Homogeneity Index,* which ranked 135 nations from the most to the least homogeneous. In the rankings, based on the years 1960–65, Mexico ranked sixty-fourth (70 percent homogeneous), the United States ranked eighty-second (50 percent), and Canada ranked one hundred-sixteenth (25 percent).

Figure 3.35 **Western United States**
Percentage of Population of Hispanic Origin by State, 1990 (Includes All Races)

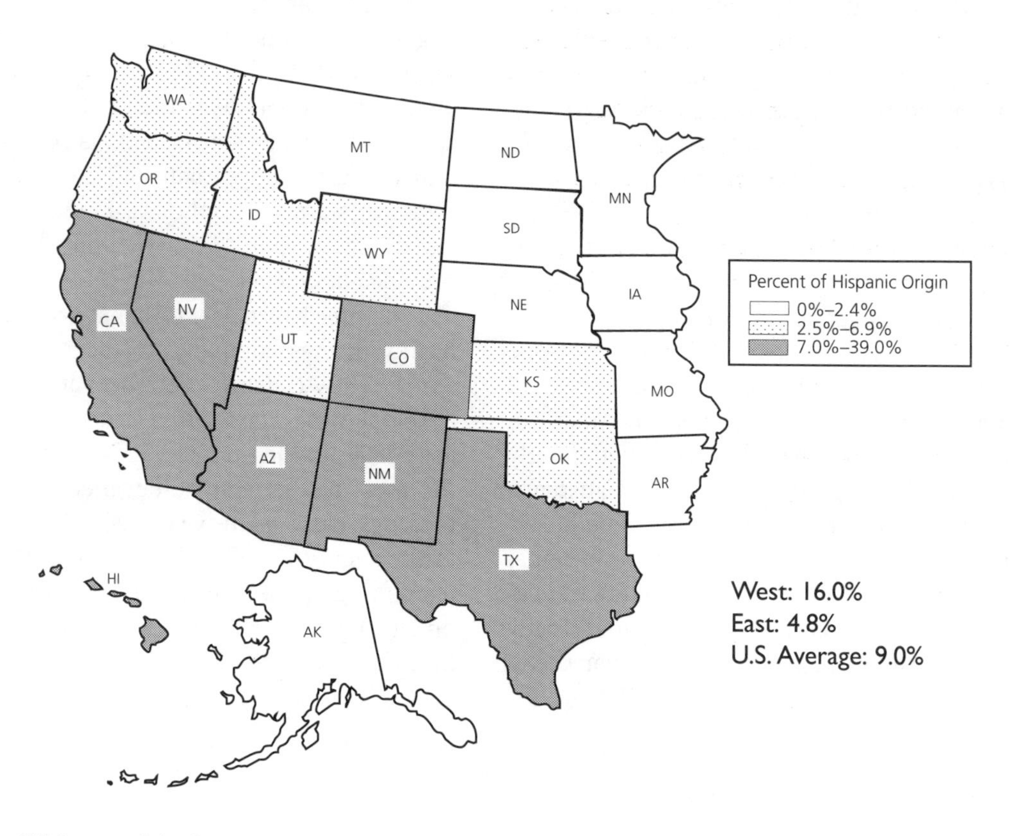

Source: U.S. Bureau of the Census

	Percent of Hispanic Origin		Percent of Hispanic Origin
New Mexico	38.2%	**Kansas**	3.8%
California	25.8%	**Alaska**	3.2%
Texas	25.6%	**Oklahoma**	2.7%
Arizona	18.8%	**Nebraska**	2.3%
Colorado	12.9%	**Montana**	1.5%
Nevada	10.4%	**Missouri**	1.2%
Hawaii	7.3%	**Minnesota**	1.2%
Wyoming	5.7%	**Iowa**	1.2%
Idaho	5.3%	**Arkansas**	0.9%
Utah	4.9%	**South Dakota**	0.8%
Washington	4.4%	**North Dakota**	0.7%
Oregon	4.0%		

Figure 3.36 **Western United States**
Distribution of Population by Race, by State, 1990

	White	Black	American Indian, Eskimo, Aleut	Asian, Pacific Islander	Other
Alaska	75.5%	4.1%	15.6%	3.6%	1.2%
Arizona	80.8%	3.0%	5.6%	1.5%	9.1%
Arkansas	82.7%	15.9%	0.5%	0.5%	0.3%
California	69.0%	7.4%	0.8%	9.6%	13.2%
Colorado	88.2%	4.0%	0.8%	1.8%	5.1%
Hawaii	33.4%	2.5%	0.5%	61.8%	1.9%
Idaho	94.4%	0.3%	1.4%	0.9%	3.0%
Iowa	96.6%	1.7%	0.3%	0.9%	0.5%
Kansas	90.1%	5.8%	0.9%	1.3%	2.0%
Minnesota	94.4%	2.2%	1.1%	1.8%	0.5%
Missouri	87.7%	10.7%	0.4%	0.8%	0.4%
Montana	92.7%	0.3%	6.0%	0.5%	0.5%
Nebraska	93.8%	3.6%	0.8%	0.8%	1.0%
Nevada	84.3%	6.6%	1.6%	3.2%	4.4%
New Mexico	75.6%	2.0%	8.9%	0.9%	12.6%
North Dakota	94.6%	0.6%	4.1%	0.5%	0.3%
Oklahoma	82.1%	7.4%	8.0%	1.1%	1.3%
Oregon	92.8%	1.6%	1.4%	2.4%	1.8%
South Dakota	91.6%	0.5%	7.3%	0.4%	0.2%
Texas	75.2%	11.9%	0.4%	1.9%	10.6%
Utah	93.8%	0.7%	1.4%	1.9%	2.2%
Washington	88.5%	3.1%	1.7%	4.3%	2.4%
Wyoming	94.2%	0.8%	2.1%	0.6%	2.3%
West	**79.2%**	**7.4%**	**1.6%**	**5.0%**	**7.5%**
East	**81.0%**	**15.2%**	**0.3%**	**1.7%**	**1.8%**
U.S. Average	**80.3%**	**12.1%**	**0.8%**	**2.9%**	**3.9%**

Source: U.S. Bureau of the Census

Figure 3.37 **Canada**
Distribution of Population by Ethnic Origin, by Province, 1991

	French	British	German	Canadian	Italian	Chinese	Aboriginal	Ukrainian	Multiple Response
Alberta	3.0%	19.6%	7.4%	3.7%	0.0%	2.8%	2.7%	4.1%	42.4%
British Columbia	2.1%	25.0%	4.8%	1.9%	1.5%	5.6%	2.3%	1.6%	39.9%
Manitoba	5.0%	17.0%	8.7%	1.4%	0.0%	0.0%	6.9%	6.9%	38.0%
New Brunswick	32.8%	33.0%	0.6%	1.3%	0.2%	0.2%	0.6%	0.0%	29.7%
Newfoundland	1.7%	78.5%	0.2%	0.2%	0.1%	0.1%	0.9%	0.0%	17.4%
Northwest Territories	2.4%	10.2%	1.5%	1.8%	0.3%	0.5%	51.2%	0.8%	27.7%
Nova Scotia	6.2%	44.0%	2.8%	1.1%	0.3%	0.0%	0.8%	0.0%	40.2%
Ontario	5.3%	25.4%	2.9%	5.3%	4.9%	2.7%	0.0%	0.0%	32.9%
Prince Edward Island	9.2%	44.0%	0.5%	0.6%	0.0%	0.0%	0.3%	0.0%	43.1%
Québec	74.6%	4.2%	0.0%	0.0%	2.6%	0.5%	1.0%	0.0%	8.4%
Saskatchewan	3.1%	16.5%	12.4%	3.0%	0.0%	0.0%	6.8%	5.7%	42.8%
Yukon	3.2%	19.2%	3.8%	2.7%	0.5%	0.0%	13.7%	1.4%	48.8%
Canada Average	**22.8%**	**20.8%**	**3.4%**	**2.8%**	**2.8%**	**2.2%**	**1.7%**	**1.5%**	**28.9%**

("British" includes English, Irish, Scottish, Welsh, and other British.
"French" includes French, Acadian, and Québécois.
"Aboriginal" includes Inuit, Métis, Aleuts, and North American Indian.
"Canadian" includes respondents who declined to identify an ethnic origin outside Canada.
Only the top eight ethnic groups are included; thus, total does not equal 100%.)

Source: The Daily

There is a great deal of variety in the ethnic makeup of the Mexican population. The overwhelming majority of Mexicans are of mixed European, mostly Spanish, and Indian blood, and are called *mestizos* north of the Isthmus of Tehuantepec, *ladinos* south of it. During the Spanish colonial rule (1521–1821), the population was divided into three groupings: pureblooded Spanish born in Mexico *(criollo)*, those of mixed descent (*mestizo*), and pureblooded natives (*indio*). Mexicans of pure European descent have mostly disappeared, excepting those who have migrated in the 20th century, and the *mestizo* category has been swollen by natives who have chosen to assimilate into the national culture.

No one knows what proportion of the population is descended from the Aztecs, Toltecs, Maya, and other peoples indigenous to Mexico before the arrival of Cortes (most educated guesses hover around 10 percent). In 1940, the Mexican census stopped classifying people according to ethnic origin. Since then, an *indio* has been defined linguistically, as someone who speaks a language other than Spanish as a primary language. In other words, any *indio* who chooses not to follow the old customs is, officially, no longer an *indio*. The proportion of the population that speaks only aboriginal languages has declined from an estimated 14.7 percent in 1921 to 6.3 percent in 1940, then increased slightly to an estimated 7.9 percent today. The proportion of the population that speaks indigenous languages as its only tongue is highest in Yucatán (46.1 percent), and lowest in Aguascalientes and Zacatecas (0.1 percent).

Figure 3.38 **Mexico**
Distribution of Population Speaking an Indigenous Language, 1990

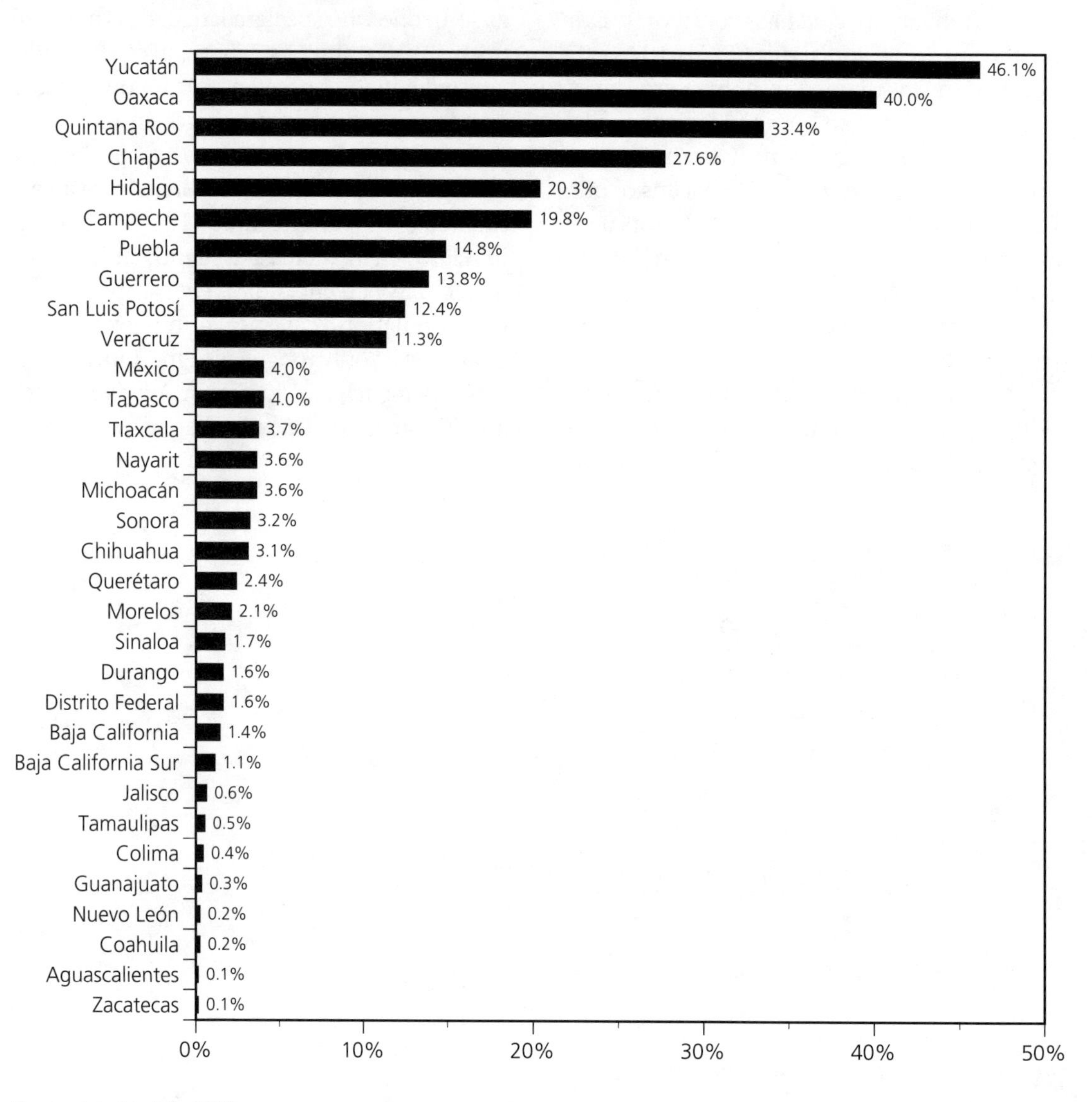

Source: Interior, July 1993

Chapter Four
Health

Infant Mortality

Infant mortality is an important measure of national health. In 1990, Canada had an infant mortality rate of only 6.8 per 1,000 live births. The United States had 10 infant deaths per 1,000 live births. Mexico had an infant mortality rate of 24, more than double the infant mortality rates of the United States and Canada.

But Mexico is showing marked improvement since 1988 when Mexico's infant mortality rate was 50 infant deaths per 1,000 live births. All three nations are well below the worldwide average of 37.2 infant deaths per 1,000 live births recorded in 1980. But none is close to the leaders in reducing infant mortality—Japan (5.2), Iceland (5.4), and Finland (5.8). Of 171 countries surveyed by the World Development Report in 1988, Canada ranked tenth in reducing infant deaths, the United States twenty-first, and Mexico eighty-fourth.

Figure 4.1 **Canada**
Infant Mortality Rate by Province, 1990

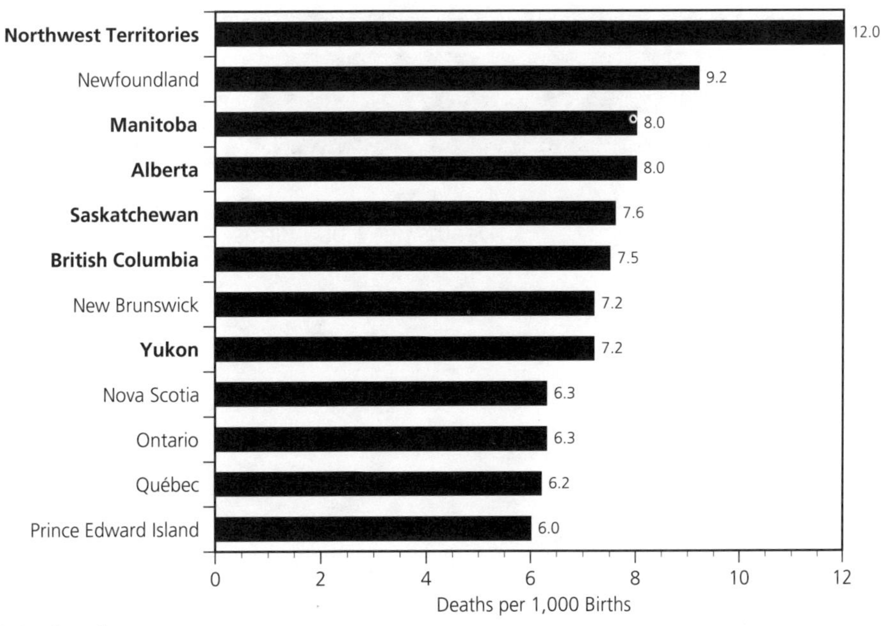

Source: Statistics Canada

Figure 4.2 **Western United States**
Infant Mortality Rate by State, 1988

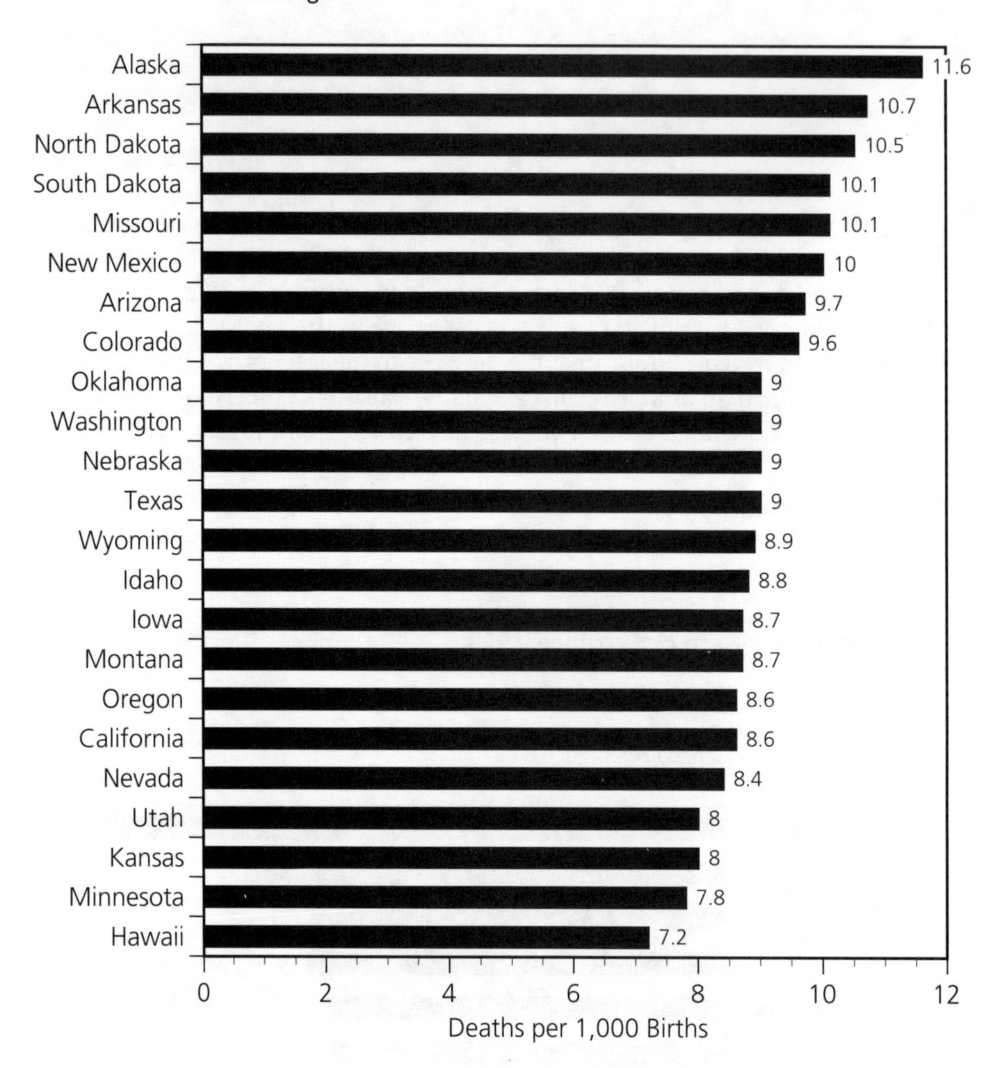

Source: U.S. National Center for Health Statistics

Figure 4.3 **Mexico**
Infant Mortality Rate by State, 1990

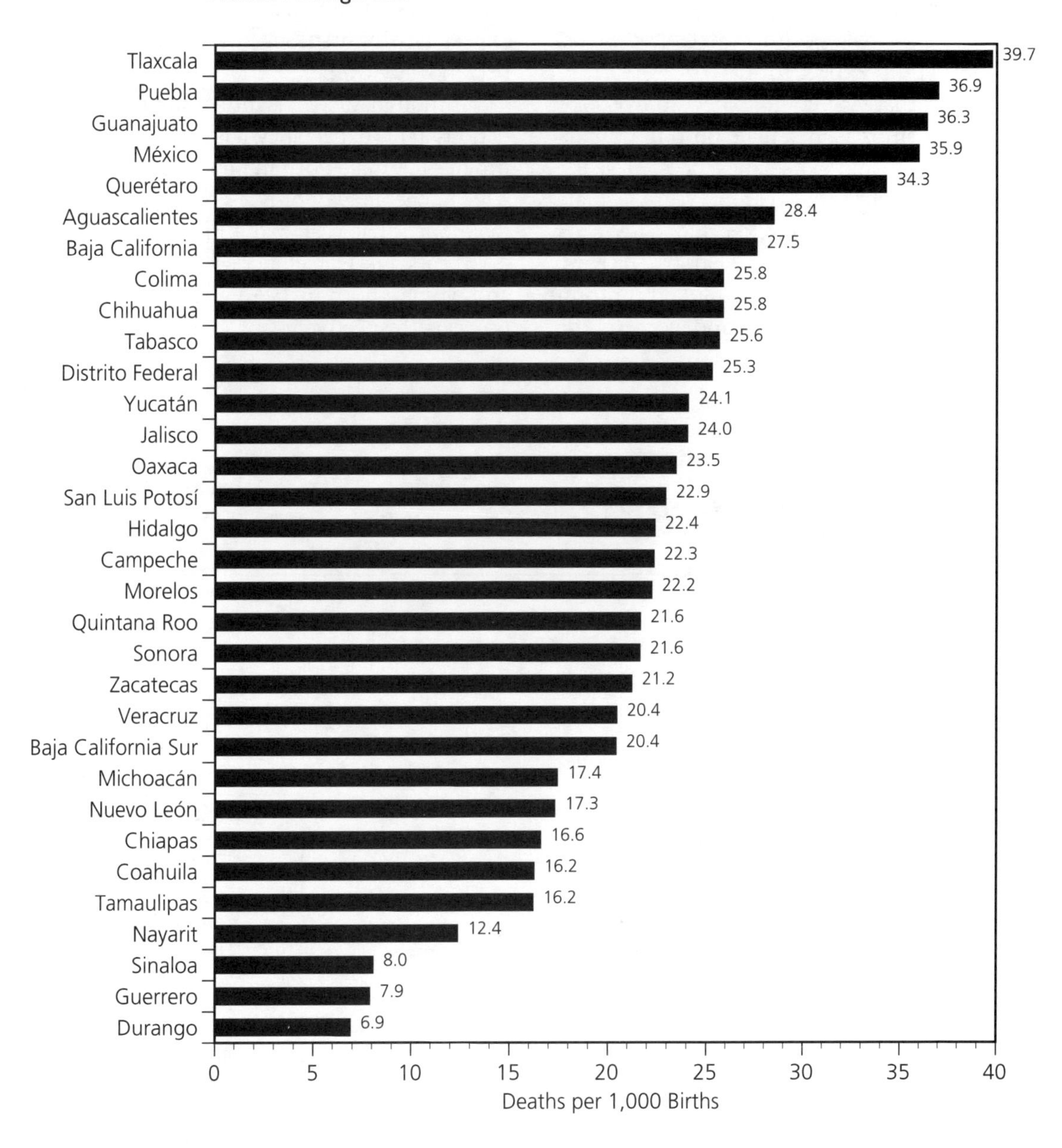

Source: INEGI

Life Expectancy

Life expectancy is another major indicator of national health, and Canadian citizens live longer than people in the United States or Mexico. A male born in the United States in 1980 could expect to live 70 years; a female born that year could expect to live 77.4 years. By comparison, a male born in Canada in 1981 has a life expectancy of 71.9 years, and a female born that year, 80 years. A male born in Mexico in 1980 is expected to live 63.7 years, a female born that year 69.9 years. Worldwide in 1988, Canada ranked ninth in life expectancy for both men and women. The United States ranked seventeenth in life expectancy for males, twelfth in life expectancy for females. Mexico ranked eighty-fifth in life expectancy for men, seventy-fourth in life expectancy for women.

Although Mexico is third among the three nations in life expectancy, it is making the most progress. A male born in Mexico in 1940 had a life expectancy of only 39.5 years, and a female born that year had a life expectancy of 41.5 years. Life expectancies for Canadian men and women born in 1940 were 63 years and 66 years, respectively.

Figure 4.4 **United States Life Expectancy, 1930–1990**

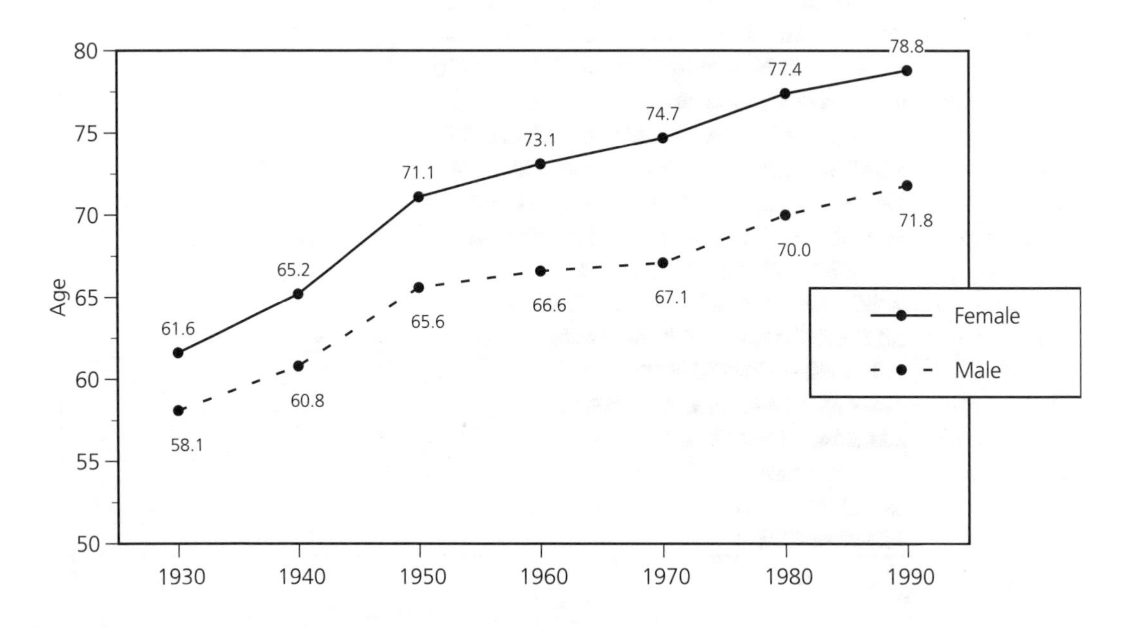

Source: U.S. National Center for Health Statistics

Figure 4.5 **Canada**
Life Expectancy, 1931–1985

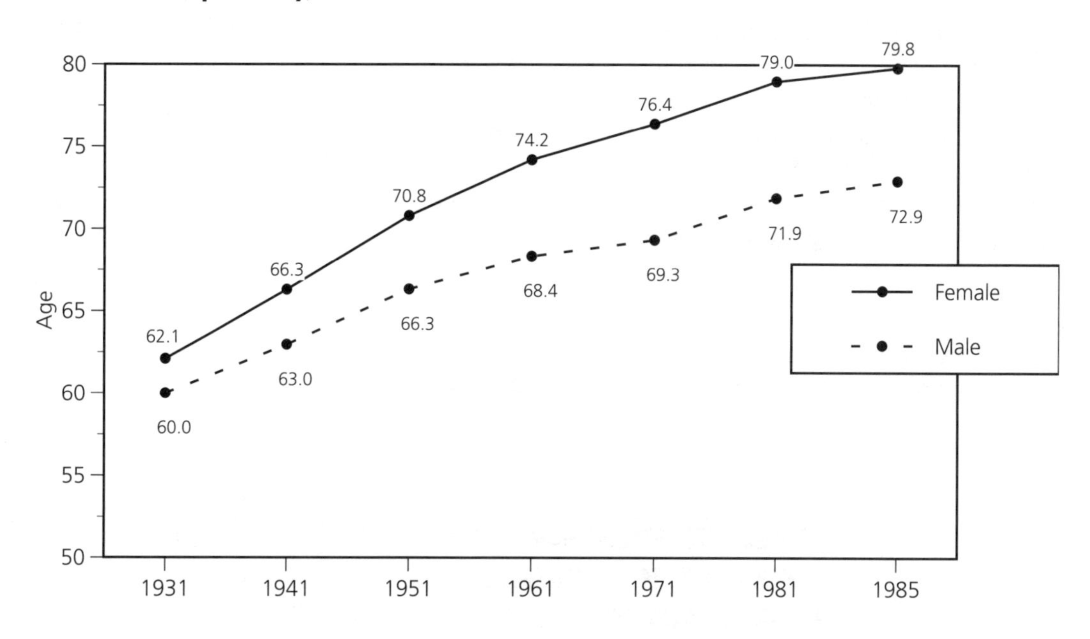

Source: Statistics Canada

Figure 4.6 **Mexico**
Life Expectancy, 1930–1988

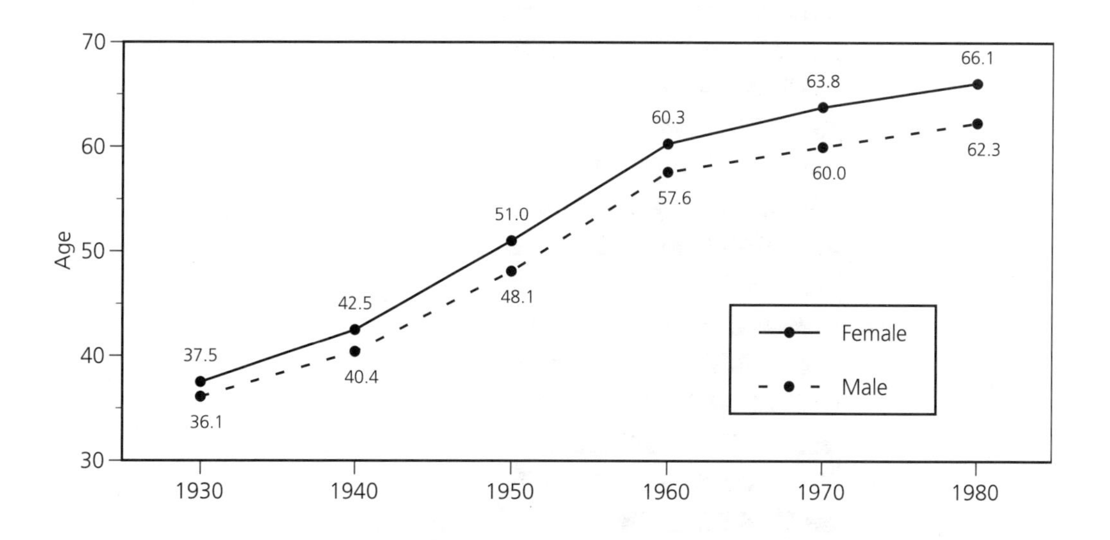

Source: Population Reference Bureau

Cancer and Heart Disease

North America's biggest killers, cancer and heart disease, cause far fewer deaths in Mexico than they do in the United States or Canada. Only 10.5 percent of deaths in Mexico in 1990 were caused by cancer, compared to 23.5 percent in the United States and 27.2 percent in Canada. This is chiefly because cancer and heart disease are ailments of old age, and many Mexicans die at a younger age than people in either the United States or Canada. In both the United States and Canada, cancer caused proportionately fewer deaths in the West than in the East.

Figure 4.7 **Western United States**
Percent of Total Deaths from Heart Disease and Cancer by State, 1990

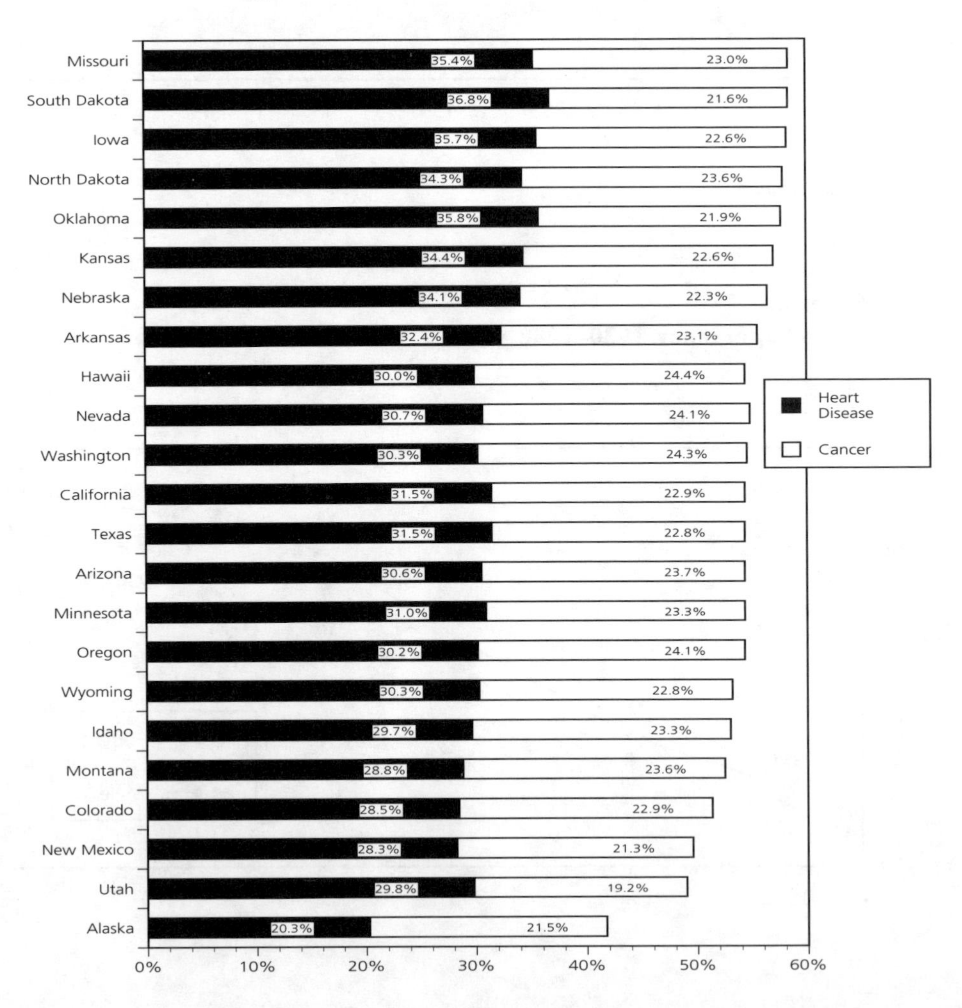

Source: U.S. National Center for Health Statistics

Figure 4.8 **Canada**
Percent of Total Deaths from Circulatory Disease and Cancer by Province, 1989

West: 39.8% Circulatory Disease/ 26.1% Cancer
East: 41.0% Circulatory Disease/ 27.6% Cancer
Canada Average: 40.7% Circulatory Disease/ 27.2% Cancer

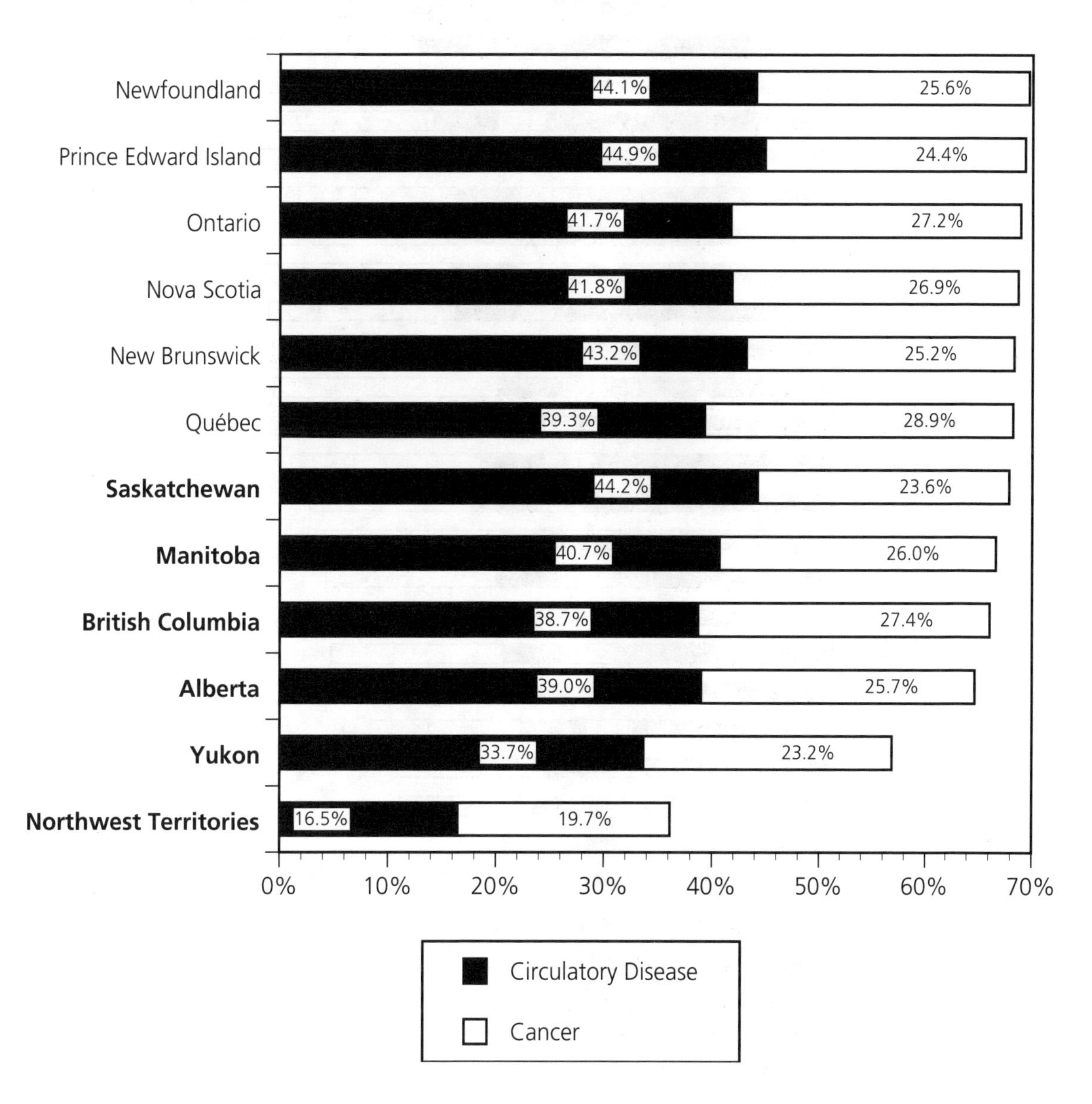

Source: Statistics Canada

Figure 4.9 **Mexico**
Percent of Total Deaths from Circulatory Disease and Cancer, 1990

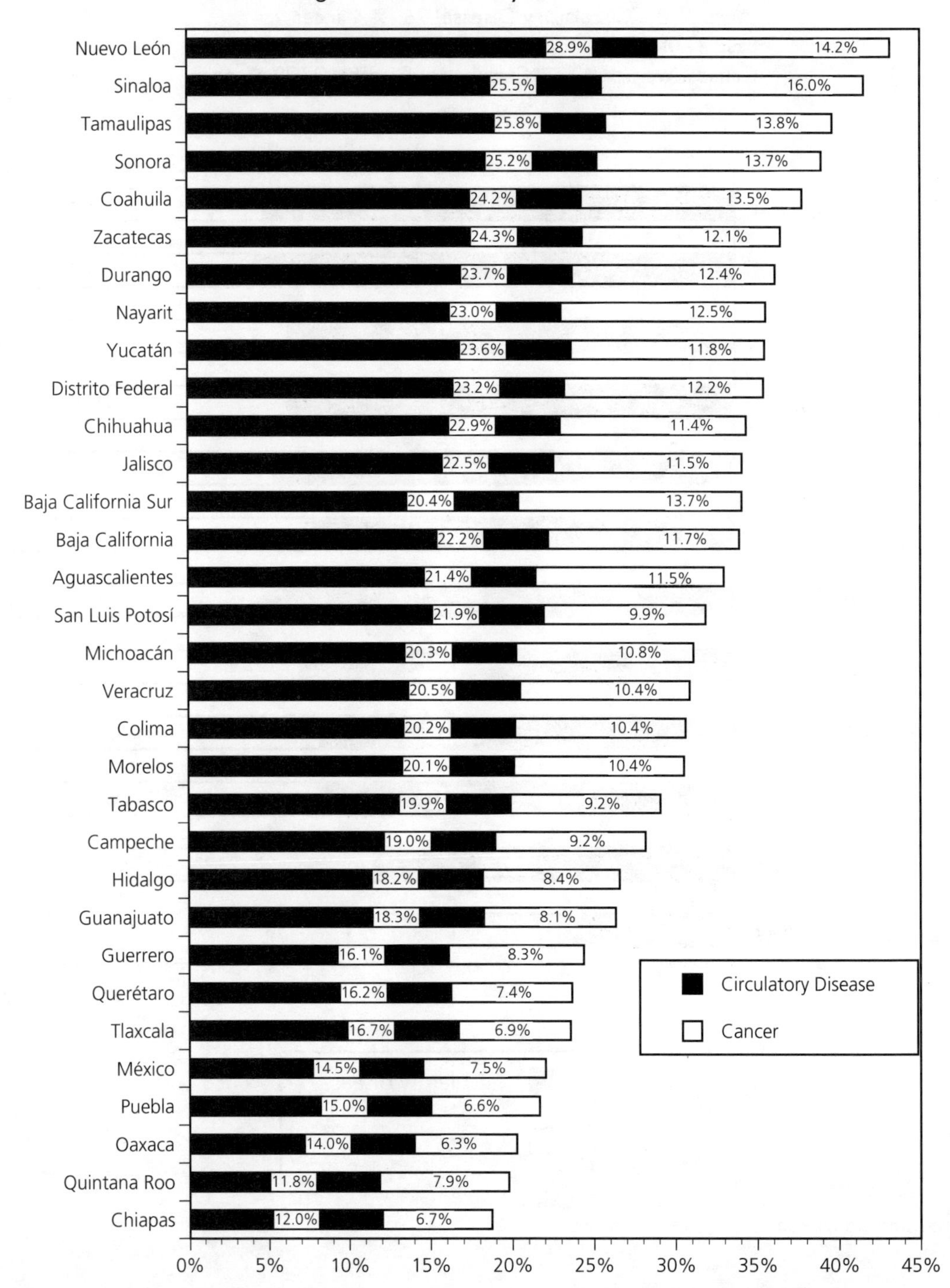

Source: INEGI, Anuario Estadistico

Physicians

Another indicator of national health is the number of physicians per capita. The United States had one physician for every 392 people in 1992. Canada had one for every 511 people in 1991, and Mexico had one for every 975 people. In 1988, the United States ranked twenty-first of 203 countries in population per physician. Canada ranked thirty-first, and Mexico ranked seventy-ninth. The world average for physicians per capita in 1980 was one per 2,000 people.

In the United States, there are more physicians per capita in the West than in the country as a whole. In Canada, the reverse is true. In Western Canada, the average physician serves 595 people, compared to a Canadian average of 511 people.

Figure 4.10 **Western United States**
Active Physicians by State, 1989
(Does not include physicians employed by the federal government.)

West: 184,653
East: 347,985
U.S. Total: 532,638

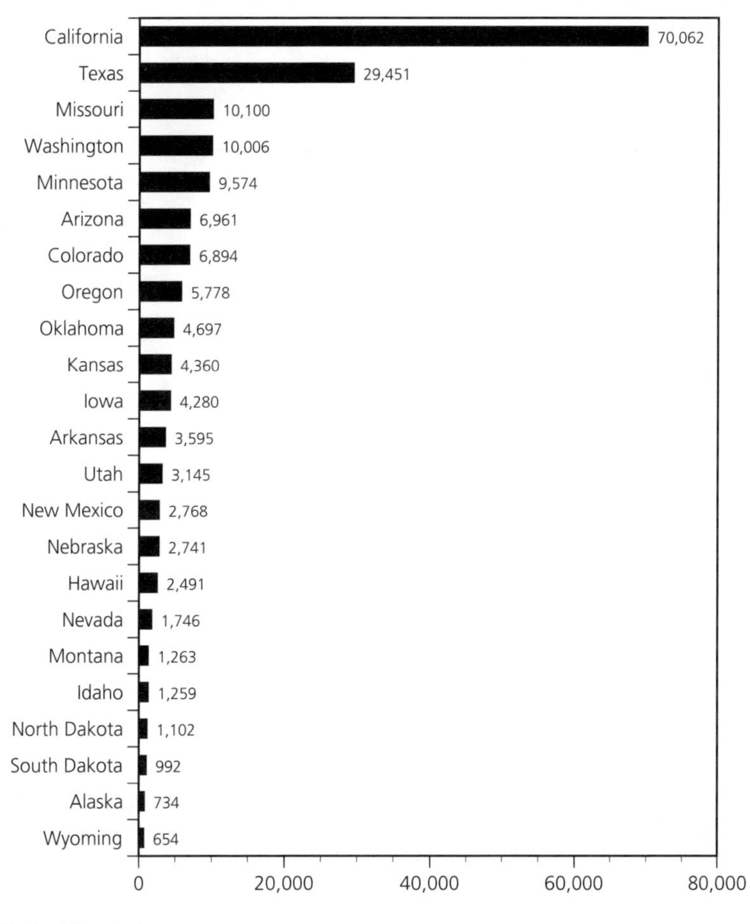

Source: American Medical Association

Figure 4.11 **Canada**
Active Physicians by Province, 1991

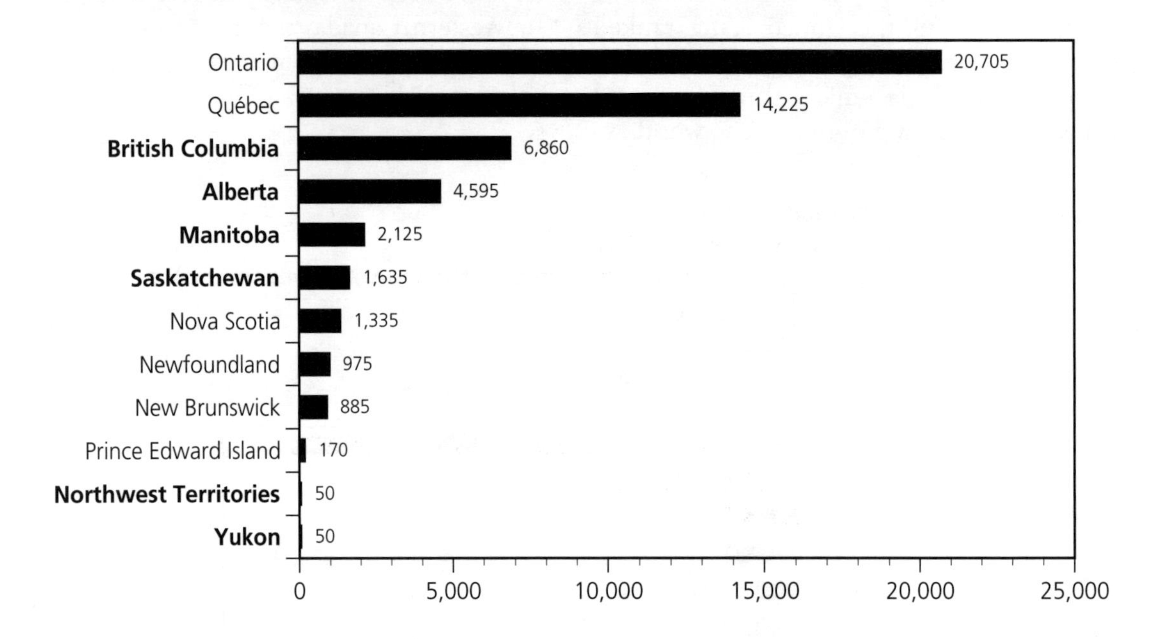

Source: Statistics Canada

Figure 4.12 **Mexico**
Active Physicians by State, 1990

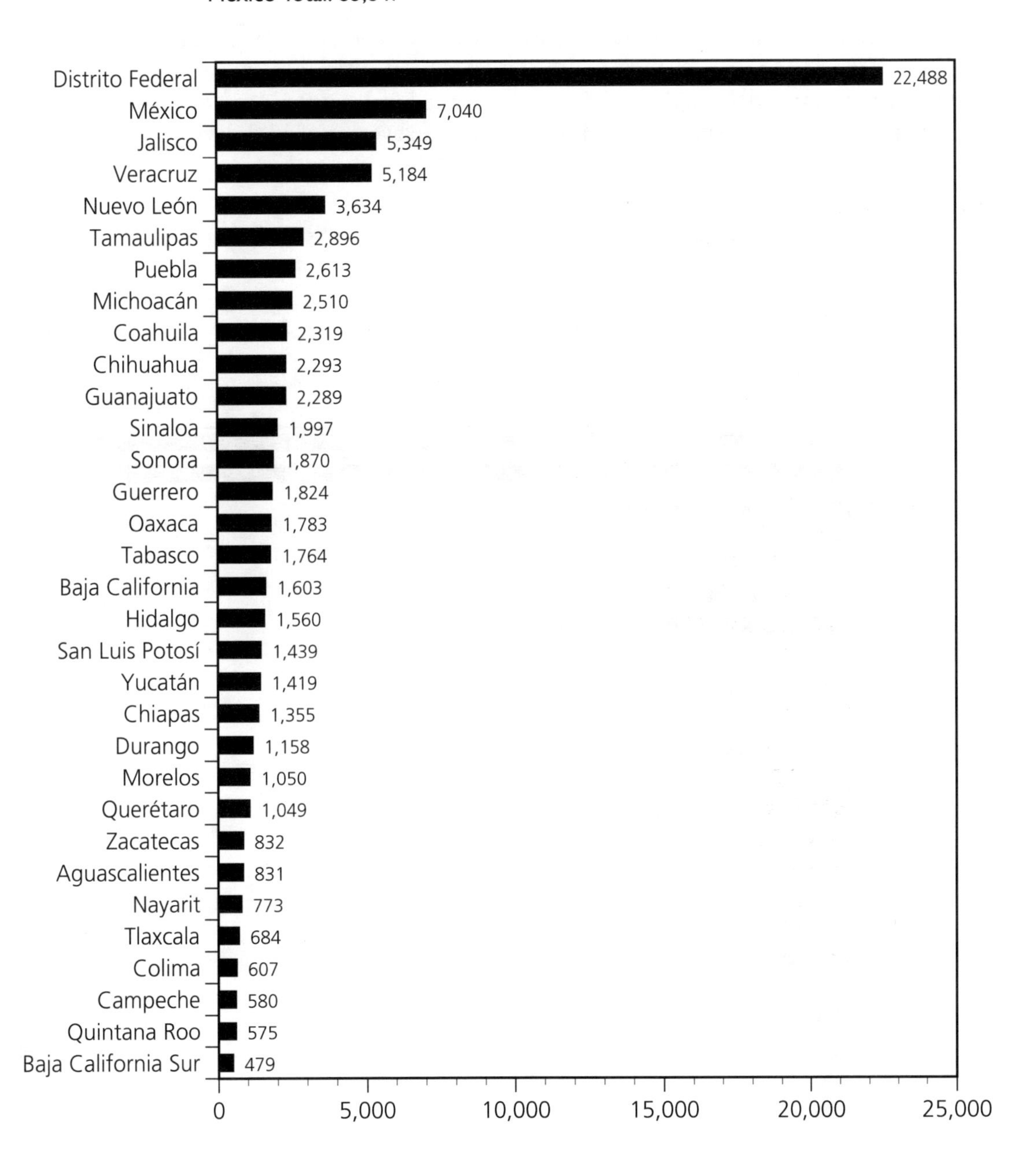

Source: INEGI

Hospitals

Canada has a hospital bed for every 114 people. The United States has a bed for every 159 people, Mexico for every 893 people. Worldwide, there is an average of one hospital bed for every 290 people. In 1980, Canada ranked fortieth worldwide; the United States sixty-sixth, and Mexico one hundred seventieth.

One reason why the United States has a lower ratio of hospital beds to population than Canada is because U.S. hospital stays tend to be shorter. The average hospital stay in the United States is 9.3 days, compared to a Canadian average of 13.2 days.

Figure 4.13 **Western United States Hospitals by State, 1990**

West: 2,868
East: 3,781
U.S. Total: 6,649

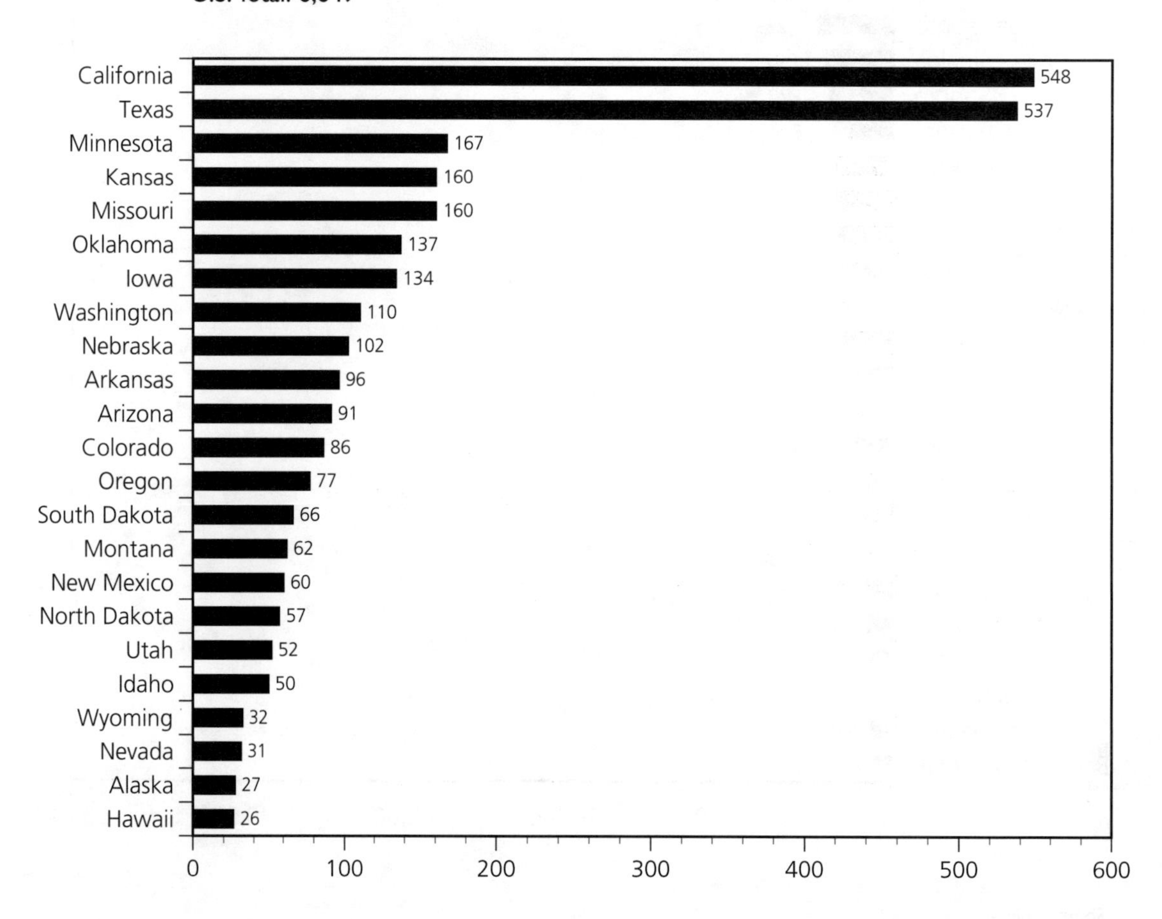

Source: American Hospital Association

Figure 4.14 **Western United States Hospital Beds by State, 1990**

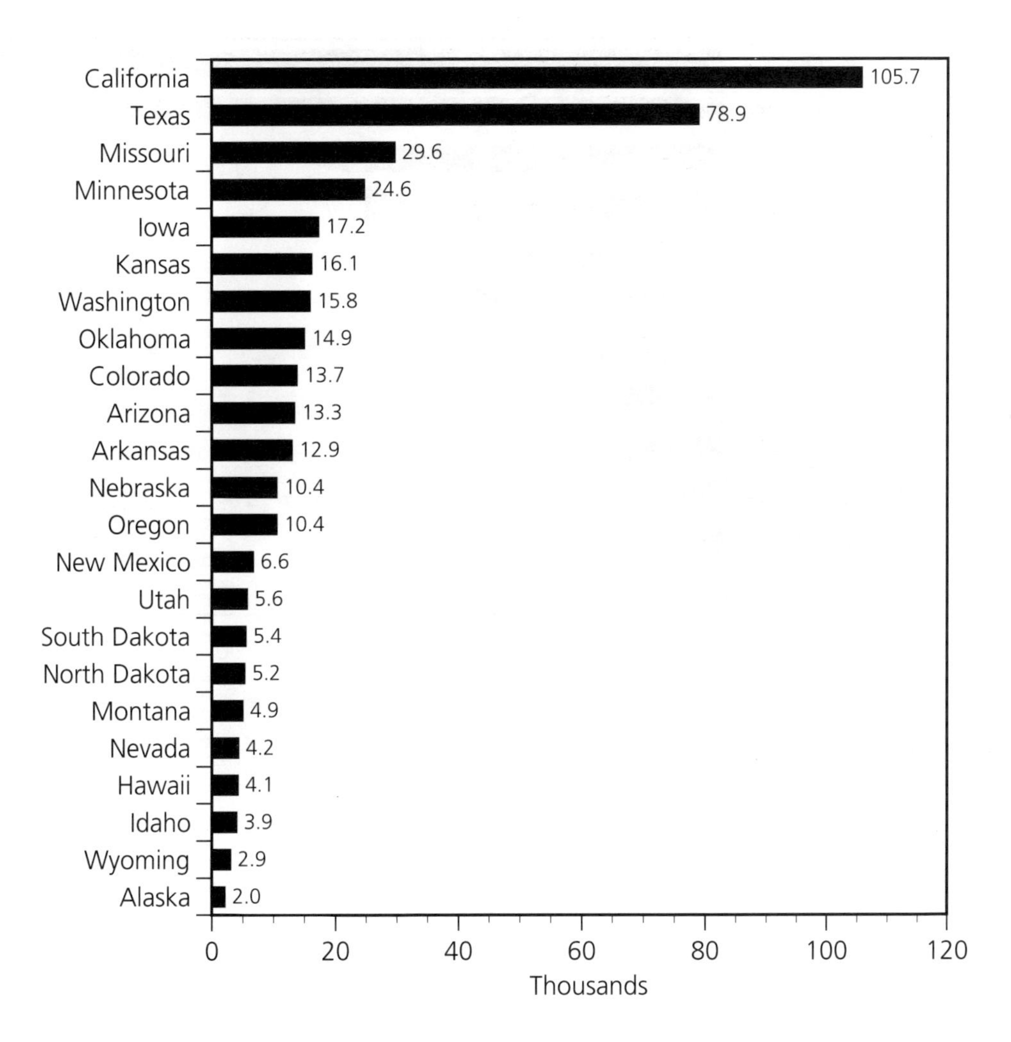

Source: American Hospital Association

Figure 4.15 **Canada**
Hospitals by Province, 1991

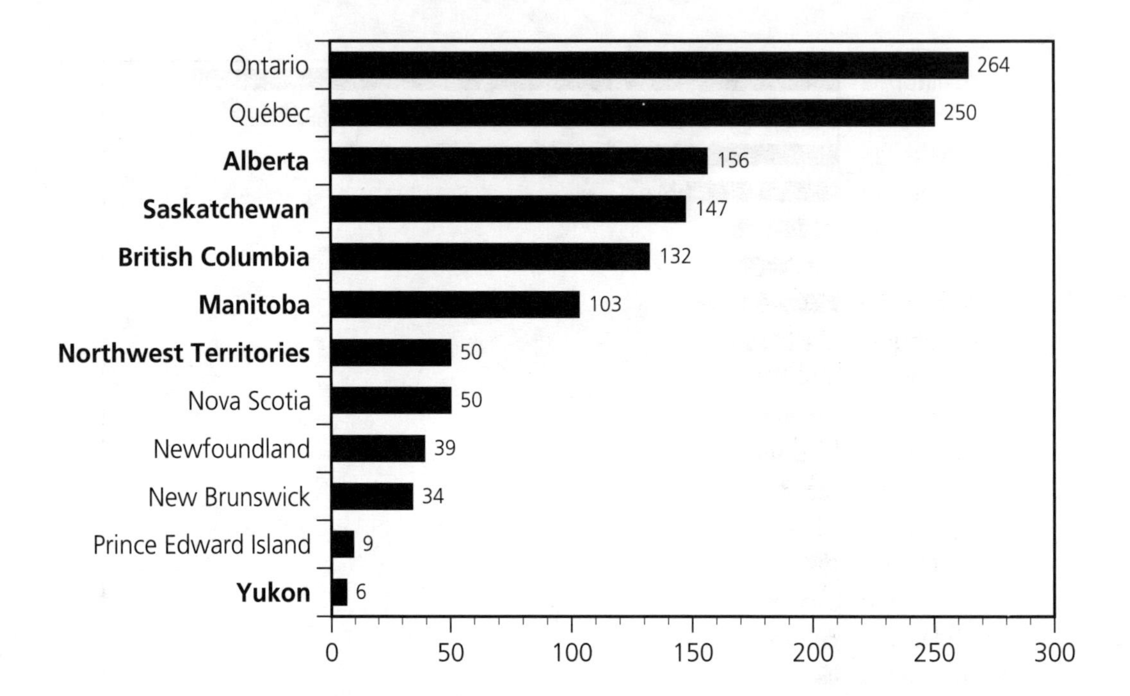

Source: Statistics Canada

Figure 4.16 **Canada**
Hospital Beds by Province, 1991

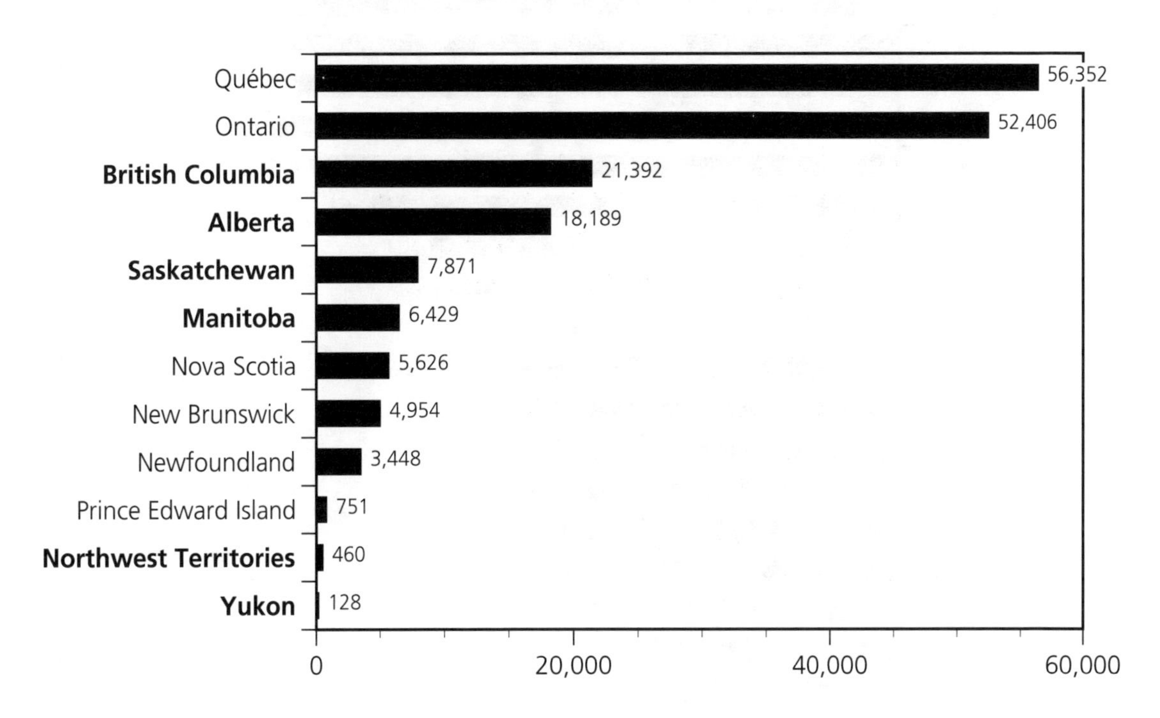

Source: Statistics Canada

Figure 4.17 **Mexico**
Hospitals and Clinics by State, 1988

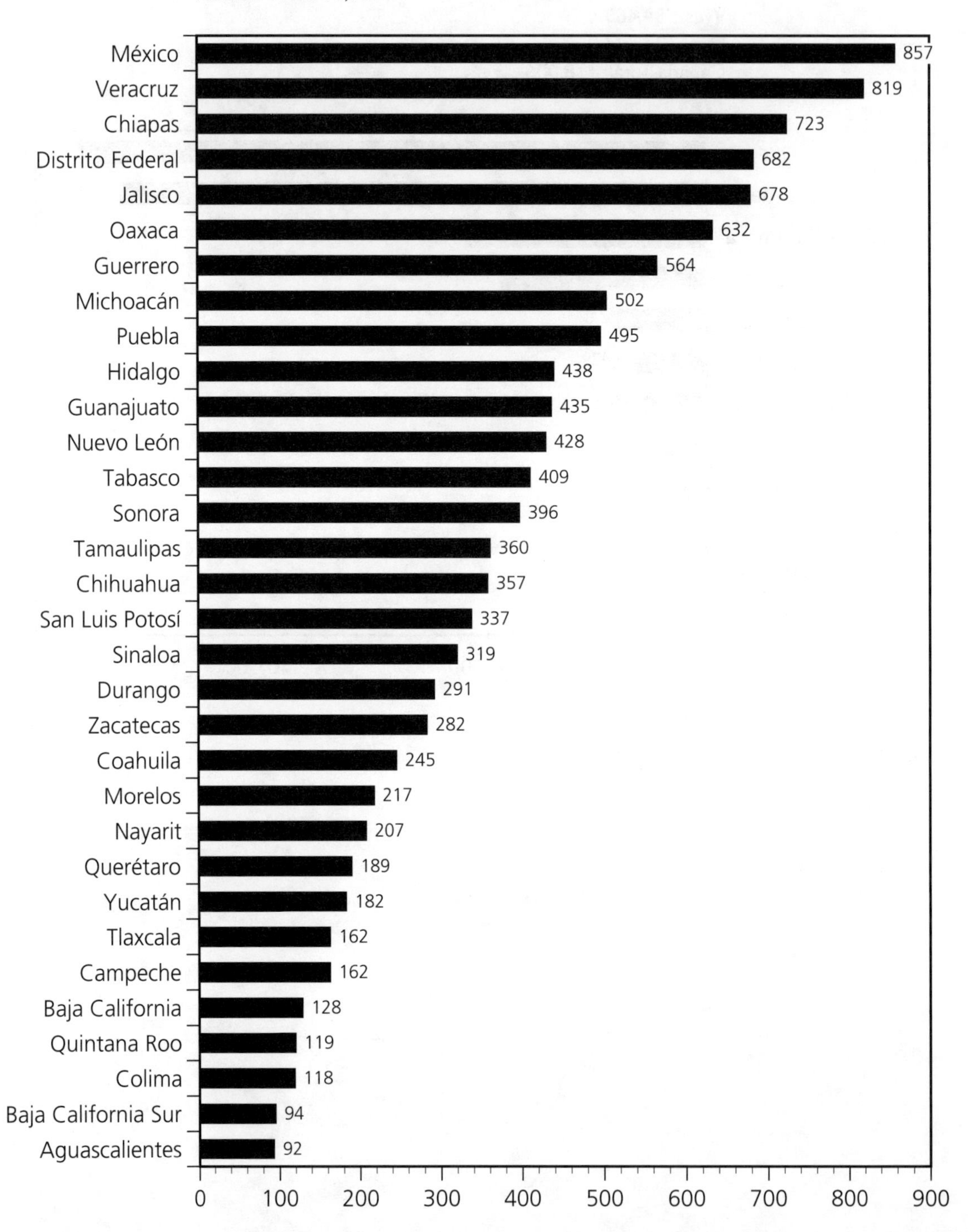

Source: INEGI

Figure 4.18 **Mexico**
Hospital Beds by State, 1990

Mexico Total: 61,251

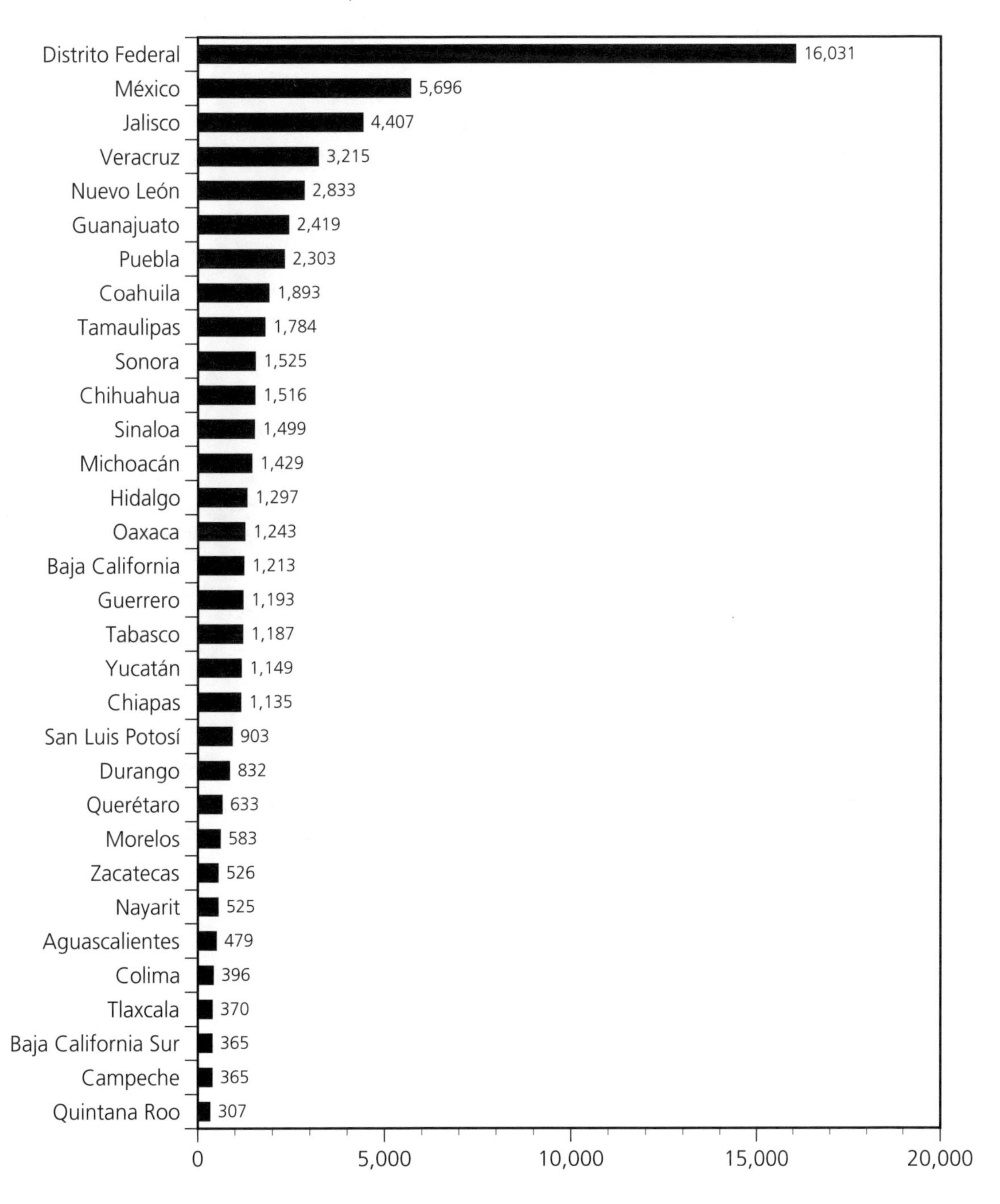

Source: INEGI

Chapter Five
Housing

Canadian homes have more rooms than either Mexican or U.S. homes. There is an average of 5.7 rooms per dwelling unit in Canada, compared to 5.1 in the United States and 2.3 in Mexico.

Just about every home in Canada and the United States has electricity. About three-quarters of Mexican homes do. Just about every Canadian home (99.5 percent in 1988) has water piped in. The United States ranked slightly lower, with 97.6 percent; less than two-thirds of Mexican homes had piped water.

In the western United States, the median price of a single family home in 1993 was highest in Honolulu ($347,000), and lowest in Amarillo, Texas ($59,700).

Figure 5.1 **Canada**
Median Sales Price of Residential Housing by Province, May 1993

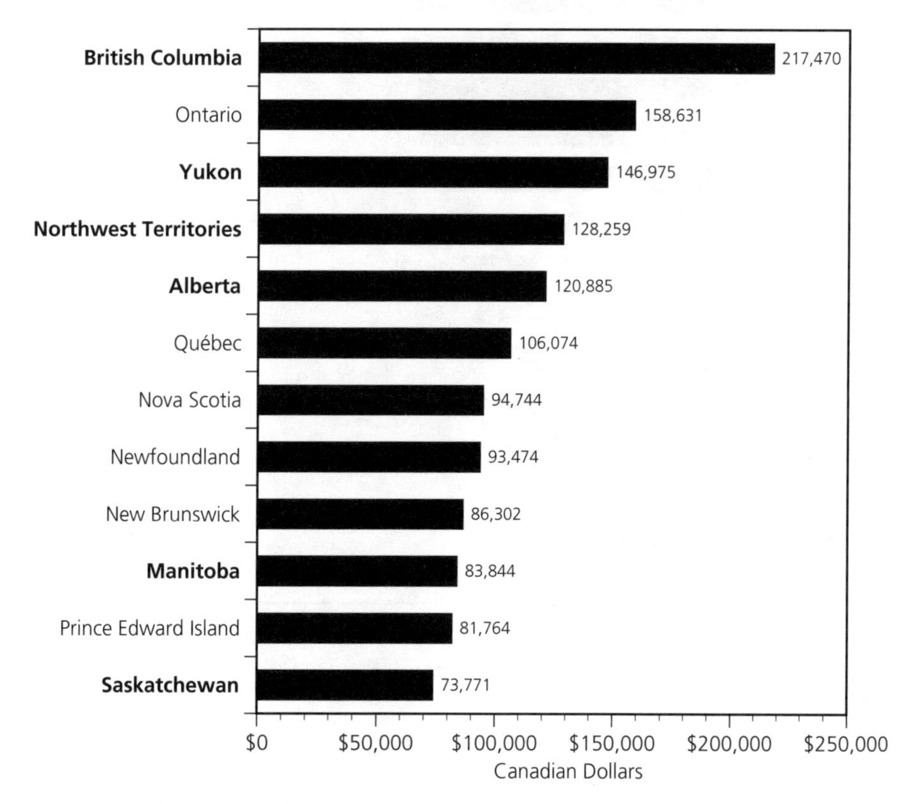

Source: The Canadian Real Estate Association

The most expensive homes in Canada are in the Victoria and Vancouver area of British Columbia and in Ontario. The least expensive homes are in Saskatchewan and Quebec.

Figure 5.2 **Canada**
Median Sales Price of Residential Housing by Major Metropolitan Areas, May 1993

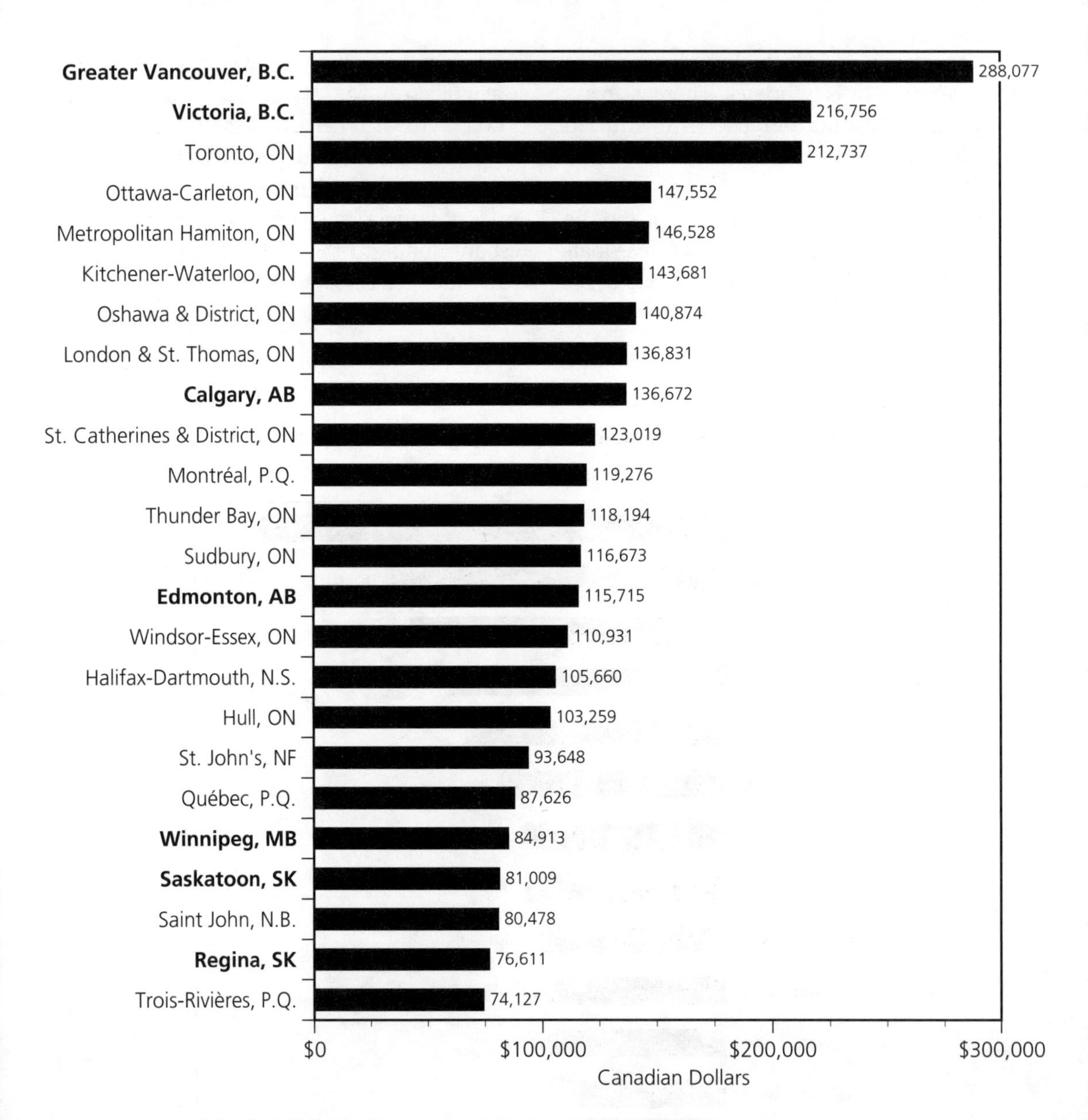

Source: The Canadian Real Estate Association

Figure 5.3 **Western United States Median Sales Price of Existing Single-Family Homes, by Major Metropolitan Areas, First Quarter 1993**

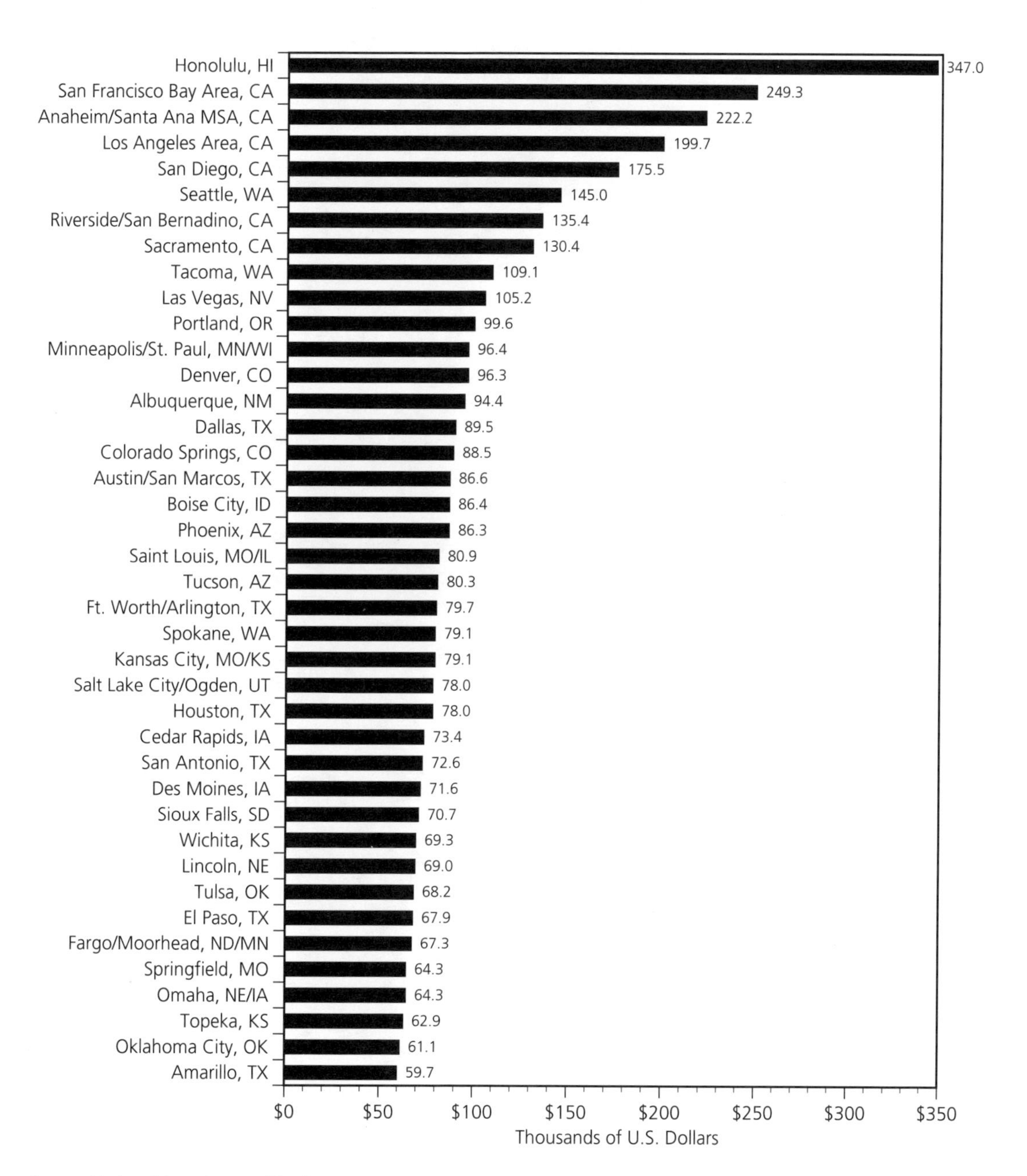

Source: National Association of Realtors

Chapter Six
Transportation

Roads

North Americans get around primarily by road. There are 3.9 million miles (6.28 million km) of roads in the United States, of which 52 percent are in the West; 549,255 miles (883,916 km) of roads in Canada; and 148,661 miles (239,344 km) of roads in Mexico.

As one might expect, most of those roads are in rural areas and between metropolitan areas. In the western United States, 312,482 miles (502,877 km) of roads are in urban areas, and 1.702 million (2.74 million km) in rural areas.

Canada does not keep statistics on urban and rural road mileage as the United States does. In Canada, 64.6 percent of roads are under municipal jurisdiction, 33.6 percent under provincial or territorial jurisdiction.

The western United States has a slightly higher proportion of rural roads (85 percent) than the nation as a whole (80 percent). In Canada, the proportion of roads under provincial or territorial jurisdiction is somewhat higher in the West than in the East.

In the western United States, total road mileage is more dependent on population than on the geographic size of the state. Texas leads in both rural and urban mileage, but Alaska has the second lowest highway mileage in the West, ahead only of Hawaii. Kansas has more road mileage than Minnesota, which has substantially more road mileage than Montana. Hawaii has by far the greatest proportion of urban mileage. North Dakota has the least.

Of those roads, 45,074 miles (72,537 km) in the United States are part of the interstate highway system. In Canada, highways are a provincial responsibility. Except in national parks and in the capital of Ottawa, there are few federal highways. Less than 1.8 percent of Canadian roads are under federal jurisdiction. Except in the Distrito Federal, only a small proportion of Mexican roads are federal highways. A substantial proportion of Mexican roads are unpaved. Virtually all road mileage in the United States and Canada is paved.

Autos and Trucks

The United States leads the world in the number of motor vehicles in relation to population, an average of 788.2 per 1,000 persons in 1990. Canada had 638 per 1,000, while Mexico had 147.3 motor vehicles per 1,000 persons in 1990. The world average is 43.6 motor vehicles per 1,000 persons.

California (21.9 million) and Texas (12.8 million) lead the West in motor vehicle registration. There are more cars and trucks in California than in all of Canada or Mexico.

Figure 6.1 **Western United States Total Road and Street Mileage by State, 1990**

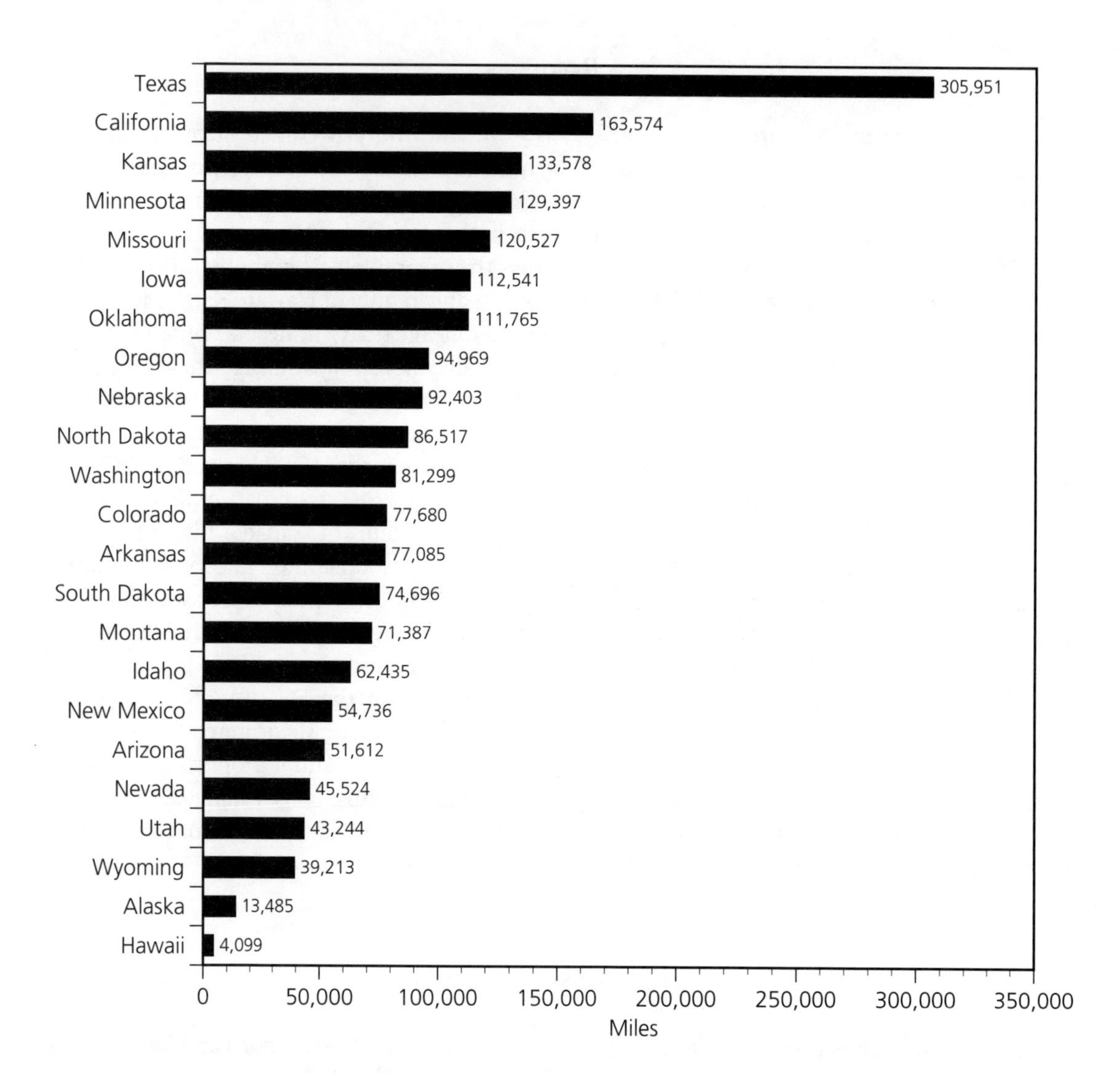

Source: U.S. Department of Transportation, Federal Highway Administration

Figure 6.2 **Canada**
Total Road and Street Mileage by Province, 1988–1989

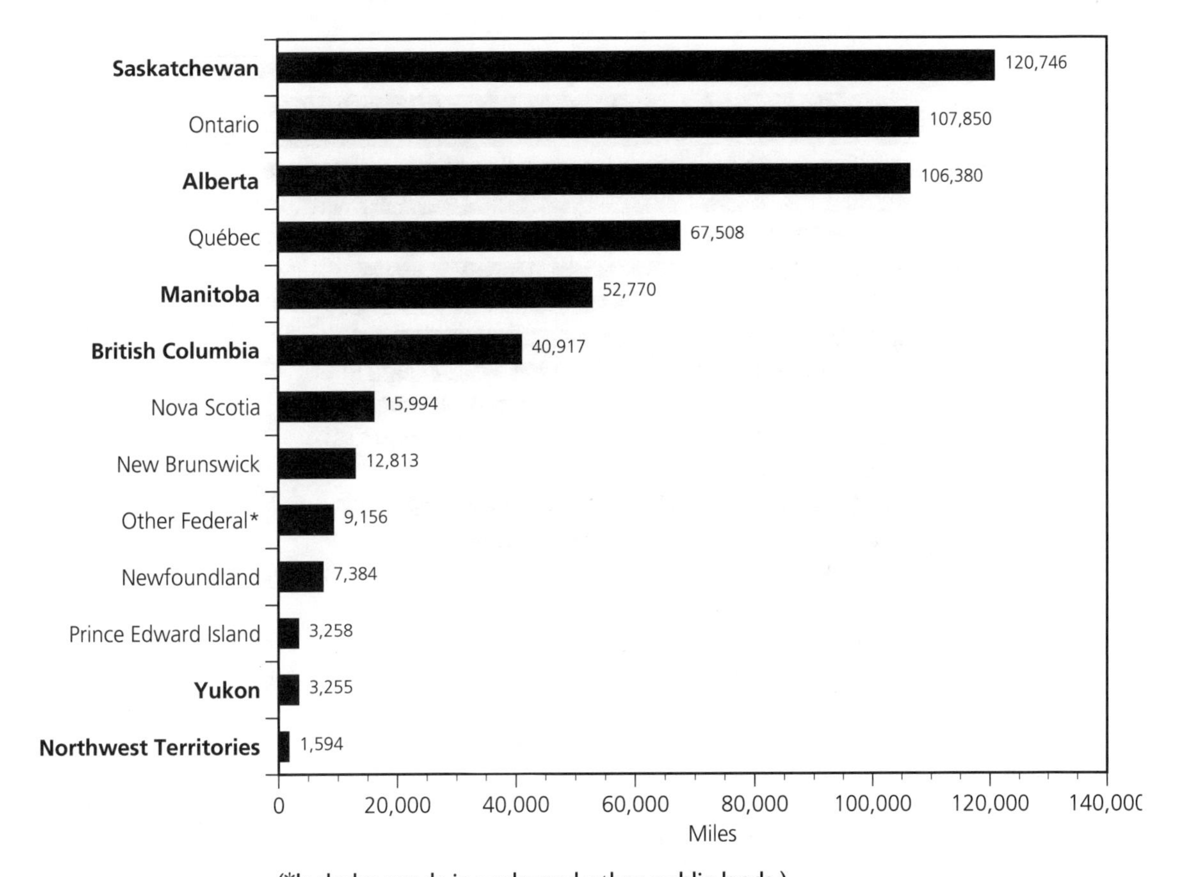

(*Includes roads in parks and other public lands.)

Source: Transportation Association of Canada

Figure 6.3 **Mexico**
Total Road and Street Mileage by State, 1990

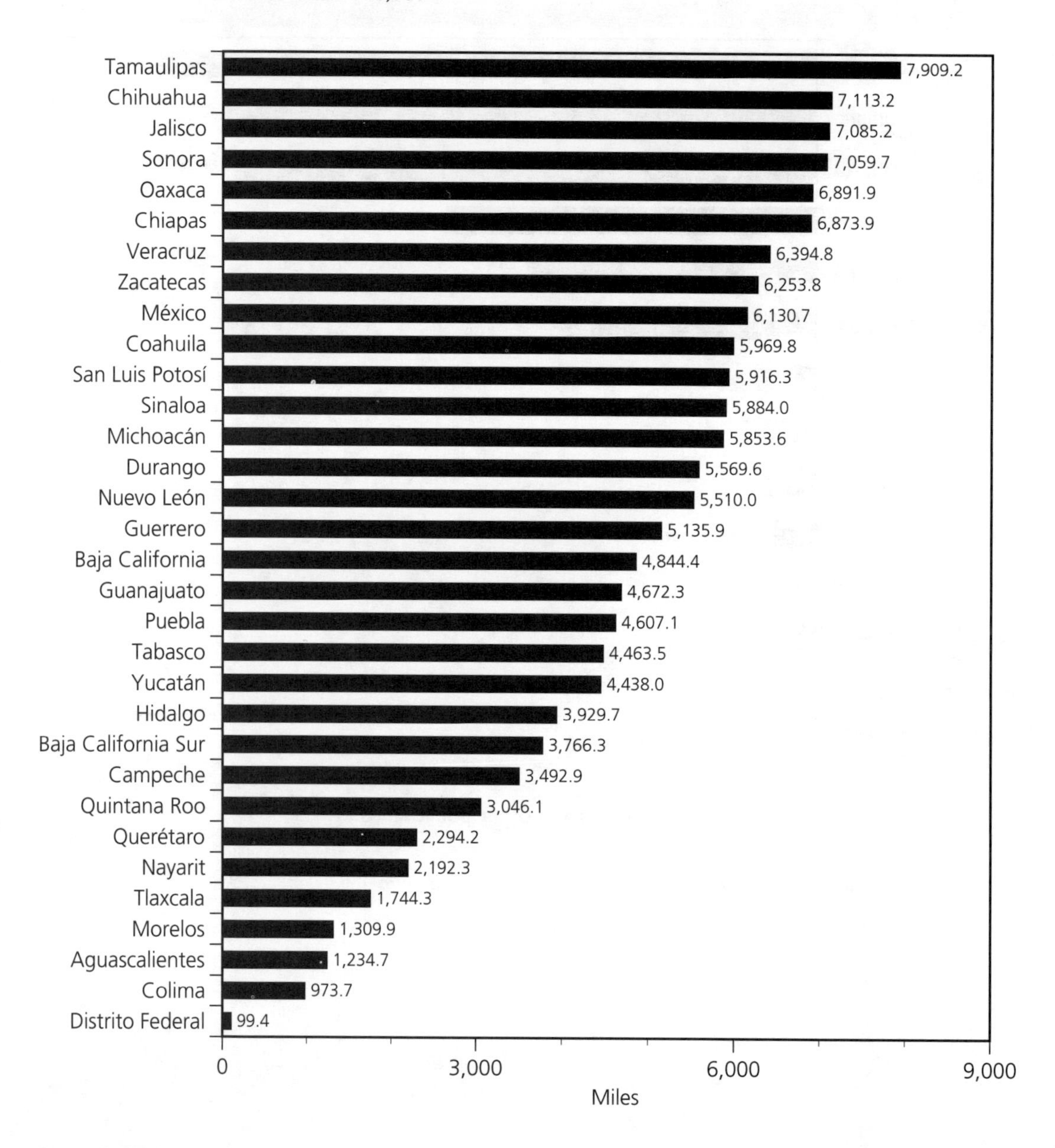

Source: INEGI, Anuario Estadistico

Figure 6.4 **Western United States Motor Vehicle Registrations by State, 1990**

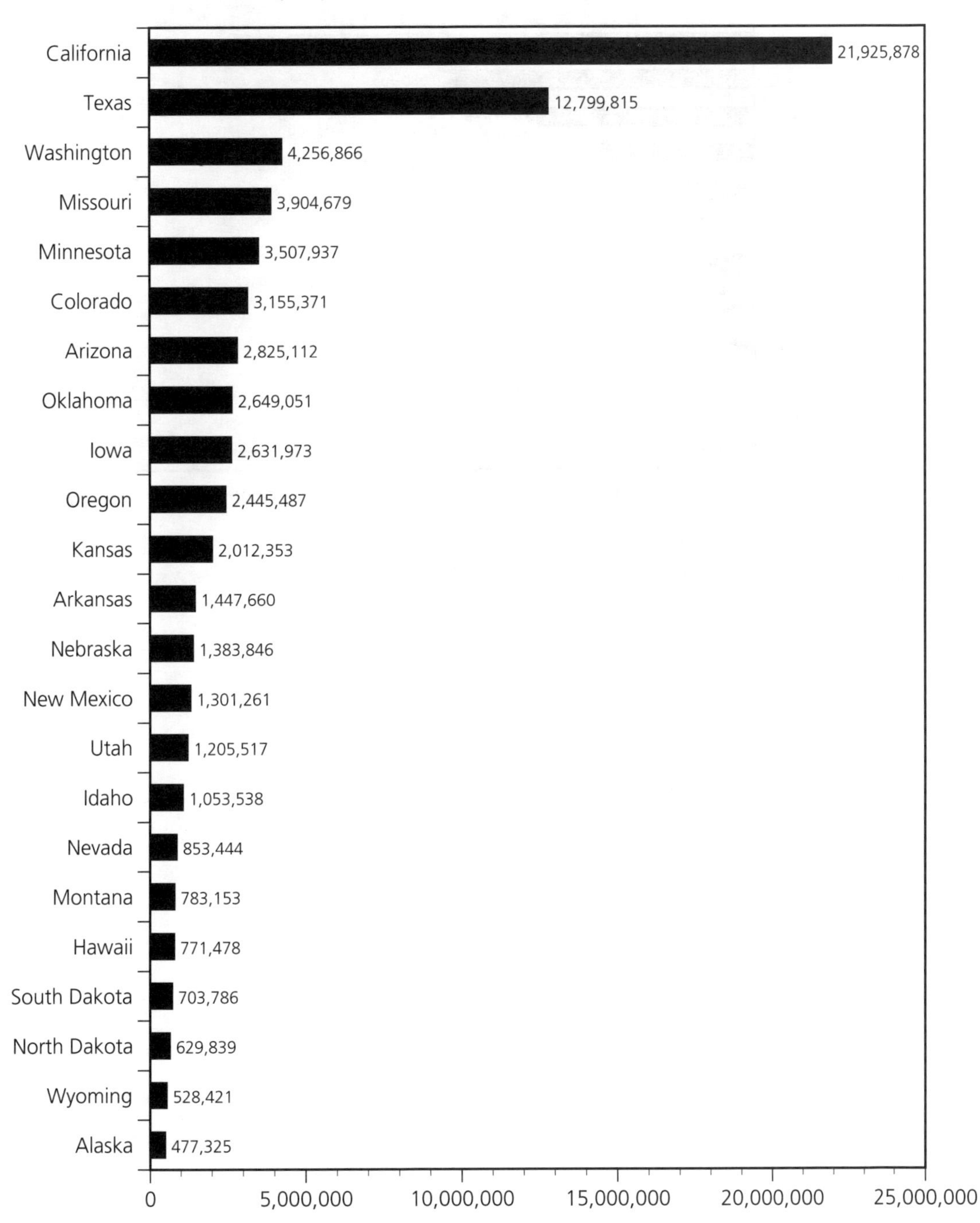

Source: U.S. Department of Transportation, Federal Highway Administration

Figure 6.5 **Canada**
Motor Vehicle Registrations by Province, 1988

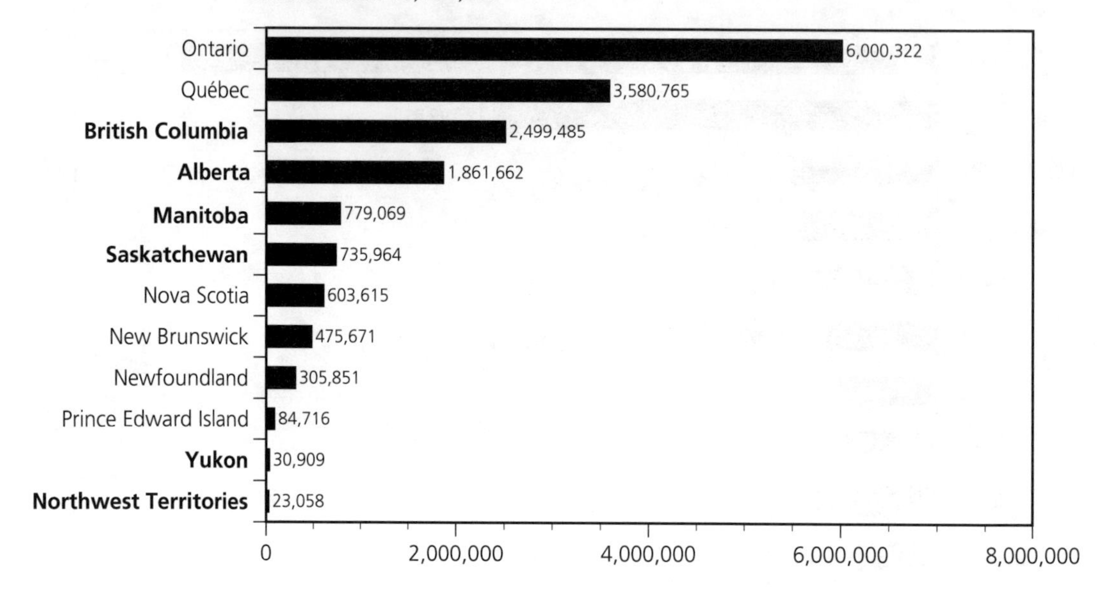

Source: Statistics Canada, Canada Yearbook

Figure 6.6 **Mexico**
Motor Vehicle Registrations by State, 1990 preliminary

Note: Does not include rental vehicles, except for D.F., which includes rental motorcycles.
Mexico Total: 11,493,706

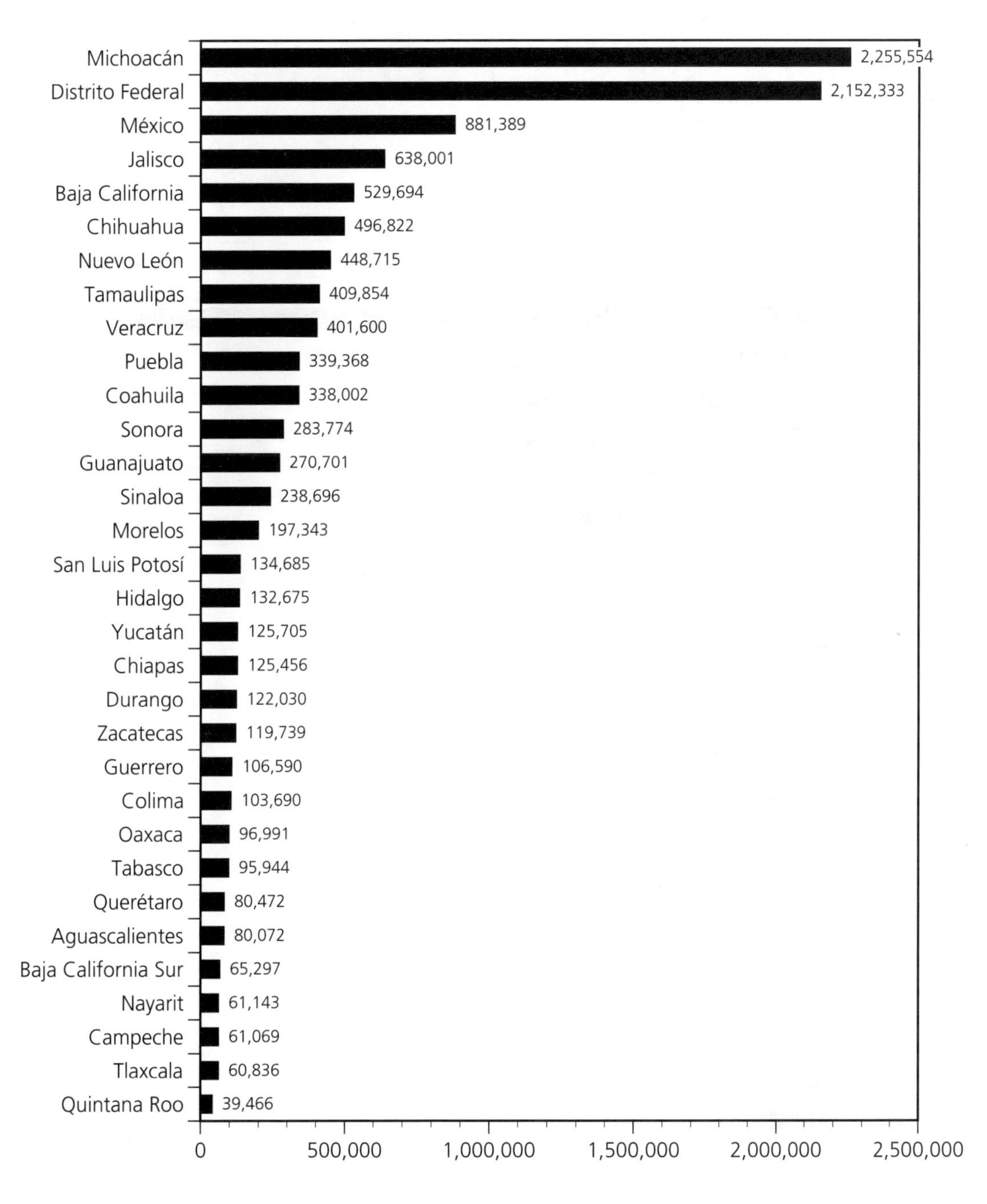

Source: INEGI, Anuario Estadistico

RAILROAD TRACKAGE

The United States has 128,939 miles (207,501 km) of railroad tracks, of which 71,039 (114,323 km), or 55 percent, are in the West. Canada has 39,519 miles (63,598 km) of track, of which 16,923 miles (27,234 km), or 43 percent, are in the Canadian West. Mexico has 16,369 miles (26,343 km) of track. Taken together, the West has 56 percent of all railroad track miles in North America.

Figure 6.7 **Western United States Railroad Mileage by State, 1991**

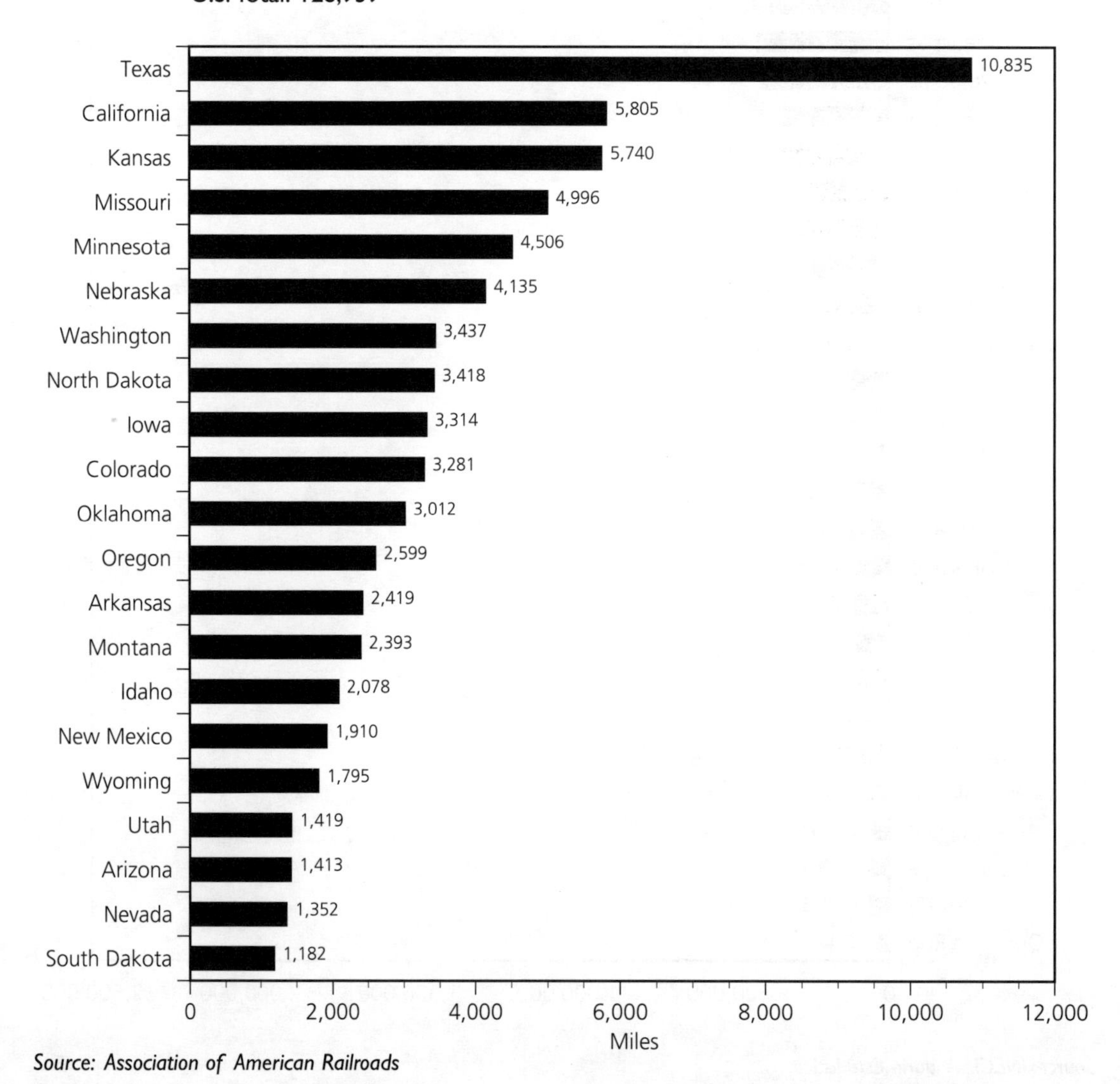

Source: Association of American Railroads

Figure 6.8 **Canada**
Railroad Mileage by Province, 1989

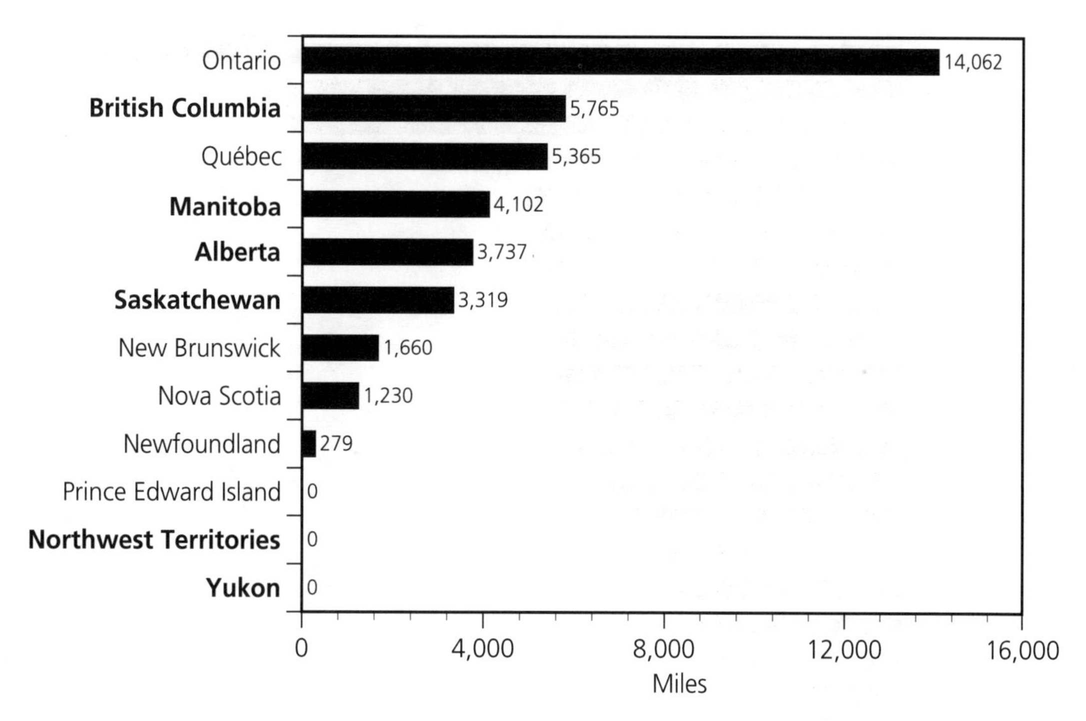

Source: Transport Canada

Figure 6.9 **Mexico**
Railroad Mileage by State, 1990

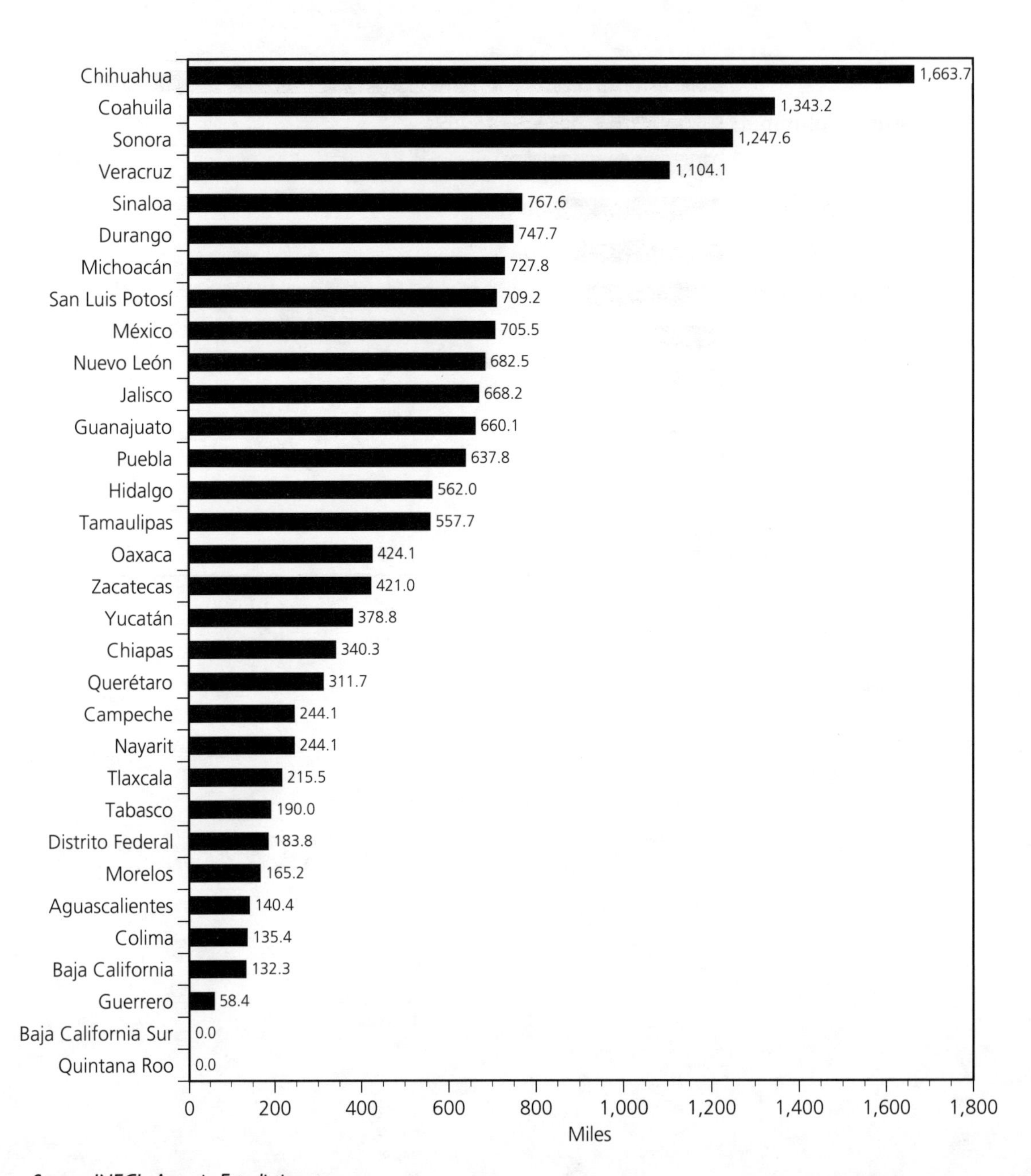

Source: INEGI, Anuario Estadistico

Airports

Four of the top ten airports in the United States measured by passenger and cargo traffic are located in the West. In 1992, Dallas/Ft. Worth ranked second in passengers and tenth in cargo. Los Angeles ranked third in passengers and second in cargo. San Francisco was fifth in passengers and seventh in cargo. Denver was sixth in passengers, and Anchorage was ninth in cargo. Chicago O'Hare is first in passengers, and Memphis is first in air cargo.

Figure 6.10 **Western United States Principal Airports and Passengers, 1992**

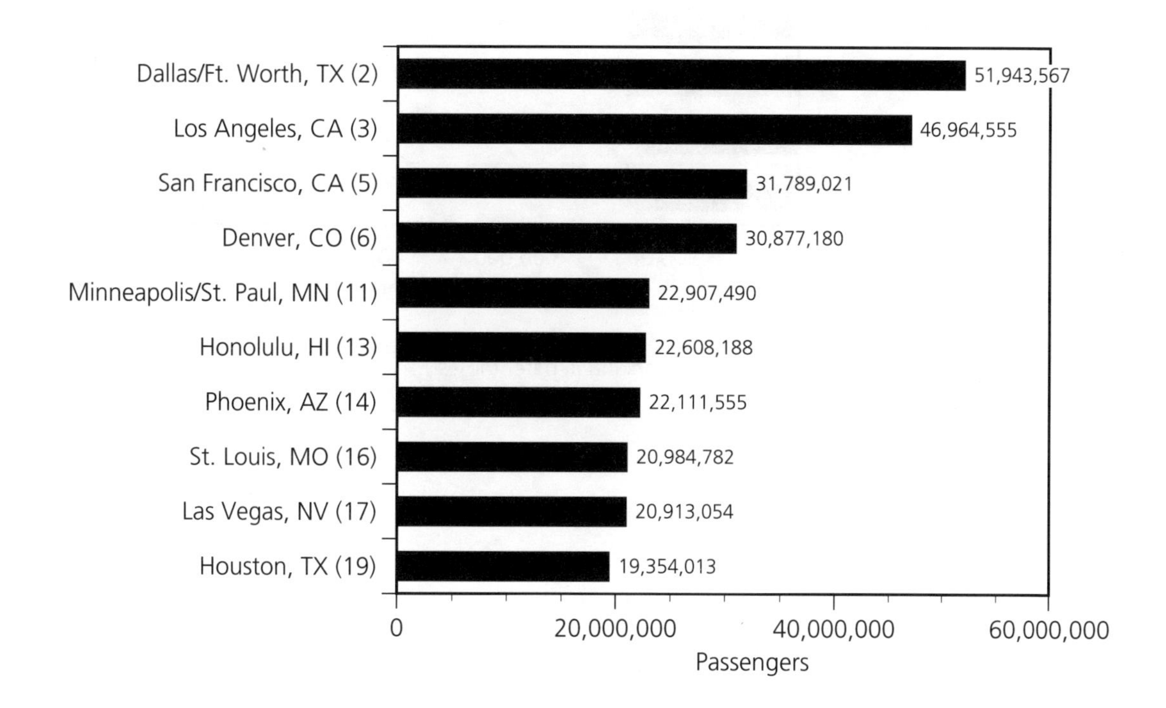

(National ranking in parentheses.)

Source: Air Transport Association

Figure 6.11 **Western United States Principal Airports and Cargo, 1992**

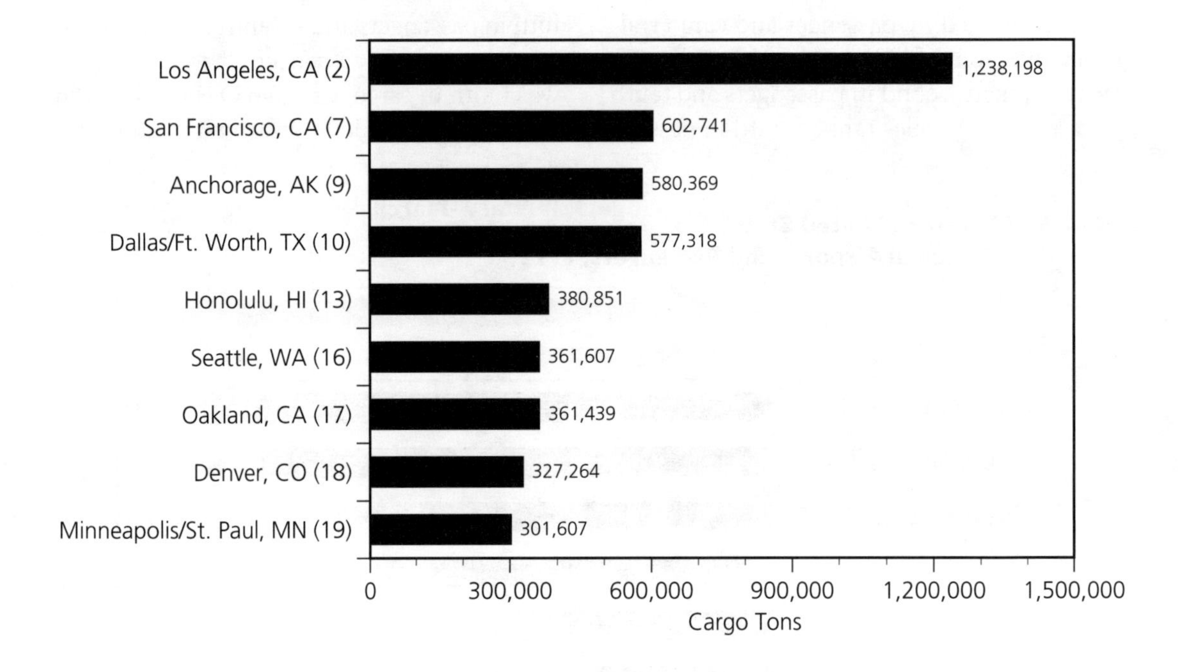

(National ranking in parentheses.)

Source: Air Transport Association

Five western Canadian airports ranked in Canada's top ten in Canadian passenger traffic in 1992. Vancouver International was second, Calgary International was fourth, Winnipeg International was seventh, and Edmonton's two airports, International and Municipal, ranked ninth and tenth. Vancouver, Calgary, Winnipeg, and Edmonton ranked second, fourth, seventh, and eighth, respectively, in Canadian cargo tonnage in 1991.

Figure 6.12 **Canada**
Principal Airports and Passengers, January–March 1992

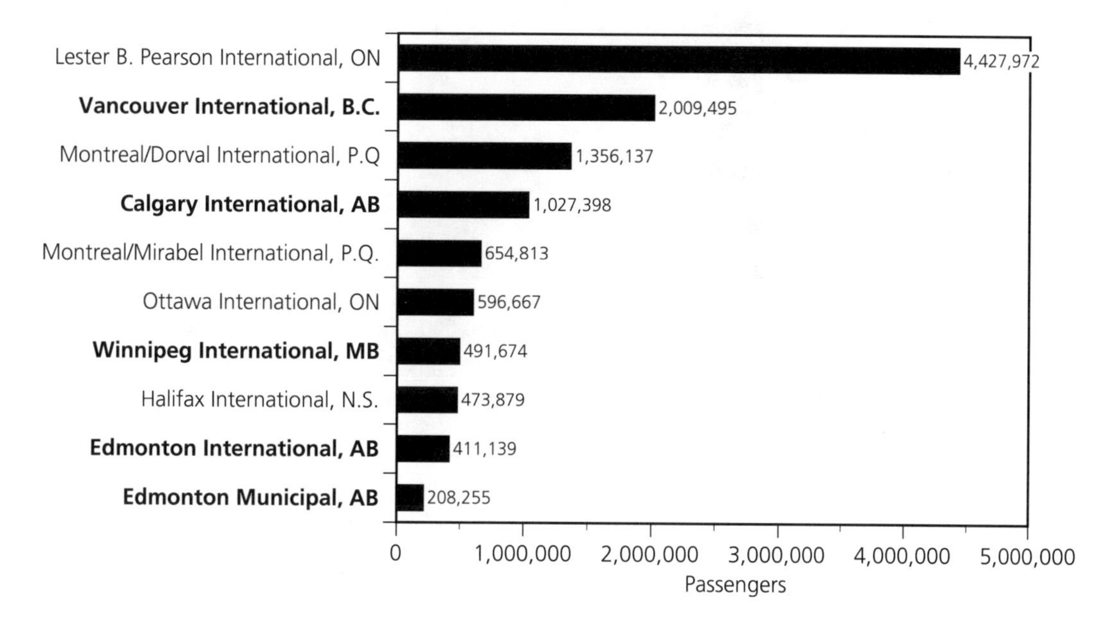

Source: Statistics Canada

Figure 6.13 **Canada**
Principal Airports and Cargo, January–March 1992

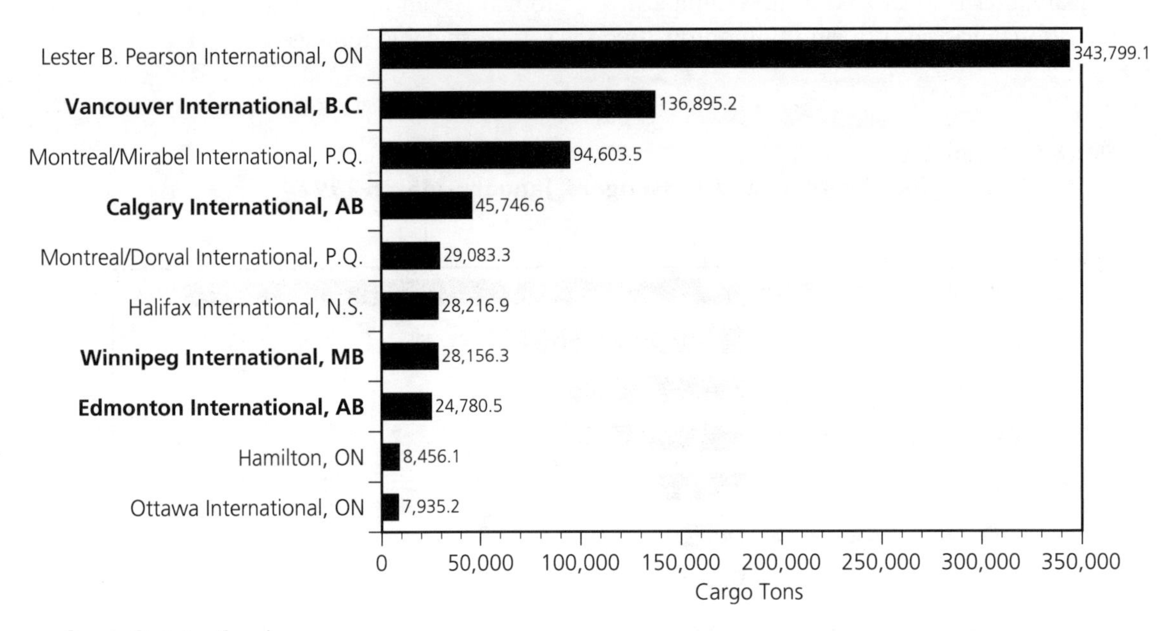

Source: Statistics Canada

Mexico City's airport was the busiest in Mexico, with 12.2 million passengers embarking and disembarking, 45 percent of the nation's total. In cargo tonnage, at 138,236 tons (125,407 mt), Mexico City's airport handled two-thirds of total Mexican cargo tonnage shipped by air.

Figure 6.14 **Mexico**
Principal Airports and Passengers, 1990 preliminary

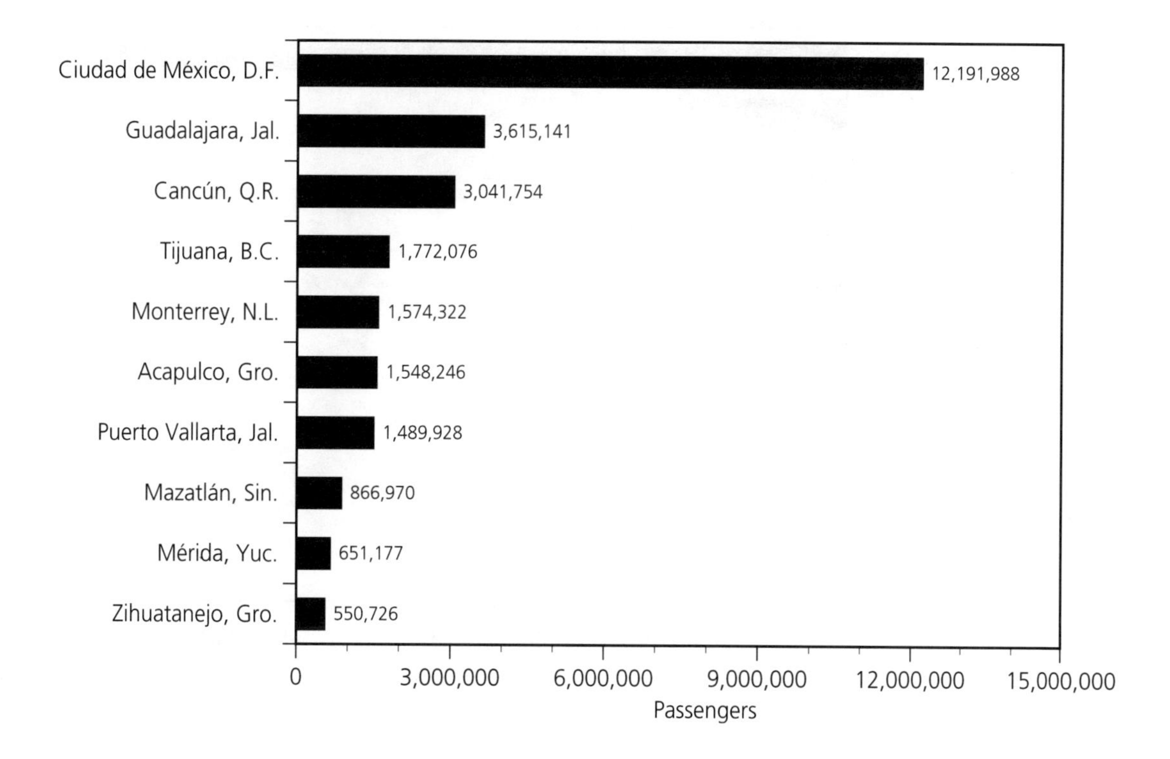

Source: INEGI

Figure 6.15 **Mexico**
Principal Airports and Cargo, 1990 preliminary

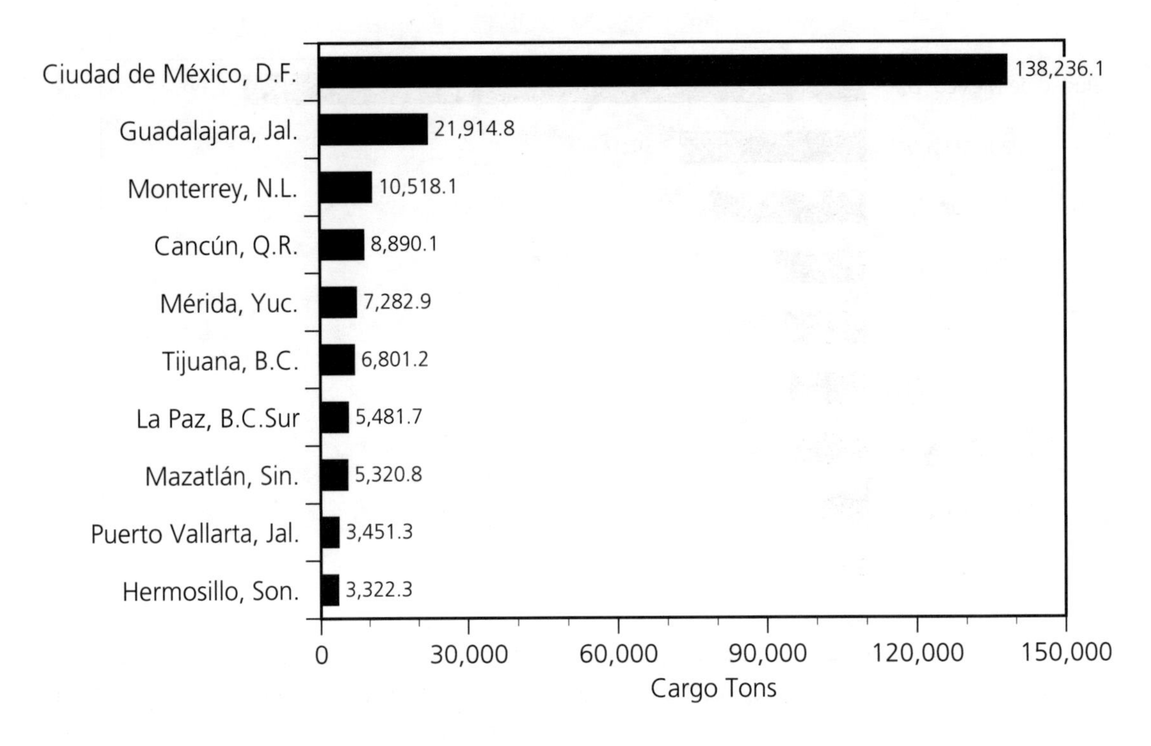

Source: INEGI

Airline passengers in North America in 1992 (Mexican figures are from 1990) totaled 468.2 million, of which the United States accounted for 94 percent.

Seaports

Western ports accounted for 38.7 percent of all tonnage handled by United States ports in 1992. Four Western ports—Houston (third), Valdez, Alaska (fourth), Corpus Christi (sixth), and Long Beach (tenth)—ranked among the top ten U.S. ports in total tonnage. The top U.S. seaports in total tonnage were the port of South Louisiana and New York.

Figure 6.16 **Western United States Principal Ports and Cargo, 1992**

Western ports accounted for 38.7% of all U.S. port tonnage in 1992.

	Millions of Tons	Percent Foreign Tonnage	National Rank
Houston, TX	126.2	49.9%	3
Valdez Harbor, AK	96.0	0.0%	4
Corpus Christi, TX	62.0	57.8%	6
Long Beach, CA	52.4	43.5%	10
Texas City, TX	48.1	59.6%	12
Los Angeles, CA	46.4	53.7%	13
Duluth-Superior, MN	40.8	14.1%	17
Port Arthur, TX	30.7	65.7%	20
Portland, OR	27.5	59.1%	21
St. Louis, Metro, MO	27.1	0.0%	22
Beaumont, TX	26.7	29.6%	23
Seattle, WA	21.6	60.7%	30
Tacoma Harbor, WA	21.4	76.7%	31
Richmond, CA	21.2	32.8%	32
Anacortes, WA	15.4	14.6%	35
Freeport, TX	14.5	49.3%	40

Source: U.S. Department of the Army, Corps of Engineers

In the value of cargo handled, the ports of Los Angeles and Long Beach, which share the same harbor, handled twice the value of cargo as all the ports of the New York Port Authority combined.

Five western Canadian ports, all in British Columbia, ranked among that nation's largest. Vancouver was the largest, handling 25 percent of all the tonnage handled by Canadian seaports.

Figure 6.17 **Canada**
Principal Ports and Cargo, 1991*

	Million of Tons	Percent International Tonnage
Vancouver, B.C.	75.4	96.7%
Port-Cartier, P.Q	25.3	75.1%
Sept-îles/Pointe-Noire, P.Q.	23.6	82.7%
Québec/Lévis, P.Q.	20.1	64.0%
Saint John, N.B.	19.0	87.2%
Thunder Bay, ON	18.6	13.3%
Montréal/Contrecoeur, P.Q.	17.9	58.5%
Halifax, N.S.	15.7	77.0%
Prince Rupert, B.C.	14.3	95.5%
Hamilton, ON	11.8	51.3%
Nanticoke, ON	9.6	63.5%
Baie-Comeau, P.Q.	9.2	63.2%
Come By Chance, NF	6.8	99.9%
Howe Sound, B.C.	6.6	0.0%
New Westminster, B.C.	5.8	47.7%
Sault Ste-Marie, ON	5.5	89.5%
Sorel, P.Q.	5.1	39.4%
Vancouver Island, East, B.C.	4.9	0.0%
Sarnia, ON	4.4	53.5%
Port Alfred, NF	4.0	96.1%

(Western ports are in bold.)

(* Ranked 1-20)

Source: Statistics Canada

Mexican ports handled 185 million tons of cargo in 1990. The busiest ports were Coatzacoalcos in Veracruz and Cayo Arcas in Campeche, both on the Mexican Gulf Coast. Each handled about 17 percent of the total tonnage handled by Mexican ports that year, more than a third of the total.

Figure 6.18 **Mexico Principal Ports and Cargo, 1990**

	Millions of Tons	Percent International Tonnage
Coatzacoalcos, Ver.	32.3	67.1%
Cayo Arcas, Camp.	32.1	99.0%
Salina Cruz, Oax.	20.7	47.6%
Dos Bocas, Tab.	20.1	91.8%
Isla de Cedros, B.C.	13.2	49.1%
Tampico, Tamps.	10.0	63.2%
Tuxpan, Ver.	7.5	28.0%
Manzanillo, Col.	7.1	45.6%
Venustiano Carranza, B.C. Sur	6.5	0.0%
Veracruz, Ver.	6.2	75.5%
Lázaro Cárdenos, Mich.	5.4	68.2%
Guaymas, Son.	5.4	46.3%
Topolobampo, Sin.	3.2	0.1%
Rosarito, B.C.	2.9	41.4%
San Marcos, B.C. Sur	2.9	100.0%
Mazatlán, Sin.	2.5	14.1%
San Miguel de Cozumel, Q.R.	2.3	84.1%
Progeso, Yuc.	1.9	36.4%
La Paz, B.C. Sur	1.8	0.1%
Lerma, Camp.	1.7	0.0%

Source: INEGI, Anuario Estadistico

Chapter Seven
Energy

Energy Production

More energy is produced in North America than anywhere else in the world. The United States is the world's largest producer of primary energy. Canada is fifth. Mexico is eighth. Primary energy includes coal, oil, natural gas, hydro, and nuclear, but not electricity which is produced from one or more primary energy sources. In 1990, from all sources the three North American nations produced 88.89 quadrillion Btus of energy, 38 percent of the energy produced by the top ten world producers that year.

Figure 7.1 **Top Ten Producers of Electric Energy, 1990**

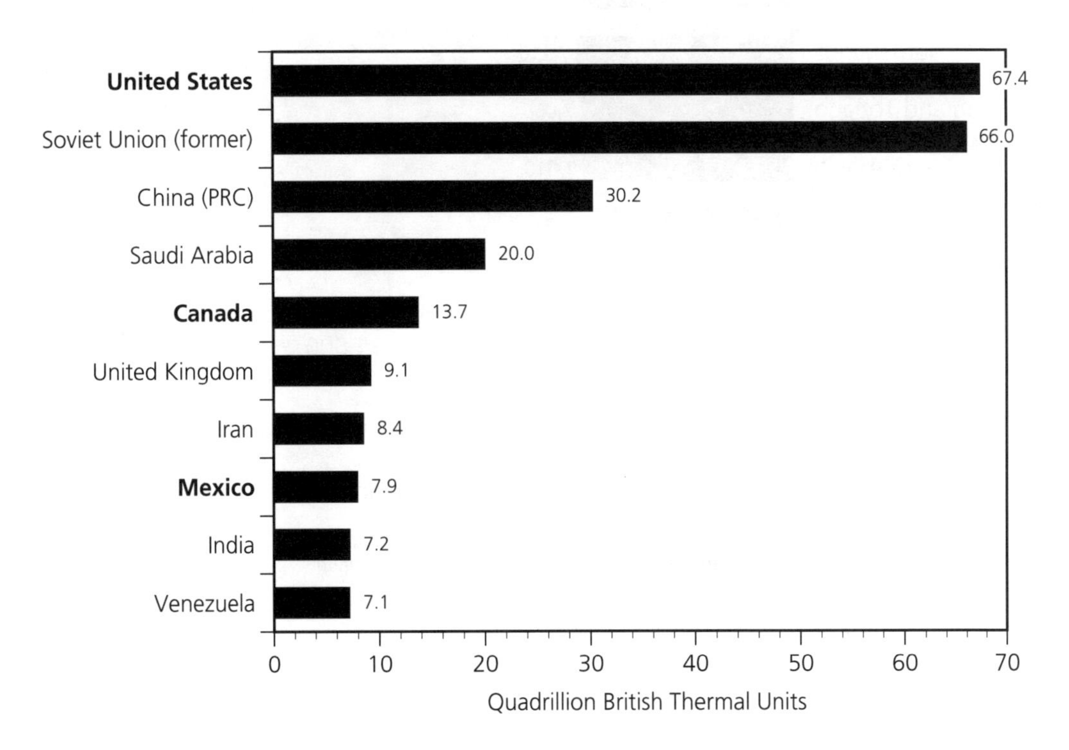

Source: Statistical Office of the United Nations, New York, NY, Energy Statistics, annual

Energy Consumption

The United States, Canada, and Mexico are also among the world's largest consumers of energy. Per capita energy consumption is higher in the West, but total energy consumption is greater in the East. In 1990, the twenty-three states west of the Mississippi accounted for 37 percent of U.S. population but 41 percent of U.S. energy consumption.

Figure 7.2 **Top Ten Consumers of Primary Energy, 1990**

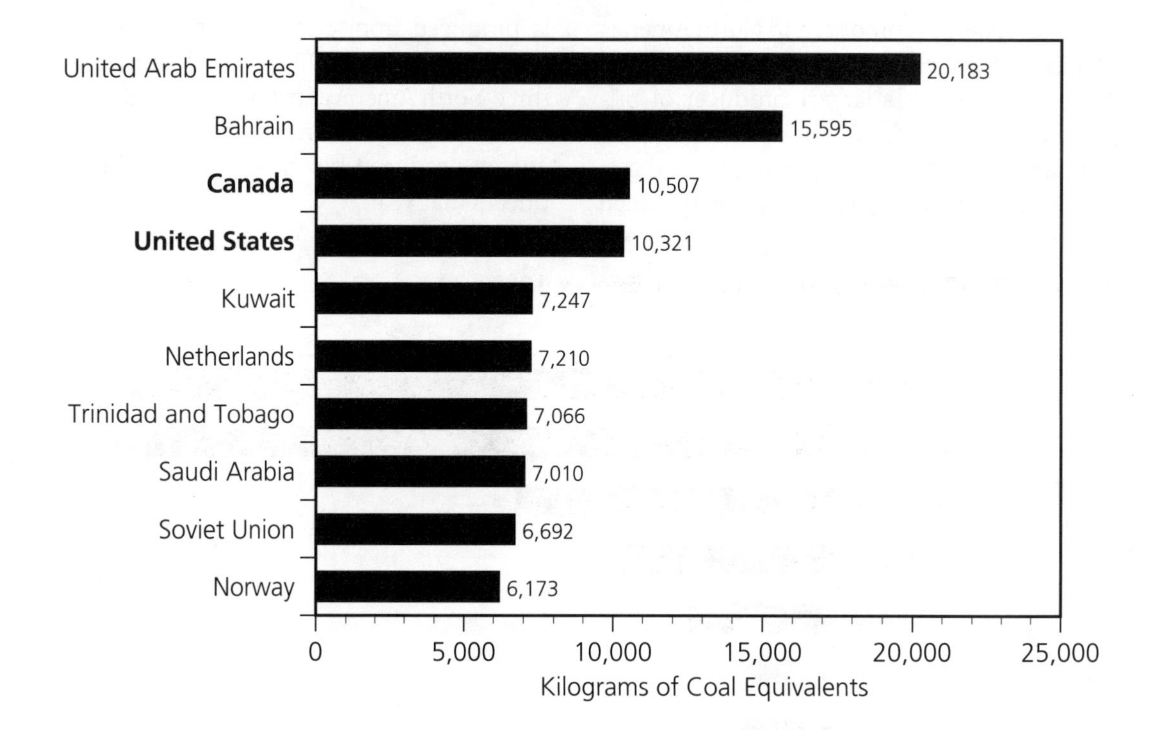

(Due to reunification, Germany is not included.)

Source: Statistical Office of the United Nations, New York, NY, Energy Statistics, annual

In the western United States and Canada, energy consumption per capita is greater in the energy-producing states. There are several reasons for this. First, energy production is an energy-intensive business: it takes energy (lots of it) to produce energy, and private, commercial, and agricultural transportation in the large and sparsely populated energy-producing states consumes a disproportionate share of energy fuels. All but one of the major energy-producing states in the western United States—Alaska, Oklahoma, Texas, and Wyoming—have rates of per capita energy consumption above the regional average. The exception is California, which recorded the lowest energy consumption per capita in the region. In Canada, in energy-rich Alberta, per capita energy consumption is higher than in every other Canadian province or territory.

In both the western United States and Canada, energy consumption tended to be greater in the colder states and provinces, as one might expect. An anomaly is South Dakota, which had one of the lowest rates of per capita energy consumption of the twenty-three states west of the Mississippi.

When an energy-producing area is in a cold climate, per capita energy consumption increases substantially. Per capita energy consumption in Alaska is nearly triple the average for the western United States. Wyoming's per capita energy consumption is more than double the regional average. Alberta has a per capita energy consumption rate double the Canadian average.

Figure 7.3 **Western United States**
Energy Consumption per Capita by State, 1990

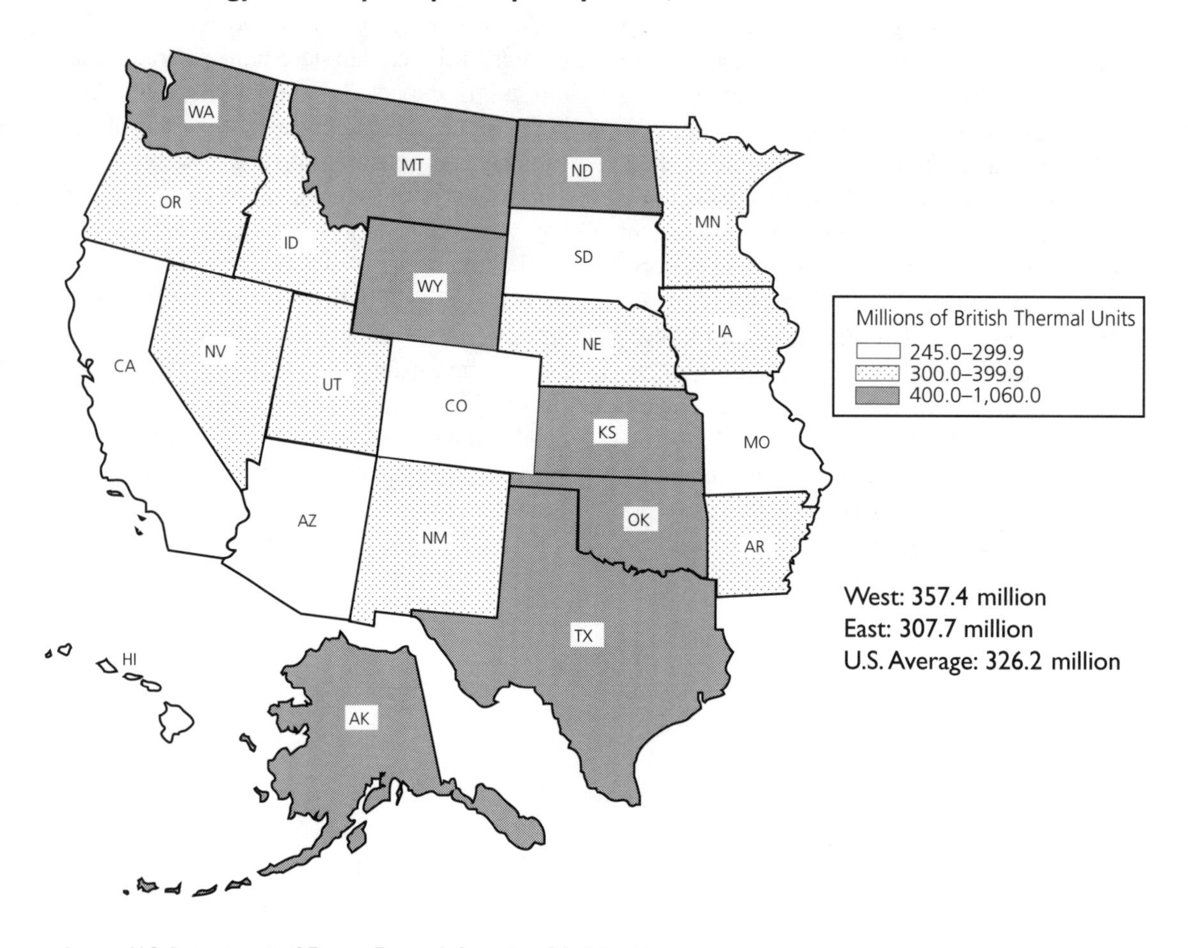

Source: U.S. Department of Energy, Energy Information Administration

	Millions of Btus		Millions of Btus
Alaska	1,057.5	**Nevada**	325.7
Wyoming	877.4	**Iowa**	323.8
Texas	576.7	**Nebraska**	321.8
North Dakota	480.1	**Utah**	315.1
Oklahoma	429.8	**Minnesota**	301.1
Montana	423.1	**South Dakota**	292.6
Kansas	415.7	**Missouri**	288.4
Washington	401.3	**Colorado**	277.2
New Mexico	394.0	**Hawaii**	269.9
Idaho	370.1	**Arizona**	250.0
Arkansas	332.5	**California**	245.5
Oregon	326.4		

Figure 7.4 **Canada**
Energy Consumption per Capita by Province, 1991

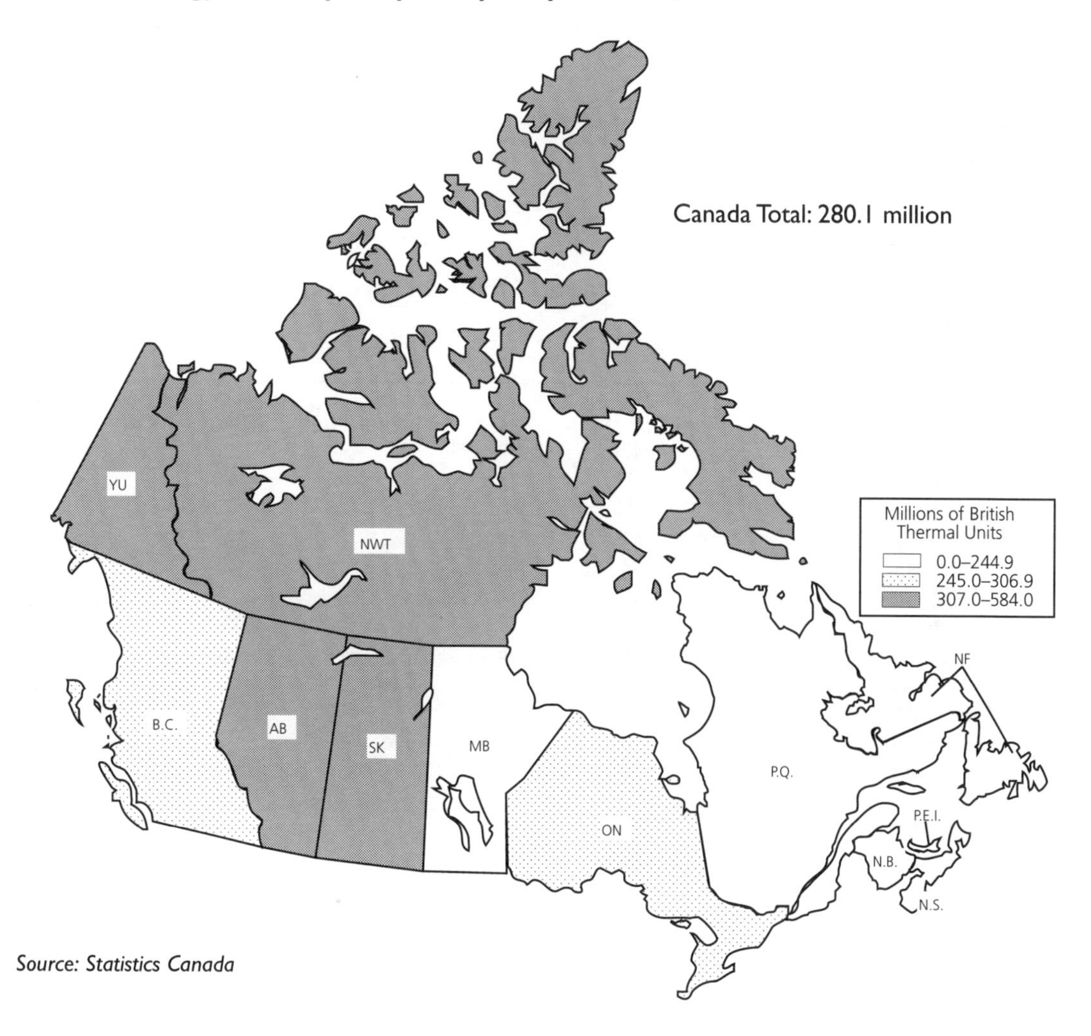

Source: Statistics Canada

	Millions of Btus
Alberta	584.0
Saskatchewan	403.6
NWT and Yukon	343.5
British Columbia	270.2
Ontario	258.4
Atlantic Provinces	244.6
Manitoba	225.8
Québec	205.3

Electric Power Generation

Coal is the primary source of electricity generation in the western United States, followed by oil and gas, hydro- and nuclear power. Canada relies primarily upon hydropower and conventional steam for electricity generation. In Mexico, too, steam and hydropower are the primary sources for electricity.

The United States generates the most electricity from every source except hydro-power.

Figure 7.5 **Western United States Sources of Energy in Million Kilowatt Hours for Electric Generation by State, 1992**

	Coal*	Fuel Oil**	Gas	Nuclear Fuel	Hydro	Other***
Alaska	290	407	2,552	-	918	-
Arizona	34,602	73	2,914	25,609	6,911	-
Arkansas	20,030	50	2,584	11,326	3,380	-
California	-	325	56,610	35,244	19,205	7,925
Colorado	30,002	39	353	-	1,505	-
Hawaii	-	6,851	-	-	10	-
Idaho	-	-	-	-	6,260	-
Iowa	24,844	36	146	3,405	981	14
Kansas	22,153	45	1,076	8,491	-	-
Minnesota	24,443	639	370	11,166	758	407
Missouri	46,830	81	183	8,084	1,450	0
Montana	17,126	16	23	-	8,223	79
Nebraska	12,402	9	145	8,748	1,075	8
Nevada	16,443	328	2,210	-	1,982	-
New Mexico	25,348	36	2,068	-	255	-
North Dakota	26,865	29	-	-	1,699	-
Oklahoma	27,666	15	15,051	-	3,210	-
Oregon	3,683	9	1,474	4,573	31,476	6
South Dakota	2,626	7	1	-	3,612	-
Texas	118,673	254	93,621	24,496	2,638	281
Utah	31,543	34	578	-	580	186
Washington	9,625	6	464	5,692	67,967	361
Wyoming	41,154	55	8	-	636	-
U.S. Total	**1,575,895**	**88,916**	**263,872**	**618,776**	**239,559**	**10,200**

* Excludes petroleum coke.

** Includes petroleum coke.

*** Includes generation by geothermal, wood, waste, wind, and solar.

Source: U.S. Department of Energy, Energy Information Administration

Figure 7.6 **Western United States Energy-Producing Installations by State, 1989**

	Coal	Petroleum	Gas	Water	Nuclear	Other
Alaska	5	347	26	43	-	5
Arizona	13	6	62	38	3	0
Arkansas	5	34	21	40	2	0
California	-	59	149	414	6	34
Colorado	31	53	38	48	-	-
Hawaii	-	73	-	4	-	-
Idaho	-	4	-	110	-	1
Iowa	53	270	52	24	1	3
Kansas	20	233	177	7	1	2
Minnesota	55	168	47	54	3	10
Missouri	47	176	80	29	1	-
Montana	6	168	3	84	-	7
Nebraska	12	106	118	20	2	-
Nevada	8	32	16	17	-	-
New Mexico	13	6	29	6	-	1
North Dakota	14	29	2	5	-	-
Oklahoma	10	26	85	37	-	-
Oregon	1	4	6	175	1	2
South Dakota	6	39	9	28	-	-
Texas	33	33	325	45	2	6
Utah	14	13	20	78	-	1
Washington	2	10	7	256	1	1
Wyoming	19	9	-	25	-	-
West	**367**	**1,898**	**1,272**	**1,587**	**23**	**73**
East	**883**	**1,659**	**720**	**1,892**	**87**	**35**
U.S. Total	**1,250**	**3,557**	**1,992**	**3,479**	**110**	**108**

Source: U.S. Department of Energy, Energy Information Administration

Figure 7.7 **Canada**
Sources of Energy in Million Kilowatt Hours for Electrical Generation by Province, 1990

	Hydro	Steam—Conventional	Steam—Nuclear	Internal Combustion	Gas Turbine
Alberta	2,059.9	38,739.8	-	16.4	2,058.3
British Columbia	57,245.3	3,196.5	-	220.0	0.4
Manitoba	19,746.9	374.9	-	27.5	-
New Brunswick	3,483.4	7,841.5	5,338.2	0.0	1.7
Newfoundland	34,622.4	1,864.8	-	85.1	13.1
Northwest Territories	257.2	-	-	214.5	-
Nova Scotia	1,149.6	8,261.0	-	-	19.0
Ontario	40,225.1	28,862.3	59,352.6	0.3	902.5
Prince Edward Island	-	76.5	-	-	4.2
Québec	129,403.9	1,654.4	4,145.8	256.5	(2.1)
Saskatchewan	4,220.2	9,269.7	-	1.9	48.2
Yukon	422.5	-	-	62.1	-
Canada Total	**125,342.9**	**47,010.2**	**26,925.5**	**405.4**	**1,243.4**

Source: Statistics Canada

Figure 7.8 **Mexico**
Electricity-Producing Installations by State, 1990

Mexico Total: 509

	Electric Generation Plants		Electric Generation Plants
México	40	**Tamaulipas**	13
Sonora	38	**Sinaloa**	11
Jalisco	35	**Nayarit**	9
Michoacán	35	**Guanajuato**	8
Puebla	33	**Querétaro**	7
Chiapas	32	**Campeche**	7
Veracruz	29	**San Luis Potosí**	7
Baja California Sur	26	**Colima**	6
Chihuahua	24	**Oaxaca**	6
Baja California	24	**Yucatán**	6
Nuevo León	21	**Distrito Federal**	4
Guerrero	21	**Zacatecas**	3
Durango	19	**Aguascalientes**	0
Quintana Roo	16	**Tlaxcala**	0
Coahuila	15	**Morelos**	0
Hidalgo	14	**Tabasco**	0

Source: INEGI

Figure 7.9 **Mexico**
Installed Electric Generation Capacity by State, 1991

(National total includes 63 Megawatts of mobile units and emergency reserves.)

	Hydro	Steam	Internal Combustion	Gas Turbo	Combined Cycle	Geothermal	Coal	Nuclear
Baja California	-	460	3	177	-	620	-	-
Baja California Sur	-	113	77	76	-	-	-	-
Campeche	-	150	-	14	-	-	-	-
Coahuila	66	-	1	90	-	-	1,200	-
Colima	-	1,900	-	-	-	-	-	-
Chiapas	3,928	-	-	-	-	-	-	-
Chihuahua	25	753	-	197	-	-	-	-
Distrito Federal	-	-	-	148	-	-	-	-
Durango	-	471	-	84	200	-	-	-
Guanajuato	-	902	-	45	-	-	-	-
Guerrero	1,638	-	-	37	-	-	-	-
Hidalgo	4	1,500	-	-	482	-	-	-
Jalisco	150	-	1	61	-	-	-	-
México	348	954	-	226	-	-	-	-
Michoacán	508	-	-	-	-	90	-	-
Nayarit	3	-	2	-	-	-	-	-
Nuevo León	-	566	-	86	678	-	-	-
Oaxaca	156	-	-	-	-	-	-	-
Puebla	431	38	-	-	-	10	-	-
Querétaro	2	-	-	-	218	-	-	-
Quintana Roo	-	-	-	258	-	-	-	-
San Luis Potosí	20	700	-	-	-	-	-	-
Sinaloa	273	656	-	57	-	-	-	-
Sonora	256	1,211	-	100	-	-	-	-
Tamaulipas	32	1,145	-	24	-	-	-	-
Veracruz	92	817	-	-	400	-	-	675
Yucatán	-	217	-	60	140	-	-	-
Zacatecas	-	-	-	15	-	-	-	-
Mexico Total	**7,932**	**12,553**	**147**	**1,755**	**1,818**	**720**	**1,200**	**675**

Megawatts*

(* 1 Megawatt - .001 million kilowatts)

Source: INEGI

Chapter Eight
Communications

Western North America is blessed with superb communications services. The United States is the world leader in communications technology, and, according to the CIA *World Factbook*, Canada has "excellent service" provided by modern media, and Mexico has a "highly developed system with extensive radio." In the western United States and Canada, virtually every home has a telephone, radio, and television set. Mexican homes have fewer telephone lines, but almost every home, including the poorest, has a radio and television set.

The telecommunications services industry in the United States is growing much more rapidly than the economy as a whole, fueled by spectacular growth in cellular mobile telephone and satellite services. Revenues overall are expected to increase 7.7 percent in 1994 from 1993, with cellular revenues expected to grow 39 percent and satellite revenues expected to grow 25 percent. The Canadian telecommunications industry also has been growing at a robust pace. With the deregulation of long distance service in Canada, hardware sales alone are predicted to grow at more than five percent per year through 1996.

Figure 8.1 **Communications, 1990**

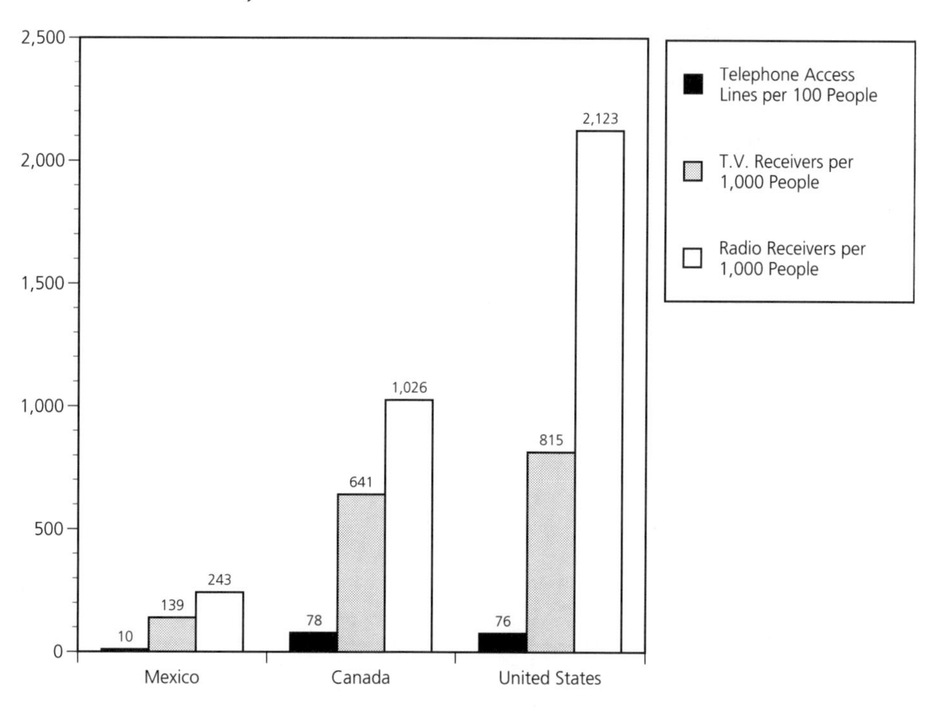

Sources: International Telecommunications Union, Geneva, Switzerland, Telecommunication Statistics; UNESCO, Paris, France, Statistical Yearbook; U.S. Statistical Abstract

Figure 8.2 **Western United States**
Number of Daily and Sunday Newspapers by State, 1991

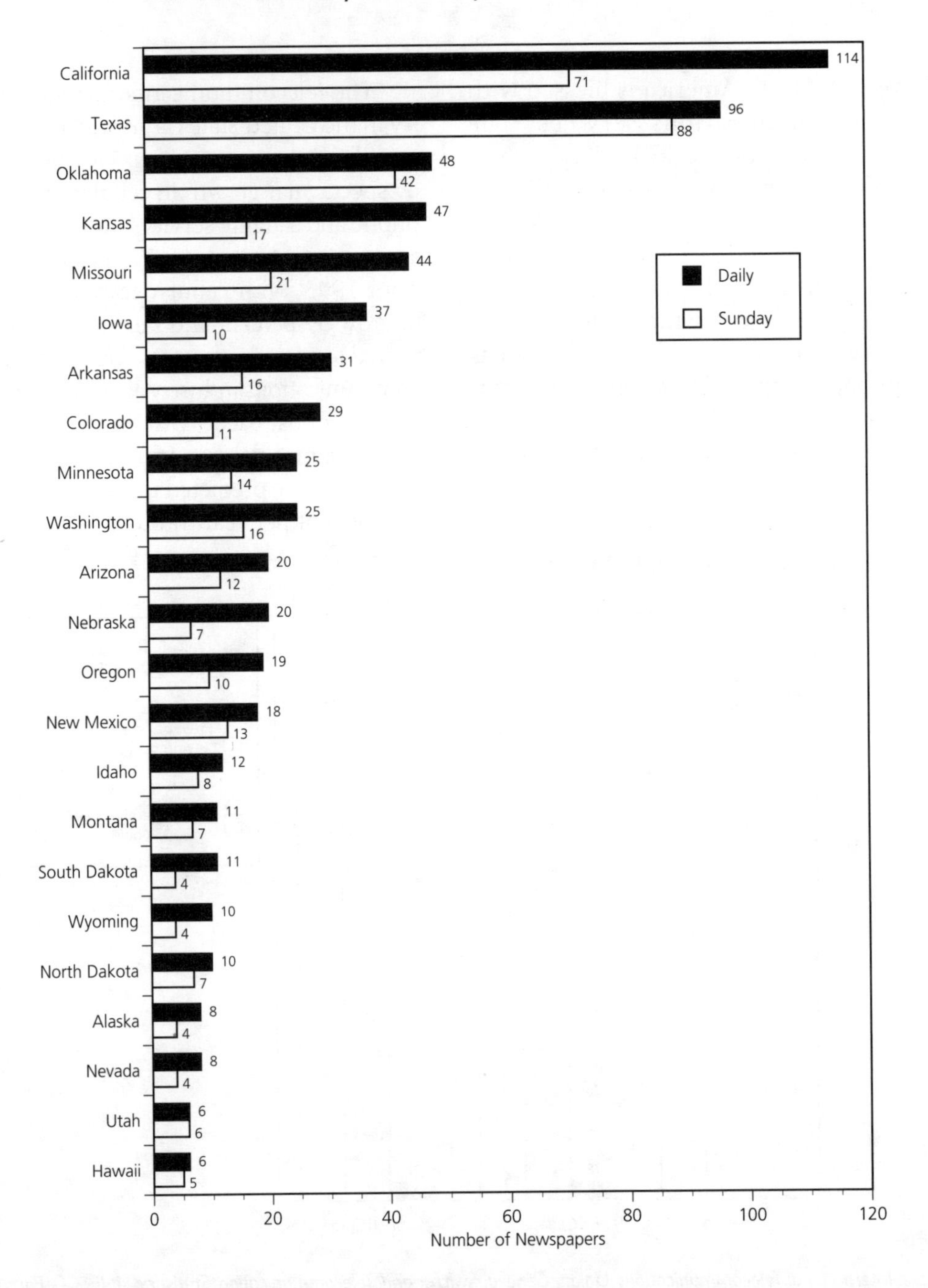

Source: Editor & Publisher Co., New York, NY, Editor and Publisher International Yearbook

The leading newspaper in the western United States is the Los Angeles *Times*, with a daily circulation in 1993 of 1.01 million, nearly as much as the circulation of the Dallas *Morning News* (575,000) and the San Francisco *Chronicle* (518,125), the second and third largest newspapers in the region, combined. Other regional leaders are the Minneapolis *Star-Tribune* (410,754), the San Diego *Union-Tribune* (383,827), Denver's *Rocky Mountain News* (348,885), and Phoenix's *Arizona Republic* (347,839). Six newspapers in California have daily circulations larger than the largest newspaper in western Canada, the Vancouver *Sun* (262,527).

There is more competition among newspapers in Mexico (40 in Mexico City alone) than in the western United States or Canada. The largest Mexican newspaper is *ESTO* (400,200). Several periodicals are published in English. The largest is a daily newspaper, *The News* (24,115). Another useful English-language source is *Business Mexico*, the Monthly magazine of the American Chamber of Commerce in Mexico. The leading Spanish-language publications in the United States are the newspapers *La Opinion* of Los Angeles (109,558) and *El Nuevo Herald* of Miami (107,384).

Figure 8.3 **Canada**
Number of Newspapers and Television Stations by Province, 1991

West: 466 Newspapers/ 46 T.V. Stations
East: 670 Newspapers/ 77 T.V. Stations
Canada Total: 1,136 Newspapers*/ 123 T.V. Stations

	Newspapers*	T.V. Stations
Alberta	140	13
British Columbia	163	10
Manitoba	55	8
New Brunswick	25	4
Newfoundland	19	4
Northwest Territories	8	2
Nova Scotia	39	6
Ontario	370	30
Prince Edward Island	5	1
Québec	212	32
Saskatchewan	96	12
Yukon	4	1

(* Includes English, French, and bilingual published at least once a month.)

Source: Corpus Almanac and Canadian Sourcebook, 1993

Figure 8.4 **Mexico**
Established Periodicals and Television Stations by State, 1989

Mexico Total: 165 Periodicals/ 439 T.V. Stations

	Periodicals	T.V. Stations
Aguascalientes	3	4
Baja California	7	9
Baja California Sur	3	24
Campeche	5	6
Coahuila	19	24
Colima	9	4
Chiapas	10	21
Chihuahua	14	31
Distrito Federal	27	8
Durango	9	41
Guanajuato	19	9
Guerrero	10	12
Hidalgo	4	10
Jalisco	10	14
México	18	8
Michoacán	21	20
Morelos	5	2
Nayarit	4	5
Nuevo León	16	9
Oaxaca	6	19
Puebla	7	4
Querétaro	6	3
Quintana Roo	2	26
San Luis Potosí	4	11
Sinaloa	15	8
Sonora	10	45
Tabasco	7	10
Tamaulipas	25	19
Tlaxcala	1	1
Veracruz	19	12
Yucatán	4	7
Zacatecas	6	13

Source: Banamex

Chapter Nine
Education

In Mexico, the federal government has primary responsibility for all levels of public education. The emphasis on education, and the federal government's leading role in it, are rooted in the Mexican Revolution of 1910. In 1921, the Secretariat of Public Education was formed and given full cabinet status. The secretariat is responsible for regulating the academic calendar, curriculum development, free textbook distribution, grading scales, and graduation requirements in all types of primary and secondary schools. In 1991, the secretariat's budget accounted for 13.5 percent of the total Mexican federal budget, up from 6 percent in 1989 and down from over 20 percent in 1955.

Elementary education in Mexico is divided between kindergarten and six years of primary education. There are two types of secondary schools: a basic three-year program, and the *bachillerato,* or *preparatoria,* which requires an additional 2–3 years of instruction for the college-bound.

Mexico has more than 260 institutions of higher learning, with enrollments ranging from less than 1,000 to more than 290,000. Nearly half of Mexico's college students are in Mexico City, chiefly at the National Autonomous University, which had 255,177 students on campuses throughout Mexico in 1992.

Literacy

Most people in the United States and Canada can read and write, but both nations rank low in the proportion of their populations that are literate. In 1988, the United States ranked forty-first of 164 countries surveyed in male literacy (95.7 percent), and thirty-sixth in female literacy (95.3 percent). Canada ranked forty-second in male literacy (95.6 percent) and thirty-fourth in female literacy (95.7 percent). Mexico wasn't included in the survey. Mexican government figures indicate a 92 percent literacy rate for the population as a whole in 1989, up from 83 percent in 1980. The world averages for literacy in 1988 were 82.6 percent for males and 74 percent for females.

High School and College Graduation

In 1990, 77.6 percent of Americans over the age of twenty-five had completed high school, and 21.3 percent had completed college. In the West, Washington state had the highest proportion of high school graduates (87.6 percent), and the college-educated population was proportionately largest in Colorado, where 28.9 percent of the population had completed four or more years of college.

Figure 9.1 **Western United States**
Years of School Completed by Persons 25 and Older by State, 1990

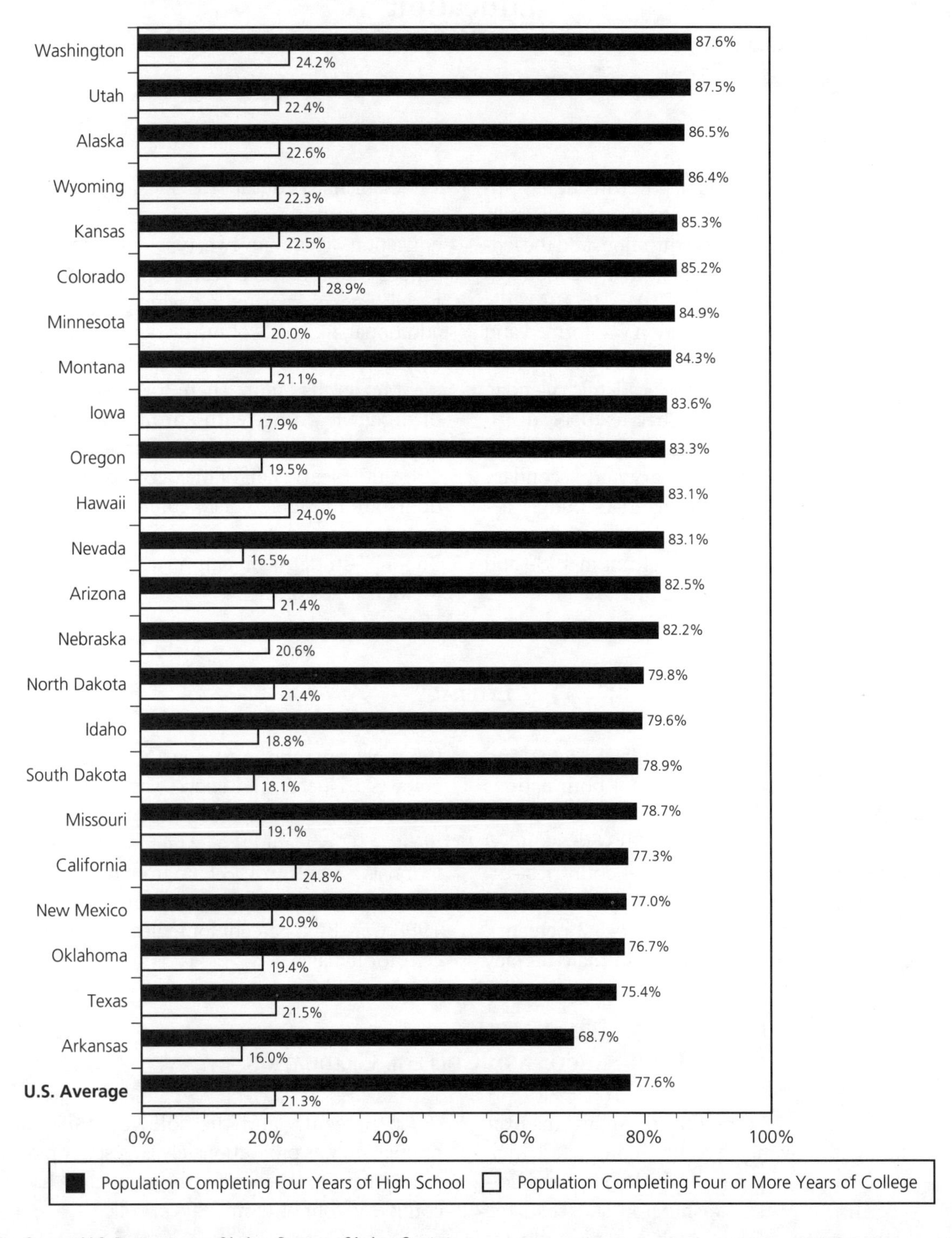

Source: U.S. Department of Labor, Bureau of Labor Statistics

In Canada in 1991, 62 percent of the population aged fifteen or older had a high school or trades diploma. The proportion of the population with high school diplomas was highest in the eastern province of Ontario, and in the Yukon and Northwest Territories. They also had the highest proportion of college graduates in the Canadian population.

Figure 9.2 **Canada**
Years of School Completed by Persons 25 and Older by Province, 1991

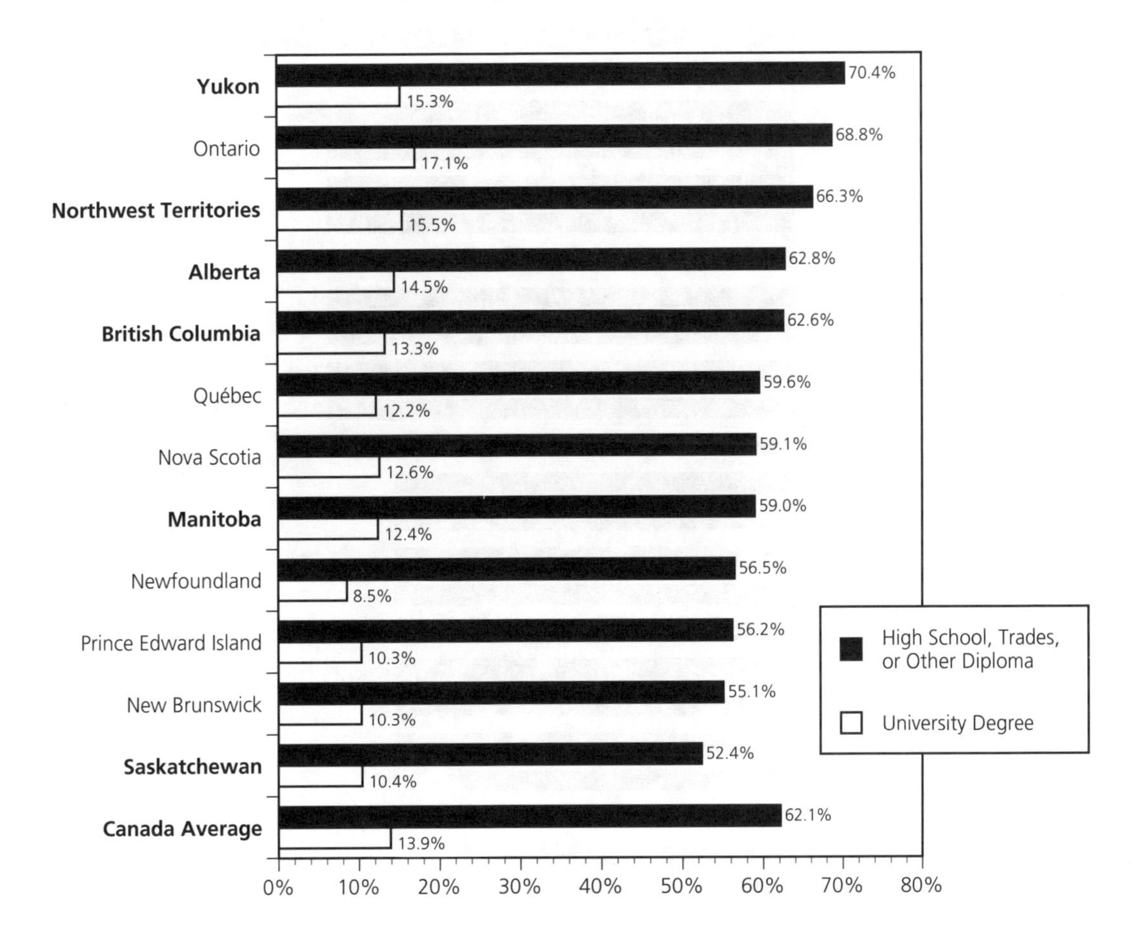

Source: Statistics Canada

In Mexico in 1989, people fifteen years old and older had an average of six years of schooling. This ranged from nine years in the Distrito Federal to four years in the southern states of Chiapas and Oaxaca.

Figure 9.3 **Mexico**
Average Years of School Completed by Persons 15 and Older by State, 1989

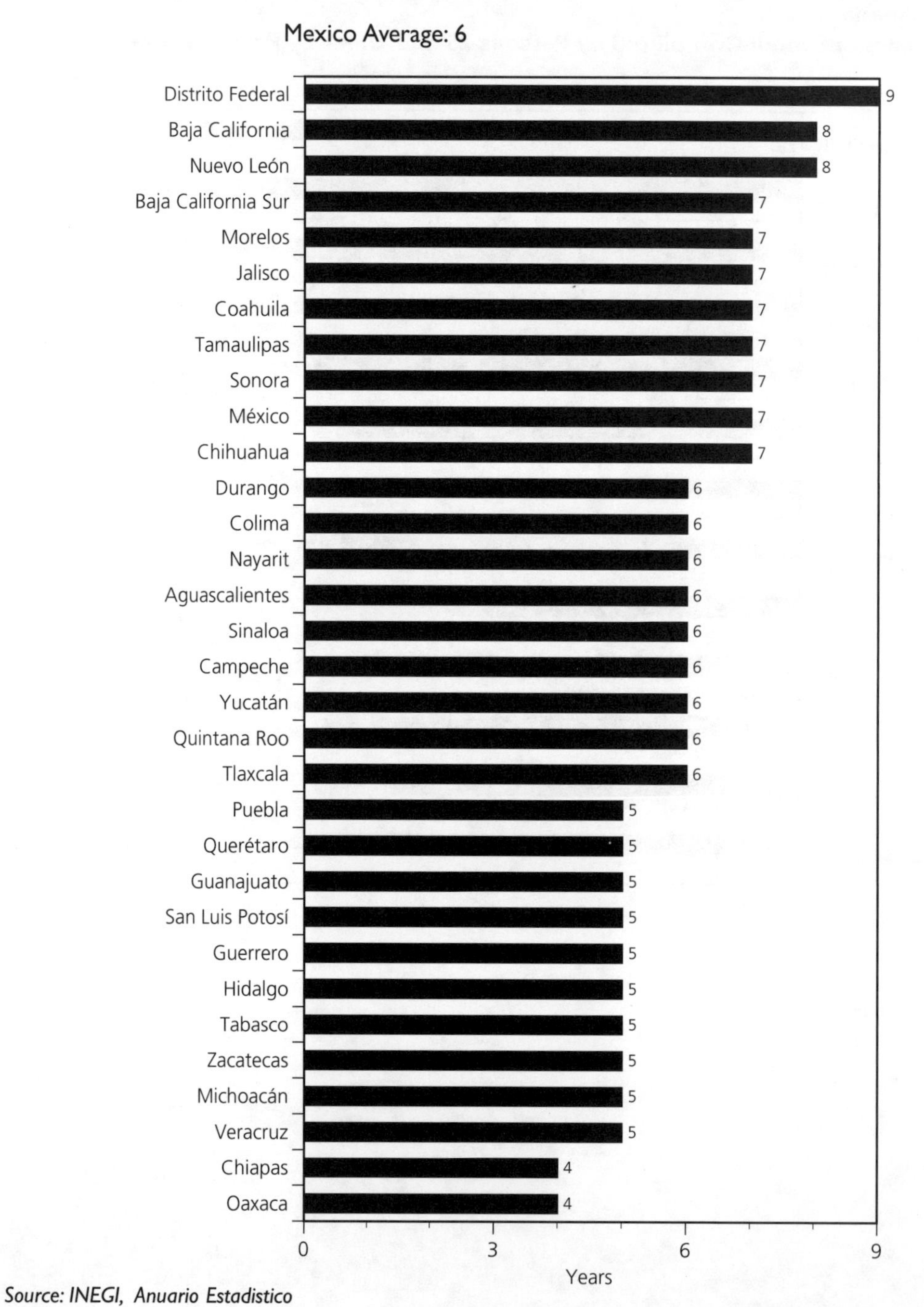

Source: INEGI, Anuario Estadistico

In 1990, 5.4 million students were enrolled in post-secondary schools in the twenty-three western U.S. states. In Canada, there were 856,000 students enrolled in universities and community colleges, with almost 200,000 enrolled in the western provinces and territories. Mexico, meanwhile, had an enrollment of 1.2 million students in its universities and teacher-colleges.

Figure 9.4 **Western United States Enrollment in Institutions of Higher Education by State, 1990**

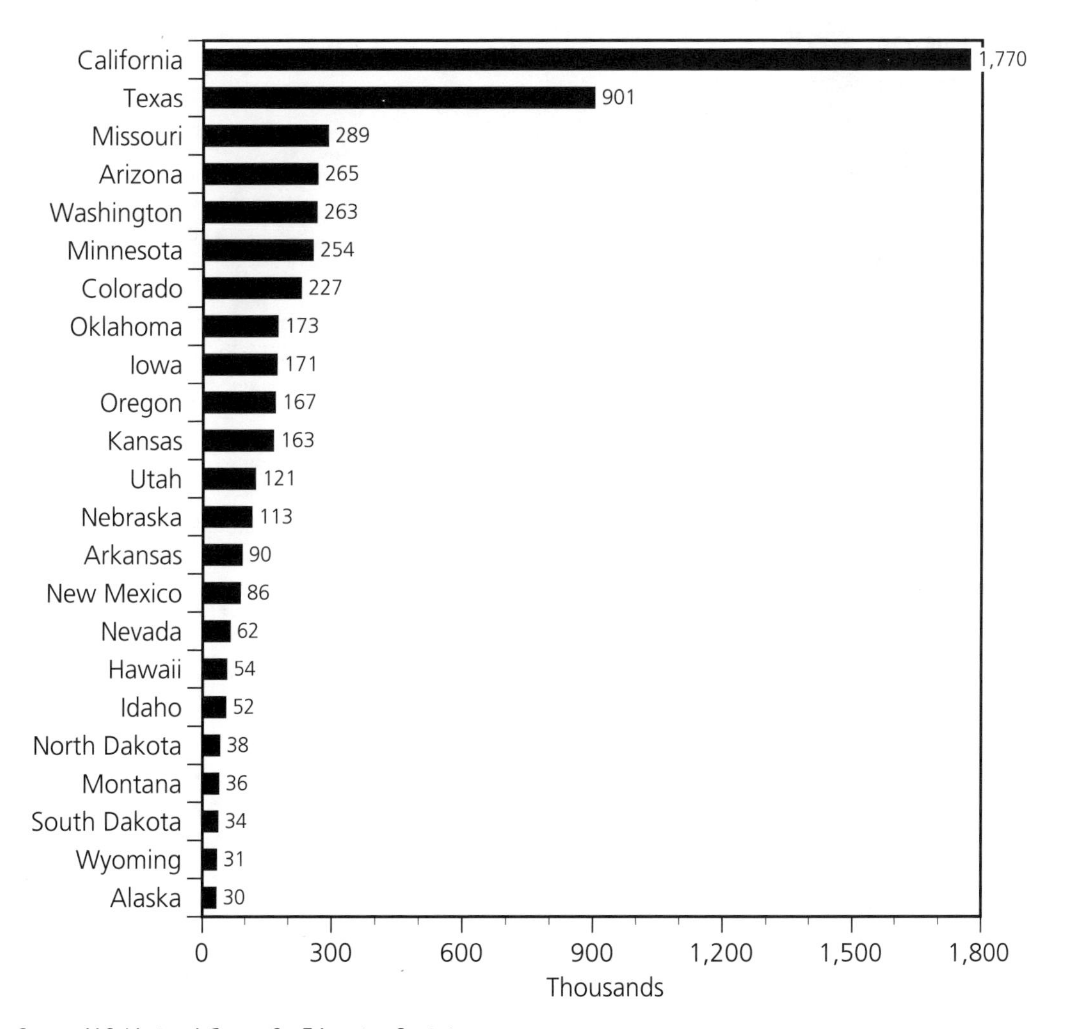

Source: U.S. National Center for Education Statistics

Figure 9.5 **Canada**
Full-Time Enrollment in Universities and Community Colleges by Province, 1991

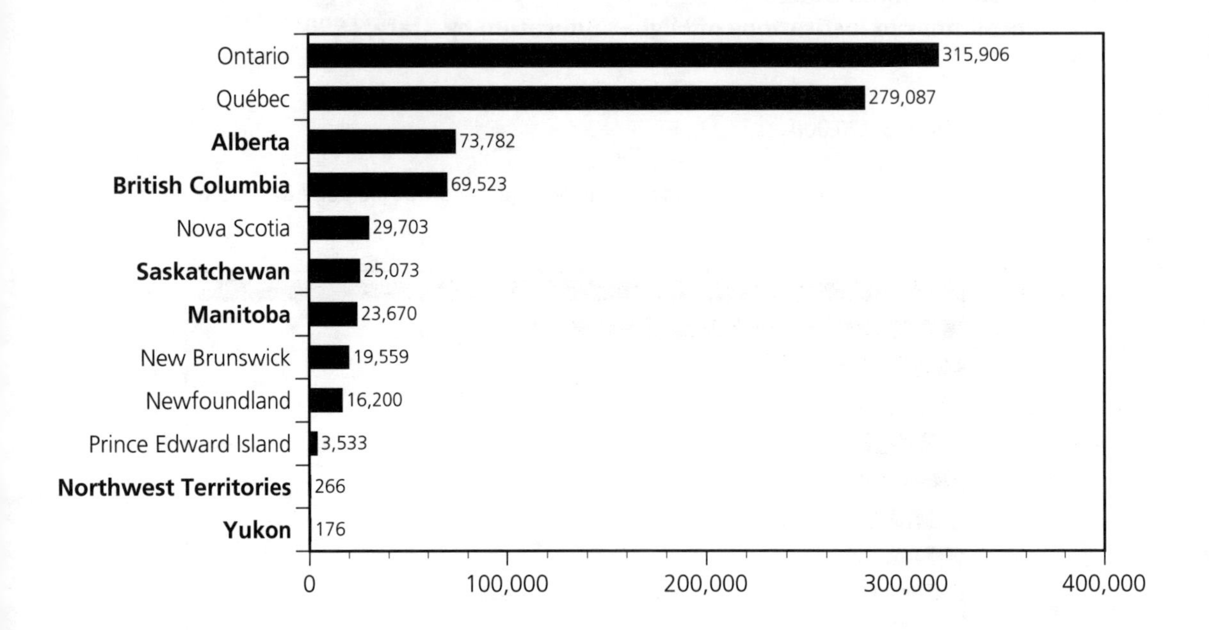

Source: Statistics Canada

Figure 9.6 **Mexico**
Enrollment in Universities, 1990–1991

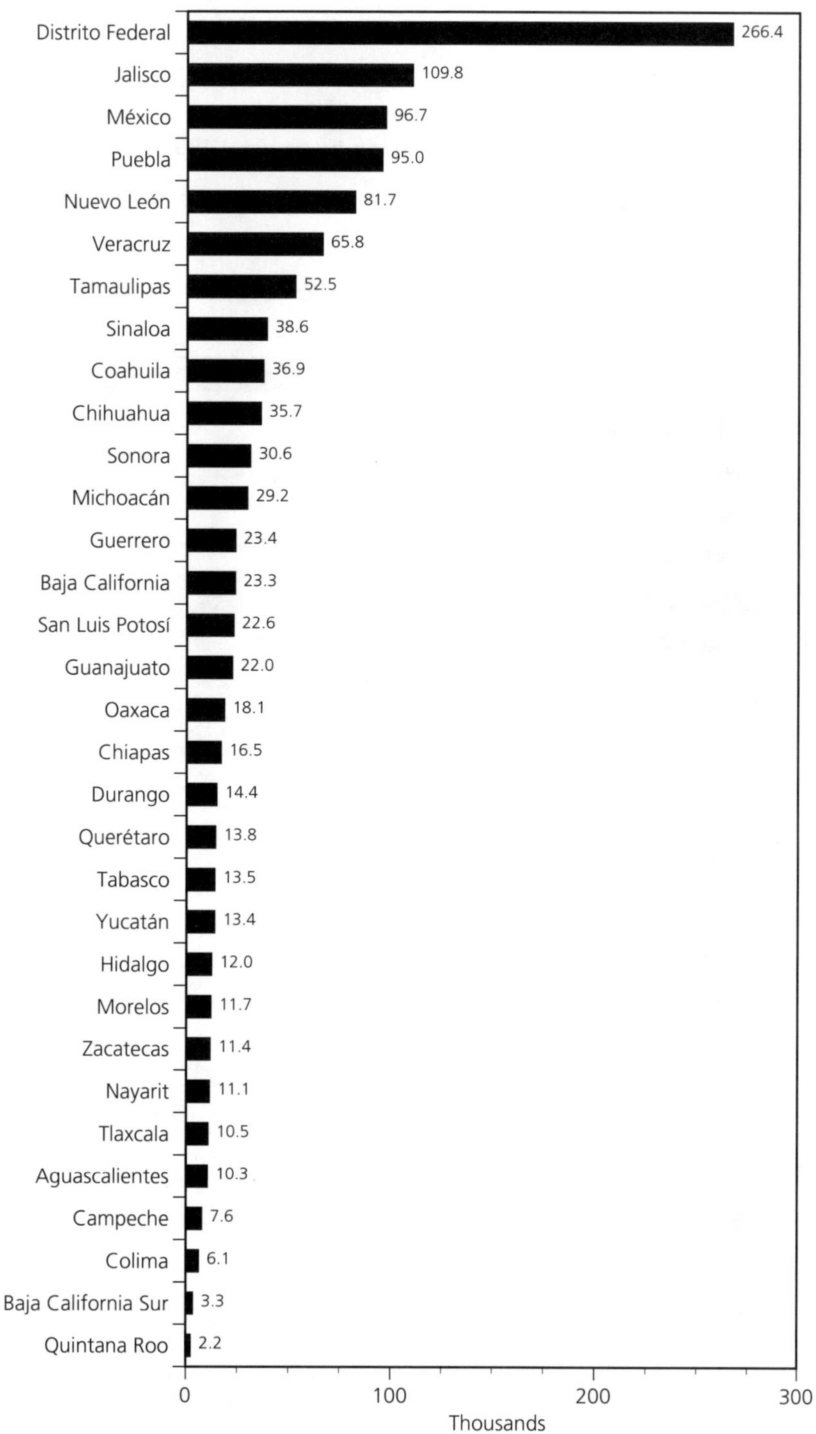

Source: Banamex

There were 1,231 institutions of higher education in the western United States in 1990. In Canada, during the 1992–93 academic year, there were sixty-nine degree-granting universities, twenty of which were in the western provinces. During the 1990–91 academic year, Mexico had 835 universities and teacher-colleges. The Canadian and Mexican figures do not include community colleges.

Figure 9.7 **Western United States**
Number of Institutions of Higher Education by State, 1990

West: 1,231
East: 2,319
U.S. Total: 3,559 (Includes military academies not included in state totals.)

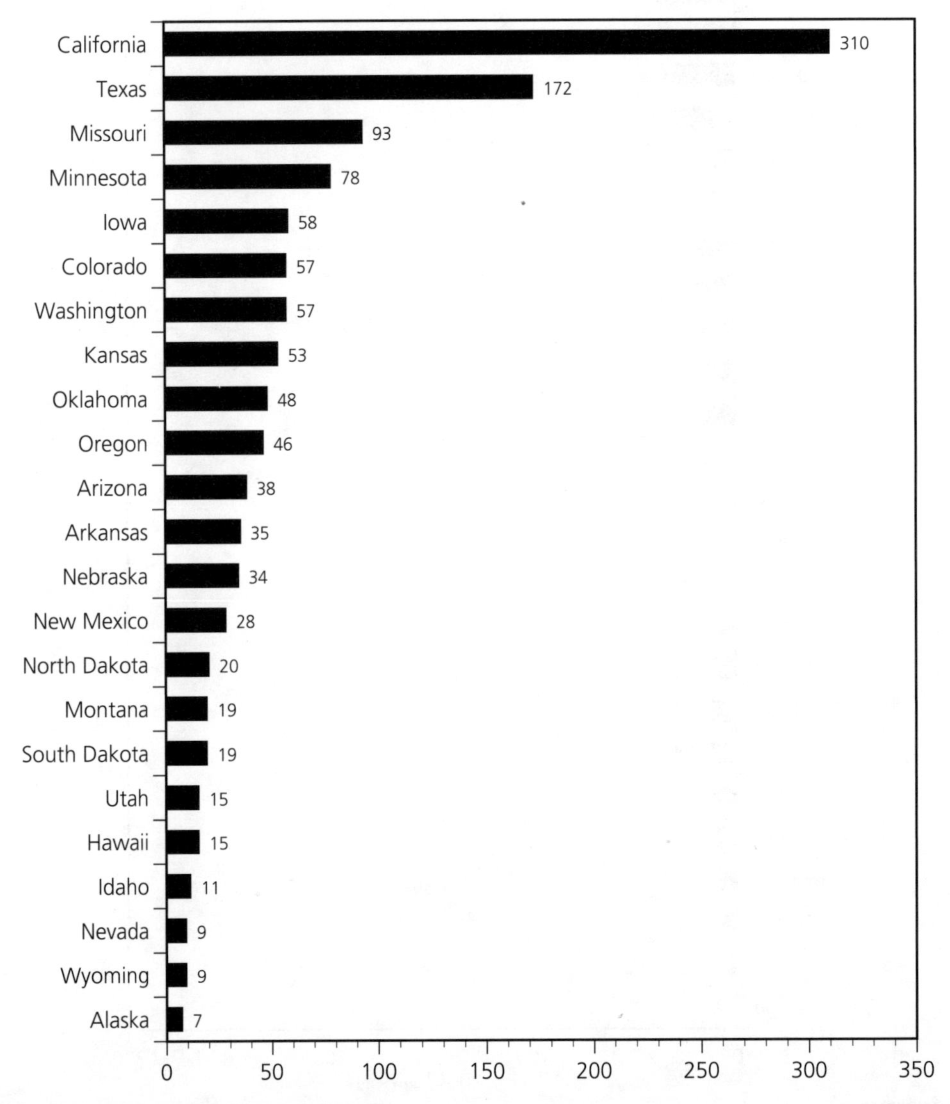

Source: U.S. National Center for Education Statistics

Figure 9.8 **Canada**
Number of Universities by Province, 1992–1993

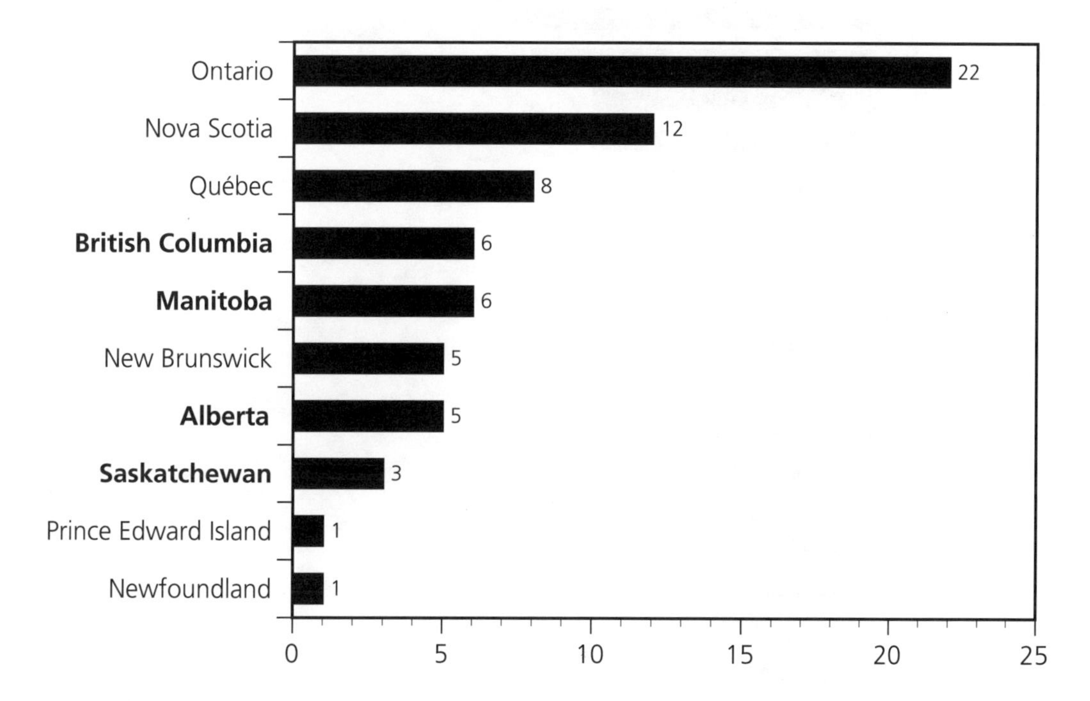

Source: Statistics Canada

Figure 9.9 **Mexico**
Number of Universities by State, 1990–1991

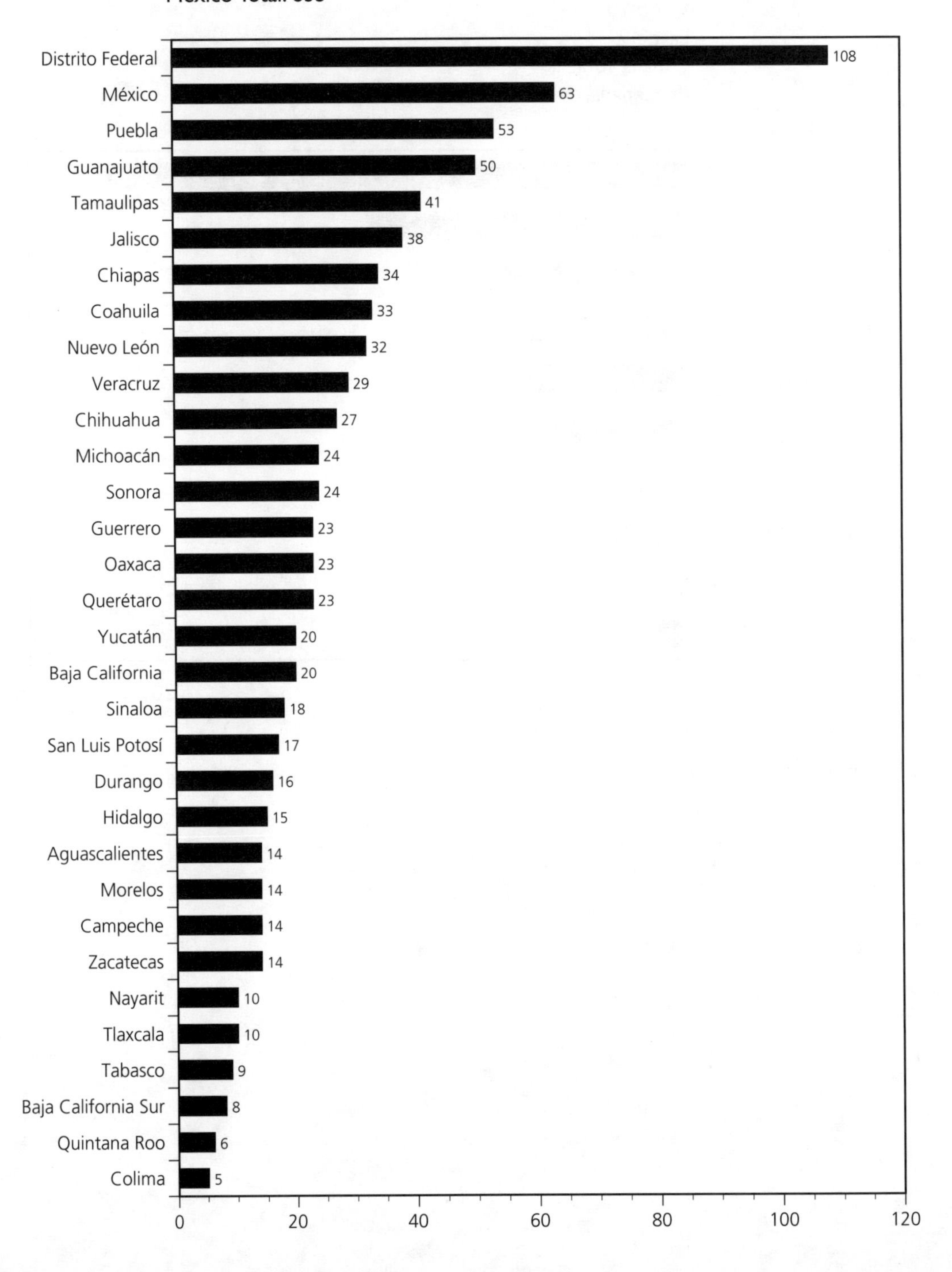

Source: Banamex

Scientists and Engineers

In the age of high technology, a nation's scientists and engineers give a nation an important economic advantage. In 1990, U.S. universities awarded 15,435 doctorates in science and engineering, 6,435, or 42 percent, by colleges and universities in the West.

Figure 9.10 **Western United States Science and Engineering Doctorates Awarded by State, 1990**

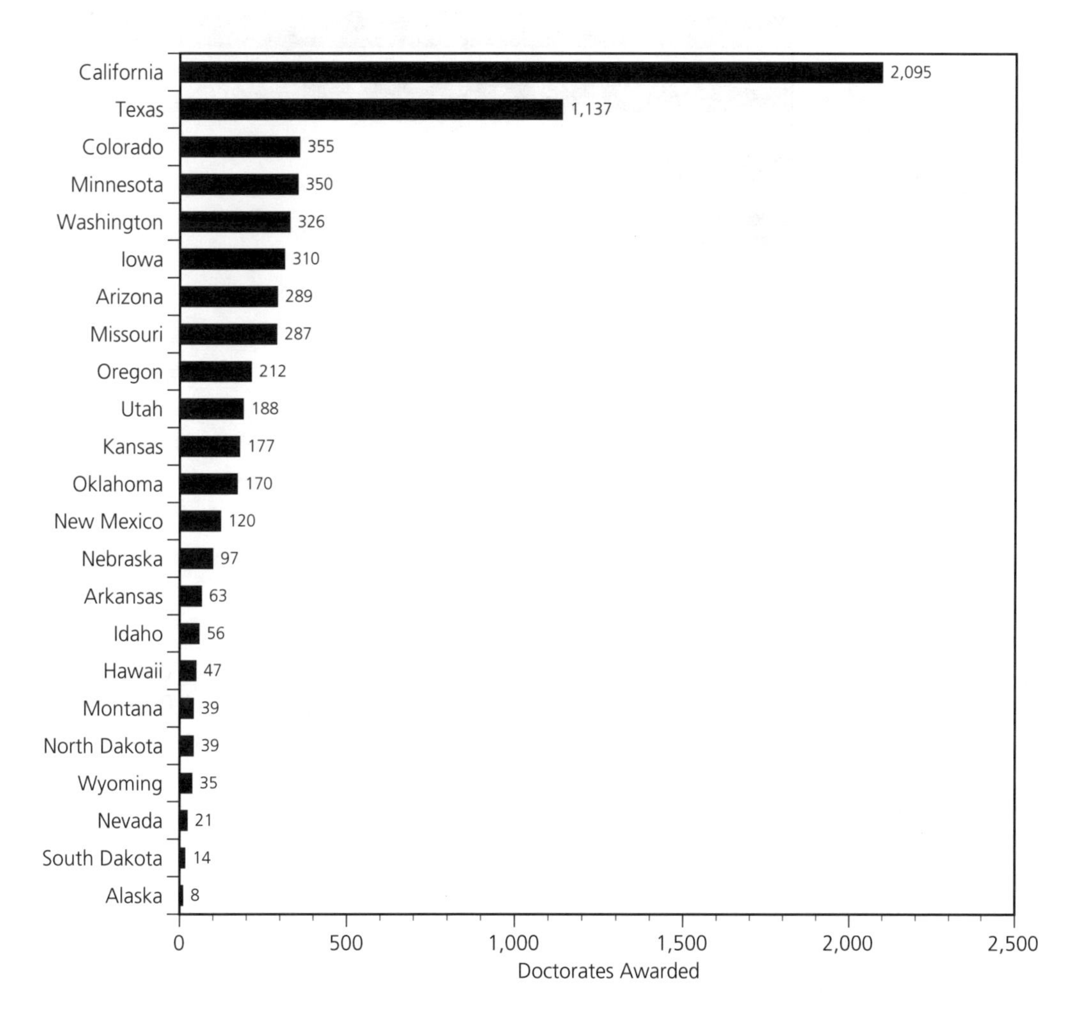

Source: National Research Council, Office of Scientific and Engineering Personnel

In 1989, Canadian colleges and universities awarded 28,337 graduate and undergraduate science and engineering degrees, 6,456, or 23 percent, from schools in the Canadian West.

Figure 9.11 Canada
Science and Engineering Degrees Awarded by Province, 1989
Includes undergraduate and graduate. (Does not include health professions.)

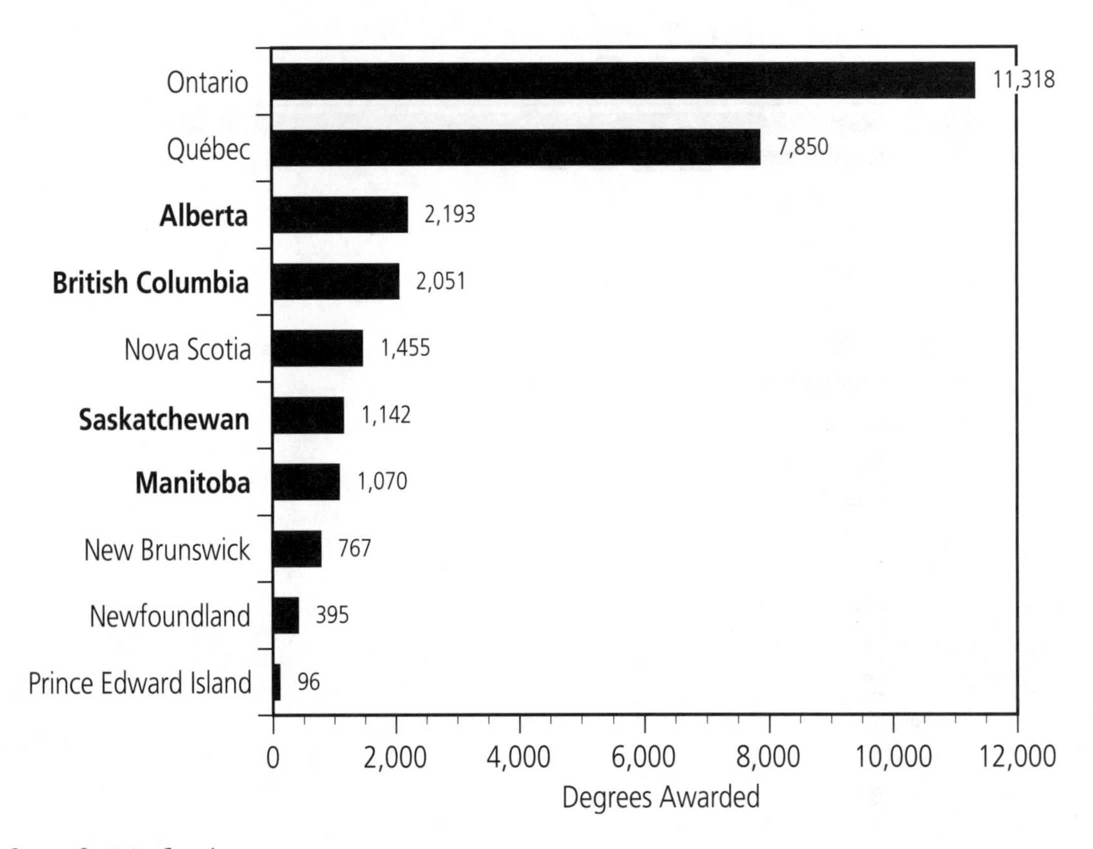

Source: Statistics Canada

Patents issued are correlated with population distribution—the more people, the more patents issued. But some states clearly produce more innovation than others. For example, California is not quite twice as large as Texas, yet its entrepreneurs claimed almost three times as many patents. Minnesota, Colorado, Utah, New Mexico, Hawaii, South Dakota, Idaho, and Nebraska issue more patents than would be expected from their population size. Missouri and Arkansas issue fewer. Per capita patents are much higher in the United States than Canada (one patent for every 4,235 Americans in 1992, compared to just one for each 22,910 Canadians). The five to one ratio favoring the United States suggests that the greater attractiveness of the U.S. market with 250 million plus consumers versus Canada's smaller market greatly fuels the patent process.

Figure 9.12 **Western United States Patents Issued by State, 1992**

West: 22,363
East: 37,367
U.S. Total: 59,760 (includes Puerto Rico)

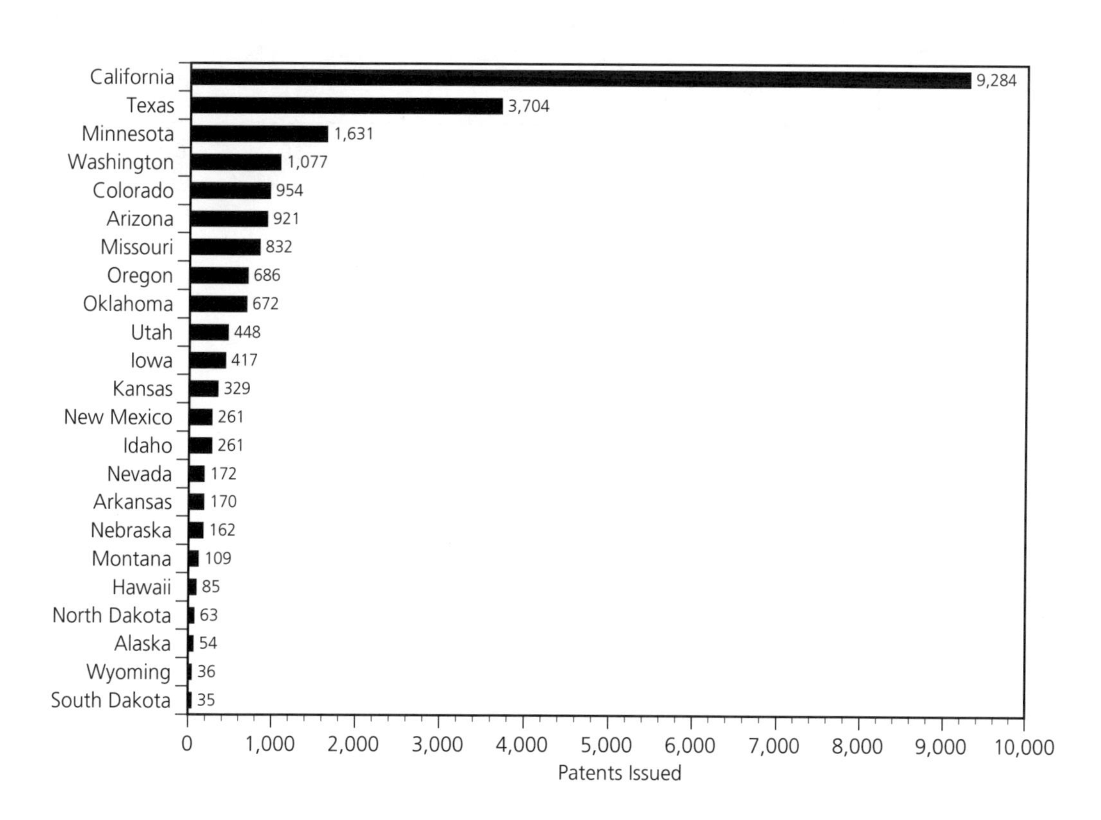

Source: U.S. Department of Commerce, Patent and Trademark Office

Figure 9.13 **Canada**
Patents Issued by Province, 1992

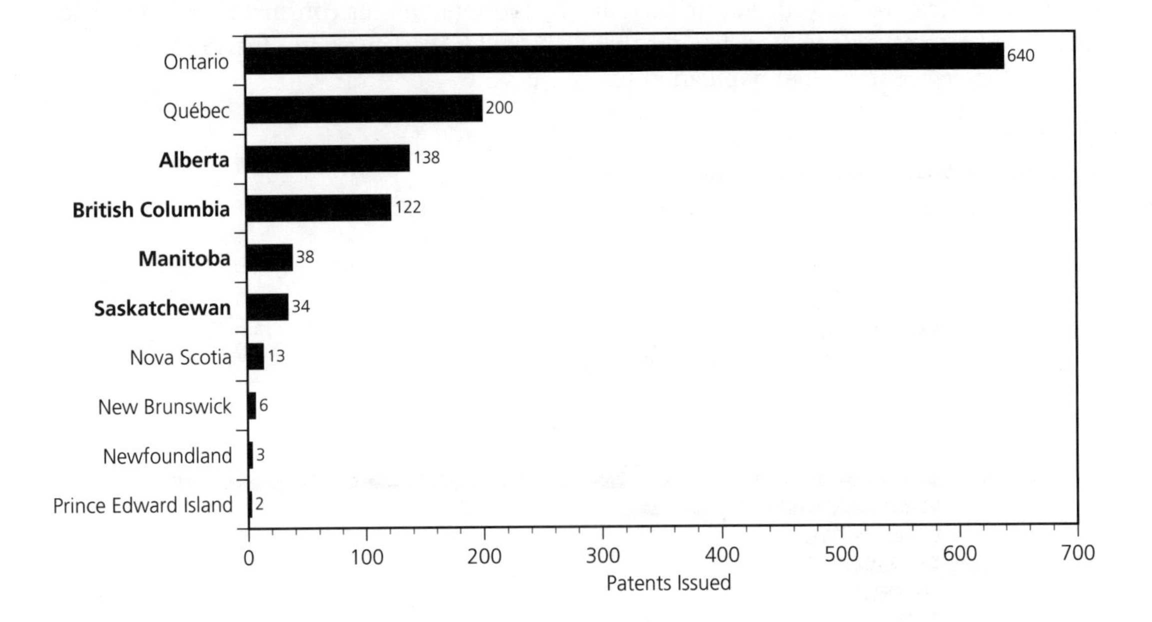

Source: Bureau of Corporate Affairs and Legislative Policy

Primary/Secondary School Enrollment

In countries with extensive systems of post-secondary education, as in all three North American countries, primary and secondary school enrollment is driven chiefly by the proportion of young people in the population. Because the Mexican population is much younger than that of Canada or the United States, it is not surprising that primary and secondary school enrollment is proportionately larger, 54 percent of the total enrolled in western North America.

Figure 9.14 **Western United States Elementary and Secondary Enrollment by State, 1992**

West: 12,442,000 Elementary/ 4,667,000 Secondary
East: 18,223,000 Elementary/ 6,923,000 Secondary
U.S. Total: 30,665,000 Elementary/ 11,590,000 Secondary

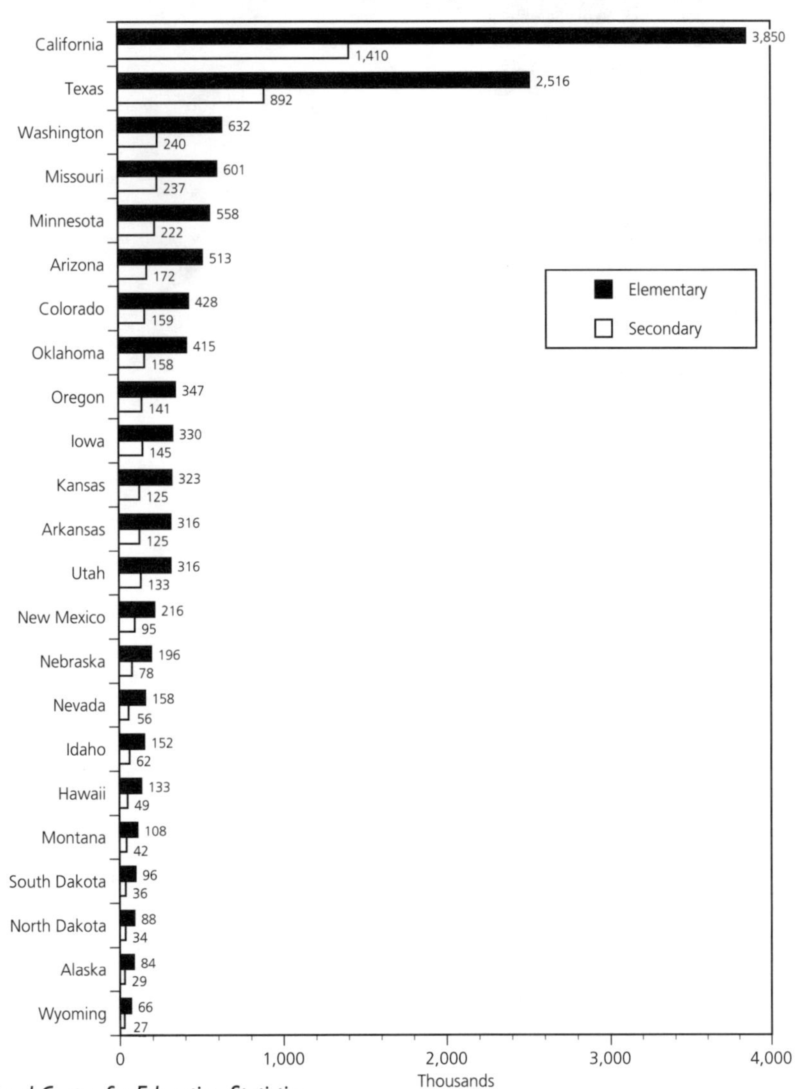

Source: U.S. National Center for Education Statistics

Figure 9.15 **Canada**
Primary and Secondary Enrollment by Province, 1990–1991

West: 730,585 Primary/ 671,273 Secondary
East: 1,643,207 Primary/ 1,616,573 Secondary
Canada Total: 2,373,792 Primary/ 2,287,846 Secondary
(Primary includes grades 1–6; secondary includes grades 7–13.)

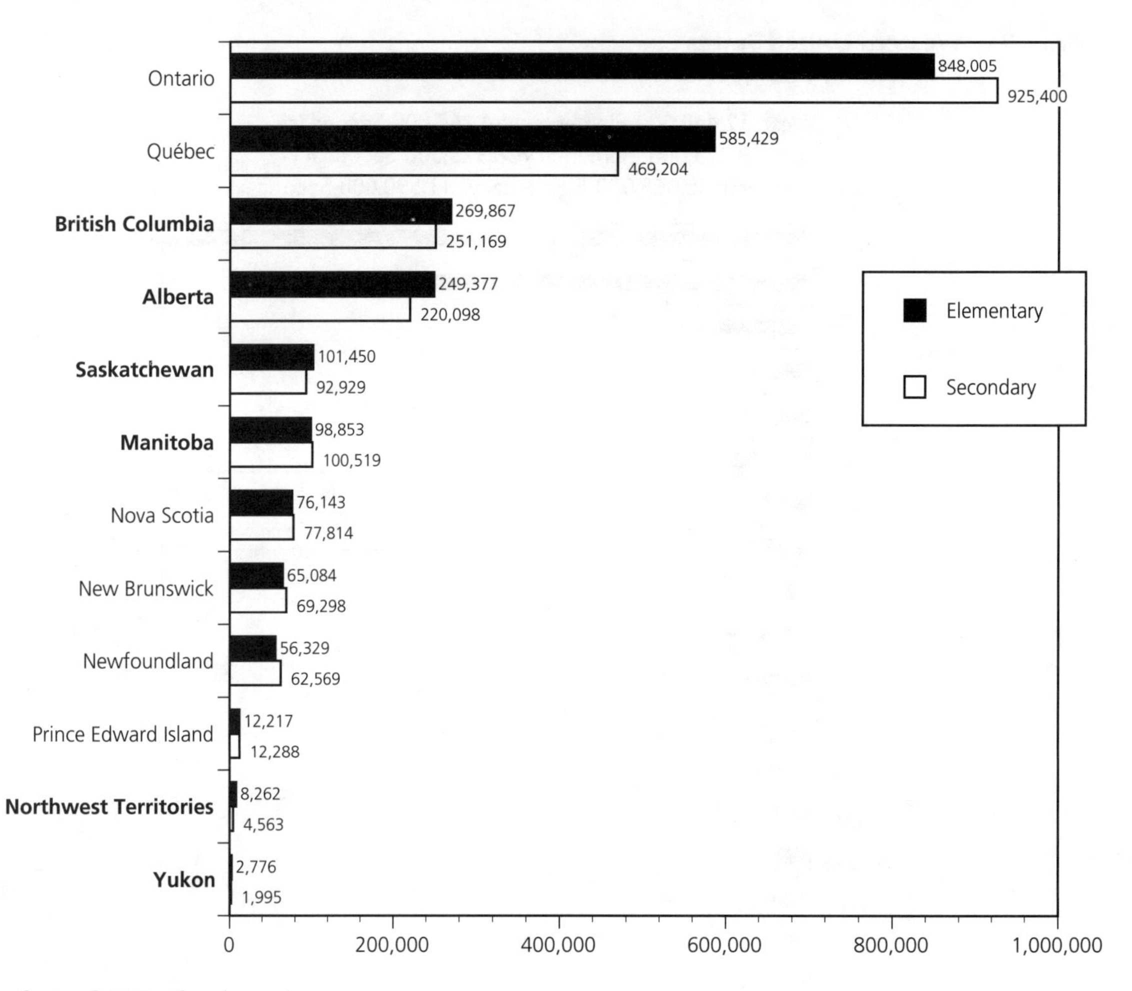

Source: Statistics Canada

Figure 9.16 **Mexico**
Primary and Secondary Education by State, 1989–1990

Mexico Total: 14,916,063 Primary/ 7,132,932 Secondary
(Primary includes grades 1–6; secondary includes grades 7–13 as well as vocational education.)

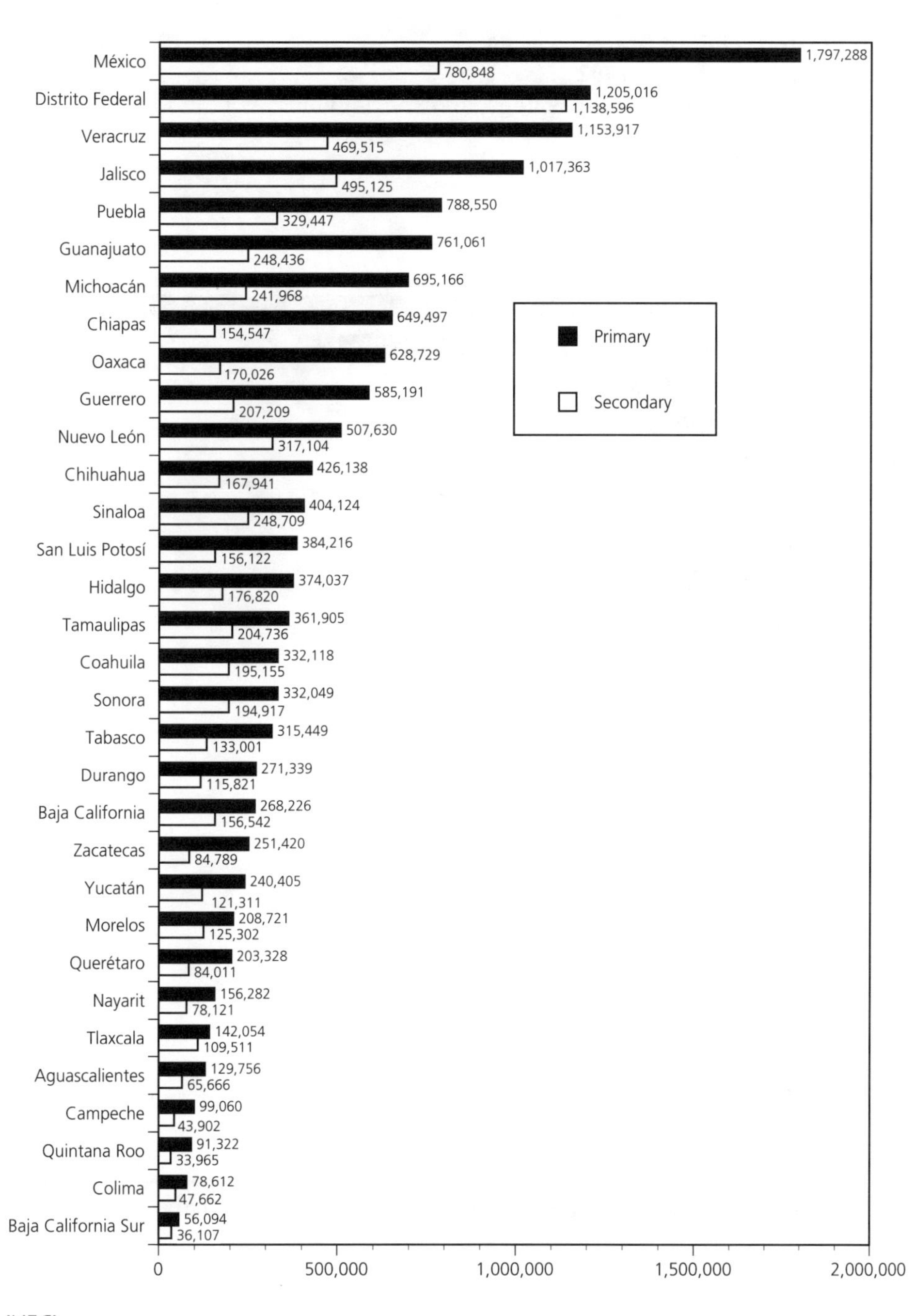

Source: INEGI

Libraries

Western North America boasts 61,515 libraries of all types, of which 84 percent are in the western United States. Included are public, school, college, and special libraries. Mexico had 7,951 libraries in 1989 (13 percent), and western Canada had 1,788 (3 percent) in 1991. California has more libraries than Mexico and western Canada combined.

Figure 9.17 **Western United States Libraries by State, 1992 (Includes public, school, college, and special libraries.)**

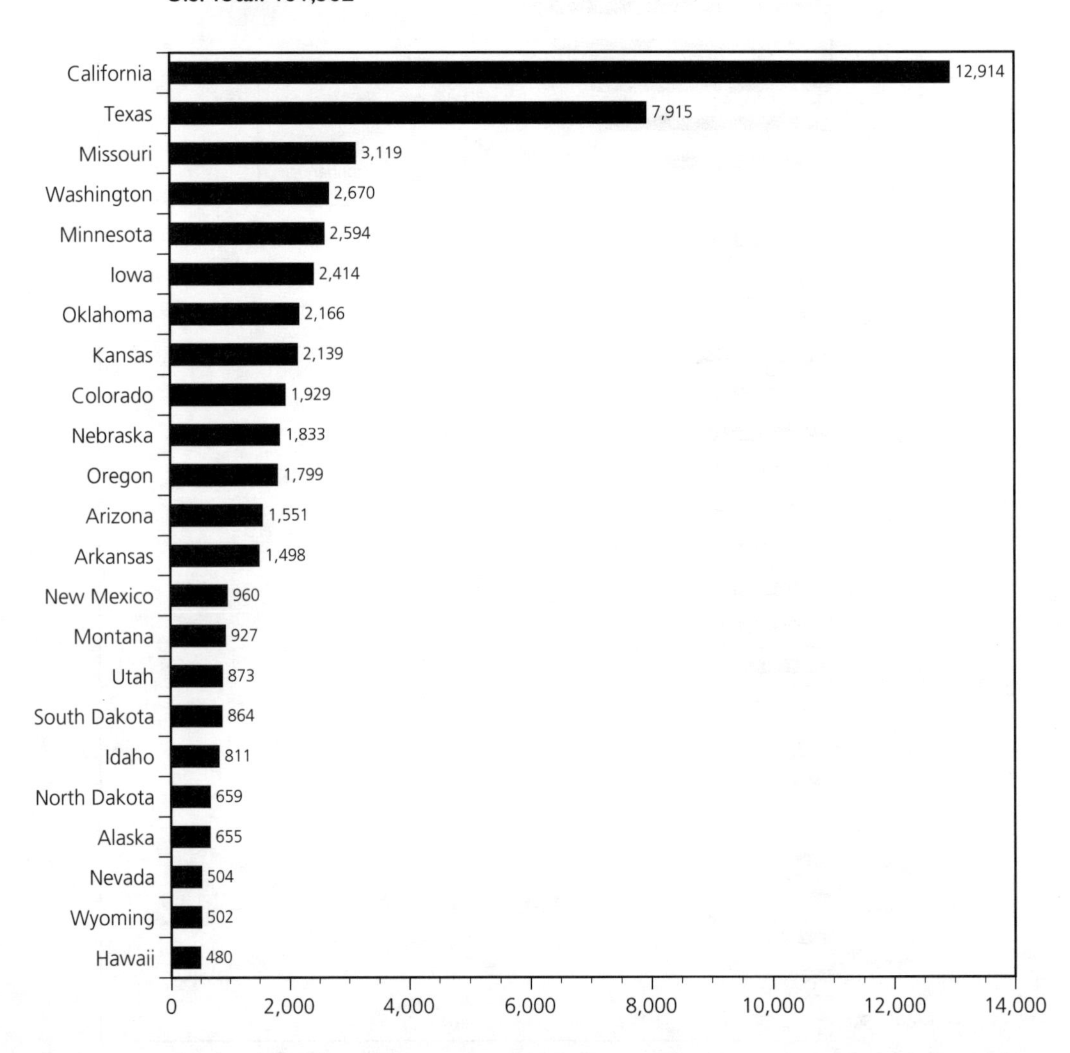

Source: American Library Association

Figure 9.18 **Canada**
Libraries by Province, 1990–1991
(Includes permanent locations, mobile stations, and mobile station stops.)

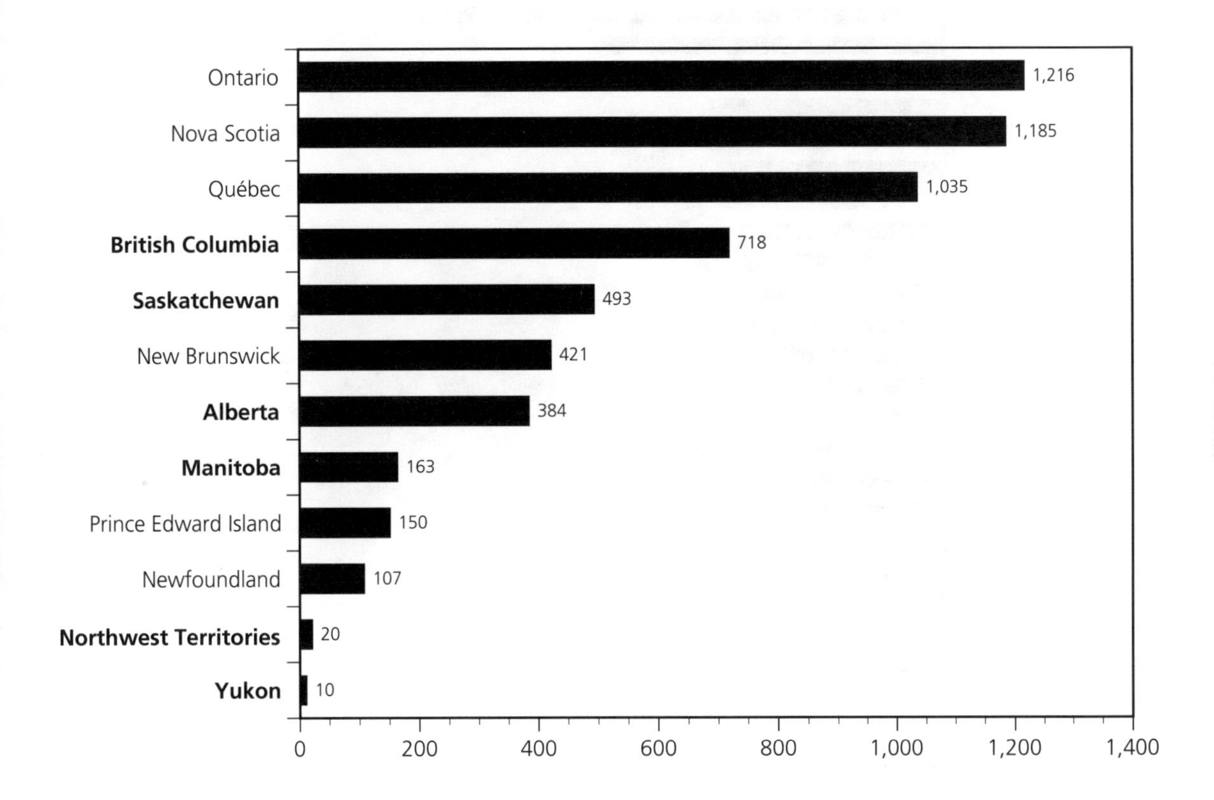

Source: Statistics Canada

Figure 9.19 **Mexico**
Libraries by State, 1989

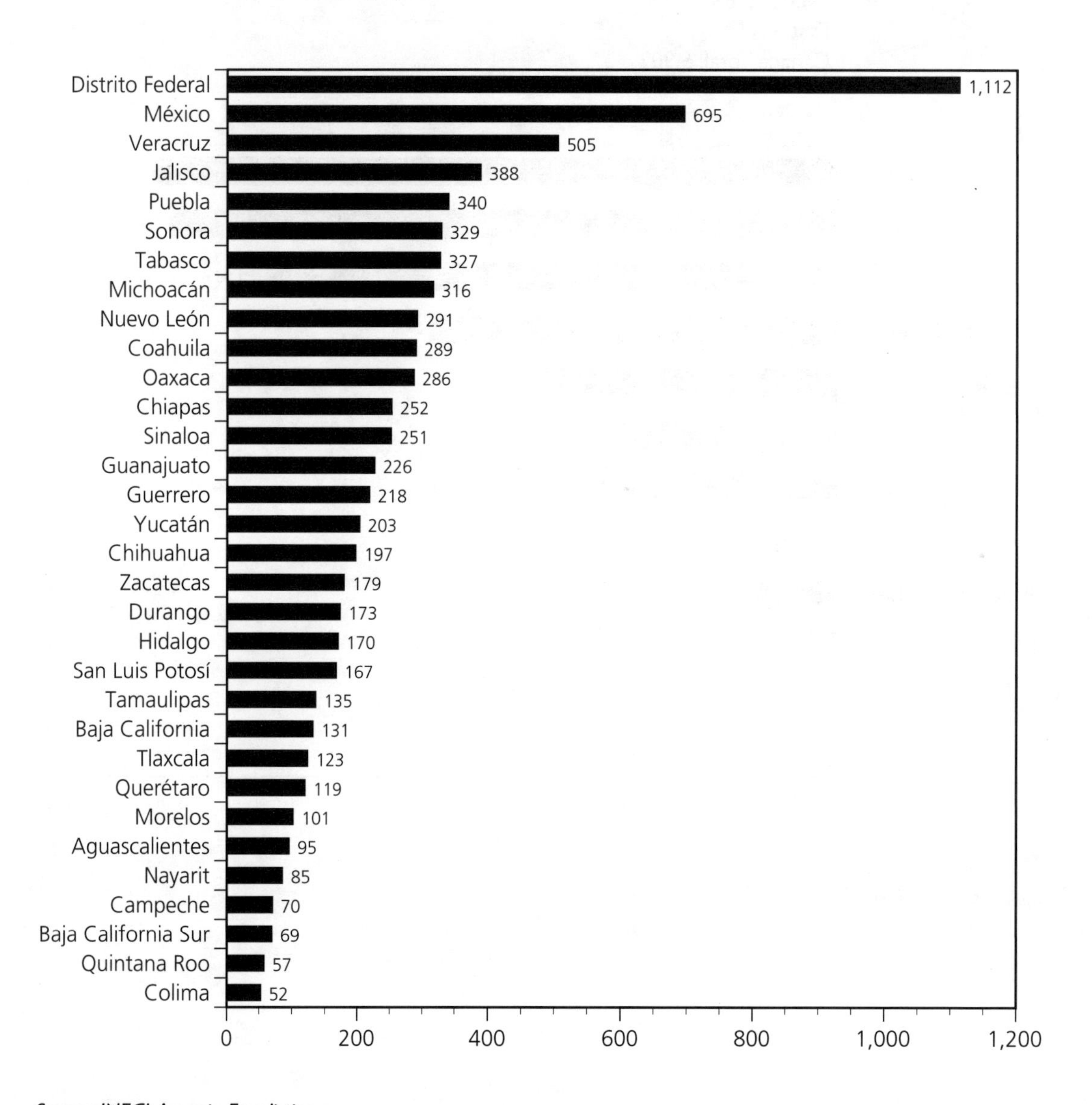

Source: INEGI, Anuario Estadistico

Chapter Ten
Public Safety

Crime Rates

In the 1880s, when guns were blazing from Tombstone, Arizona, to the Black Hills of South Dakota, and the U.S. West was truly wild, the Canadian West was relatively placid. The Royal Canadian Mounted Police saw to that. Stories are told of the Montana cowboys on a cattle drive to Alberta being met at the border by a single Mountie, who ordered them to put their pistols in their saddlebags and keep them there. The Royal Canadian Mounted Police treatment of the Plains Indians was everything that U.S. cavalry behavior was not, with the result that the Indian wars that plagued the American frontier never occurred in Canada.

But times change. Crime rates today are higher in western Canada than in the western United States. And the proportion of violent crimes to total crimes is higher in Canada than it is in the western United States—although the murder rate in the United States is higher.

In terms of reported crimes, Canada ranked fifth in the world in 1988, and the United States ranked nineteenth. (Mexico was not among the seventy-seven countries from which Interpol gathered crime statistics that year.)

The murder rate per 100,000 population was 7.91 in the United States (fifth in the world) in 1988. In Canada, the rate was 6.33 (twentieth in the world).

The rates of theft for Canada and the United States in 1988 were virtually identical. They ranked sixth and seventh in the world that year. Canada ranked tenth in the world in the number of sex crimes, and the United States was twenty-third.

The Northwest Territories, the Yukon, and the provinces of British Columbia and Alberta all had more crimes per 100,000 population than the most crime-infested western state, Arizona. And seven Canadian provinces and territories had rates of violent crime greater than the most violent western U.S. state, California. Overall, violent crimes accounted for 26 percent of all crimes committed in Canada, and 10 percent in the western United States.

In the United States, crime rates are lower in the West and higher in the East. In Canada, crime rates are higher in the West, and lower in the East.

Figure 10.1 **Western United States**
Violent Crime and Property Crime by State, 1990

U.S. Total: 732 Violent crimes/ 5,088 Property crimes per 100,000
(Violent crime includes murder, non-negligent manslaughter, rape, robbery, and aggravated assault; property crime includes burglary, larceny, and motor vehicle theft.)

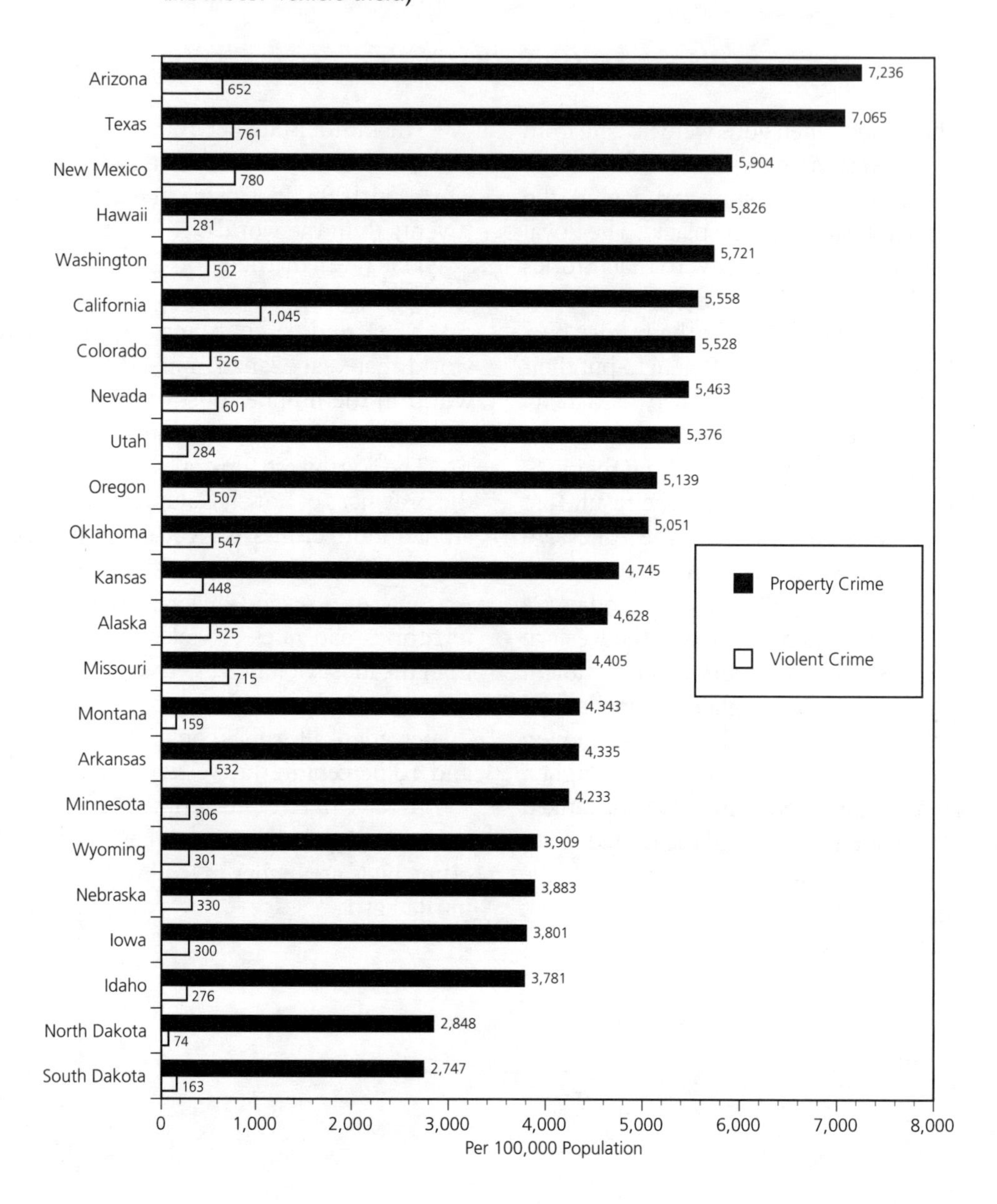

Source: U.S. Federal Bureau of Investigation

Figure 10.2 **Canada**
Violent Crime and Property Crime by Province, 1991

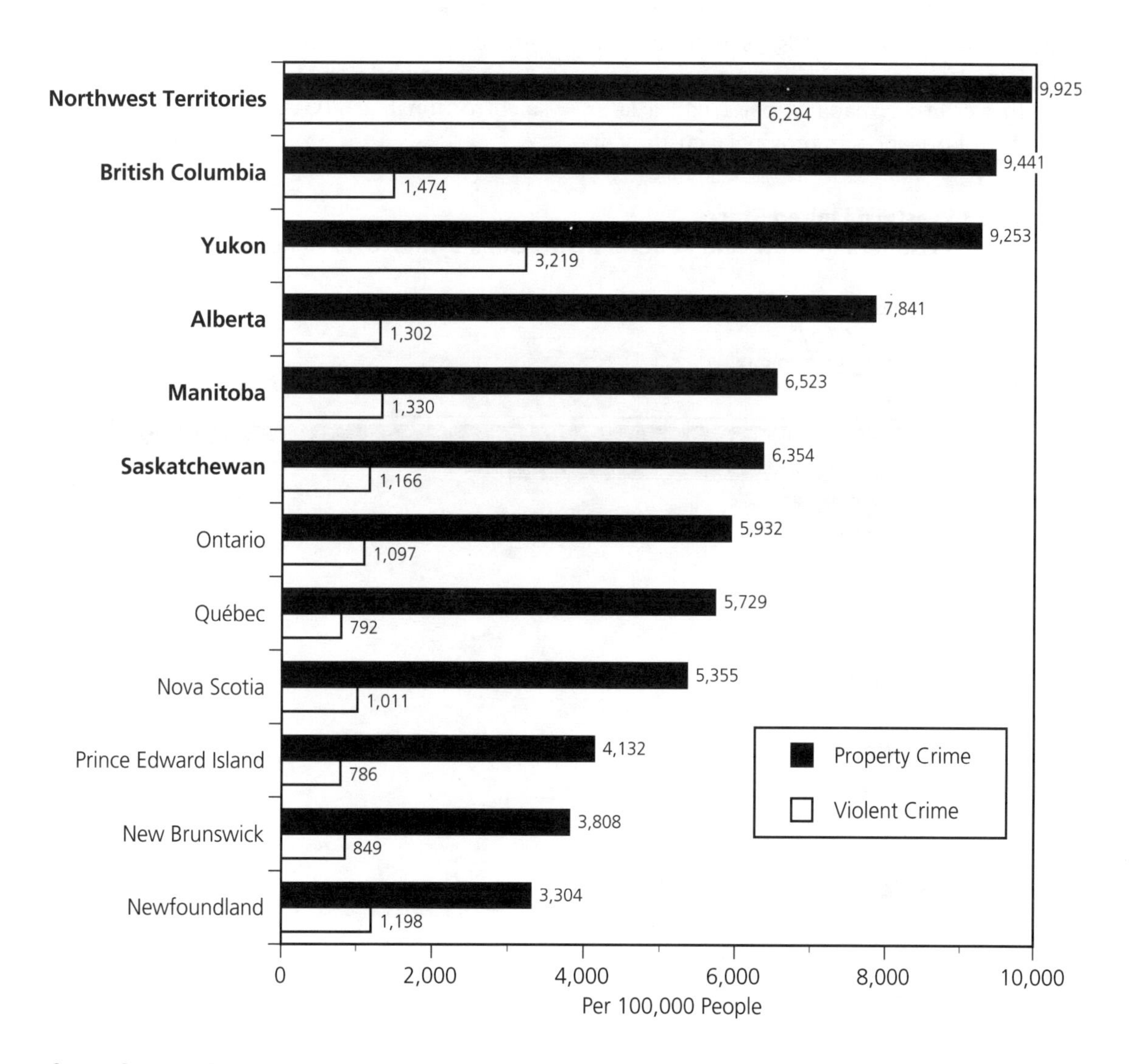

Source: Canadian Center for Justice Statistics

Prisoners in Correctional Facilities

Although the western United States has lower rates of crime and violent crime than Canada, there are far more prisoners in U.S. correctional facilities than in either Canada or Mexico. California had nearly as many prisoners as Mexico, and almost as many as western Canada.

In western Canada, Alberta had a large number of prisoners in provincial custody in proportion to population. Alberta had more prisoners than Québec, which has nearly three times Alberta's population, and more than twice as many prisoners as British Columbia, which has only 29 percent more people. Approximately one of every 233 Canadians, 327 Americans, and 818 Mexicans is a prisoner in a correctional facility.

Figure 10.3 **Western United States**
Prisoners in State and Federal Institutions by State, 1990 advance figures

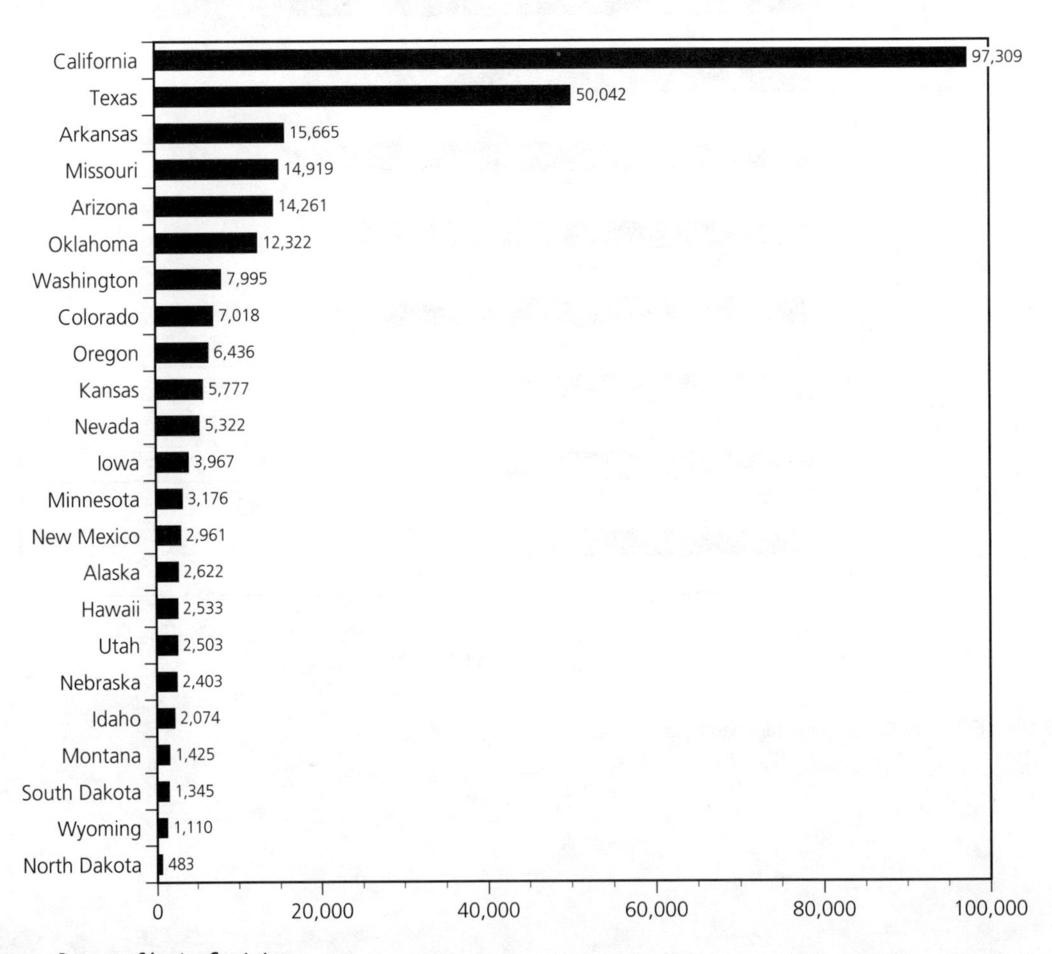

Source: Bureau of Justice Statistics

Figure 10.4 **Canada**
Number of Prisoners Sentenced to Custody by Province, 1990–1991

West: 43,620
East: 71,214
Canada Total: 114,834
(Alberta has a federal penitentiary.)

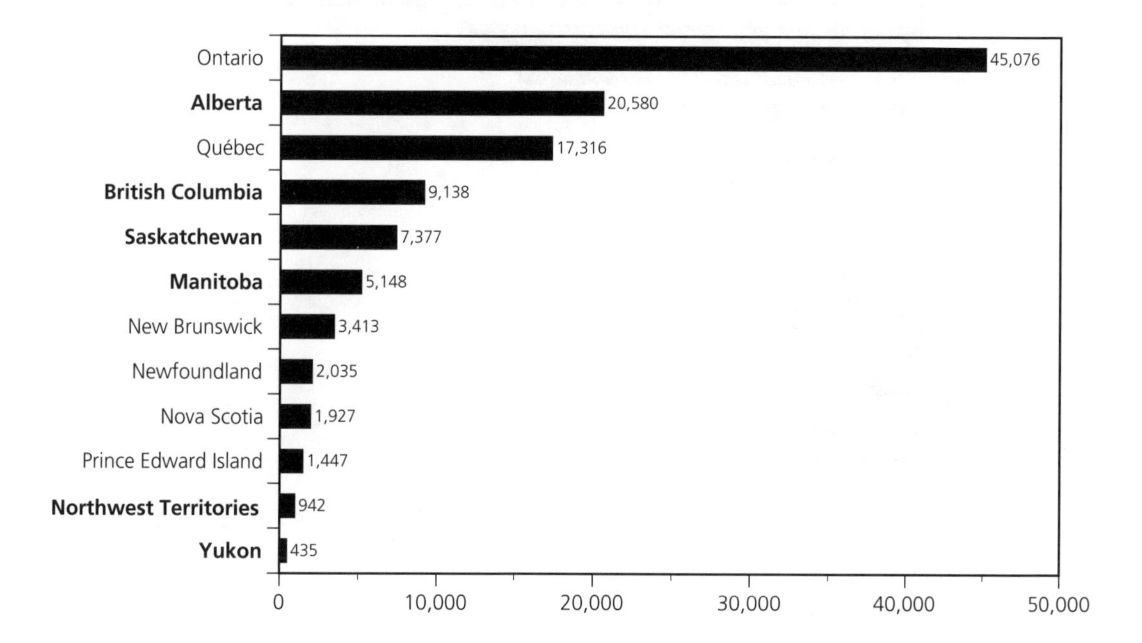

Source: Statistics Canada

Figure 10.5 **Mexico**
Number of Sentenced Prisoners in Federal and Common Courts by State, 1988

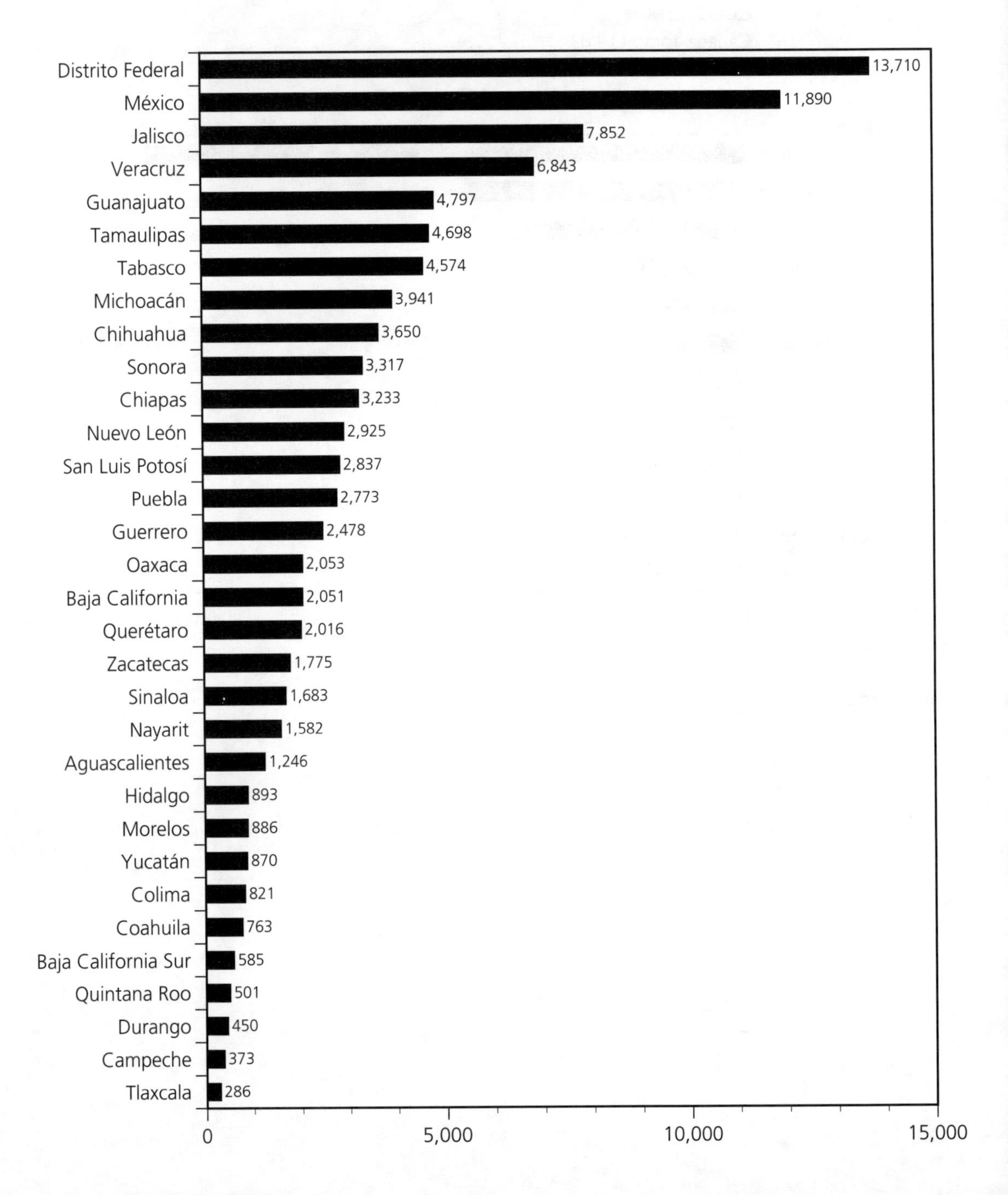

Source: INEGI, Anuario Estadistico

Motor Vehicle Fatalities

In 1990, there were 17,529 motor vehicle fatalities in the twenty-three western U.S. states, twice as many as in Mexico, and five times as many as in Canada. Mexican drivers were the most likely to have fatal accidents, with one fatality for every 1,300 registered vehicles. Canadians were the least likely, with one fatality for every 4,900 vehicles. In the United States there was one traffic fatality for every 1,900 motor vehicles.

Figure 10.6 **Western United States Motor Vehicle Fatalities by State, 1990**

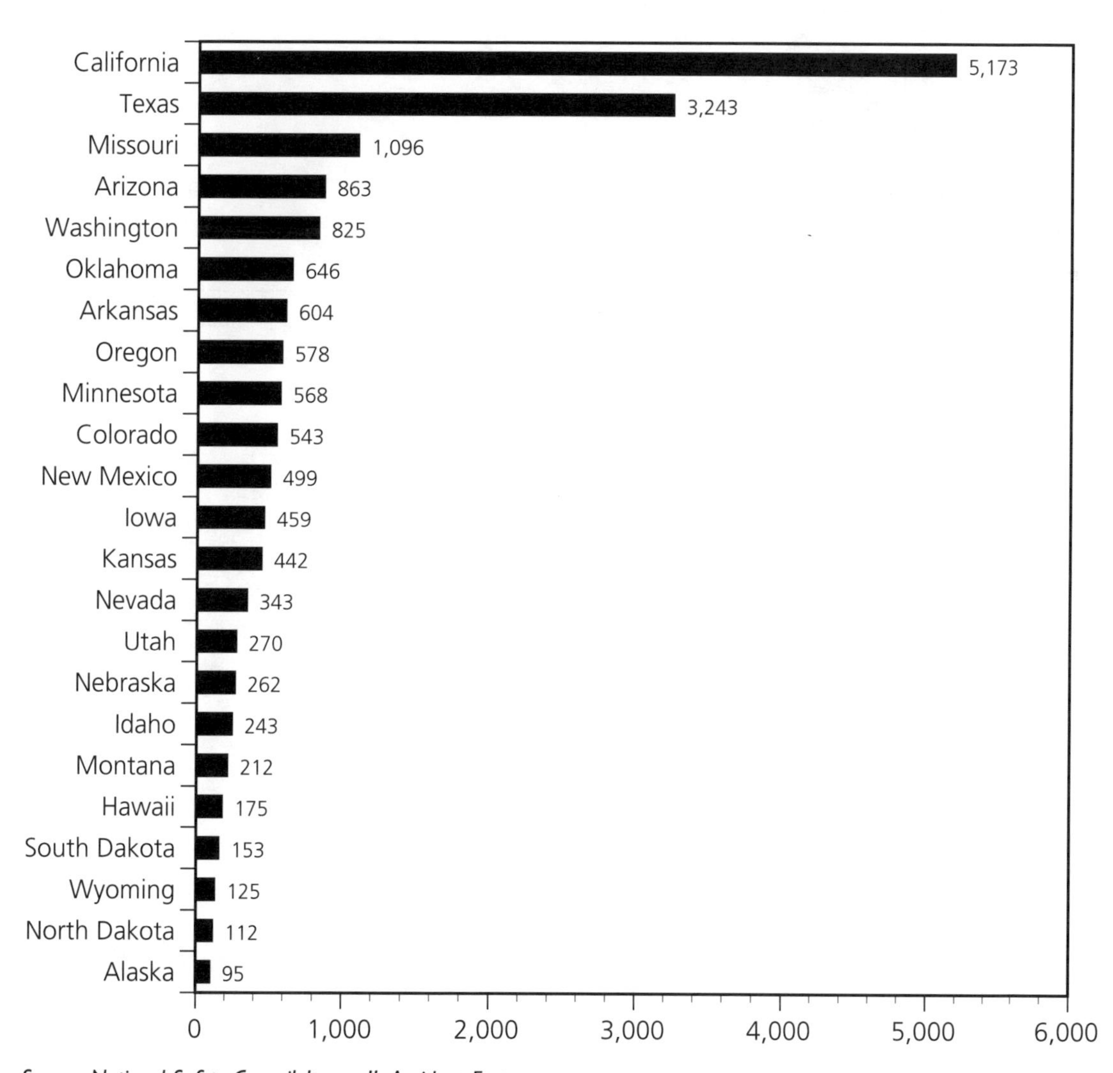

Source: National Safety Council, Itasca, IL, Accident Facts

Figure 10.7 **Canada**
Motor Vehicle Fatalities by Province, 1992

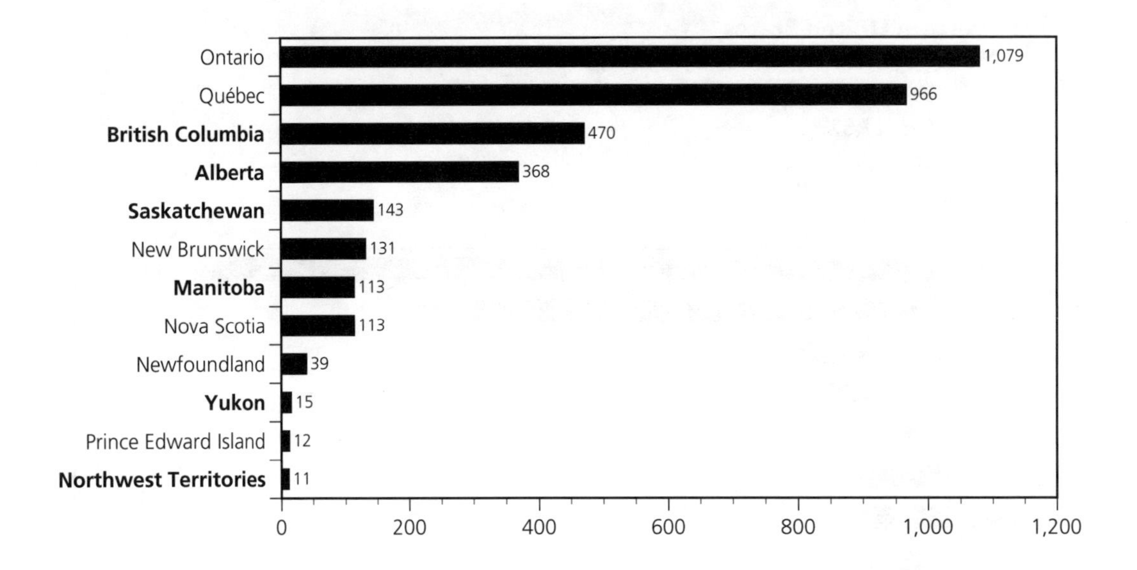

Source: Transport Canada

Figure 10.8 **Mexico**
Motor Vehicle Fatalities by State, 1990

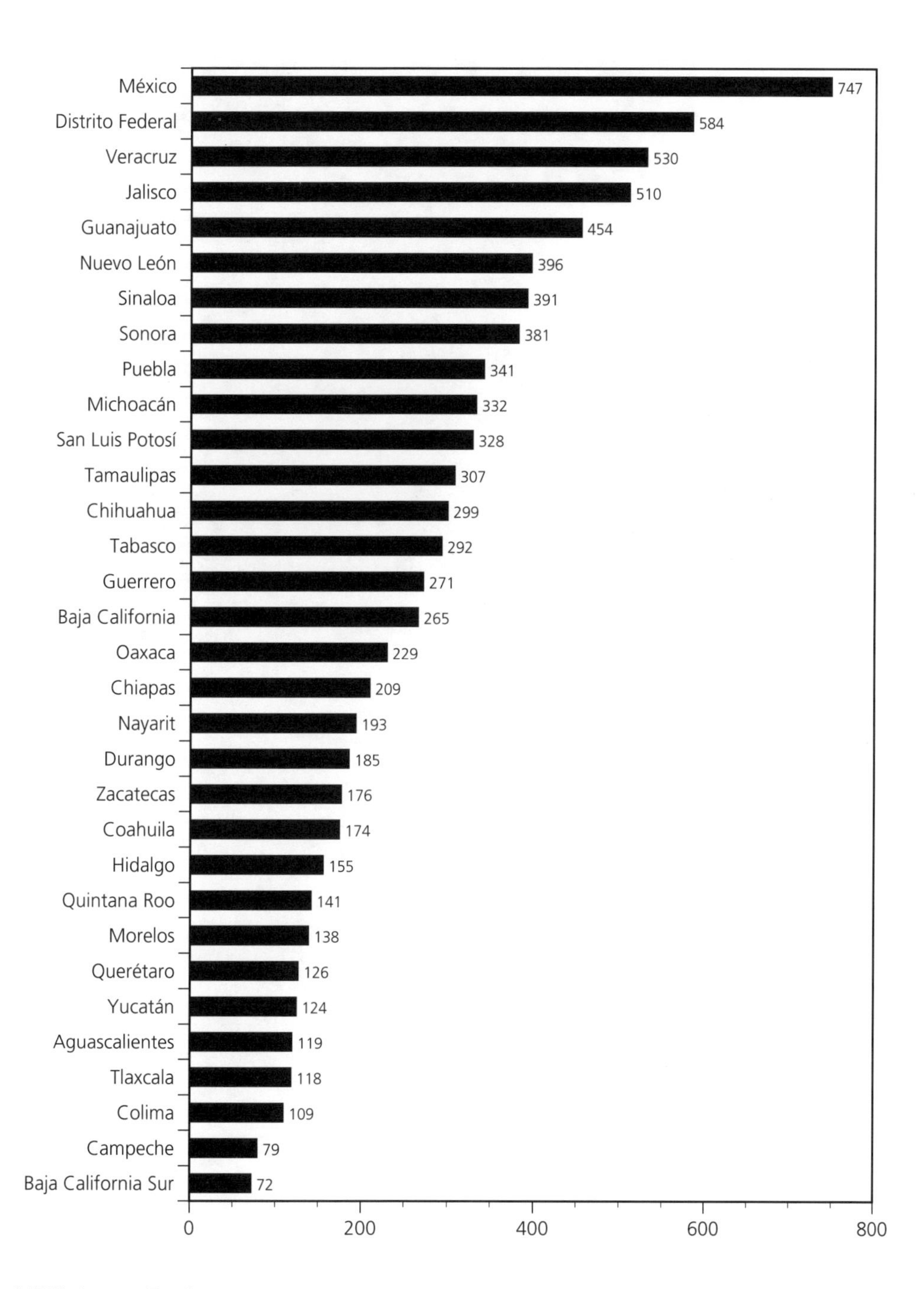

Source: INEGI, Anuario Estadistico

Chapter Eleven
Environment

Air Pollution and Hazardous Waste Disposal, United States

Although the United States is the world leader in efforts to stem air pollution, many urban areas fail to meet national standards for smog—the number one environmental problem in the country. Overall, western cities represented 20.4 percent of the metropolitan areas in the United States exceeding ozone standards, and 74 percent of the areas exceeding carbon monoxide standards. Los Angeles had the worst standard for both in the country. Air quality improved in most Western cities between 1988 and 1990. Significant exceptions were Anchorage, Alaska, (emissions of carbon monoxide) and Houston and San Diego (emissions of ozone).

Western states contain 383 of the 1,270 hazardous waste sites on the National Priority List for cleanup.

Figure 11.1 **Western United States Metropolitan Areas Failing to Meet National Ambient Air Quality Standards for Carbon Monoxide, 1989–1990**

(Western cities accounted for 74% of the cities exceeding the standards.)

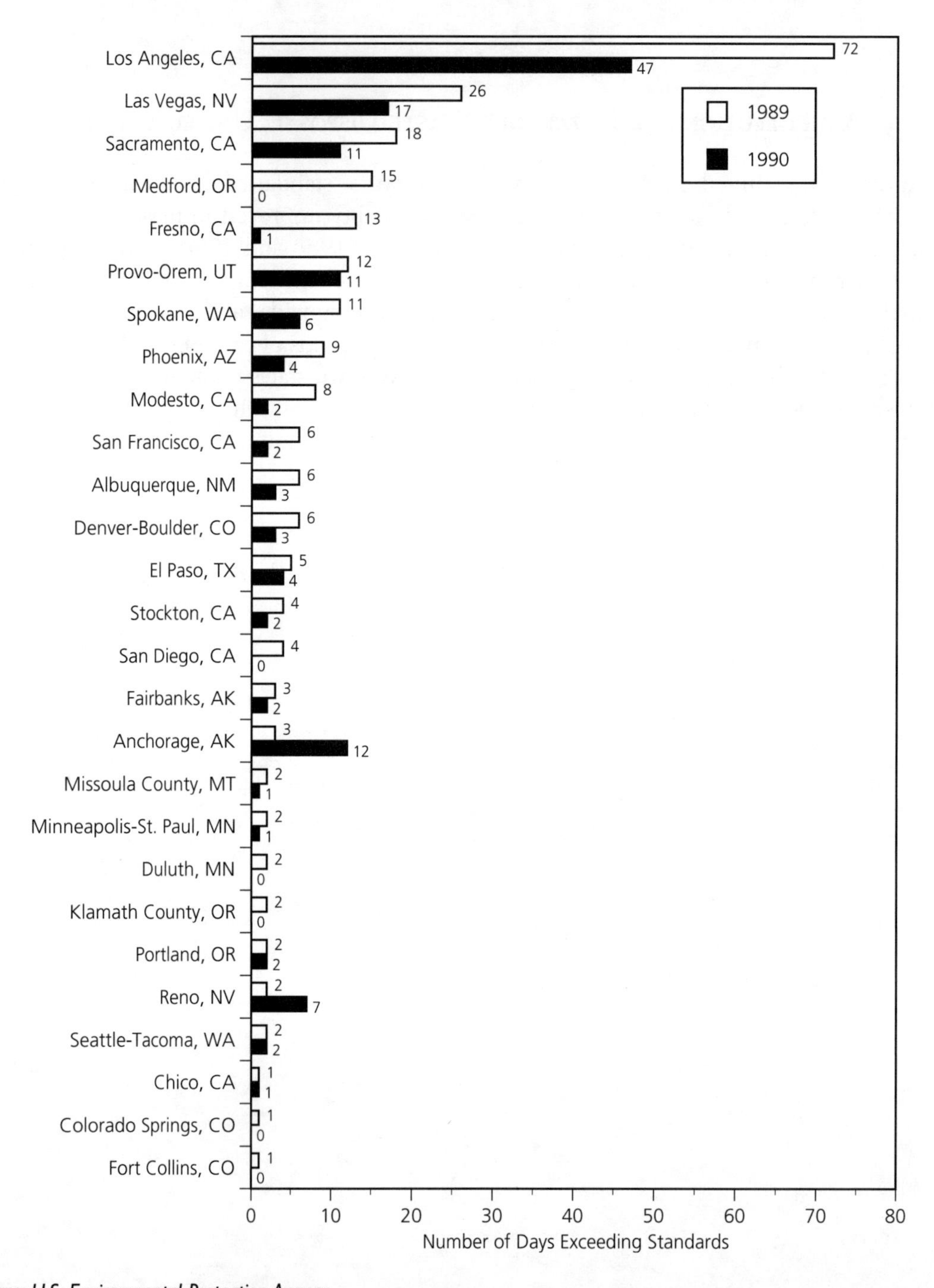

Source: U.S. Environmental Protection Agency

Figure 11.2 **Western United States**
Metropolitan Areas Failing to Meet National Ambient Air Quality Standards for Ozone, 1988–1990

(Western cities accounted for 20.4% of the areas exceeding the standards.)

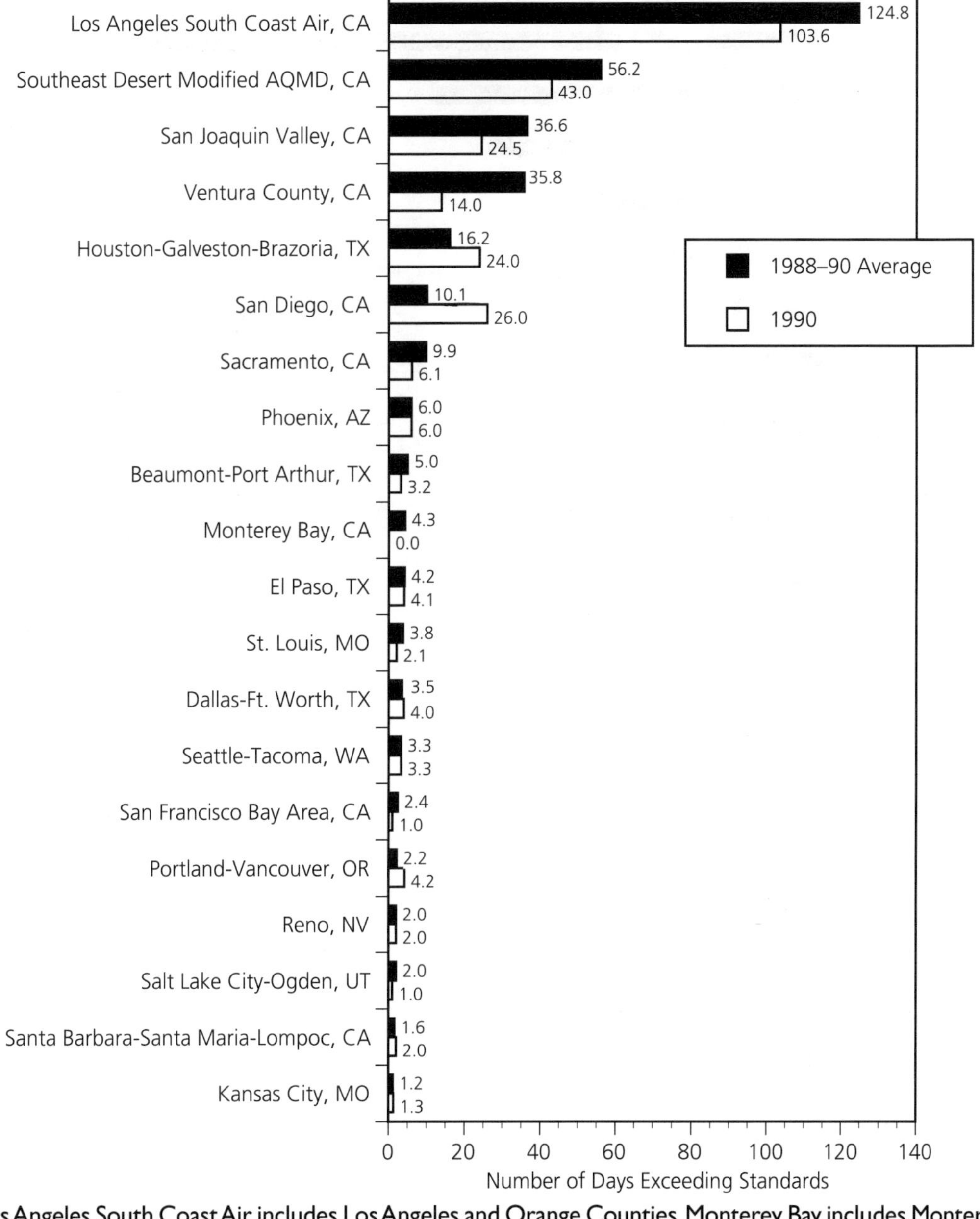

(Los Angeles South Coast Air includes Los Angeles and Orange Counties. Monterey Bay includes Monterey, Santa Cruz, and San Benito Counties. Southeast Desert Modified includes San Joaquin, Turlock, Merced, Madera, Fresno, Kings, Tulare, and Kern Counties)

Source: U.S. Environmental Protection Agency

Figure 11.3 **Western United States Hazardous Waste Sites on the National Priority List, 1991**

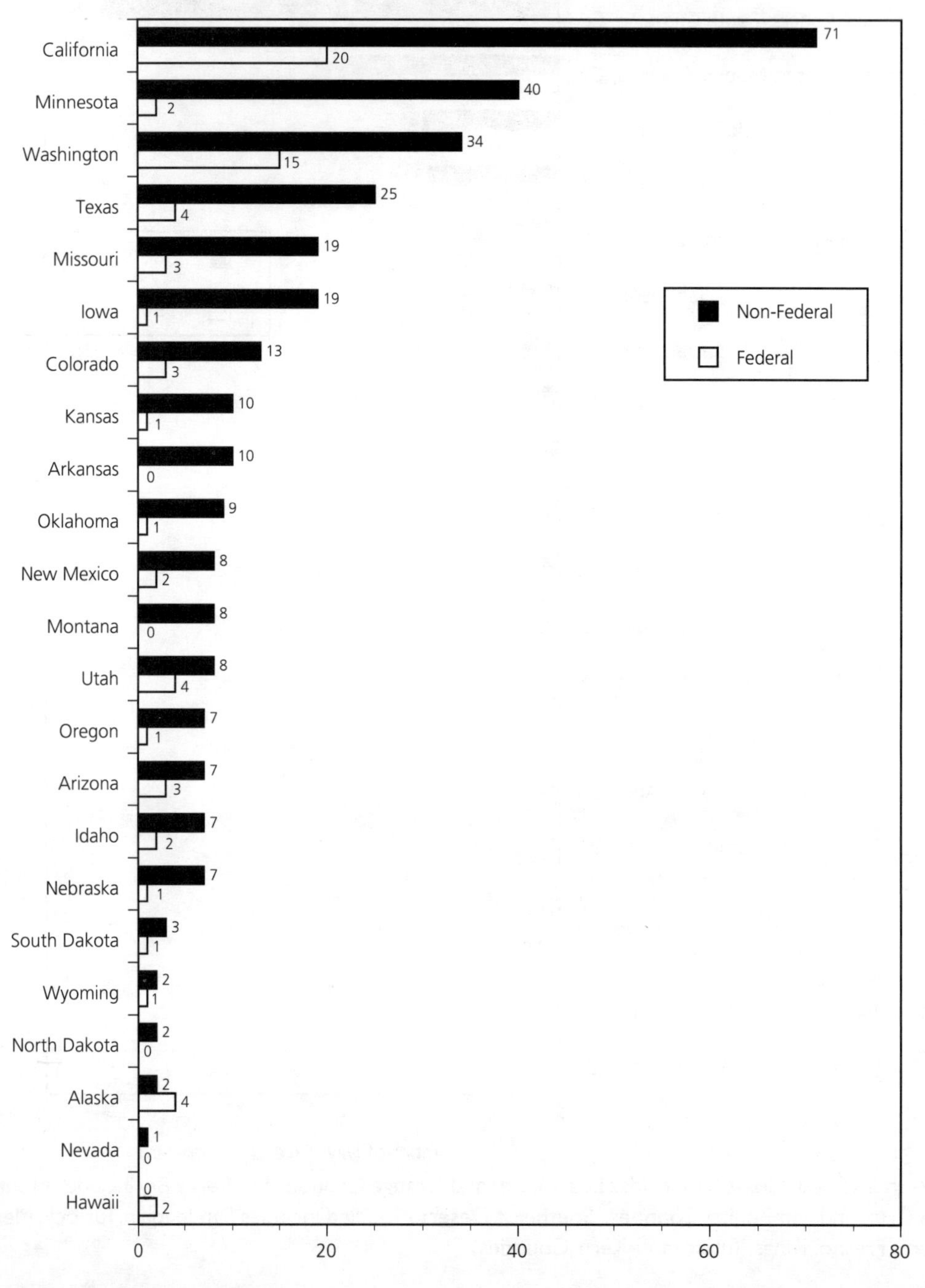

Source: Environmental Protection Agency

Air Pollution, Canada

Acid rain and marine-habitat degradation are Canada's primary environmental problems. Lakes and forests have been harmed by smelting, coal burning, and vehicle emissions, and contaminated water has closed shellfish-growing areas on both coasts.

Air quality in most Canadian cities is very good. No western Canadian city exceeded acceptable limits for either carbon monoxide or ozone emissions, and only Calgary and Vancouver exceeded the desirable limit for CO_2. In the West, emissions were greatest in British Columbia and Alberta.

Figure 11.4 **Canada**
Average Urban Air Quality for Carbon Monoxide (Eight Hour) for Selected Cities, 1985–1989

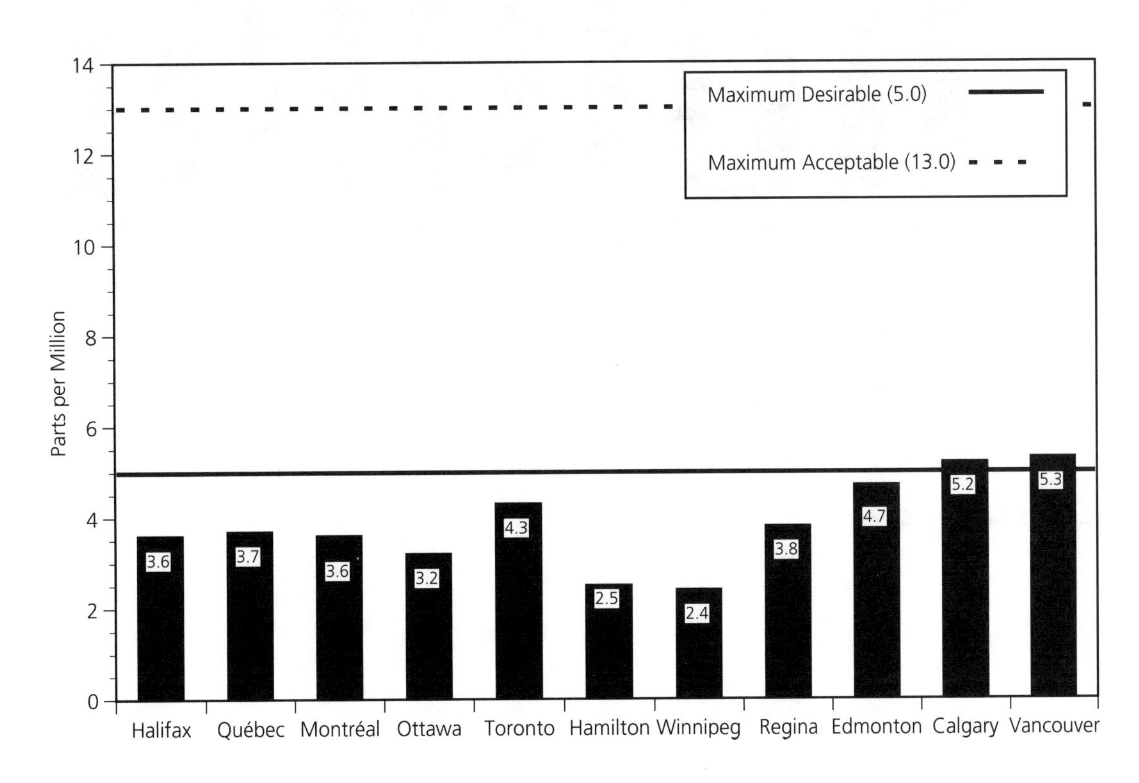

Source: Environment Canada

Figure 11.5 **Canada**
Average Urban Air Quality for Ozone (One Hour) for Selected Cities, 1985–1989

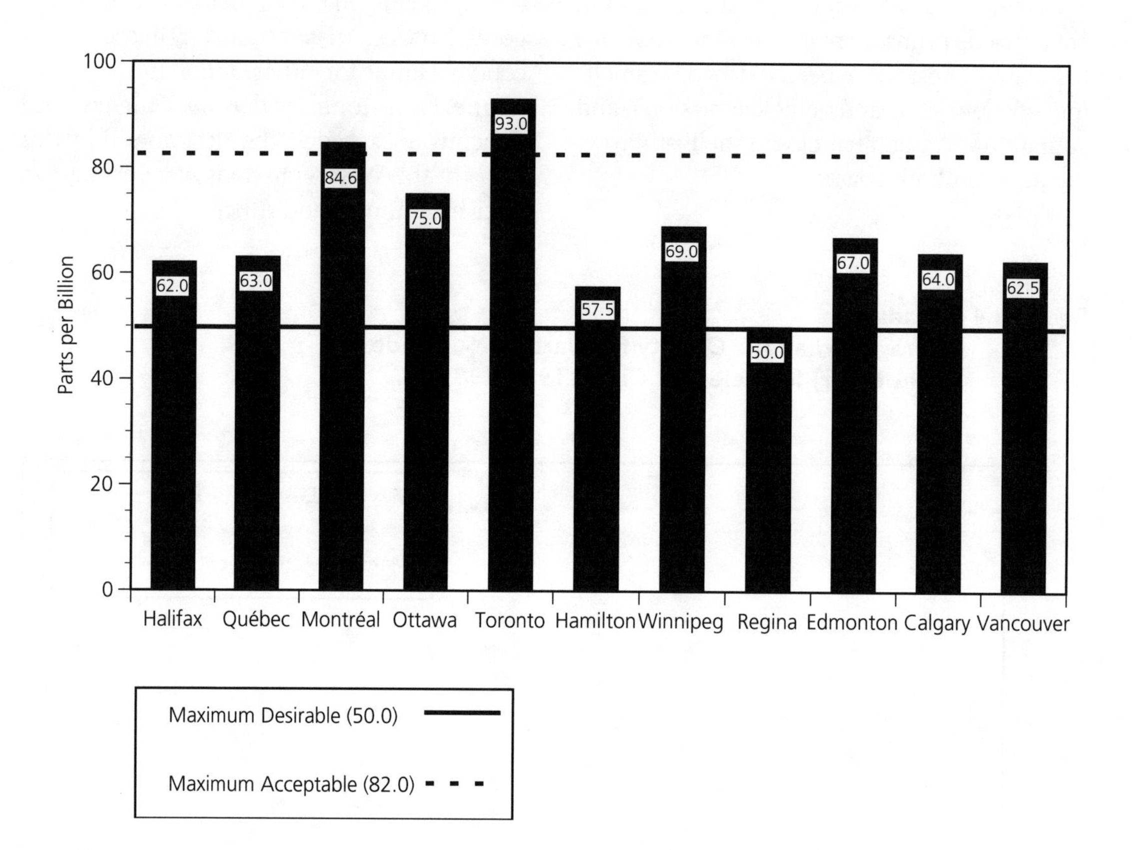

Source: Environment Canada

Figure 11.6 **Canada**
Summary of Emissions by Province, 1985

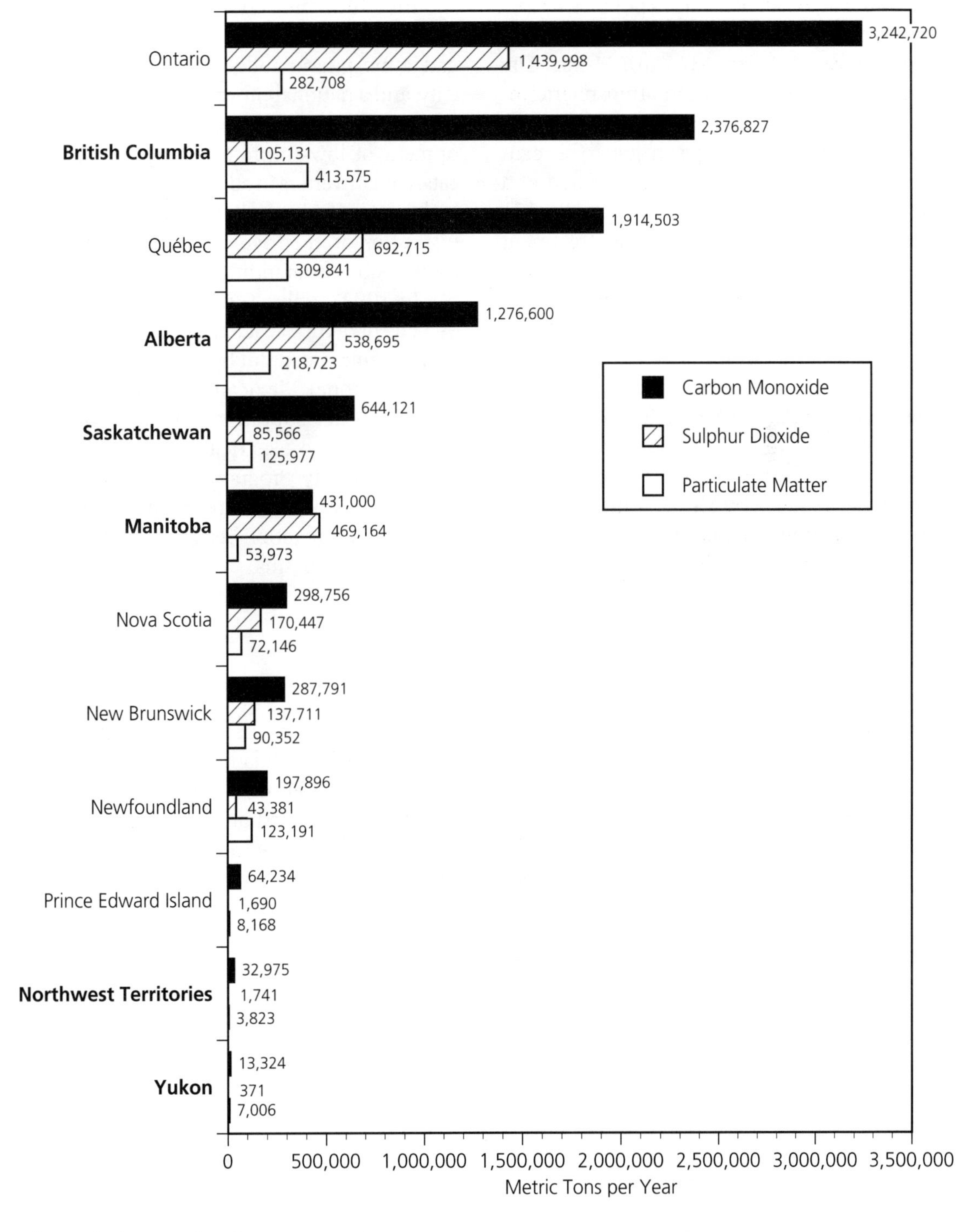

Source: Environment Canada

In Mexico, air pollution (especially in Mexico City), deforestation, and a scarcity of water are among the most serious problems.

More than half of Mexico's population and industry and much of its cropland are at altitudes above 1,600 feet (487.7 m). Mexico City and its environs suffer from atmospheric inversions that trap smog and toxic emissions from industries that have just begun to be regulated. Mexican forests are being cleared at a rate estimated at 2,375 square miles (6,151 sq. km) per year. Lack of adequate treatment of sewage and industrial effluents flowing into Mexican rivers and the Rio Grande along the U.S. border make the problem worse.

As Mexican income has risen, however, so have efforts to control pollution. In 1983, Mexico and the United States signed the Agreement of La Paz to find solutions to pollution problems in seventeen border cities. Among the measures taken by Mexico pursuant to its treaty obligations are increased investment in water treatment, streamlining of passenger and freight traffic, establishing land reserves, and stricter environmental requirements for new industries.

In 1990, the federal government announced a comprehensive program to mitigate atmospheric pollution in metro Mexico City, and a national program for environmental protection. These programs call for development of low-sulphur fuels, placement of catalytic converters in all gasoline-powered vehicles, increased use of public transportation, institution of "no drive" days, tighter controls on emissions from industry, and reforestation.

In 1992, only 79 percent of urban and 49 percent of rural residents had access to running potable water. But by 1993, thirteen million additional Mexicans gained access to drinking water supplies, and eleven million gained access to electricity through the National Solidarity Program. Thirty additional sewage and waste water treatment plants were built, and regeneration began on the country's principal river basins.

Protected Areas

The United States has set aside proportionately twice as much land in protected areas than have either Canada or Mexico. The United States has designated 379,698 square miles (983,417 sq. km), or 10.5 percent, of its land area as wilderness, parkland, or other protected areas. The vast majority of that land is in the West. Canada has set aside 190,935 square miles (494,521 sq. km) of land, or 5 percent, in protected areas. Mexico has protected 36,369 square miles (94,195 sq. km), or 4.8 percent.

Figure 11.7 **United States, Canada, and Mexico National Protected Areas, ca. 1990**

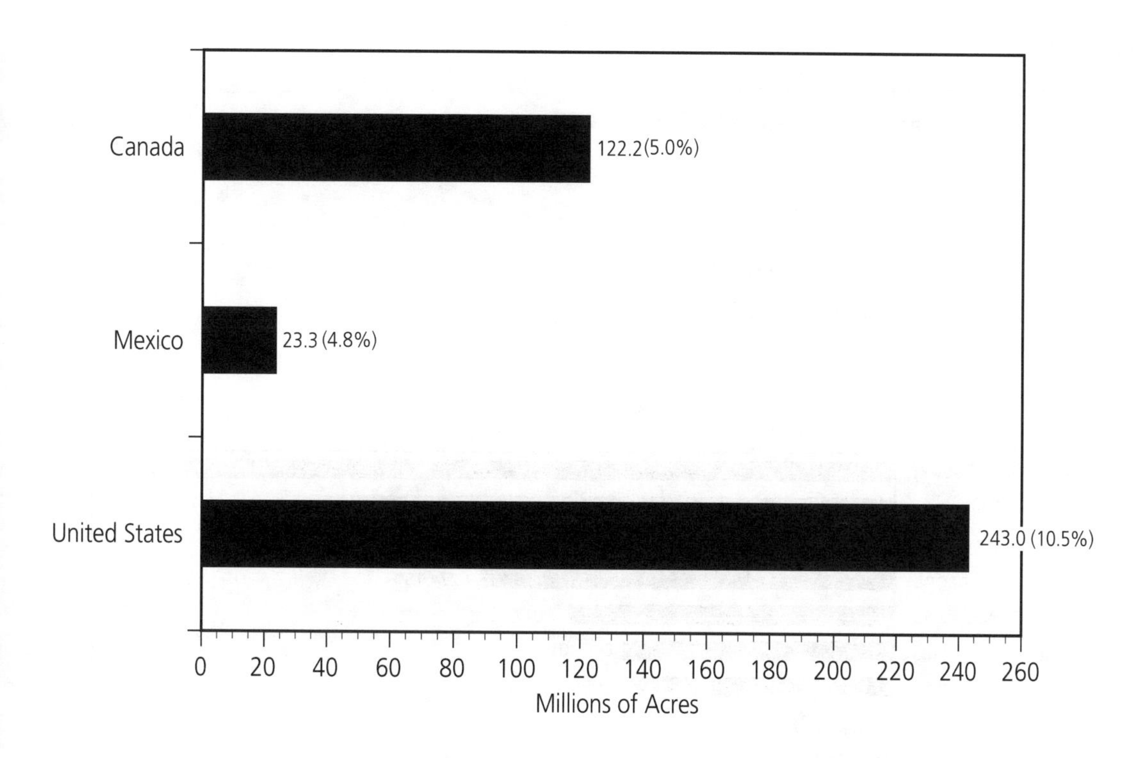

(Number in parentheses is percent of total land area.)

Source: U.N. Environment Programme

More than 90 percent of U.S. parkland is in the West. Alaska alone accounts for 69 percent of the U.S. total. California and Washington have the most federal parkland in the contiguous United States. Nearly 97 percent of Canada's parkland is in the West.

Figure 11.8 Western United States National Park Acreage by State, 1992

West: 43,171,644
East: 3,038,089
U.S. Total: 46,209,733

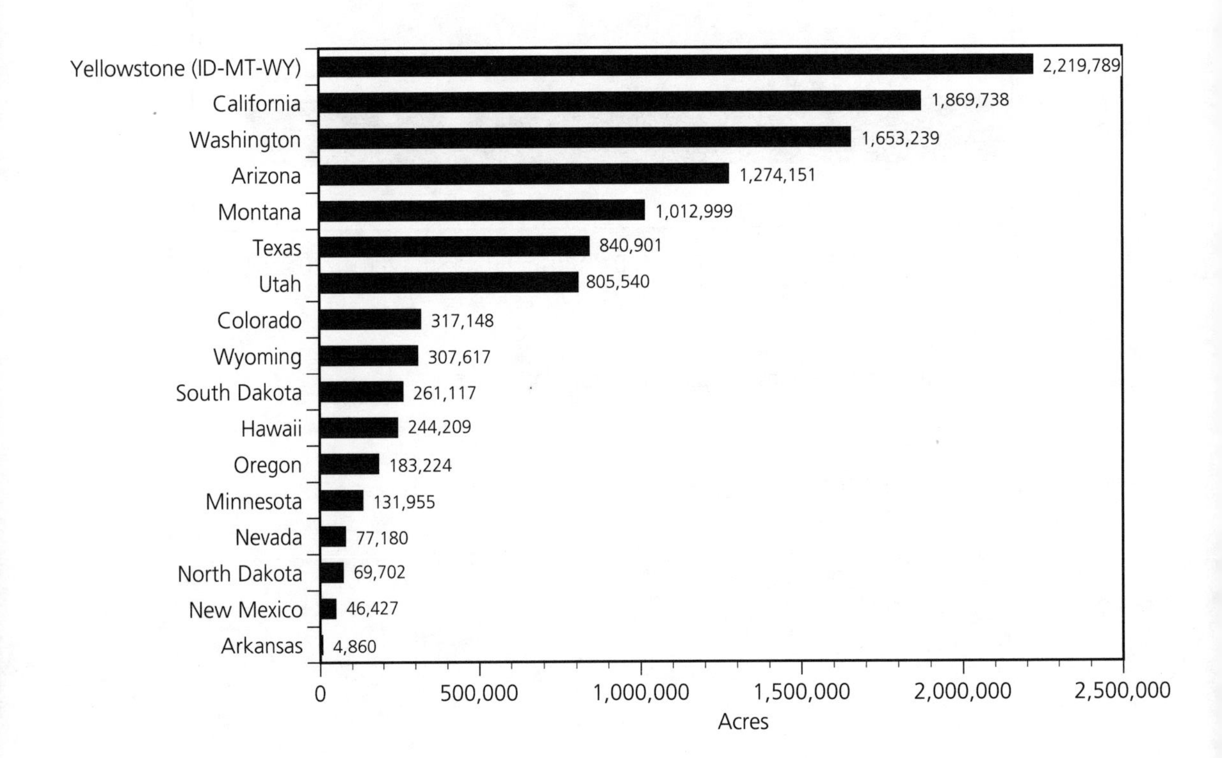

Source: National Park System

Figure 11.9 **Western United States**
National Park Acreage, Continental United States and Alaska, 1992

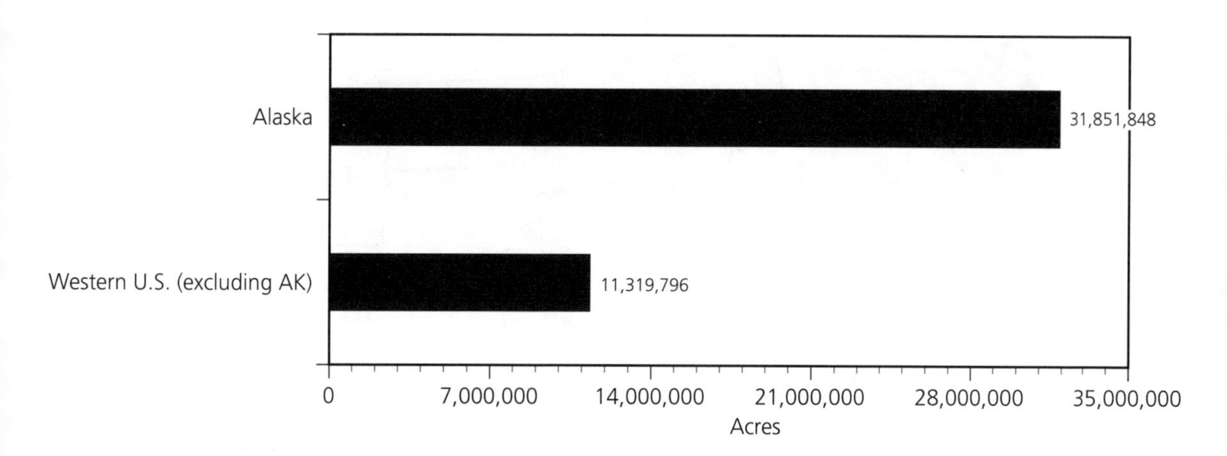

Source: National Park System

Figure 11.10 **Western United States National Park Acreage, 1992***

(* See chart, page 211)

Source: National Park System

	State	Acres		State	Acres
Denali (1)	AK	4,724,114	Yellowstone (22)	ID–MT–WY	2,219,789
Gates of the Arctic (2)	AK	7,281,655	Voyageurs (23)	MN	131,955
Glacier Bay (3)	AK	3,224,938	Glacier (24)	MT	1,012,999
Katmai (4)	AK	3,575,000	Teddy Roosevelt (25)	ND	69,702
Kenai Fjords (5)	AK	649,946	Carlsbad Caverns (26)	NM	46,427
Kobuk Valley (6)	AK	1,726,463	Great Basin (27)	NV	77,180
Lake Clark (7)	AK	2,573,724	Crater Lake (28)	OR	183,224
Wrangell–St. Elias (8)	AK	8,096,008	Badlands (29)	SD	232,822
Hot Springs (9)	AR	4,860	Wind Cave (30)	SD	28,295
Grand Canyon (10)	AZ	1,180,618	Big Bend (31)	TX	764,608
Petrified Forest (11)	AZ	93,533	Guadalupe Mountains (32)	TX	76,293
Channel Islands (12)	CA	64,255	Arches (33)	UT	66,344
Kings Canyon (13)	CA	461,845	Bryce Canyon (34)	UT	35,833
Lassen Volcanic (14)	CA	106,367	Canyonlands (35)	UT	337,570
Redwood (15)	CA	75,442	Capitol Reef (36)	UT	222,753
Sequoia (16)	CA	402,299	Zion (37)	UT	143,040
Yosemite (17)	CA	759,530	Mount Rainier (38)	WA	235,613
Mesa Verde (18)	CO	51,891	North Cascades (39)	WA	504,555
Rocky Mountain (19)	CO	265,257	Olympic (40)	WA	913,071
Haleakala (20)	HI	26,911	Grand Teton (41)	WY	307,617
Hawaii Volcanoes (21)	HI	217,298			

(Number in parentheses refers to location on map on page 210.)

Canada has 84,598 square miles (219,110 sq. km) of national parkland, of which 67,329 square miles (174,383 sq. km), or 80 percent, is in the West. Nearly 44 percent of Canada's parkland is in the Yukon and Northwest Territories. Alberta, with 24,332 square miles (63,020 sq. km) of parkland, has more parkland than the rest of the provinces combined.

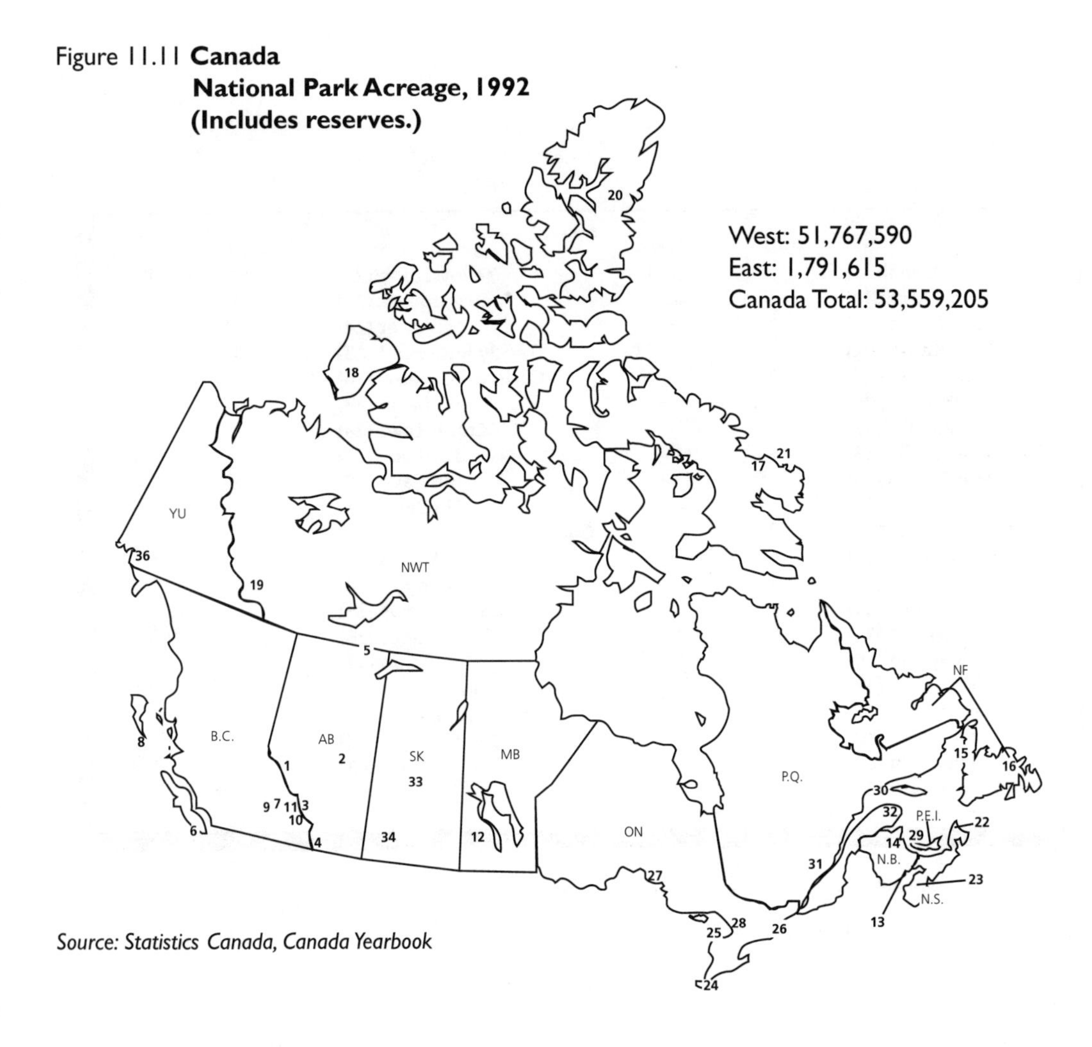

Figure 11.11 **Canada**
National Park Acreage, 1992
(Includes reserves.)

Source: Statistics Canada, Canada Yearbook

	Province	Acres
Jasper (1)	AB	2,688,012.3
Elk Island (2)	AB	48,012.6
Banff (3)	AB	1,640,977.4
Waterton Lakes (4)	AB	129,928.0
Wood Buffalo (5)	AB, NWT	11,072,050.8
Pacific Rim (6)	B.C.	123,453.8
Glacier (7)	B.C.	333,444.0
Gwaii Haanas (8)	B.C.	363,343.8
Mount Revelstoke (9)	B.C.	64,889.9
Kootenay (10)	B.C.	340,486.5
Yoho (11)	B.C.	324,474.1
Riding Mountain (12)	MB	735,360.9
Fundy (13)	N.B.	50,879.0
Kouchibougac (14)	N.B.	59,008.8
Gros Morne (15)	NF	480,002.2
Terra Nova (16)	NF	97,977.3
Auyuittuq (17)	NWT	5,305,624.3
Aulavik (18)	NWT	3,033,218.5
Nahanni (19)	NWT	1,177,605.4
Ellesmere Island (20)	NWT	9,760,662.5
Northern Baffin Island (21)	NWT	5,498,588.9
Cape Breton Highlands (22)	N.S.	234,873.7
Kejimkkujik (23)	N.S.	94,270.7
Point Pelee (24)	ON	3,830.1
Bruce Peninsula (25)	ON	66,718.5
St. Lawrence Islands (26)	ON	1,013.1
Pukaskwa (27)	ON	464,014.5
Georgian Bay Islands (28)	ON	3,508.9
Prince Edward Island (29)	P.E.I.	4,472.6
Mingan Archipelago (30)	P.Q.	37,238.8
La Mauricie (31)	P.Q.	134,400.6
Forillon (32)	P.Q.	59,404.1
Prince Albert (33)	SK	950,762.7
Grasslands (34)	SK	224,001.0
Ivvavik (–)	YU	2,512,666.3
Kluane (36)	YU	5,440,024.9

(Number in parentheses refers to map location.)

Mexico has sixty-eight protected areas, of which forty-four are national parks, eight are biosphere reserves, fourteen are special biosphere reserves, one is a wildlife protection area, and one is a national monument. Altogether, these areas cover 14 million acres (5.67 million hectares)—about the size of West Virginia.

Forty-four percent of state parks in the western United States were in Alaska in 1990. California, second in state park acreage, drew most of the nearly 310 million people that visited state parks in the West that year.

The four western Canadian provinces bordering the United States had 34,205 square miles (88,591 sq. km) of provincial parkland in 1988, of which 60 percent was in British Columbia.

National parks in British Columbia and Alberta receive the vast majority of visitors.

Figure 11.12 **Western United States State Park Acreage by State, January–June 1990**

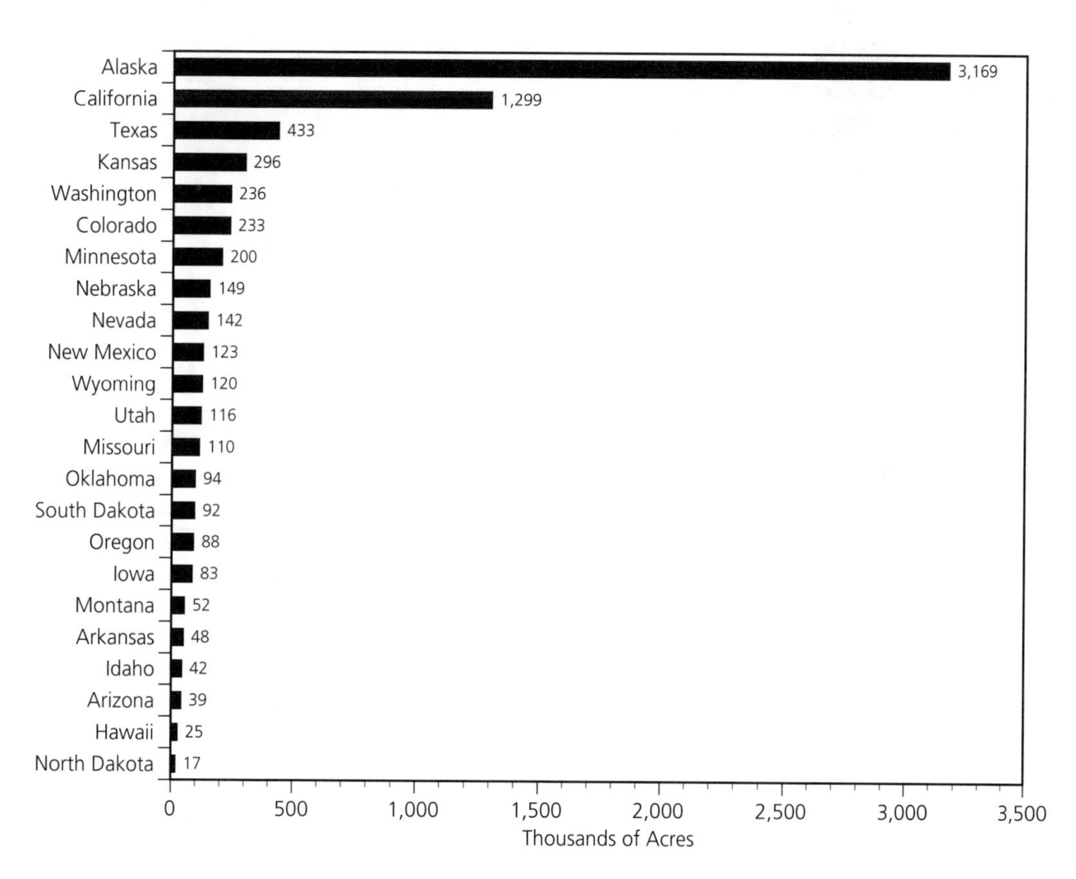

Source: National Association of State Park Directors

Figure 11.13 **Western United States**
State Park Visitors by State, January–June 1990
(Includes overnight visitors.)

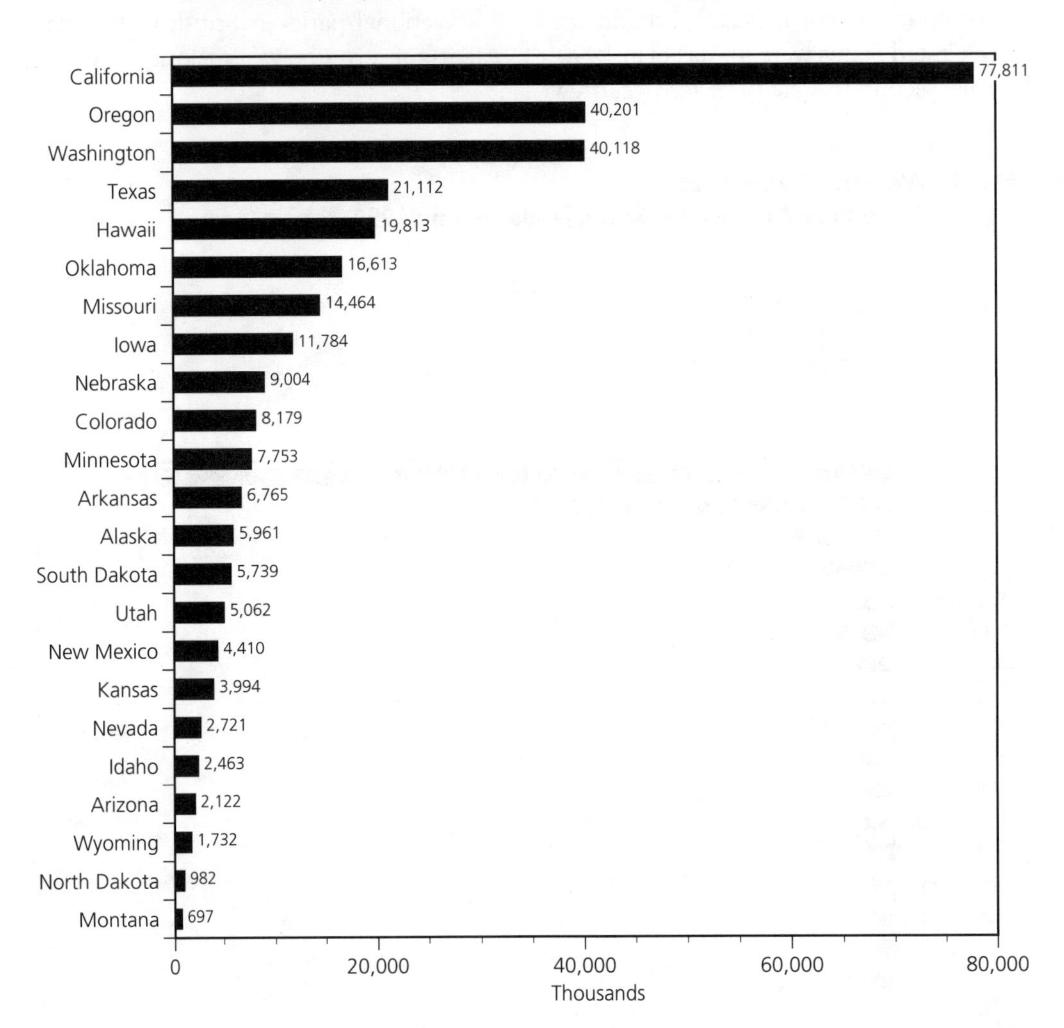

Source: National Association of State Park Directors

Figure 11.14 **Canada**
Provincial Park Acreage, 1988

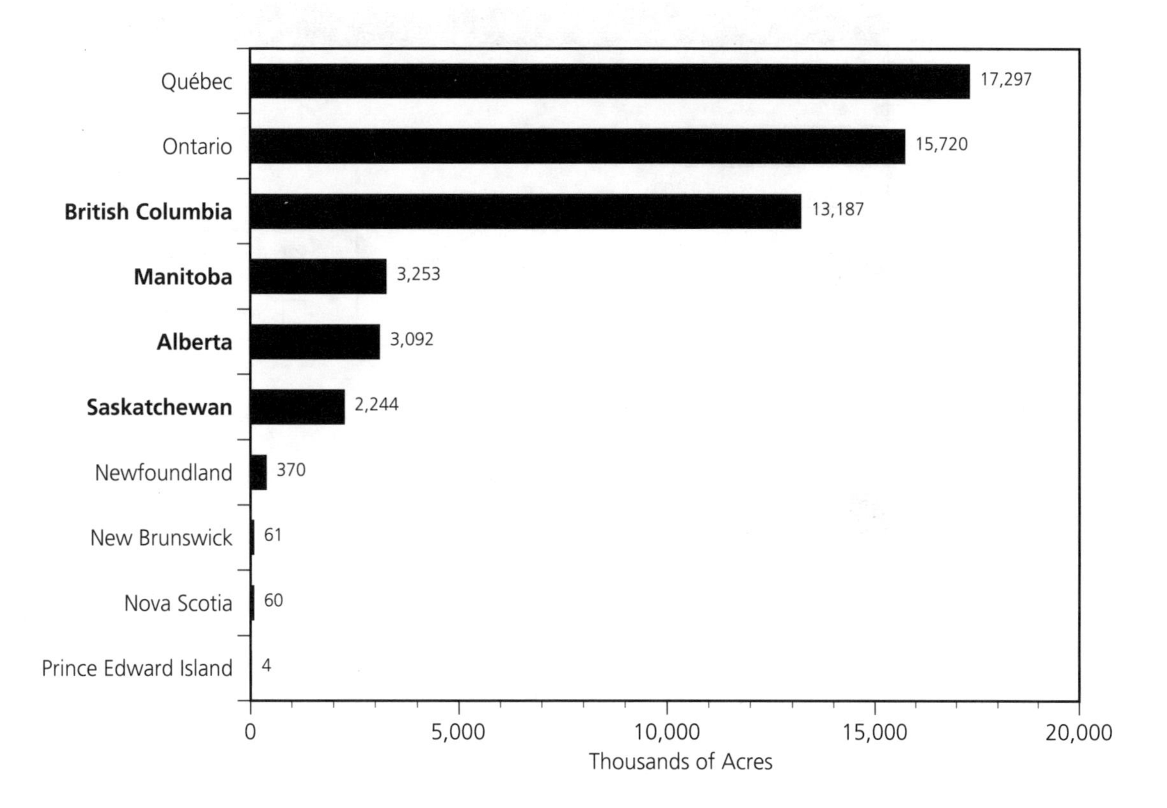

Source: Statistics Canada, Canada Yearbook

Figure 11.15 **Canada**
Person-Visits to National Parks and Historic Sites, 1991–1992

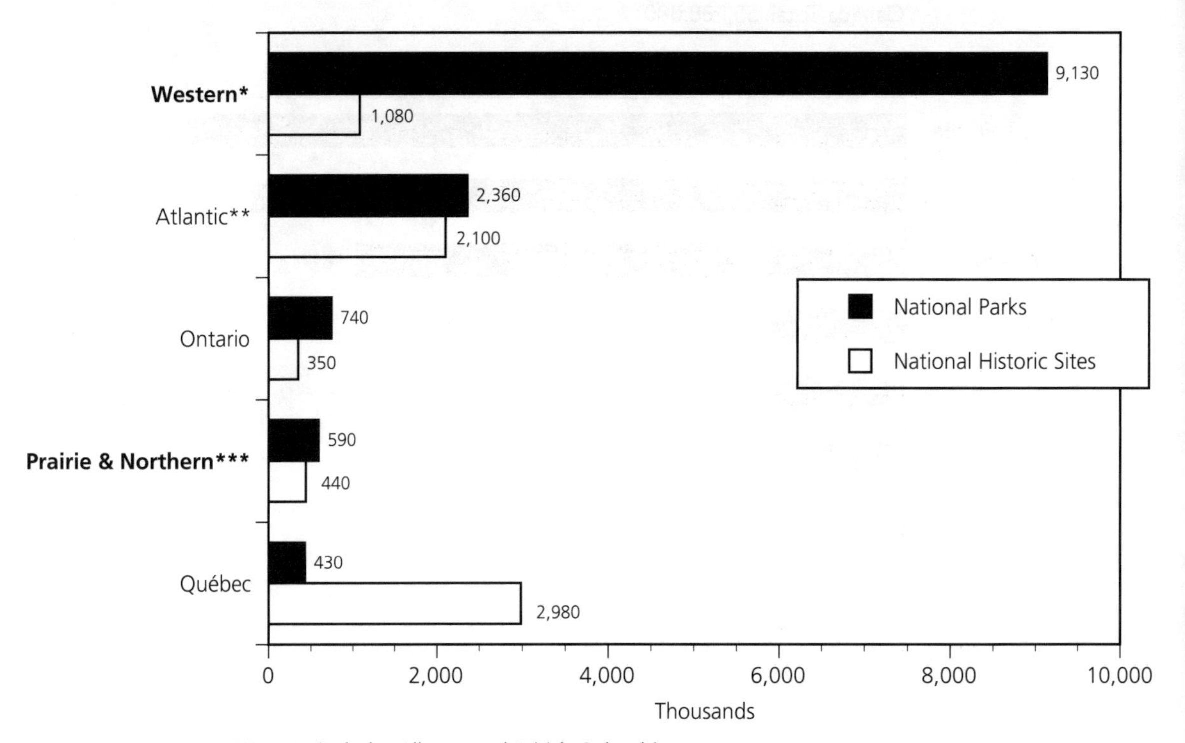

* Western includes Alberta and British Columbia.

** Atlantic includes New Brunswick, Newfoundland, Nova Scotia, and Prince Edward Island.

*** Prairie & Northern includes Manitoba, Northwest Territories, Saskatchewan, and Yukon.

Source: Parks Canada

In Mexico, the National System of Protected Areas defines national parks as "all those natural spaces that boast one or more outstanding ecosytems in terms of historical, scientific, or aesthetic value, and the presence of nationally meaningful animals and plants." There are 1,693,111.7 acres (685,470 hectares) of national parks in Mexico.

Mexico also boasts many archaeological sites and museums that display its rich cultural past.

Figure 11.16 **Mexico**
National Park Acreage, 1992

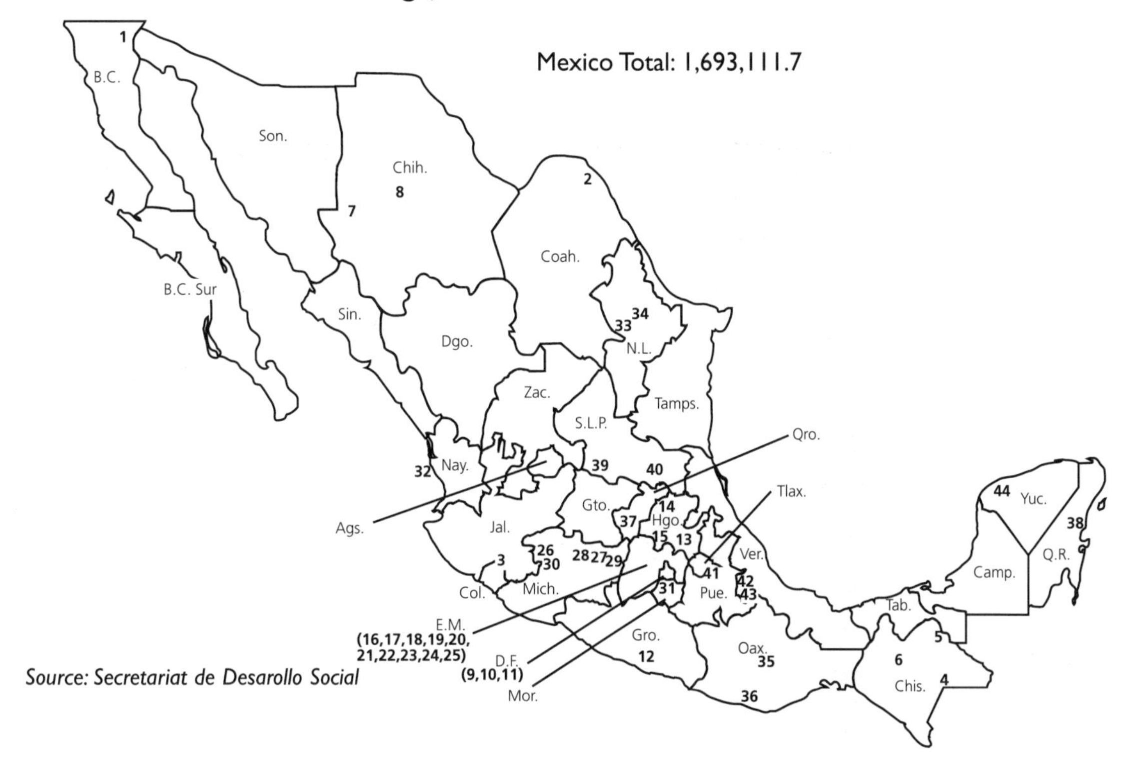

Source: Secretariat de Desarollo Social

	State	Acres		State	Acres
Constitución de 1857 (1)	B.C.	103.8	**El Sacromonte (23)**	Mex.	111.2
Balnerio los Novillos (2)	Coah.	103.8	**Lagunas de Zempoala (24)**	Mex., Mor.	11,537.4
Volcán Nevado de Colima (3)	Col., Jal.	54,363.2	**Zoquiapan y Anexas (25)**	Mex.	47,982.9
Lagunas de Montebello (4)	Chis.	14,880.7	**Lago de Camécuaro (26)**	Mich.	22.2
Palenque (5)	Chis.	4,376.2	**Cerro de Garnica (27)**	Mich.	2,392.0
Cañon del Sumidero (6)	Chis.	53,841.8	**Insurgente José María Morelos y Pávon (28)**	Mich.	4,480.0
Cascada de Basaseáchic (7)	Chih.	14,337.1	**Rayón (29)**	Mich.	61.8
Cumbres de Majalca (8)	Chih.	11,791.9	**Pico de Tancítaro (30)**	Mich.	72,441.4
Cumbres del Ajusco (9)	D.F.	2,273.4	**El Tepozteco (31)**	Mor., D.F.	59,305.3
Cerro de la Estrella (10)	D.F.	2,718.2	**Isla Isabel (32)**	Nay.	479.4
El Tepeyac (11)	D.F.	746.3	**Cumbres de Monterrey (33)**	N.L.	609,114.8
El Veladero (12)	Gro.	7,806.1	**El Sabinal (34)**	N.L.	19.8
El Chico (13)	Hgo.	6,768.2	**Benito Juárez (35)**	Oax.	6,763.3
Los Mármoles (14)	Hgo.	57,204.9	**Lagunas de Chacahua (36)**	Oax.	35,056.8
Tula (15)	Hgo.	244.6	**El Cimatario (37)**	Qro.	6,046.7
Bosencheve (16)	Mex., Mich.	37,065.8	**Tulum (38)**	Q.R.	1,640.8
Desierto del Carmen (17)	Mex.	1,307.2	**El Gogorrón (39)**	S.L.P.	61,776.3
Insurgente Miguel Hidalgo y Castilla (18)	Mex., D.F.	4,324.3	**El Potosí (40)**	S.L.P.	4,942.1
Iztaccíhuatl-Popocatépetl (19)	Mex., Mor., Pue.	63,454.2	**La Malinche (41)**	Tlax., Pue.	112,927.2
Molino de Flores Nezahuacóyotl (20)	Mex.	135.9	**Pico de Orizaba (42)**	Ver., Pue.	48,803.3
Nevado de Toluca (21)	Mex.	126,023.7	**Cañon del Río Blanco (43)**	Ver.	137,613.0
Los Remedios (22)	Mex.	988.4	**Dzibilchaltún (44)**	Yuc.	1,331.9

(Number in parentheses refers to map location.)

Figure 11.17 **Mexico**
Archaeological Sites and Museums by State, 1986

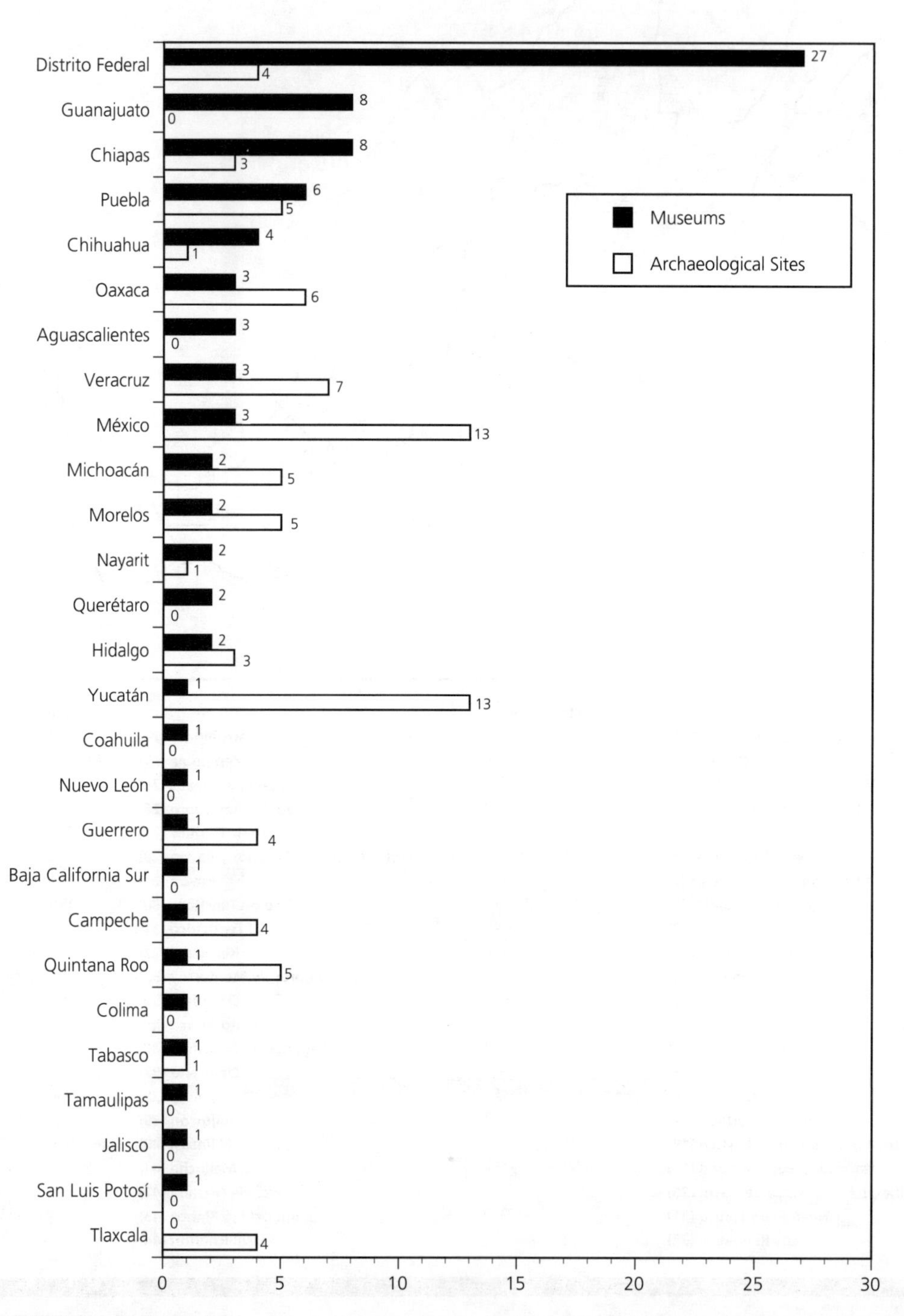

Source: INEGI, Anuario Estadistico

Figure 11.18 **Mexico**
Person-Visits to Museums, Archaeological Sites, and Historic Monuments by State, 1990 preliminary

Mexico Total: 14,545,700

	Thousands
Distrito Federal	8,441.7
México	1,436.9
Yucatán	822.7
Quintana Roo	667.9
Puebla	615.3
Oaxaca	561.9
Veracruz	383.9
Hidalgo	232.1
Morelos	209.5
Zacatecas	166.2
Chiapas	164.7
Guanajuato	119.9
Michoacán	97.8
Querétaro	84.0
Tlaxcala	79.9
Campeche	78.8
Tabasco	66.9
Guerrero	59.4
Nuevo León	47.3
Chihuahua	31.9
Colima	31.2
Coahuila	25.7
Aguascalientes	25.3
Jalisco	22.1
Baja California Sur	20.7
Nayarit	18.7
Sonora	16.1
San Luis Potosí	13.4
Tamaulipas	3.8

Source: INEGI, Anuario Estadistico

ENDANGERED SPECIES

The United States has identifie 162 species of plants, animals, and birds as endangered species. Canada has identified 23. Mexico has identified 194.

Figure 11.19 Canada, Mexico, and United States Percent of Total Known Species at Risk, 1990

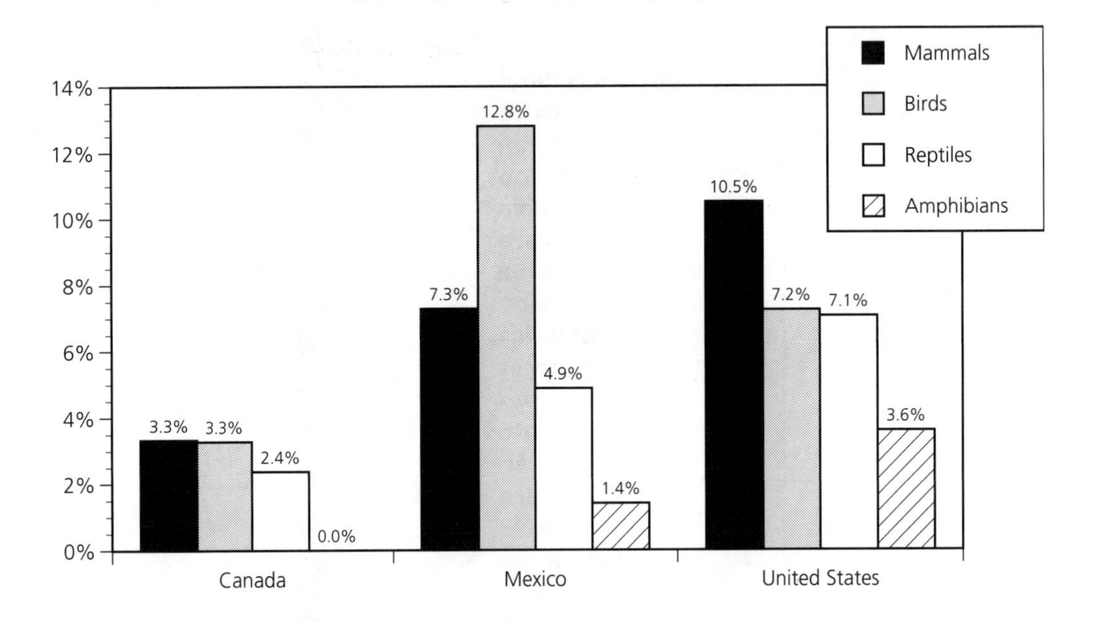

Source: U.N. Environment Programme

Emissions of Greenhouse Gases

North America is responsible for 21.5 percent of the world's emissions of greenhouse gases, which are thought to contribute to global warming. The vast majority of these emissions, 83 percent, are from the United States. Overall, the United States ranks sixth in the world in greenhouse gas emissions on a per capita basis. Canada ranks eighth. Mexico ranks sixty-second.

Figure 11.20 **Canada, Mexico, and United States**
Carbon Dioxide Emissions From Industrial Processes, 1989 and 1991

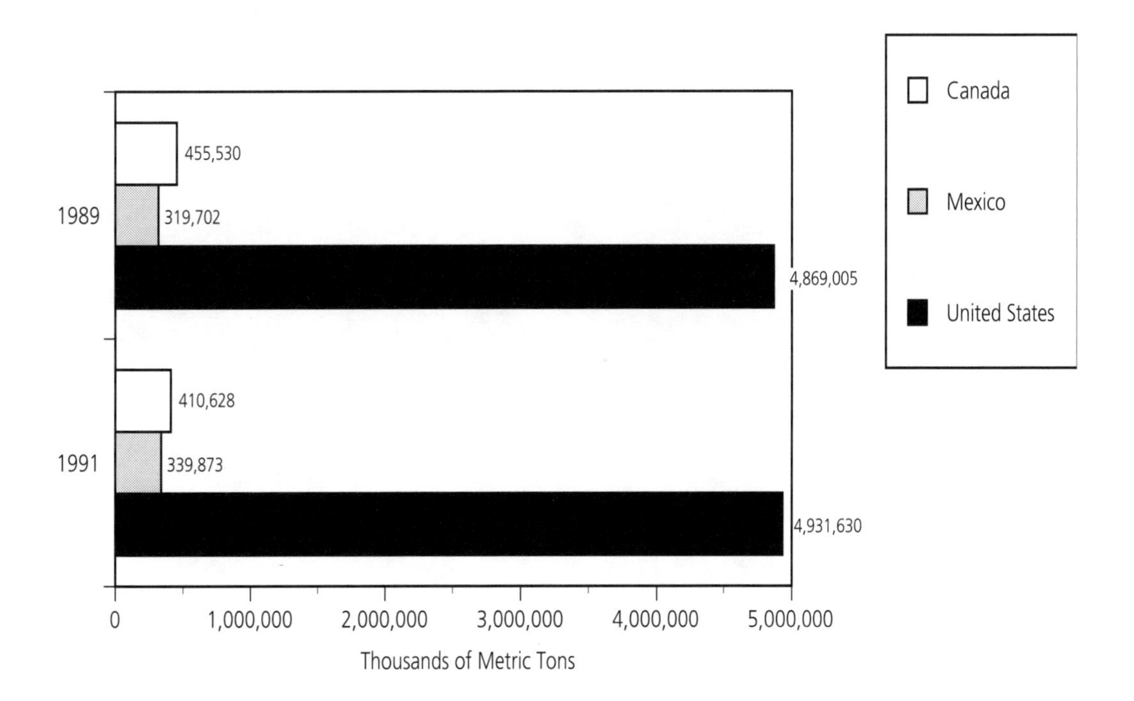

Source: Carbon Dioxide Information Analysis Center

Chapter Twelve
Economics, Finance, and Trade

Economic growth and public policies with regard to the economy in Canada and the United States are going one way, and Mexico another, but the courses are converging.

In the early 1990s, both Canada and the United States were experiencing sluggish economic growth and rising public debt. In Mexico, the economy was growing rapidly, and the federal budget deficits were coming down. In both Canada and the United States, the proportion of gross domestic product attributed to government has been growing, and government regulation of private business has been increasing. In Mexico, both have been shrinking.

The turnaround in Mexico's economic fortunes since 1986 is the most dramatic global success story since the German economic "miracle" of the Adenauer-Erhard years in the 1950s or since Mexico's economic miracle of the 1960s. In 1986, Mexico's own rate of economic growth was zero. Annual inflation was in the triple digits. The federal deficit approached 18 percent of gross domestic product, and foreign debt totaled more than 60 percent of gross domestic product. By 1991, the Mexican economy grew faster (4 percent) than its population (2.3 percent) for the first time in ten years. Inflation was reduced to 13 percent, the deficit was down to 1.1 percent of a much larger gross domestic product, and foreign debt had been cut by two-thirds, to 20 percent of gross domestic product.

Credit for the turnaround belongs primarily to Mexican President Carlos Salinas de Gortari , who assumed office in 1988, and to his predecessor Miguel de la Madrid, who began those changes in 1986. The dramatic liberalization of Mexico's economy included an extensive program of privatization and deregulation, and Salinas slashed both government spending and tax and tariff rates. Tariffs declined from an average of 100 percent to an average of about 10 percent.

Since the Revolution of 1910, Mexico has a mixed economy, with government ownership of the oil industry, many businesses, and collective farms. In 1960, the electric utility industry was nationalized, as were the banks in 1982. The thrust of "Salinastroika" has been to reverse these policies, turning Mexico into a democratic capitalist welfare state, more like Canada and the United States. President Salinas calls this new idea *liberalismo social* (social liberalism).

Mexico remains the most heavily regulated of the North American economies; the United States, the least—but the differences continue to shrink.

International trade is a substantial and rapidly rising source of new wealth and jobs in Canada, the United States, and Mexico. Their principal trading partners are each other.

All three North American nations have been experiencing wrenching political change. But whereas public confidence in the national government seems to be falling in Canada and in the United States, it appears to be increasing in Mexico, although factionalism, symbolized by the 1994 Chiapas uprising, has produced uncertainty about future stability.

Stark differences in economic power characterize North America. Canada's economy is twice the size of Mexico's, and California's economy is bigger than Canada's. The gross domestic product of Texas is greater than the gross domestic product of Mexico. The U.S. economy as a whole is more than seven times as large as the combined economies of Canada and Mexico.

Productivity

The United States has one of the world's most productive economies. In 1991, U.S. gross domestic product was the largest in the world. Canada ranked seventh that year. Mexico ranked eleventh.

Figure 12.1a **Largest Gross Domestic Product, 1991***

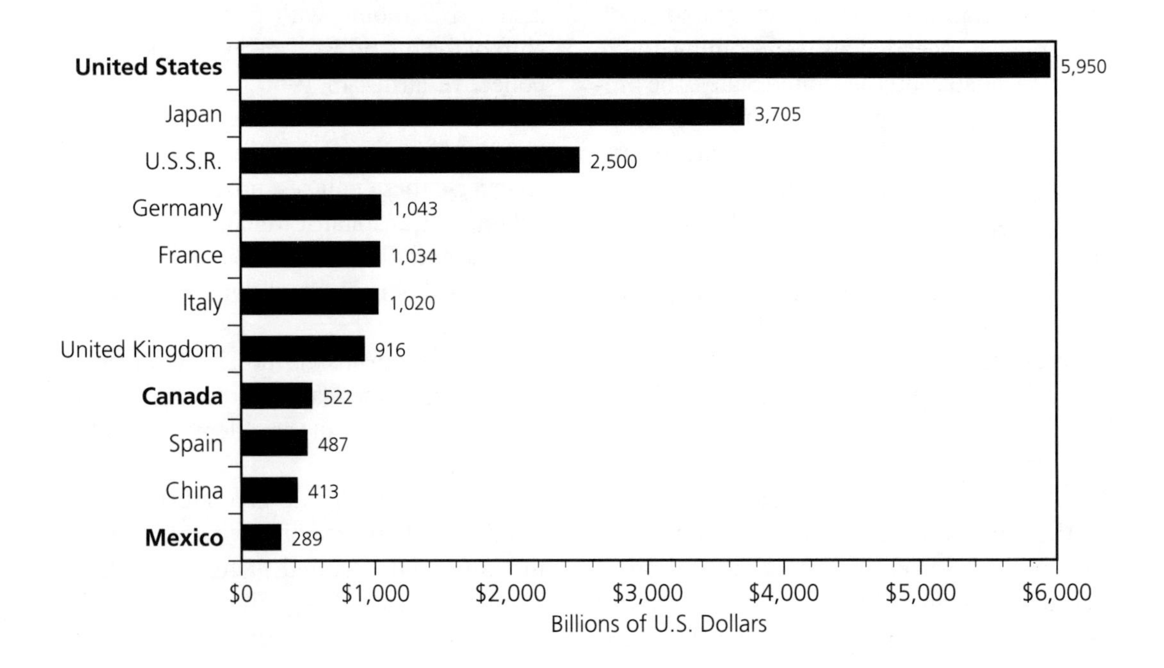

(* U.S.S.R. Information from 1988; China from 1989.)

Source: Information Please Almanac, 1994

Figure 12.1b **Largest Purchasing Power Parity per Capita, 1991**

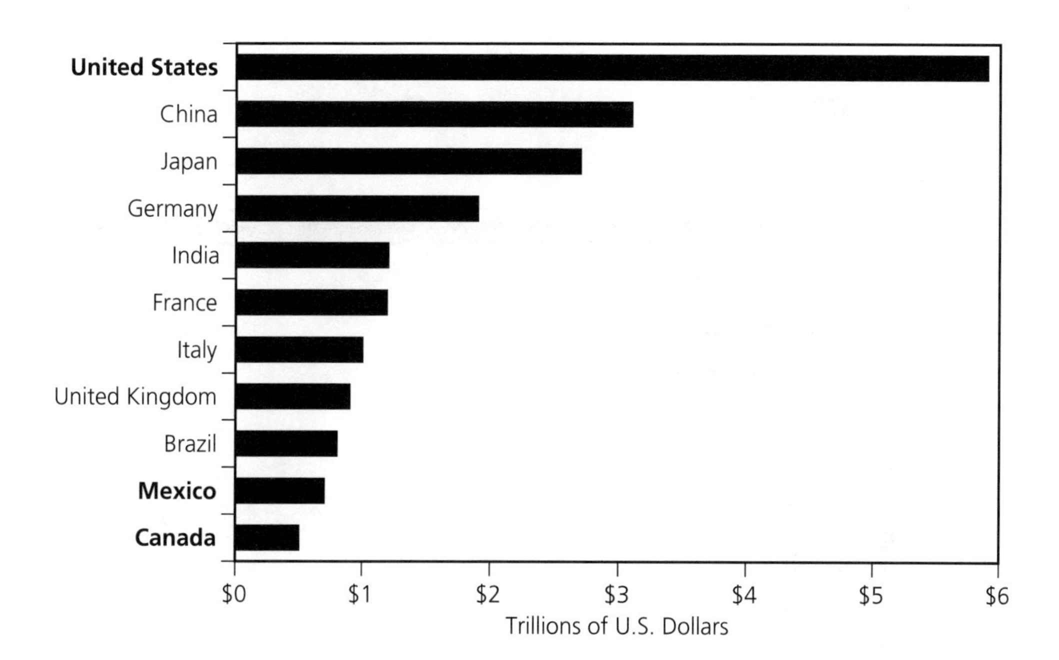

Source: United Nations, World Development Report

A Country's GDP generally is used as a measure of the size of its economy. However, GDP is computed in local currency that is then translated into market exchange rates. Thus GDP figures do not reflect barter, a very real component of many economies, nor do they reflect prices of items that are produced and consumed locally, such as food and housing ("non-traded goods"). These items generally comprise the greatest expenditures for individuals, but the tremendous discrepancies in their prices from country to country greatly distort relative income levels when GDP is used as a measuring tool.

The International Comparison Program (ICP) of the United Nations came up with purchasing power parities (PPPs) as a way to measure the relative standard of living from country to country, accounting for non-traded goods as well as for certain commodities that may have internal price supports. Purchasing power parities translate GDP into purchasing power in that particular economy. The result is a huge jump in the relative income of developing nations.

Figure 12.2a **Western United States**
Gross State Product by State, 1989 (excluding California)

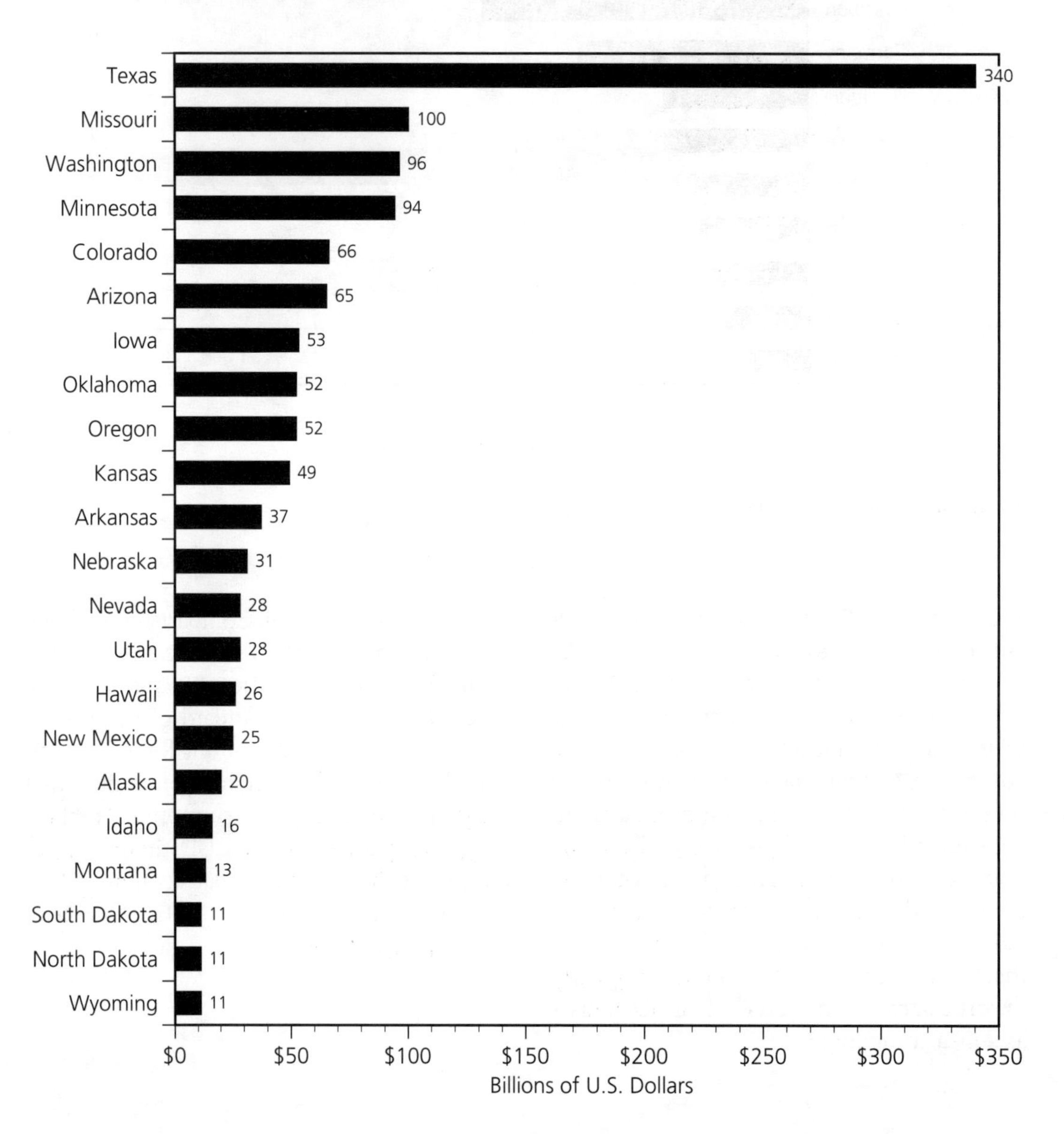

Source: U.S. Bureau of Economic Analysis

People in the western United States produced more goods and services than all the people in Canada and Mexico combined. And California and Texas account for nearly 84 percent of the region's production.

Figure 12.2b **Western United States Gross State Product, 1989**

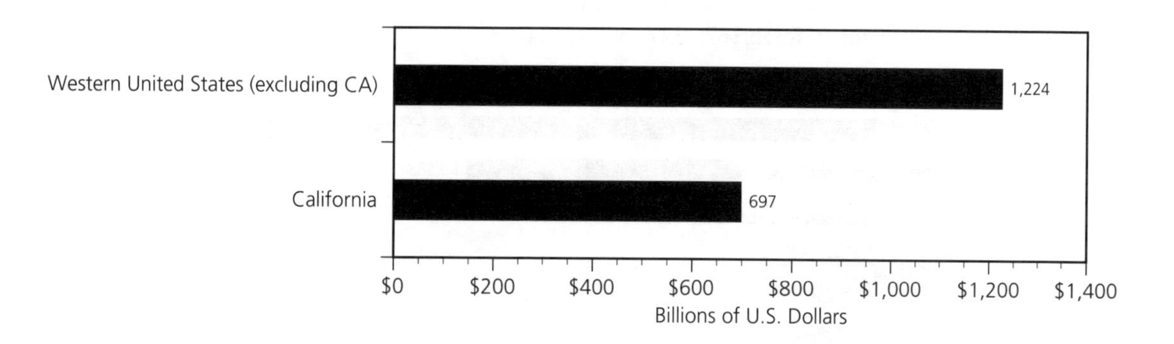

Source: U.S. Bureau of Economic Analysis

The province of Ontario accounts for more of Canada's gross domestic product than all the western provinces and territories combined. British Columbia leads western Canada in gross domestic product, but Alberta leads all of Canada in gross domestic product per capita.

Figure 12.3a **Canada**
Gross Domestic Product by Province, 1991

West: $201,167 million, Canadian
East: $474,160 million, Canadian
Canada Total: $675,327 million, Canadian

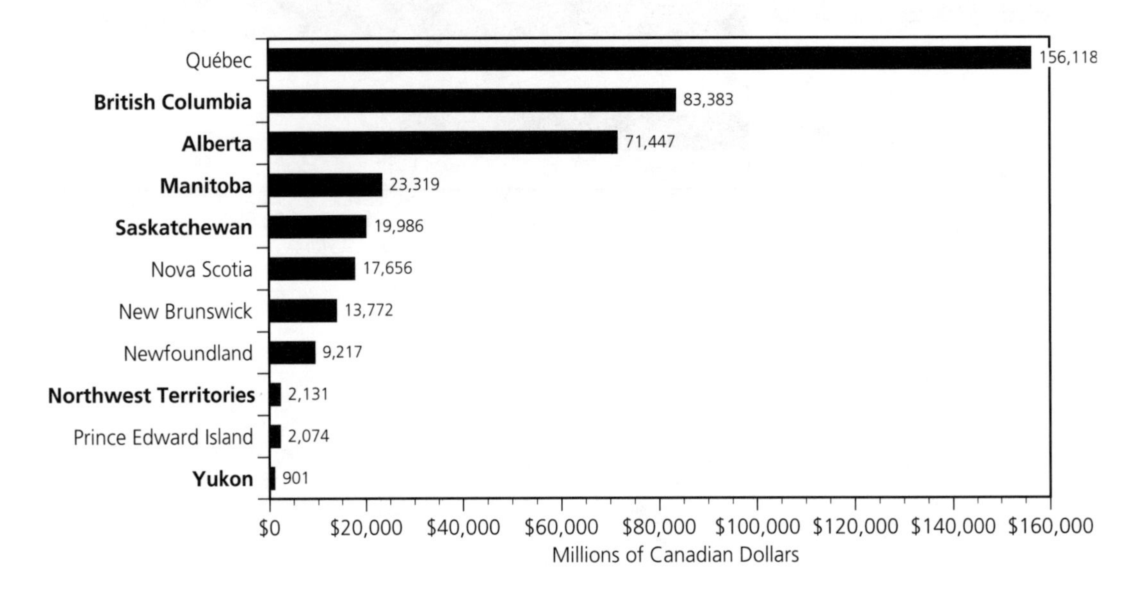

Source: Statistics Canada

Figure 12.3b **Canada**
Gross Domestic Product, 1991

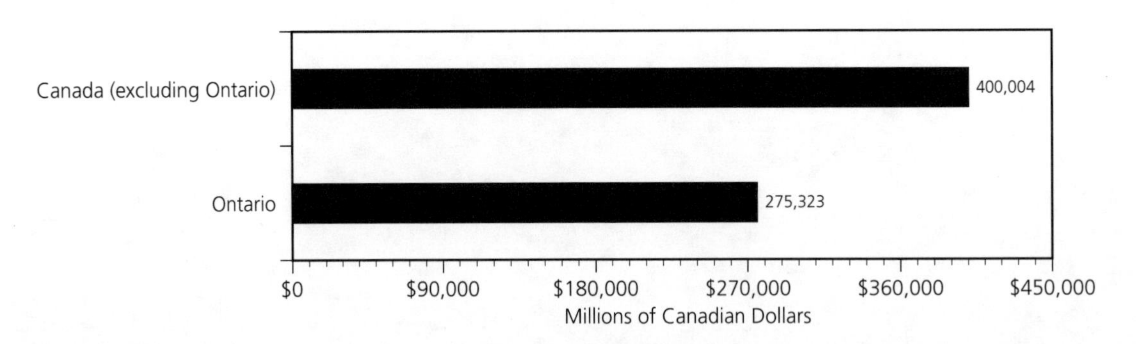

Source: Statistics Canada

Mexican gross domestic product is concentrated in Mexico City and the state surrounding it, (Mexico) as well as in the states of Nuevo León and Jalisco.

Figure 12.4 Mexico
Gross National Product Participation Rate, 1990

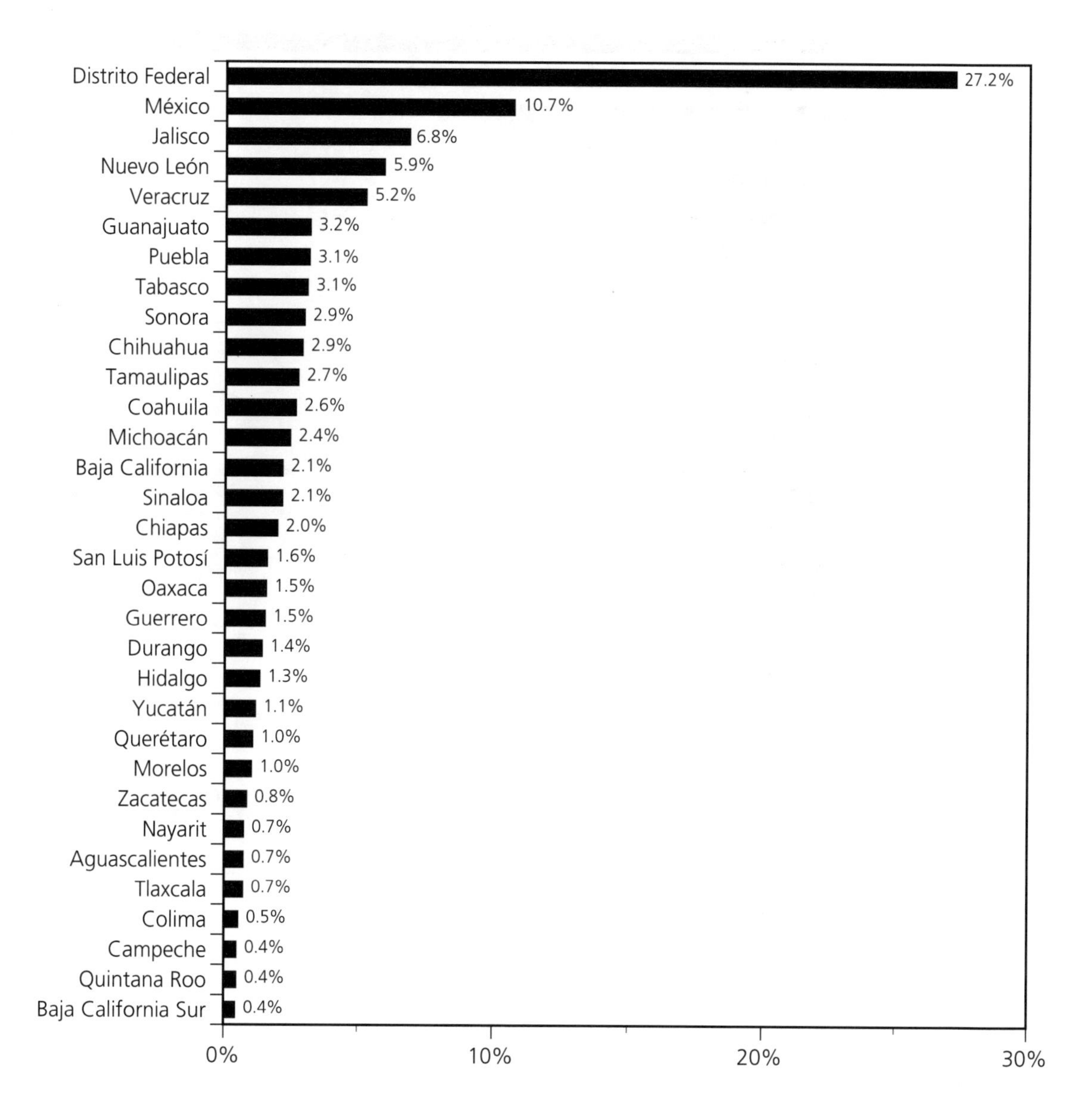

Source: Banamex, Mexico Today

Figure 12.5a **Mexico**
Gross National Product by State, 1990

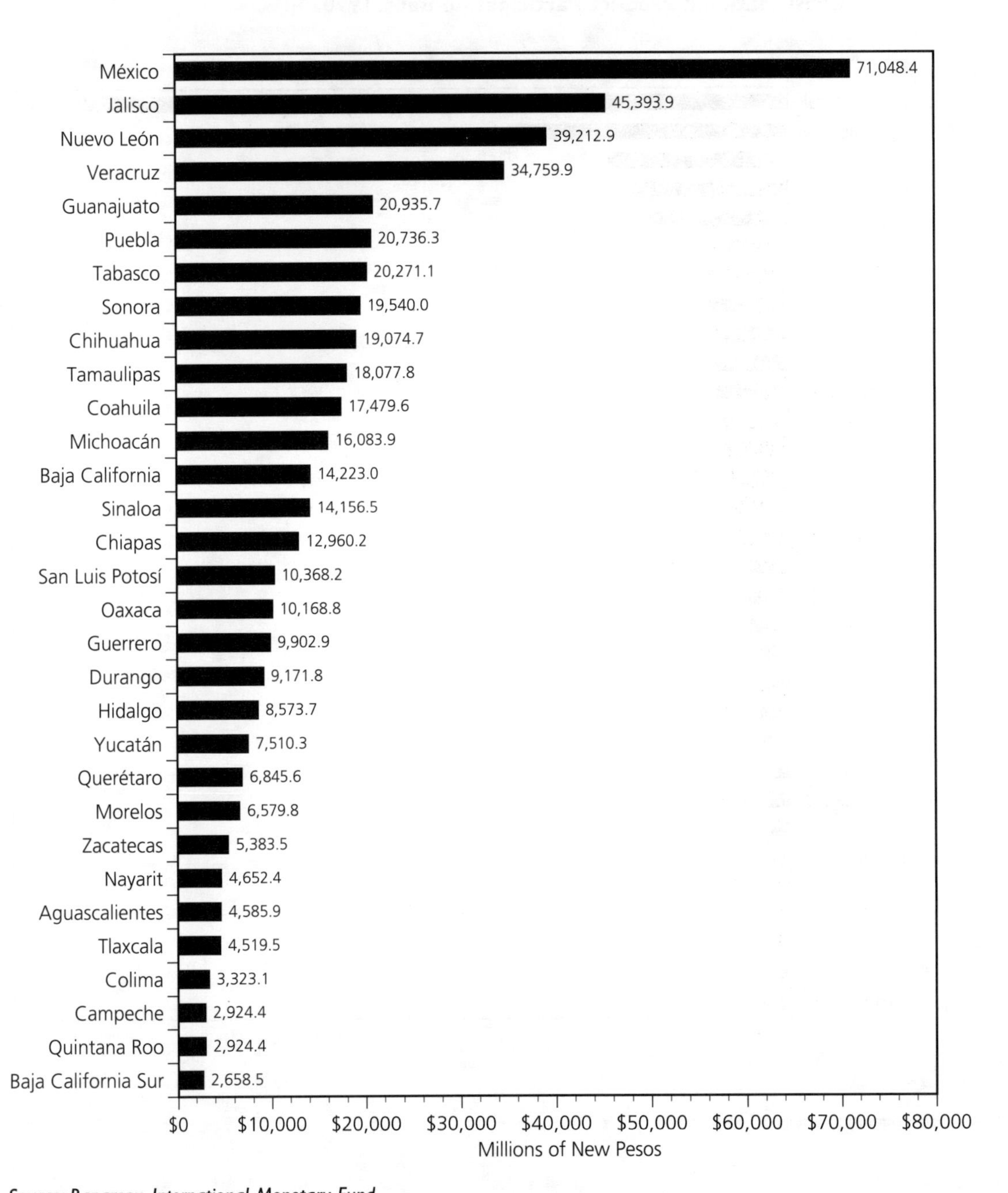

Source: Banamex, International Monetary Fund

Figure 12.5b **Mexico Gross National Product, 1990**

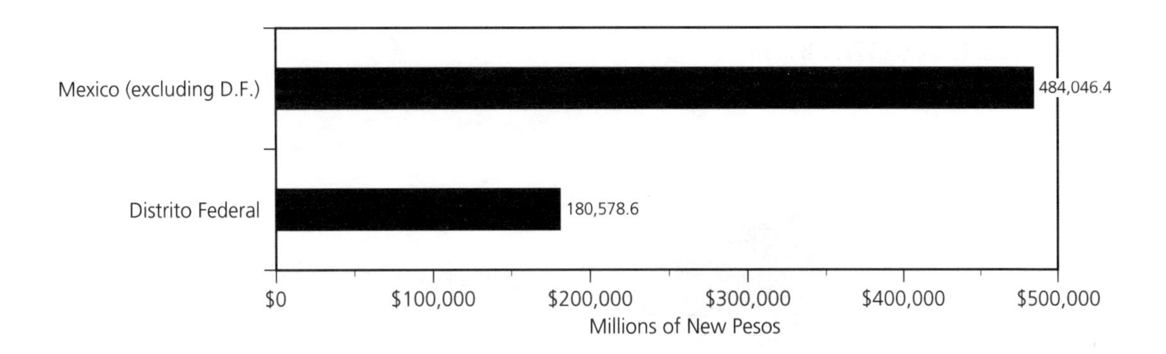

Source: Banamex, International Monetary Fund

The U.S. economy is the largest in the world for many reasons: large population, stable political system, extensive natural resources, entrepreneurial culture, limited government, and the world's highest worker productivity. In 1991, per capita gross domestic product in the United States was $22,560. Canada's per capita gross domestic product was $19,500. Mexico, with a per capita gross domestic product of $2,680, ranked forty-fifth that year. The U.S. Bureau of Labor Statistics has a productivity index based on gross domestic product per employed person. On that index, the United States ranks first, at 100. Canada is second, with a productivity of 92.9 percent of the U.S. workers. Mexican workers were not ranked in the Bureau of Labor Statistics survey. Private business measurements estimate that current productivity of Mexican workers at about one-third that of U.S. workers.

Government spending accounts for a smaller proportion of U.S. gross domestic product than it does of either Canadian or Mexican gross domestic product.

Figure 12.6 **United States Gross Domestic Product, Third Quarter 1992 (Annualized figure)**

U.S. Total: **$5,978.5** billion

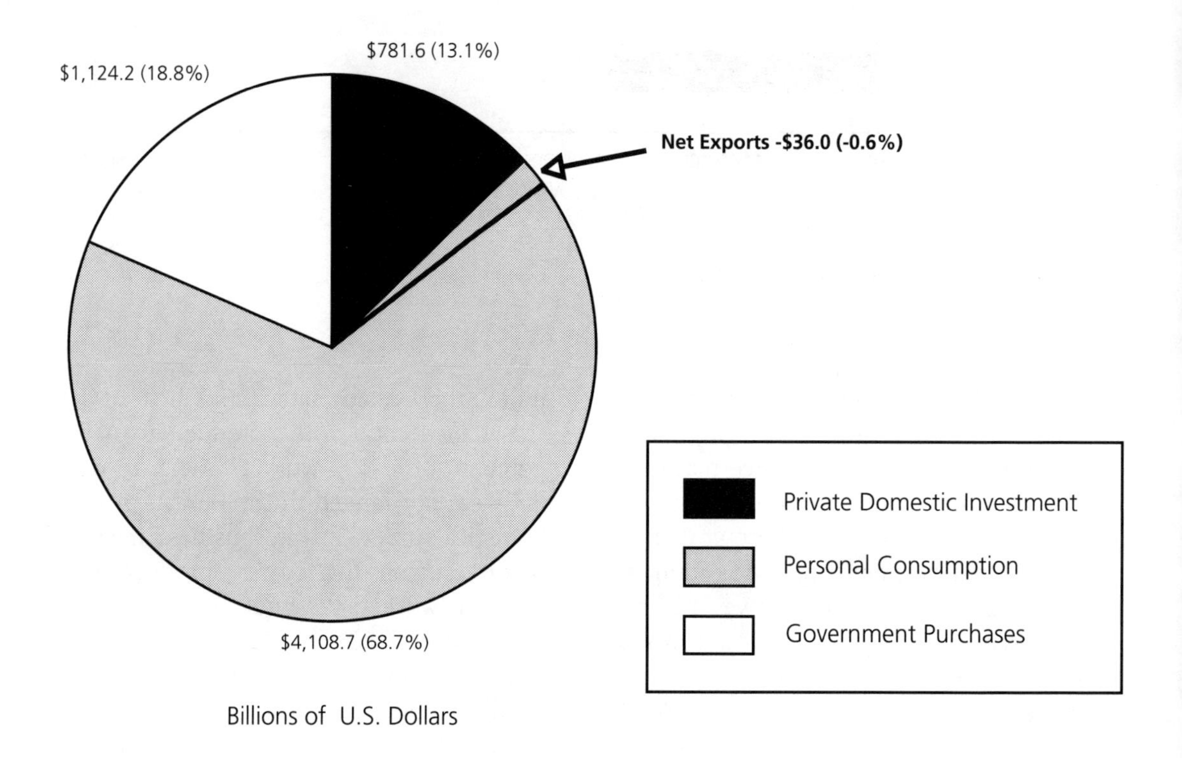

Source: Economic Report of the President, 1993

Figure 12.7 **Canada**
Gross Domestic Product, 1991 preliminary

Canada Total: **$676,251** million, Canadian

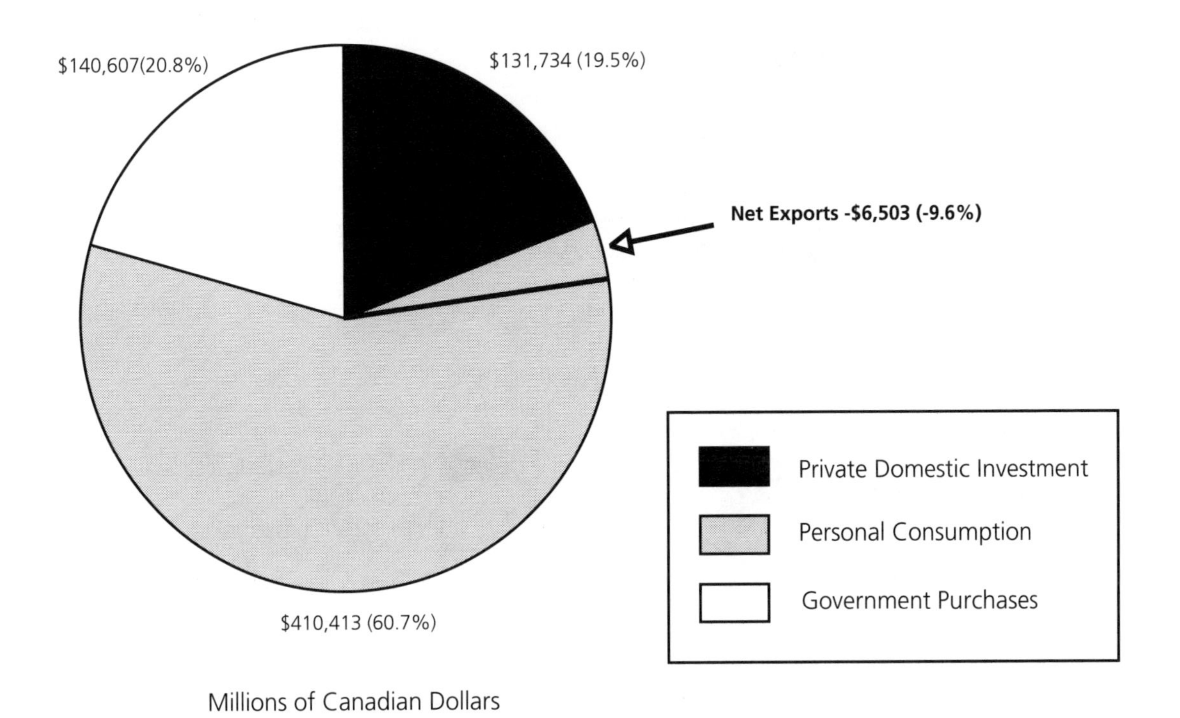

Millions of Canadian Dollars

Source: Europa Yearbook

Figure 12.8 **Mexico**
Gross Domestic Product, 1991

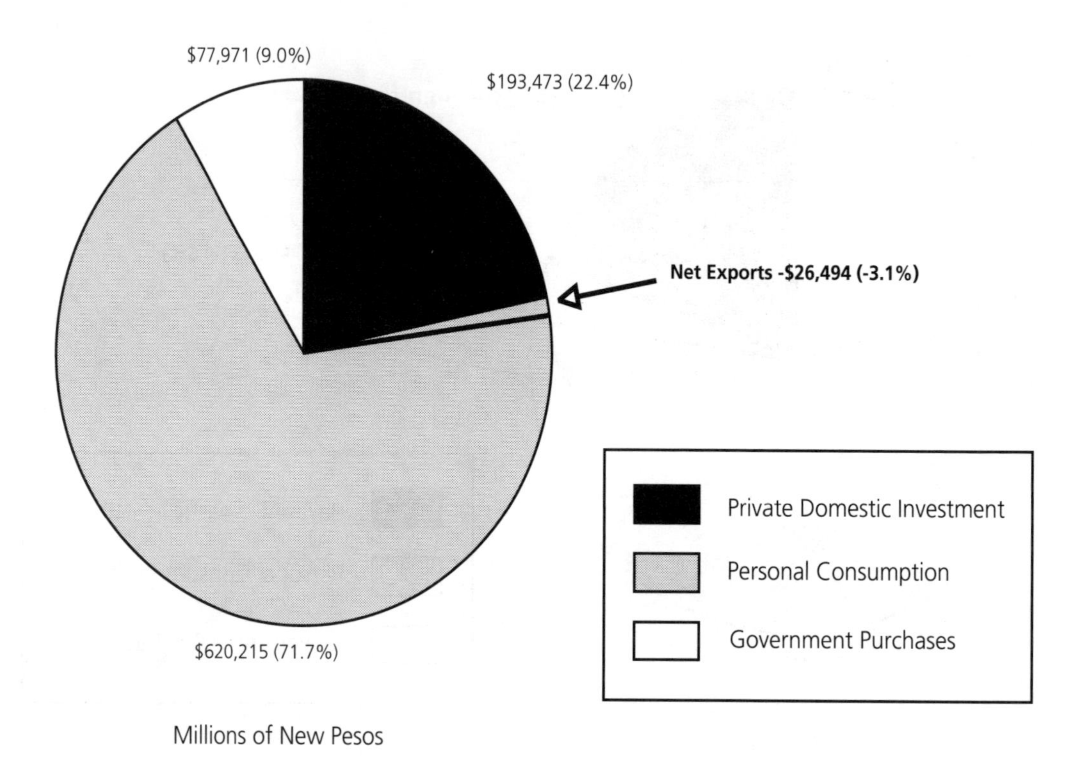

Source: International Monetary Fund, National Product Accounts

Economic and Productivity Growth

After rapid growth in the 1980s, the growth of both the U.S. and Canadian economies slowed in the 1990s. Mexican growth since 1986 has been spectacular, similar to that experienced in Mexico between 1940 and 1980, when Mexico's annual growth averaged about 6 percent.

Figure 12.9a **United States**
Gross Domestic Product, 1960–1992 (through third quarter)

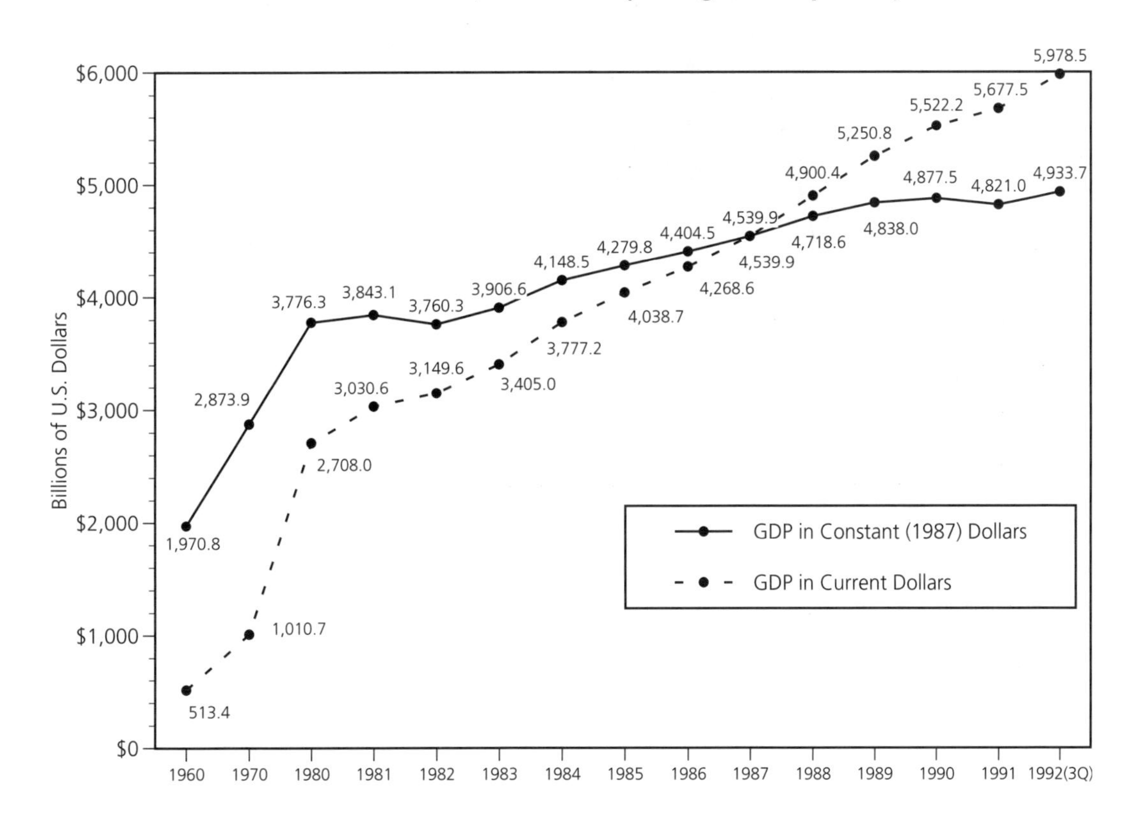

Source: Economic Report of the President, 1993

Figure 12.9b **United States**
Percent Change in Gross Domestic Product, 1980–Third Quarter 1992

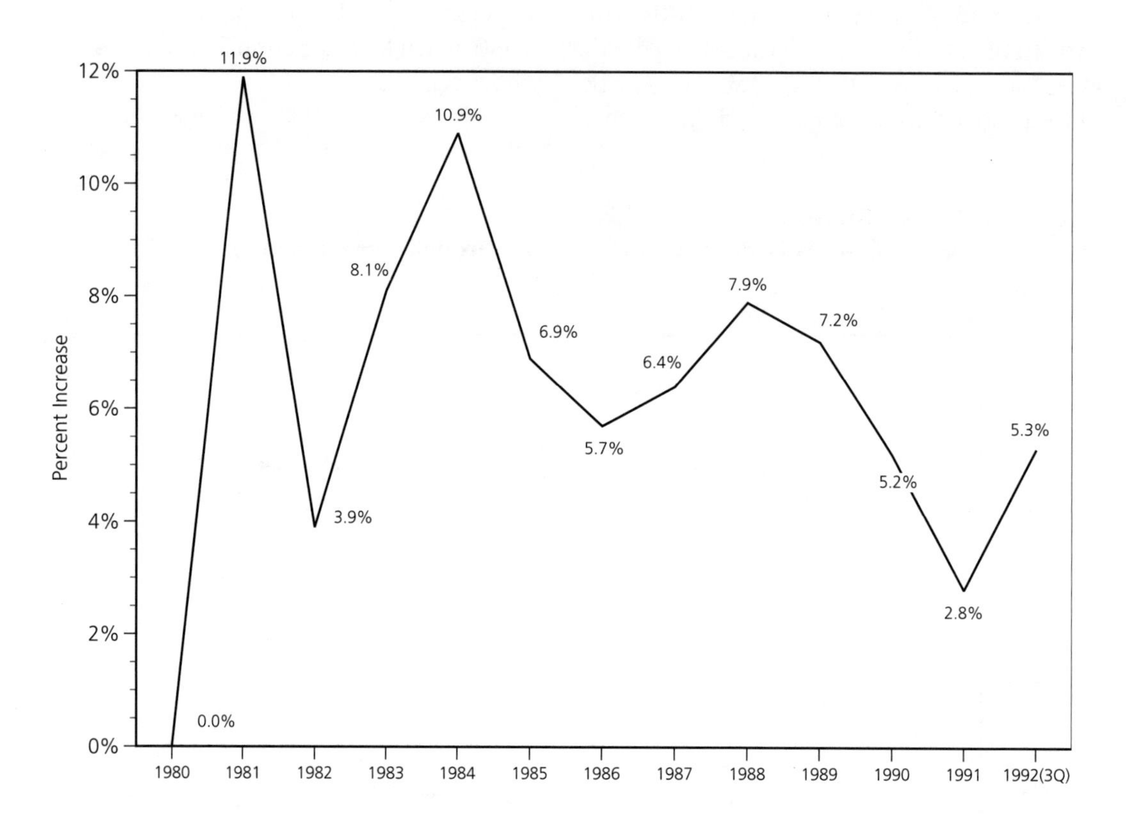

Source: Economic Report of the President, 1993

Figure 12.10a **Canada**
Gross Domestic Product, 1972–1992

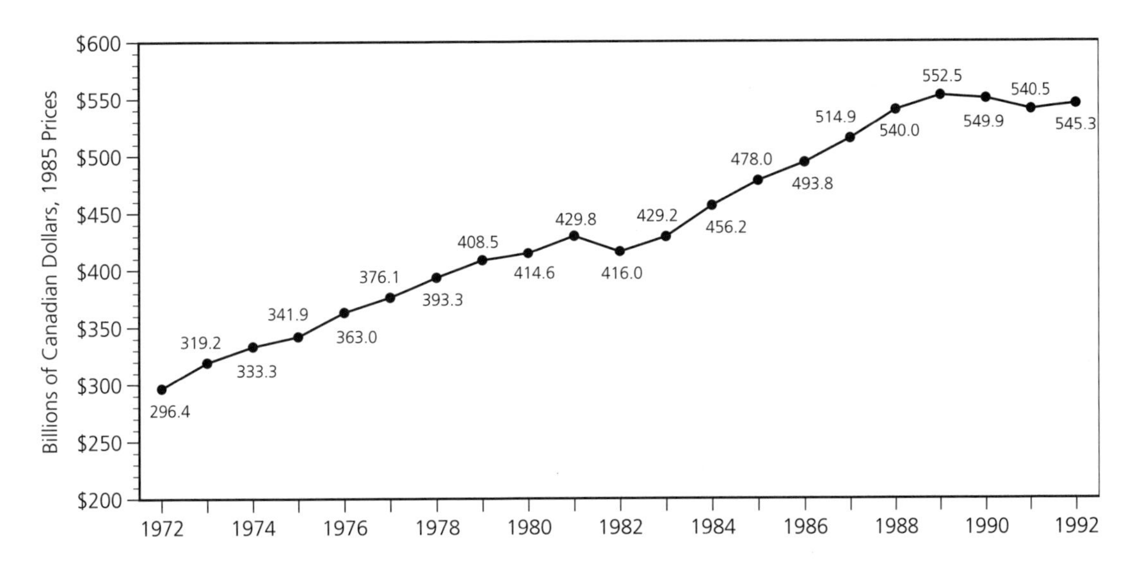

Source: International Monetary Fund, National Product Accounts

Figure 12.10b **Canada**
Percent Change in Gross Domestic Product, 1972–1992

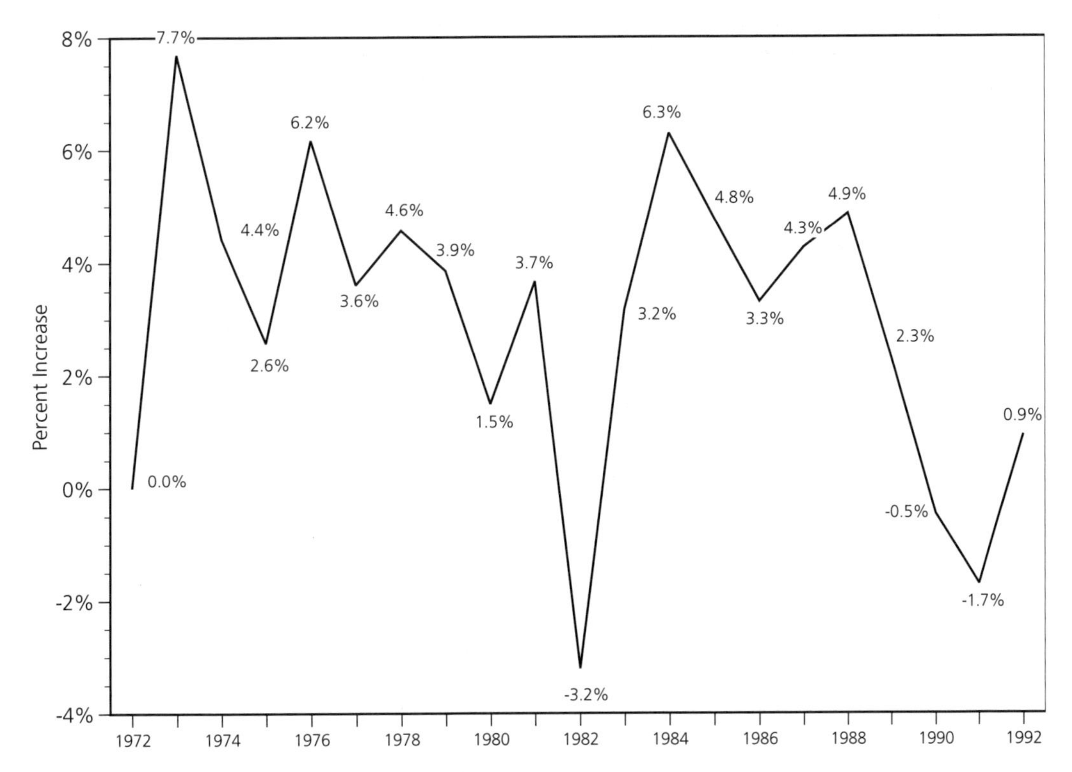

Source: International Monetary Fund, National Product Accounts

Figure 12.11a **Mexico**
Gross Domestic Product, 1977–1991

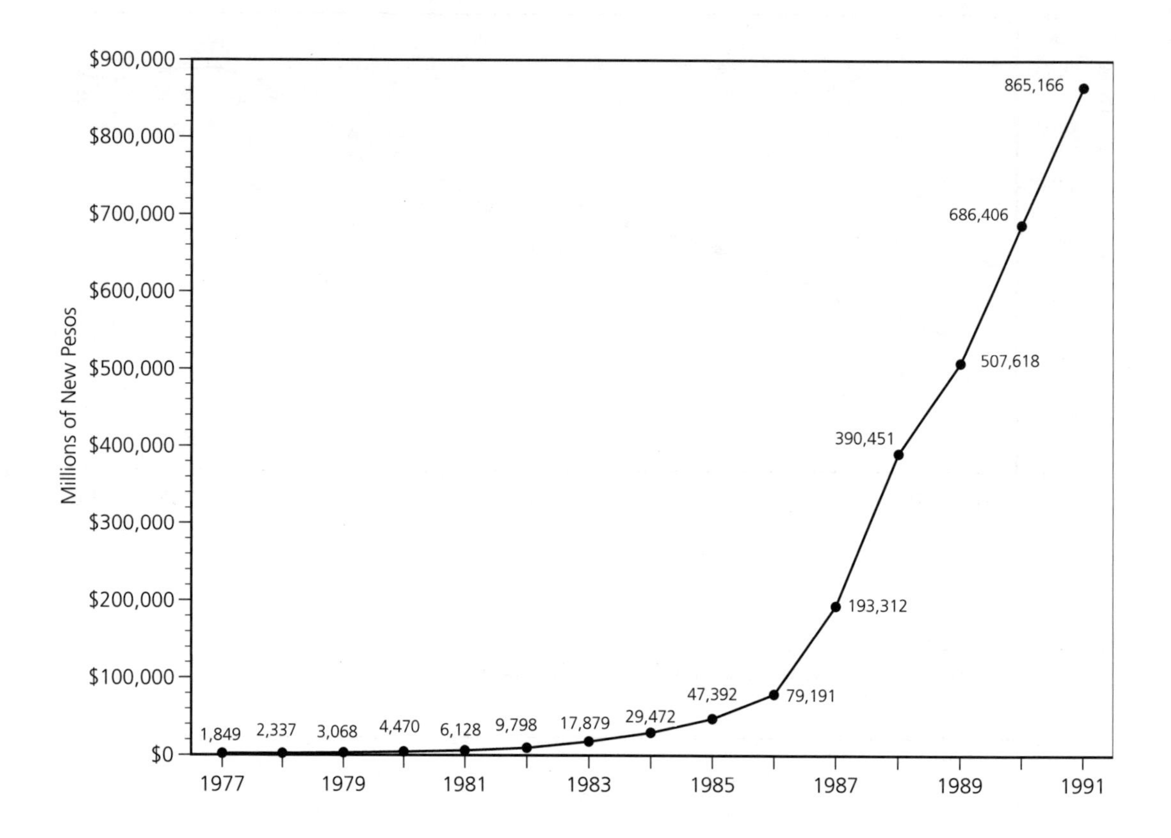

Source: International Monetary Fund

Figure 12.11b **Mexico**
Percent Change in Gross Domestic Product, 1977–1991

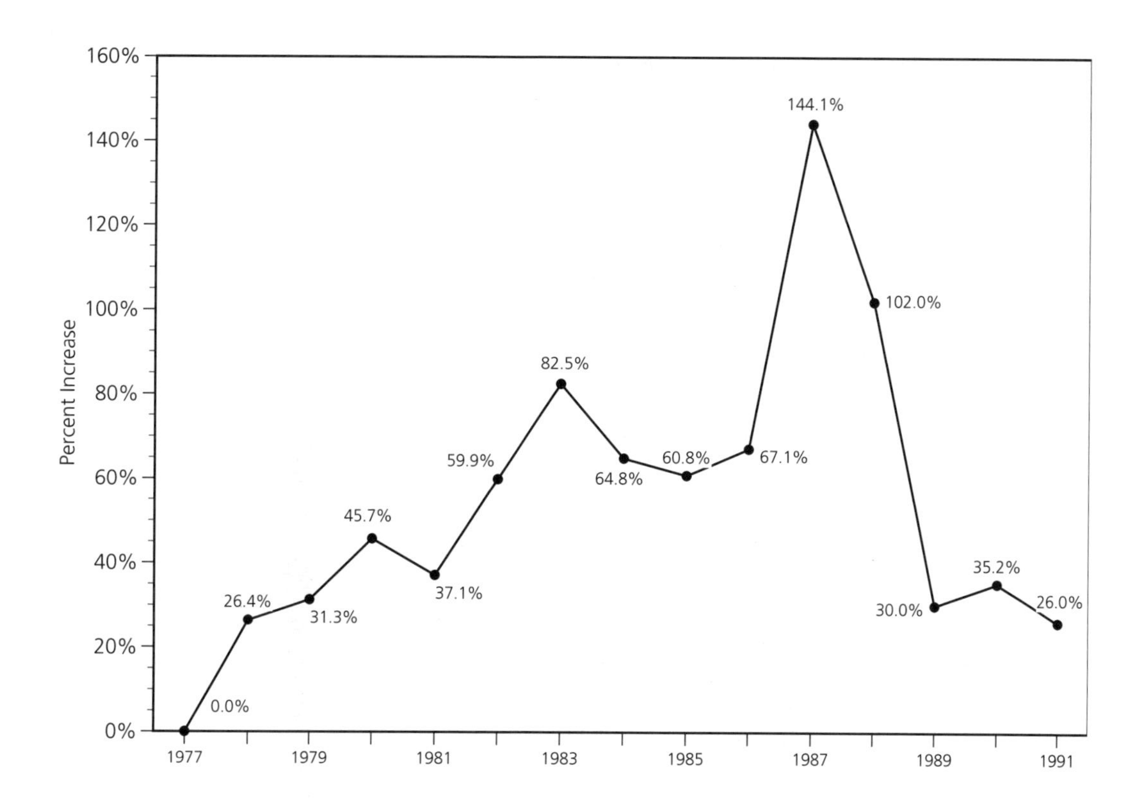

Source: International Monetary Fund

INCOME

In 1991, the U.S. per capita income was $19,082. Canadians that year had a per capita income of $19,663. Per capita income in Mexico was $2,373.

Alaska, California, and Hawaii led the western United States in per capita and household incomes. The region's lowest incomes were in Arkansas and Montana.

Personal incomes in Canada tend to be higher in the West than in the East. Family incomes are greatest in Ontario.

Figure 12.12 **Western United States Per Capita Personal Income by State, 1980–1990**

(Current Dollars)

	1980	1990	Percent Change
Alaska	$13,835	$21,646	56.5%
California	$11,603	$20,689	78.3%
Hawaii	$10,617	$20,361	91.8%
Nevada	$11,421	$19,049	66.8%
Colorado	$10,598	$18,860	78.0%
Washington	$10,725	$18,777	75.1%
Minnesota	$10,062	$18,731	86.2%
Kansas	$9,941	$18,104	82.1%
Nebraska	$9,274	$17,490	88.6%
Missouri	$9,298	$17,479	88.0%
Iowa	$9,537	$17,301	81.4%
Oregon	$9,866	$17,182	74.2%
Texas	$9,798	$16,717	70.6%
Wyoming	$11,339	$16,283	43.6%
Arizona	$9,172	$16,006	74.5%
South Dakota	$8,217	$15,890	93.4%
Oklahoma	$9,393	$15,451	64.5%
North Dakota	$8,538	$15,355	79.8%
Montana	$8,924	$15,304	71.5%
Idaho	$8,569	$15,250	78.0%
New Mexico	$8,169	$14,254	74.5%
Arkansas	$7,465	$14,176	89.9%
Utah	$7,952	$13,985	75.9%
U.S. Average	**$9,919**	**$18,696**	**88.5%**

Source: U.S. Bureau of Economic Analysis

Figure 12.13 **Western United States**
Median Household Income by State in Constant 1987 Dollars, 1990

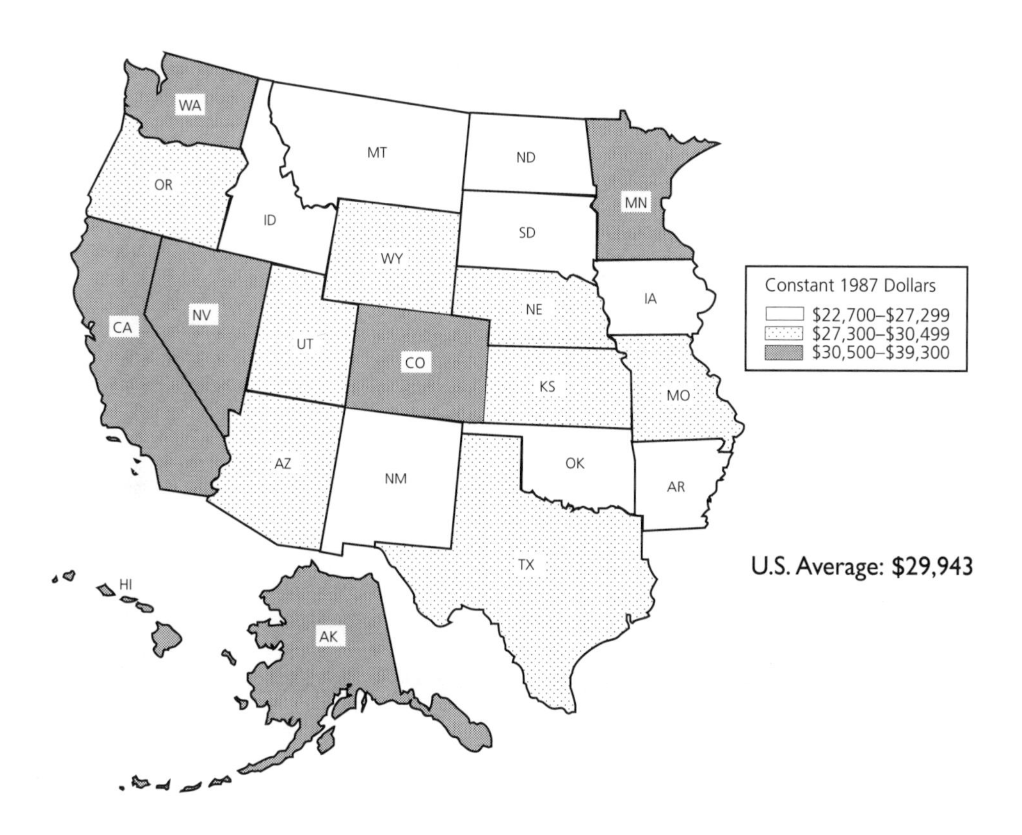

Source: U.S. Bureau of the Census

	Constant 1987 Dollars		Constant 1987 Dollars
Alaska	$39,298	**Texas**	$28,228
Hawaii	$38,921	**Nebraska**	$27,482
California	$33,290	**Missouri**	$27,332
Washington	$32,112	**Iowa**	$27,288
Nevada	$32,023	**Idaho**	$25,305
Minnesota	$31,465	**North Dakota**	$25,264
Colorado	$30,733	**New Mexico**	$25,039
Utah	$30,142	**South Dakota**	$24,571
Kansas	$29,917	**Oklahoma**	$24,384
Wyoming	$29,460	**Montana**	$23,375
Oregon	$29,281	**Arkansas**	$22,786
Arizona	$29,224		

Figure 12.14 **Canada**
Per Capita Personal Income by Province, 1981–1991

(Current U.S. Dollars)

	1981	**1991**	**Percent Change**
Northwest Territories	$8,997	$23,061	156.3%
Yukon	$12,474	$22,614	81.3%
Ontario	$10,455	$22,156	111.9%
British Columbia	$11,362	$20,031	76.3%
Alberta	$11,821	$19,615	65.9%
Québec	$9,330	$18,318	96.3%
Manitoba	$9,117	$16,820	84.5%
Nova Scotia	$7,912	$16,217	105.0%
Saskatchewan	$9,820	$15,658	59.5%
New Brunswick	$7,459	$15,516	108.0%
Prince Edward Island	$7,251	$14,737	103.2%
Newfoundland	$6,881	$14,434	109.8%
Canada Average	**$10,029**	**$19,663**	**96.1%**

Source: Statistics Canada

Figure 12.15 **Canada**
Average Family Income, 1991*

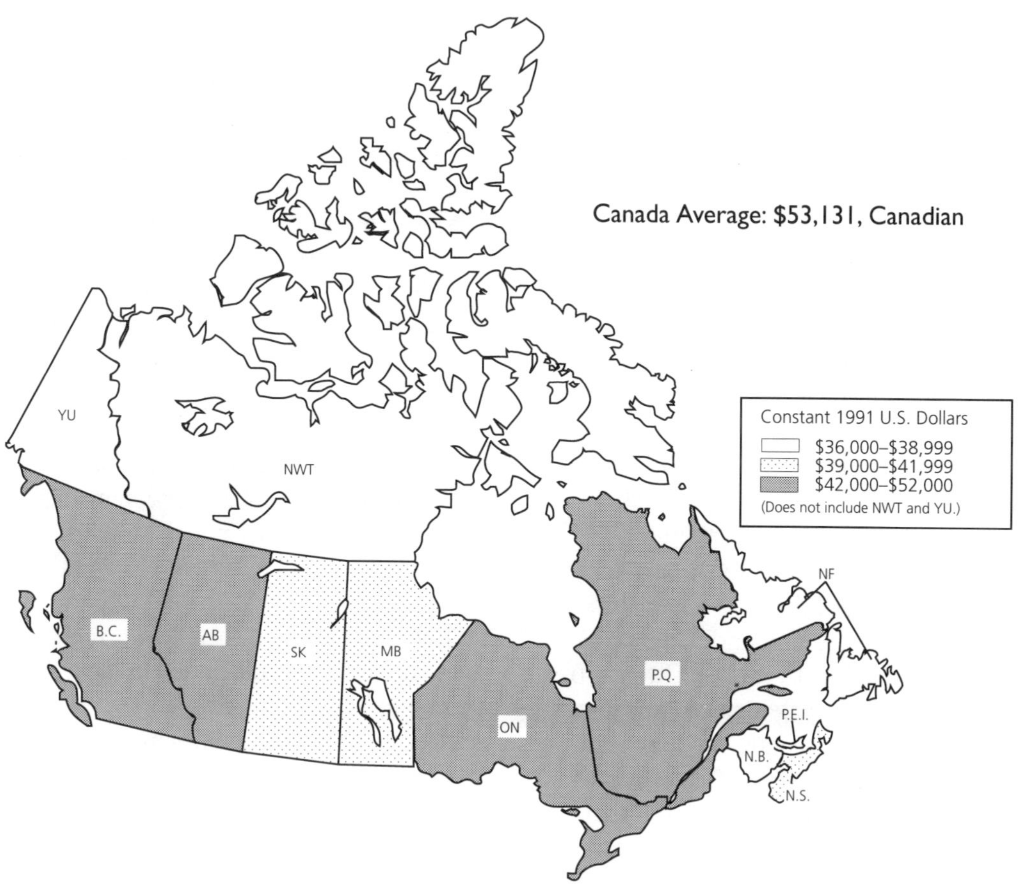

(* Constant 1991 Canadian dollars were converted to U.S. dollars using the average exchange rate for 1991.)

Source: Statistics Canada, Canada Yearbook

	Constant 1991 U.S. Dollars
Ontario	$51,164
Alberta	$48,475
British Columbia	$47,901
Québec	$42,438
Manitoba	$40,682
Saskatchewan	$40,079
Nova Scotia	$39,380
New Brunswick	$38,676
Prince Edward Island	$37,329
Newfoundland	$36,347

Income differences among Mexican states are far more profound than between states in the western United States or Canadian provinces. Generally speaking, incomes are greatest in Mexico City, its surrounding state, and in the states bordering the United States. Mexican income is lowest in its deep South.

The Minimum Salary (M.S.) varies depending on geographic area. The M.S. averaged N$9.14 (new pesos) per day between January 1 and November 15. Between November 15 and December 31, the average per day was N$10.79. Minimums for the geographic areas between November 15 and December 31 were: Area A, 11.90; Area B, 11.0 ; and Area C, 9.92. The Areas contain the following municipalities:

Area A: Baja California (all); Baja California Sur (all); Chihuahua (Guadalupe, Juárez, Praxedis G., and Guerrero); Distrito Federal (all); Guerrero (Acapulco de Juárez); México (Atizapan de Zaragoza, Coacalco, Cuatitlán, Cuatitlán Izcalli, Ecatepec, Naucalpan de Juárez, Tlalnepantla de Baz, and Tultitlán); Sonora (Agua Prieta, Cananea, Naco, Nogales, Plutarco Elias Calles, Puerto Peñasco, San Luis Río Colorado, and Santa Cruz); Tamaulipas (Camargo, Guerrero, Gustavo Díaz Ordaz, Matamoros, Mier, Miguel Alemán, Nuevo Laredo, Reynosa, Rio Bravo, San Fernando, and Valle Hermoso); and Veracruz (Agua Dulce, Coatzacoalcos, Cosoleacaque, Las Choapas, Ixhuatlán del Sureste, Minatitlán, Moloacán, Nznchital de Lázaro, and Cárdenas del Rio).

Area B: Jalisco (Guadalajara, El Salto, Tlajomulco, Tlaquepaque, Tonala, and Zapopan); Nuevo León (Apodaca, Garza Garcia, General Escobedo, Guadalupe, Monterrey, San Nicolás de los Garza, and Santa Cantarina); Sonora (Altar, Atil, Bácum, Benjamin Hill, Caborca, Cajeme, Carbó, La Colrada, Cucurpe, Empalme, Etchojoa, Guaymas, Hermosillo, Imuris, Magdalena, Navojoa, Opodepe, Oquitoa, Pitiquito, San Miguel de Horcasitas, Santa Ana, Sáric, Suaqui Grande, Trincheras, Tubutama, and Huatabampo); Tamaulipas (Aldama, Altamira, Antiguo Morelos, Cd. Madero, Gómez Farias, González, Mante, Nuevo Morelos, Ocampo, Tampico, and Xicoténcatl); and Veracruz (Coatzintla, Poza Rica de Hidalgo, and Tuxpan).

Area C: All municipalities in the following: Aguascalientes, Campeche, Coahuila, Colima, Chiapas, Durango, Guanajuato, Hidalgo, Michoacán, Morelos, Nayarit, Oaxaca, Puebla, Querétaro, Quintana Roo, San Luis Potosí, Sinaloa, Tabasco, Tlaxcala, Yucatán, and Zacatecas, and all municipalities not in A or B in Chihuahua, Guerrero, Jalisco, México, Nuevo León, Sonora, Tamaulipas, and Veracruz.

Figure 12.16 **Mexico**

Distribution by Salary Level among Employed Population by State, 1990

(Based on 1 Minimum Salary (M. S.); see page 244 for Minimum Salary by geographic location.)

	No Income	Less than 1 M.S.	1 to 2 M.S.	More than 2 up to 5 M.S.	More than 5 M.S.
Aguascalientes	4.2%	14.8%	45.6%	27.8%	7.6%
Baja California	1.4%	8.5%	32.1%	43.4%	14.6%
Baja California Sur	2.8%	13.4%	40.1%	35.5%	8.3%
Campeche	8.9%	26.3%	36.8%	22.4%	5.6%
Chiapas	19.8%	41.6%	22.1%	12.7%	3.7%
Chihuahua	5.8%	9.7%	40.1%	32.9%	11.5%
Coahuila	2.6%	16.4%	44.2%	28.5%	8.4%
Colima	3.8%	10.9%	37.4%	38.4%	9.4%
Distrito Federal	1.1%	19.5%	41.8%	27.2%	10.4%
Durango	12.1%	18.4%	39.8%	23.6%	6.0%
Guanajuato	8.4%	18.6%	38.0%	27.0%	8.0%
Guerrero	15.7%	24.8%	31.9%	22.8%	4.8%
Hidalgo	9.4%	31.8%	36.1%	18.4%	4.3%
Jalisco	5.6%	14.3%	37.9%	32.4%	9.7%
México	3.8%	16.8%	44.4%	26.6%	8.4%
Michoacán	12.9%	18.7%	34.4%	26.4%	7.5%
Morelos	5.3%	13.2%	42.9%	30.6%	7.9%
Nayarit	9.1%	13.8%	34.0%	35.5%	7.6%
Nuevo León	2.2%	13.9%	44.7%	27.9%	11.2%
Oaxaca	25.9%	29.5%	26.8%	14.5%	3.3%
Puebla	13.1%	27.0%	35.2%	19.1%	5.6%
Querétaro	8.0%	16.6%	38.2%	27.5%	9.8%
Quintana Roo	8.3%	15.1%	29.6%	35.9%	11.0%
San Luis Potosí	11.5%	26.7%	37.0%	19.2%	5.7%
Sinaloa	4.8%	11.3%	42.7%	32.2%	9.0%
Sonora	1.9%	10.4%	42.5%	34.0%	11.1%
Tabasco	11.4%	26.9%	30.9%	24.0%	6.7%
Tamaulipas	3.6%	20.6%	39.7%	29.1%	7.0%
Tlaxcala	9.3%	23.4%	42.1%	20.9%	4.4%
Veracruz	10.6%	27.1%	36.7%	20.6%	5.0%
Yucatán	6.2%	33.7%	35.9%	19.4%	4.9%
Zacatecas	18.6%	22.0%	36.1%	18.5%	4.8%
Mexico Average	**7.5%**	**20.2%**	**38.4%**	**26.0%**	**7.9%**

(Does not include responses of "not specific.")

Source: INEGI

Income Growth

In the western United States between 1980 and 1990, real per capita income grew the most in Hawaii, Arkansas, and Nebraska. It fell in Wyoming and Alaska.

Figure 12.17 **Western United States Per Capita Personal Income by State, 1980–1990**

(Constant 1987 U.S. Dollars)

	1980	1990	Percent Change
Alaska	$19,056	$18,823	-1.2%
California	$15,982	$17,990	12.6%
Hawaii	$14,624	$17,705	21.1%
Nevada	$15,731	$16,564	5.3%
Colorado	$14,598	$16,400	12.3%
Washington	$14,773	$16,328	10.5%
Minnesota	$13,860	$16,288	17.5%
Kansas	$13,693	$15,743	15.0%
Nebraska	$12,774	$15,209	19.1%
Missouri	$12,807	$15,199	18.7%
Iowa	$13,136	$15,044	14.5%
Oregon	$13,590	$14,941	9.9%
Texas	$13,496	$14,537	7.7%
Wyoming	$15,618	$14,159	-9.3%
Arizona	$12,634	$13,918	10.2%
South Dakota	$11,318	$13,817	22.1%
Oklahoma	$12,938	$13,436	3.8%
North Dakota	$11,760	$13,352	13.5%
Montana	$12,292	$13,308	8.3%
Idaho	$11,803	$13,261	12.4%
New Mexico	$11,252	$12,395	10.2%
Arkansas	$10,282	$12,327	19.9%
Utah	$10,953	$12,161	11.0%
U.S. Average	**$13,663**	**$16,257**	**19.0%**

Source: U.S. Bureau of Economic Analysis

The U.S. national recession began officially in the fourth quarter of 1990. It began earlier in the West. Between 1988 and 1990, household income declined in ten western U.S. states.

Figure 12.18 **Western United States Median Income of Households, 1988–1990**

(Constant 1990 U.S. Dollars)

	1988	1990	Percent Change
Alaska	$36,573	$39,298	7.5%
Hawaii	$36,486	$38,921	6.7%
Washington	$35,175	$32,112	-8.7%
California	$33,462	$33,290	-0.5%
Minnesota	$32,136	$31,465	-2.1%
Nevada	$30,916	$32,023	3.6%
Oregon	$30,656	$29,281	-4.5%
Arizona	$29,206	$29,224	0.1%
Wyoming	$29,188	$29,460	0.9%
Utah	$29,071	$30,142	3.7%
Colorado	$28,962	$30,733	6.1%
Kansas	$28,246	$29,917	5.9%
Nebraska	$27,796	$27,482	-1.1%
Texas	$27,580	$28,228	2.3%
Iowa	$26,853	$27,288	1.6%
North Dakota	$26,617	$25,264	-5.1%
Oklahoma	$26,148	$24,384	-6.7%
Idaho	$25,908	$25,305	-2.3%
Missouri	$25,900	$27,332	5.5%
South Dakota	$24,631	$24,571	-0.2%
Montana	$24,561	$23,375	-4.8%
Arkansas	$22,286	$22,786	2.2%
New Mexico	$21,319	$25,039	17.4%
U.S. Average	**$30,079**	**$29,943**	**-0.5%**

Source: U.S. Bureau of the Census

In the United States, real disposable per capita income has declined since 1990.

In Canada, personal income grew the most in the Northwest Territories, and the least in Saskatchewan and Alberta.

Figure 12.19 **United States Per Capita Disposable Personal Income, 1960–1992**

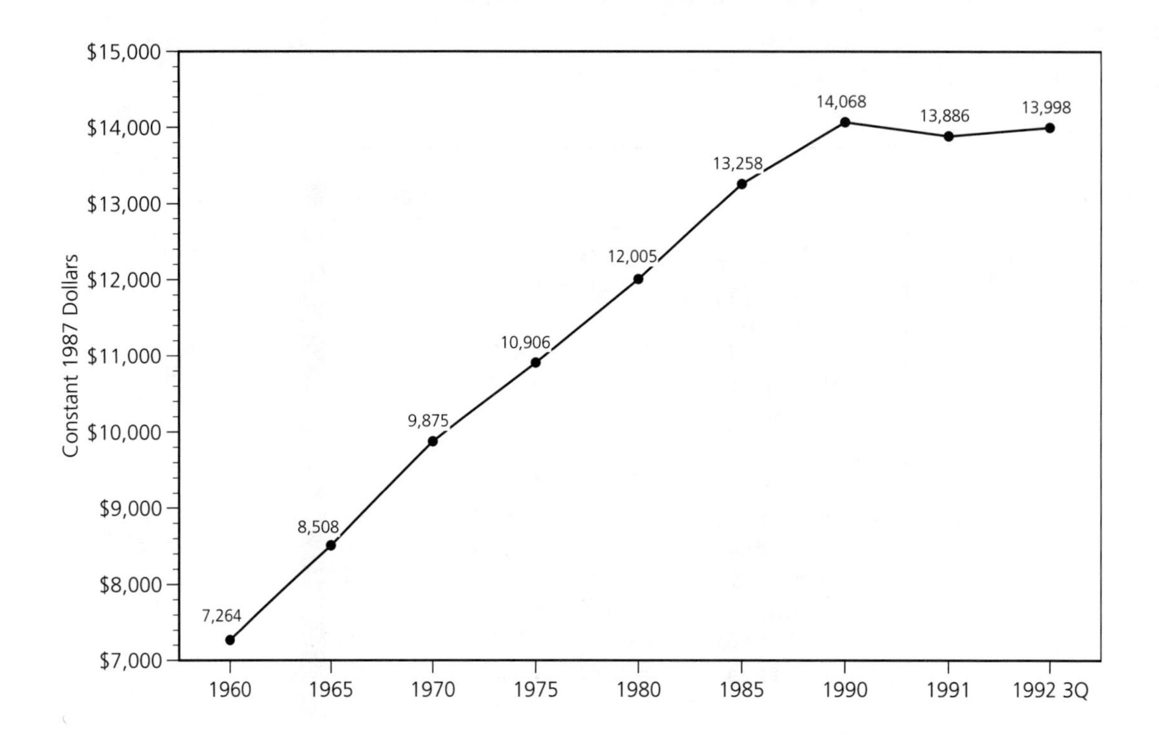

Source: Economic Report of the President, 1993

Figure 12.20 **Canada**
Average Family Income by Province, 1981–1989 (Constant 1989 Canadian Dollars) and 1991

(Current Canadian Dollars)

	1981 (Constant 1989 Dollars)	1989 (Constant 1989 Dollars)	Percent Change	1991 (Current Dollars)
Ontario	$48,806	$57,330	17.5%	$58,634
Alberta	$54,781	$49,734	-9.2%	$55,552
British Columbia	$50,867	$49,442	-2.8%	$54,895
Québec	$43,138	$44,860	4.0%	$48,634
Manitoba	$43,195	$46,551	7.8%	$46,621
Saskatchewan	$46,168	$42,978	-6.9%	$45,930
Nova Scotia	$37,542	$43,123	14.9%	$45,130
New Brunswick	$37,158	$40,670	9.5%	$44,323
Prince Edward Island	$35,417	$38,726	9.3%	$42,779
Newfoundland	$39,064	$39,648	1.5%	$41,654
Canada Average	**$46,769**	**$50,083**	**7.1%**	**$53,131**

Source: Statistics Canada, Canada Yearbook

Cost of Living

Income alone cannot adequately measure how well people live. Income has to be measured in relation to what the money can buy. When the cost of living is taken into consideration, many communities in the western United States with lower per capita incomes offer a higher standard of living.

Figure 12.21 **Western United States Cost of Living Index for Selected Cities, September 1991**

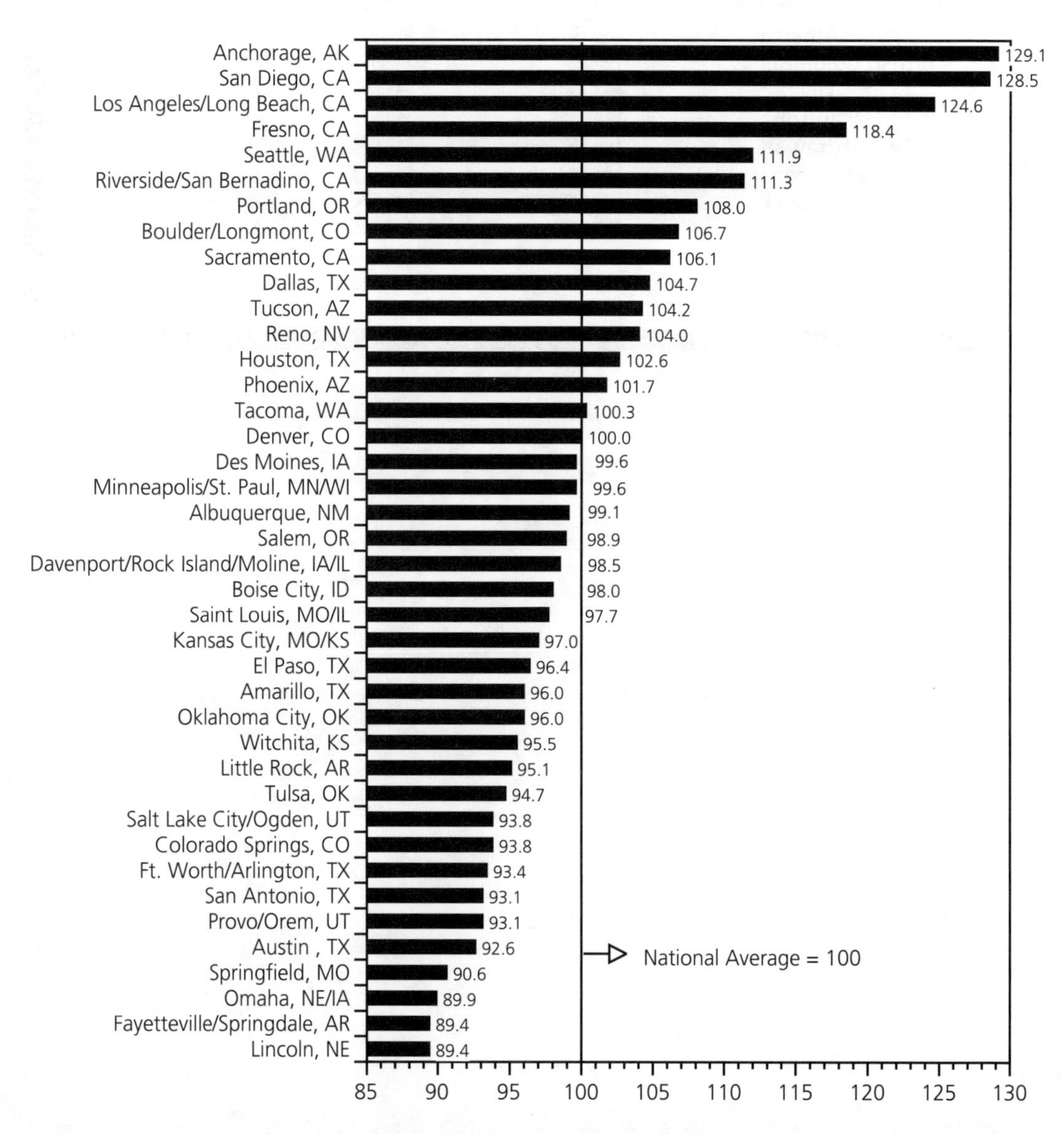

Source: American Chamber of Commerce Researchers Association, Alexandria, VA, Cost of Living Index

In Canada, living costs generally are lower in the West than they are in the East.

The cost of living in Mexico City is lower than in many other cities throughout the country because many of the goods and services are subsidized. This situation has begun to change dramatically since 1990.

Figure 12.22 **Canada**
Cost of Living Index by Province, October 1992*
(1986=100)

(* Information not available for Yukon and Northwest Territories.)

Source: Statistics Canada

Figure 12.23 **Canada**
Cost of Living Index for Selected Cities, July 1991
(1986=100)

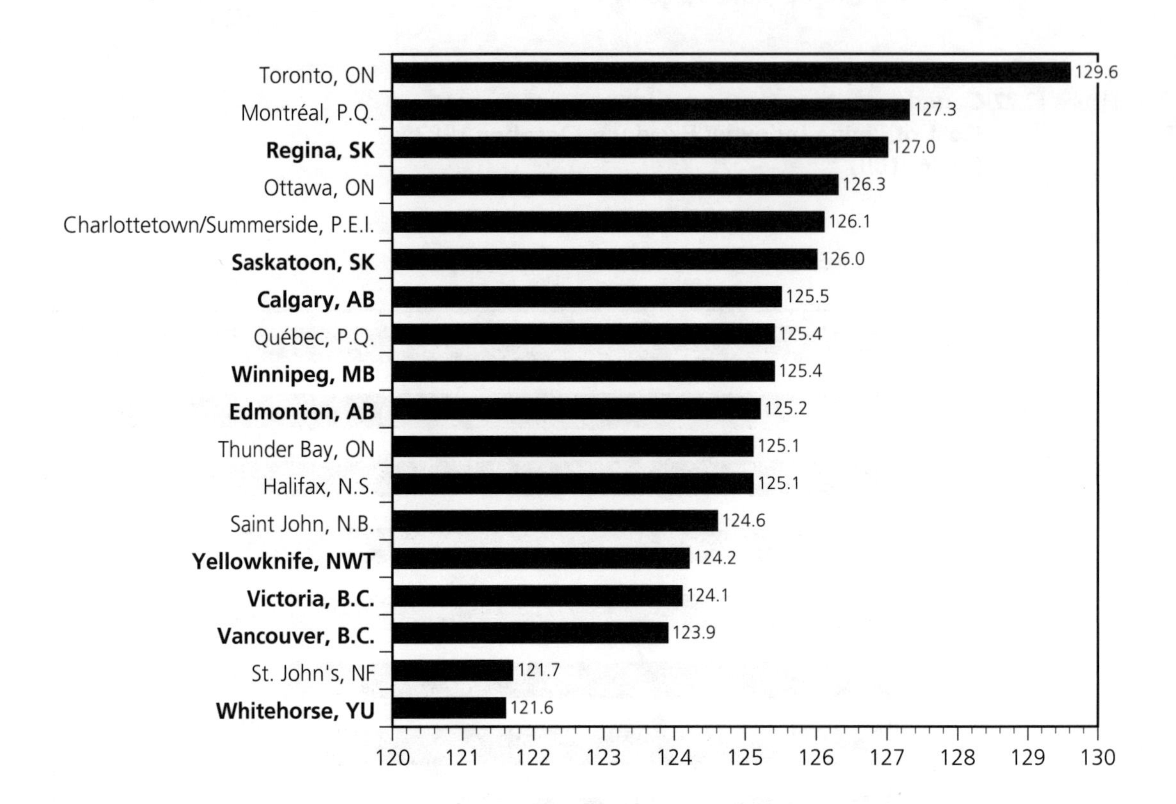

Source: Statistics Canada

Figure 12.24 **Mexico**
Cost of Living Index, January 1993

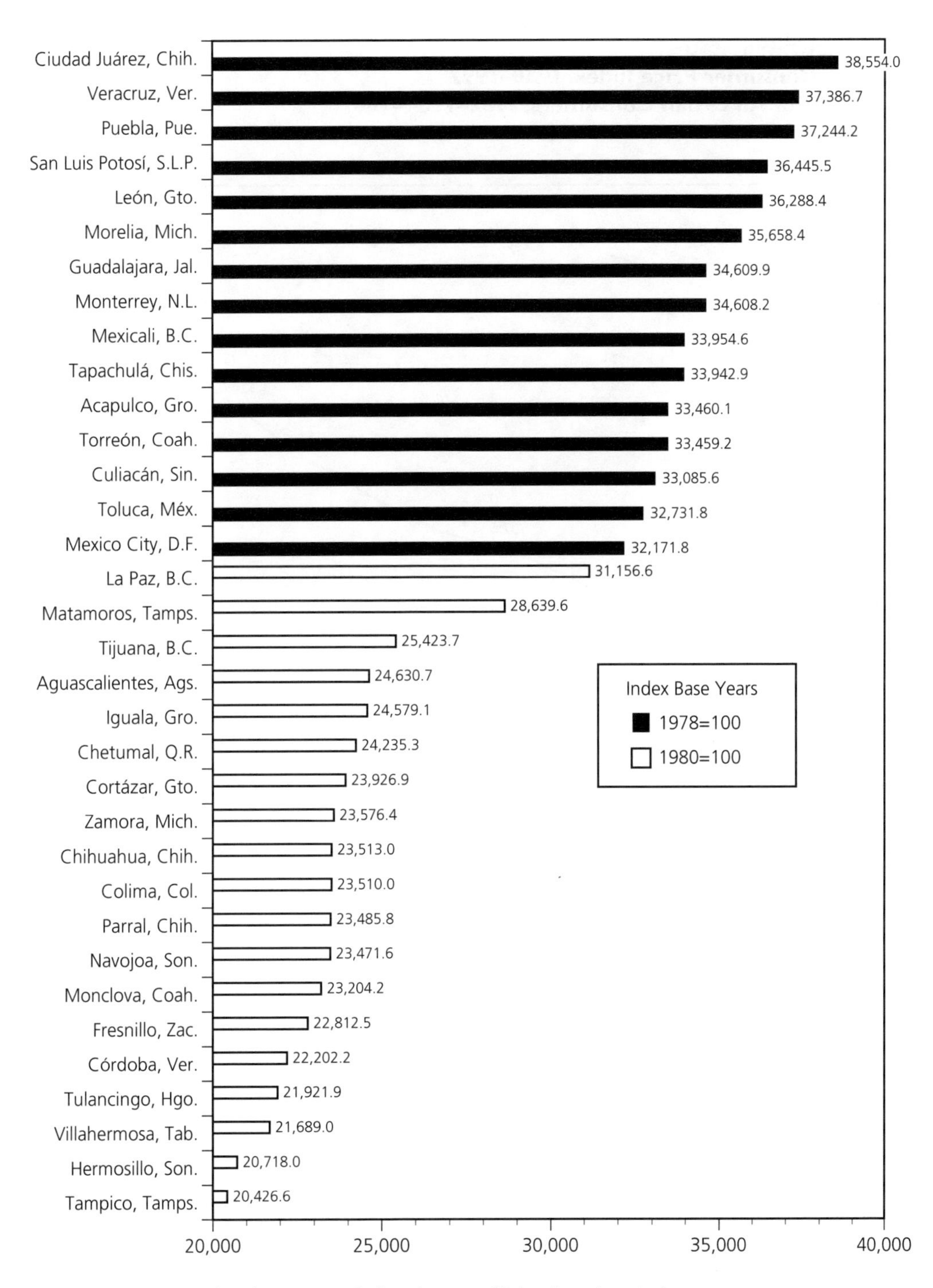

(Mexico uses two indices for cost of living, based on city.)

Source: INEGI

Canada and Mexico experienced substantial increases in the cost of living in the 1980s, while prices in the United States were relatively stable.

Figure 12.25a **United States**
Consumer Price Index, 1980–1992
(For All Urban Consumers; 1982–1984=100)

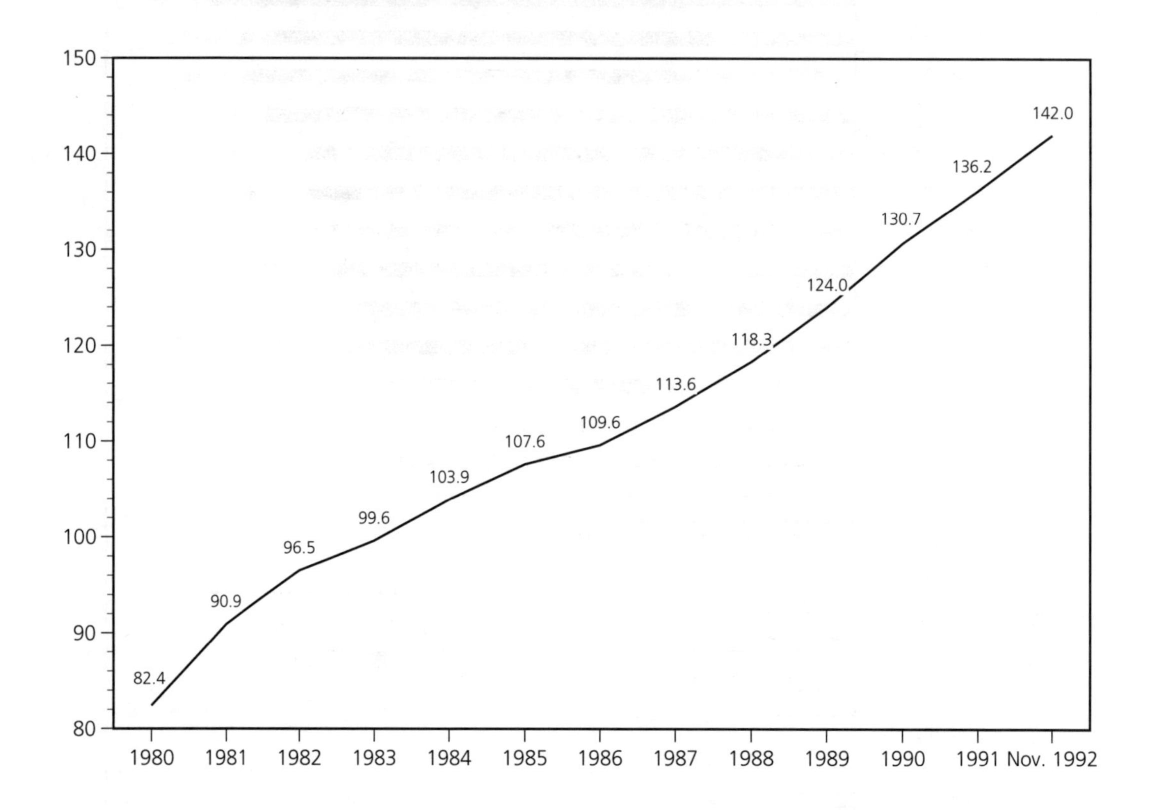

Source: Economic Report of the President, 1993

Figure 12.25b **United States**
Percent Change in the Consumer Price Index, 1980–1992
(for All Urban Consumers; 1982–1984=100)

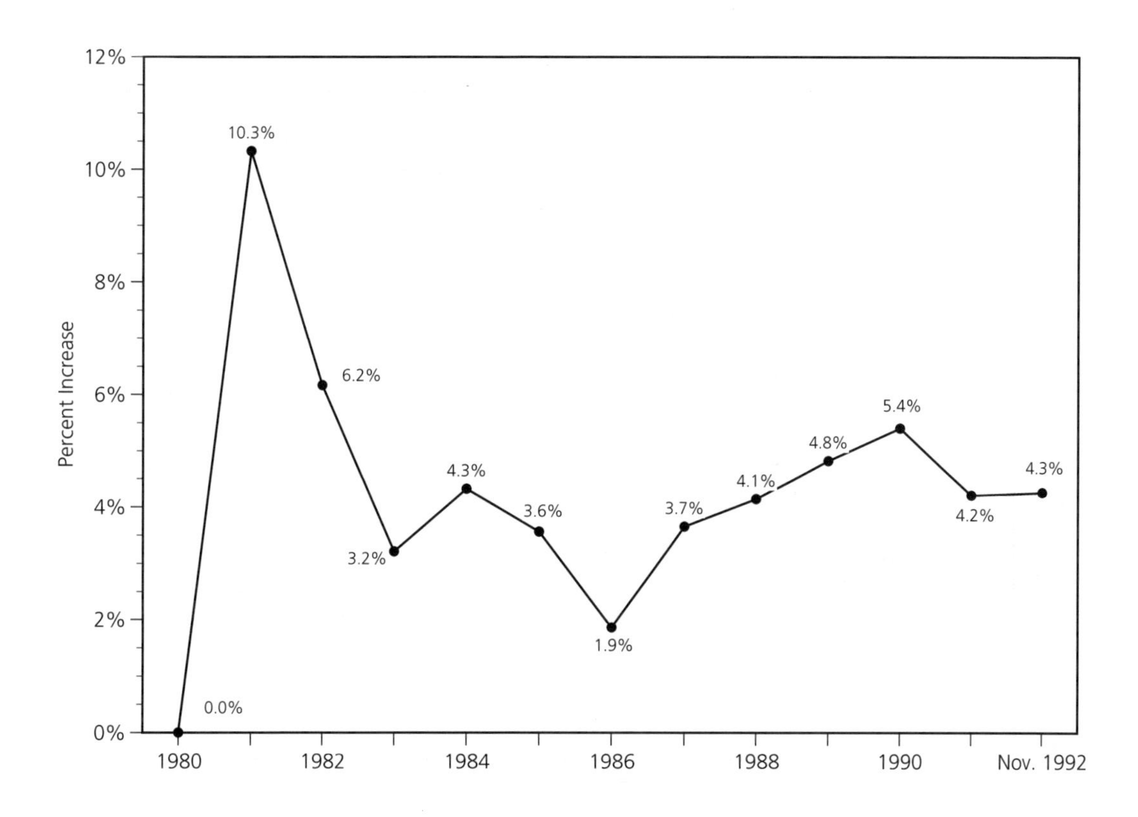

Source: Economic Report of the President, 1993

Figure 12.26a **Canada**
Consumer Price Index, 1980–1993
(1986=100)

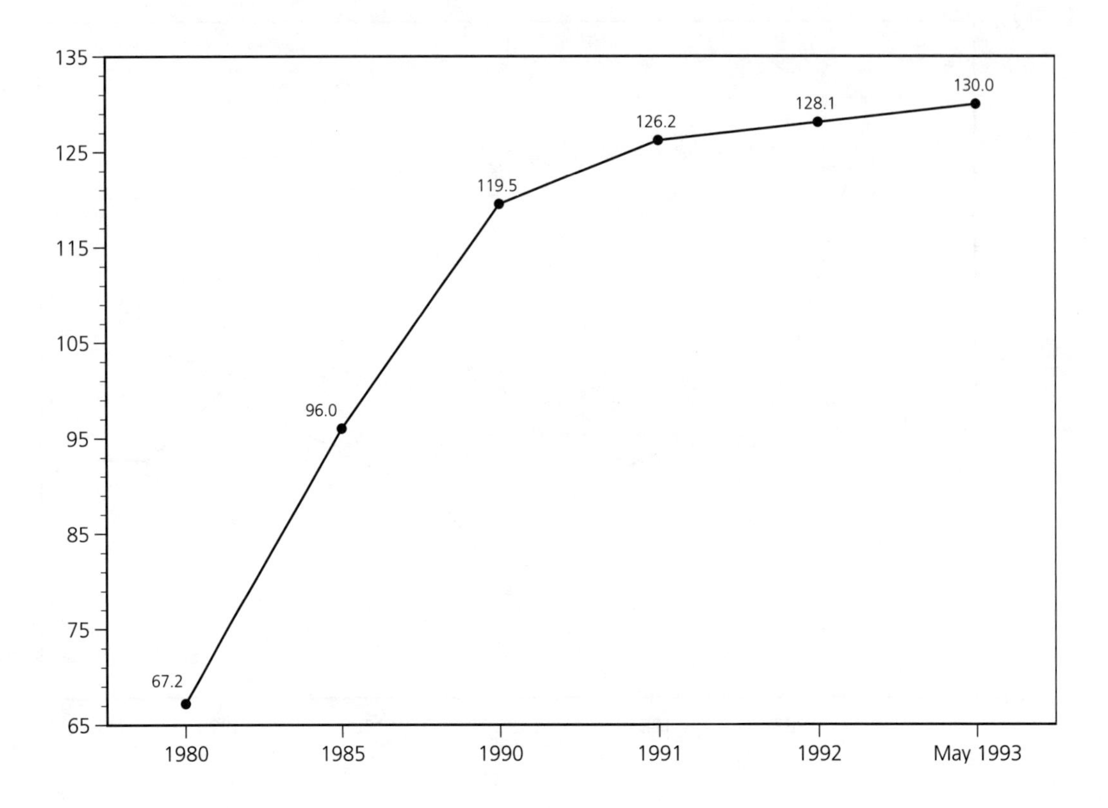

Source: Bank of Canada

Figure 12.26b **Canada**
Percent Change in the Consumer Price Index, 1980–1993 (1986=100)

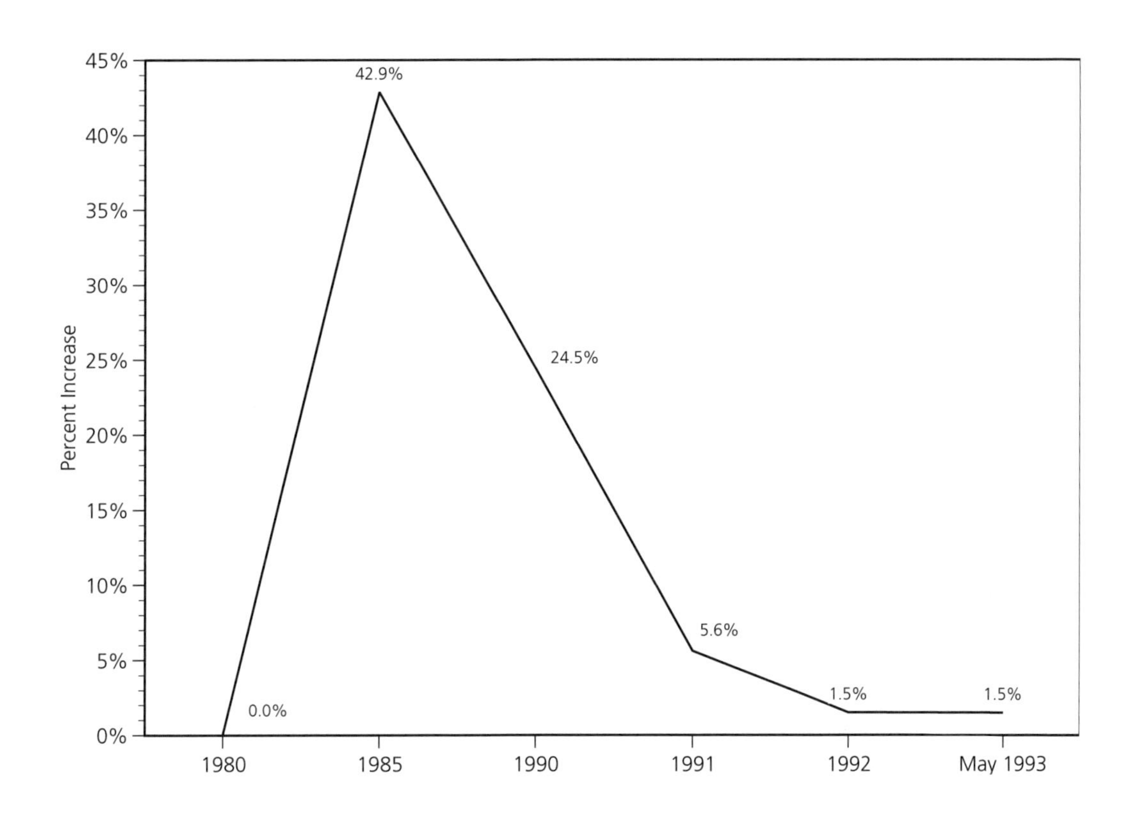

Source: Bank of Canada

Figure 12.27a **Mexico**
Consumer Price Index, 1981–1993
(1978=100)

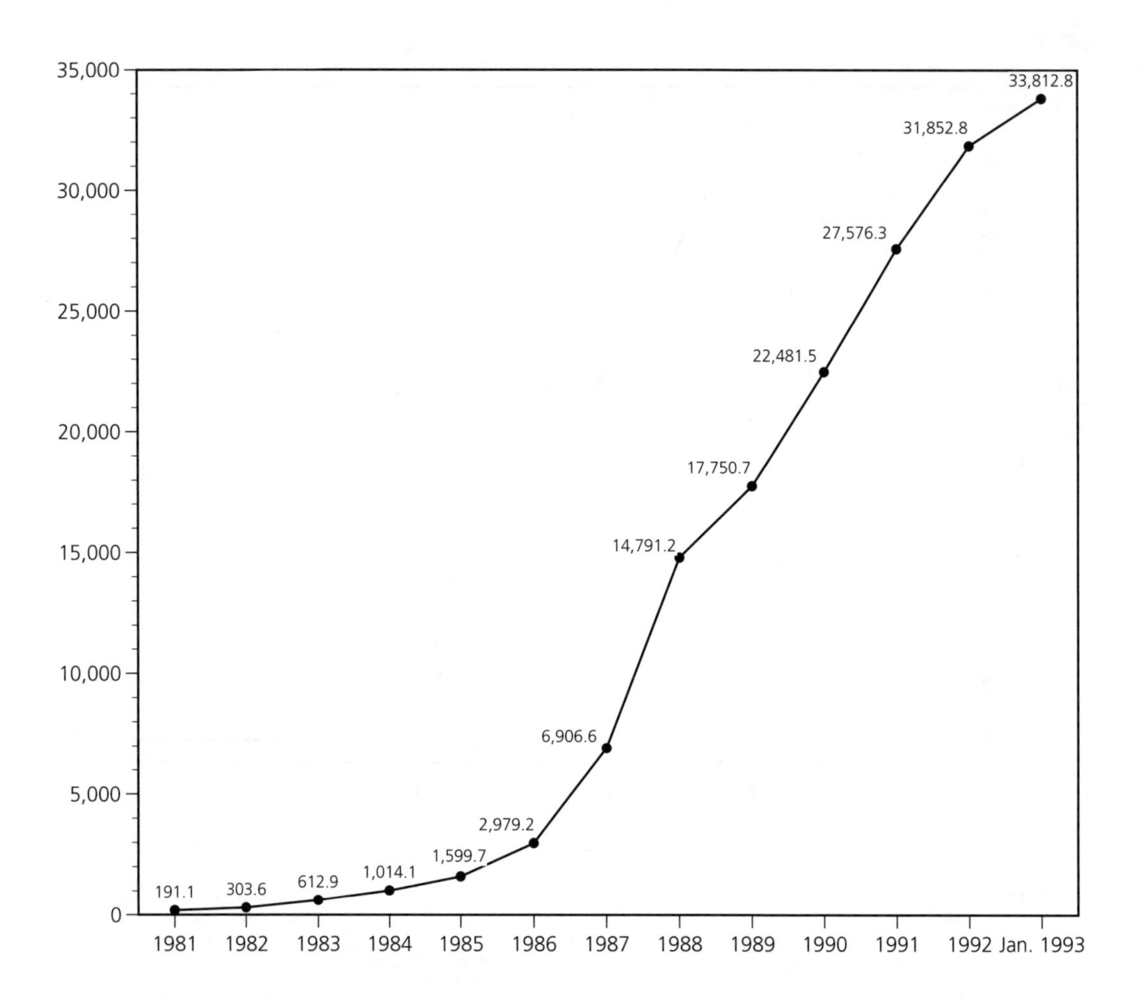

Source: Banamex

Figure 12.27b **Mexico**
Percent Change in the Consumer Price Index, 1981–1993 (1978=100)

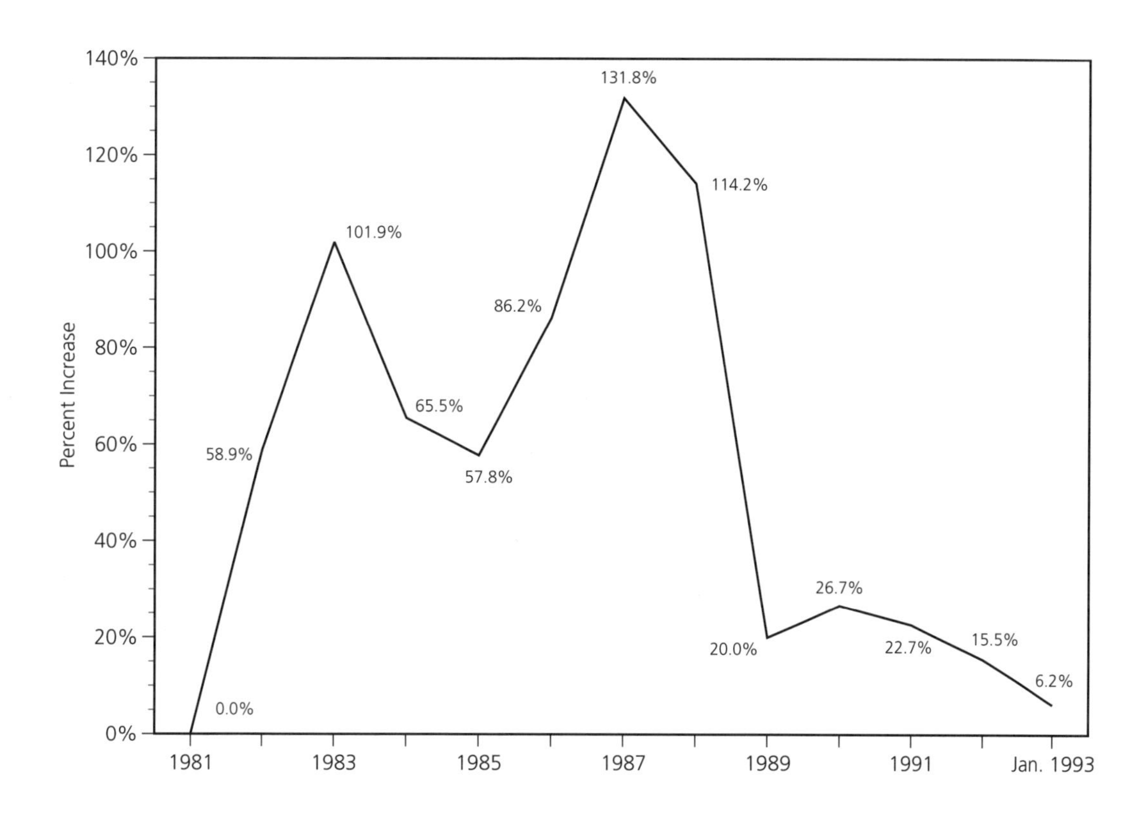

Source: Banamex

Chapter Thirteen
Business and Employment

UNEMPLOYMENT

Unemployment is a more serious problem in Canada and Mexico than it is in the western United States. For 1991, the United States had an unemployment rate of 6.7 percent. Unemployment in Canada for that year averaged 10.3 percent. In Mexico in 1991, the unemployment rate was between 14 percent and 17 percent.

In the western United States in 1991, Alaska and California had the highest unemployment rates. Nebraska had the lowest.

Figure 13.1 **Western United States Unemployment Rate by State, 1980 and 1991**

	Unemployment Rate, 1980	Unemployment Rate, 1991
Alaska	9.7%	8.5%
California	6.8%	7.5%
Arkansas	7.6%	7.3%
New Mexico	7.5%	6.9%
Montana	6.1%	6.9%
Oklahoma	4.8%	6.7%
Texas	5.2%	6.6%
Missouri	7.2%	6.6%
Washington	7.9%	6.3%
Idaho	7.9%	6.1%
Oregon	8.3%	6.0%
Arizona	6.7%	5.7%
Nevada	6.2%	5.5%
Minnesota	5.9%	5.1%
Wyoming	4.0%	5.1%
Colorado	5.9%	5.0%
Utah	6.3%	4.9%
Iowa	5.8%	4.6%
Kansas	4.5%	4.4%
North Dakota	5.0%	4.1%
South Dakota	4.9%	3.4%
Hawaii	4.9%	2.8%
Nebraska	4.1%	2.7%
U.S. Average	**7.1%**	**6.7%**

Source: U.S. Bureau of Labor Statistics

In Canada 1992 unemployment was highest in Newfoundland, lowest in Saskatchewan. British Columbia had the highest rate of unemployment in the Canadian West, but British Columbia's rate of unemployment was lower than for any of the eastern Canadian provinces except Ontario.

Figure 13.2 **Canada**
Unemployment Rate by Province, 1982 and 1992

	Average Unemployment, 1982	Average Unemployment, 1992
Newfoundland	16.7%	20.2%
Prince Edward Island	12.9%	17.7%
Northwest Territories*	11.3%	14.5%
Yukon*	11.1%	13.9%
Nova Scotia	13.1%	13.1%
Québec	13.8%	12.8%
New Brunswick	14.1%	12.8%
Ontario	9.7%	10.8%
British Columbia	12.1%	10.4%
Manitoba	8.5%	9.6%
Alberta	7.7%	9.5%
Saskatchewan	6.1%	8.2%
Canada Average	**11.0%**	**11.3%**

* Data for 1981 and 1st three quarters of 1991.

Source: Statistics Canada

Mexico doesn't keep unemployment data in a form comparable to that of the United States and Canada, but the flow of migrant workers, legal and illegal, from Mexico to the United States is evidence of a serious and chronic problem of unemployment and underemployment in Mexico.

Figure 13.3 **Mexico Unemployment Rate by State, 1980 and 1990**

	Unemployment Rate, 1980	Unemployment Rate, 1990
Zacatecas	20.1%	4.0%
Guerrero	14.2%	4.0%
Tlaxcala	10.0%	3.6%
Tamaulipas	7.4%	3.6%
Durango	13.3%	3.5%
Coahuila	5.4%	3.2%
Morelos	7.0%	3.2%
Tabasco	4.2%	3.1%
Querétaro	7.1%	3.1%
Guanajuato	8.8%	3.1%
Michoacán	9.7%	3.1%
Hidalgo	7.4%	3.0%
México	6.4%	3.0%
Chihuahua	9.5%	3.0%
Veracruz	5.6%	2.8%
Oaxaca	9.9%	2.8%
Distrito Federal	4.4%	2.6%
Sonora	6.6%	2.6%
Nuevo León	5.3%	2.6%
Puebla	6.5%	2.4%
San Luis Potosí	9.1%	2.4%
Chiapas	7.2%	2.3%
Aguascalientes	6.1%	2.2%
Jalisco	7.2%	2.2%
Baja California	5.4%	2.2%
Baja California Sur	4.8%	2.1%
Nayarit	6.2%	2.1%
Sinaloa	8.6%	2.0%
Colima	3.6%	1.9%
Campeche	4.9%	1.9%
Yucatán	5.4%	1.5%
Quintana Roo	5.1%	1.4%
Mexico Average	**7.2%**	**2.7%**

Source: INEGI, Anuario Estadistico and Atlas of Mexico

Labor force participation is affected both by economic opportunities and by cultural considerations. Labor force–participation rates have been higher in the United States than in Canada or Mexico, in part because more jobs have been available in the United States, and in part because it is more acceptable, culturally, for women to work outside the home. Minnesota, the western U.S. state with the highest labor force–participation rate, has a labor force participation substantially above that of the highest ranking Canadian province, Alberta, or the highest ranking Mexican state, Quintana Roo. Some of the difference is due to different methods of calculating labor force participation.

U.S. employment growth in the 1980s was roughly half again as much as employment growth in Canada and Mexico.

Figure 13.4 **Western United States**
Labor Force–Participation Rate by State, 1991
(Civilian Noninstitutional Population 16 and Older)

(Participation rate is defined as the percentage of persons either employed or actively seeking employment.)
U.S. Average: Male: 75.5%/ Female: 57.3%

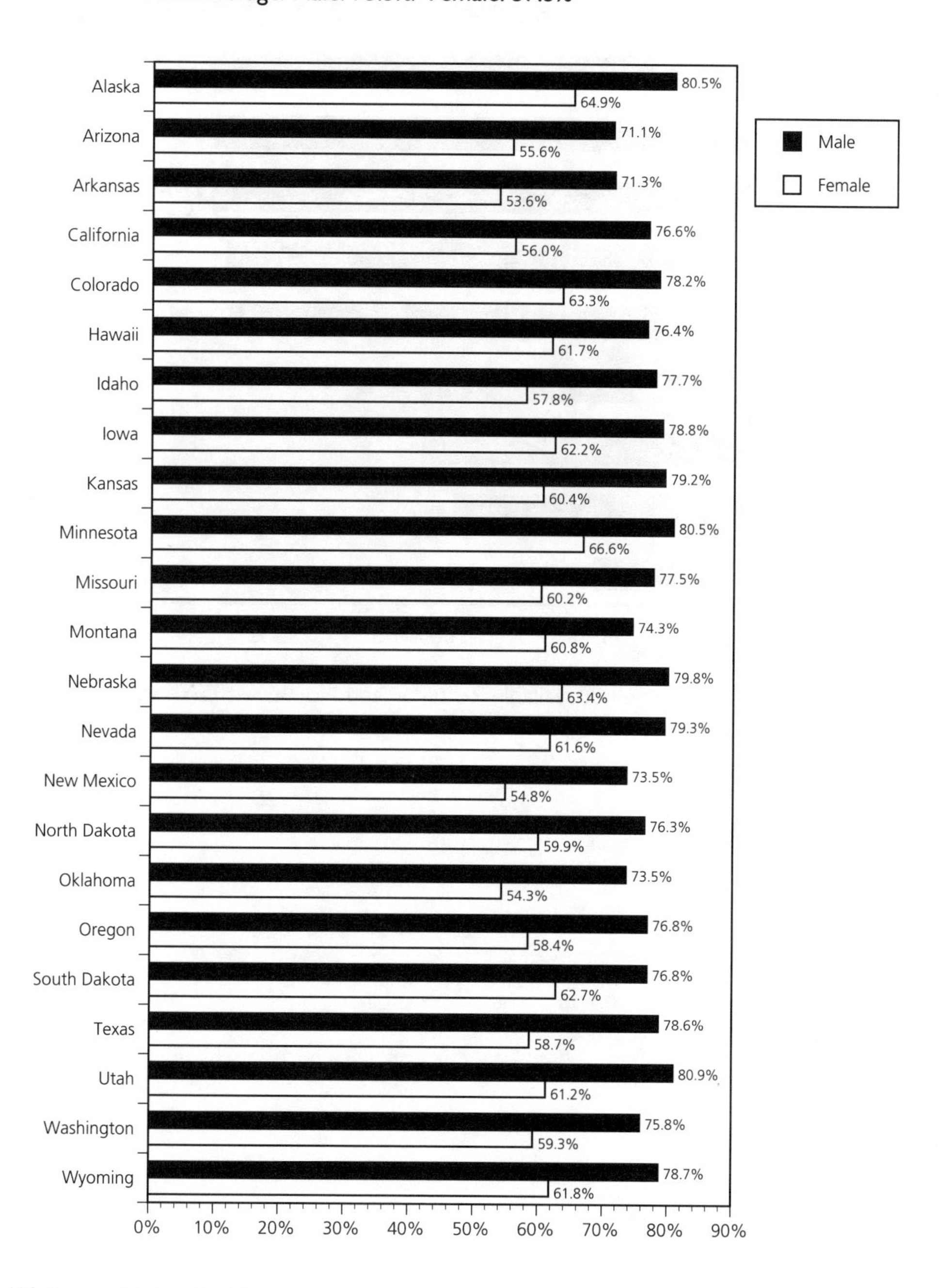

Source: U.S. Bureau of Labor Statistics

Figure 13.5 **Canada**
Labor Force–Participation Rate by Province, November 1992

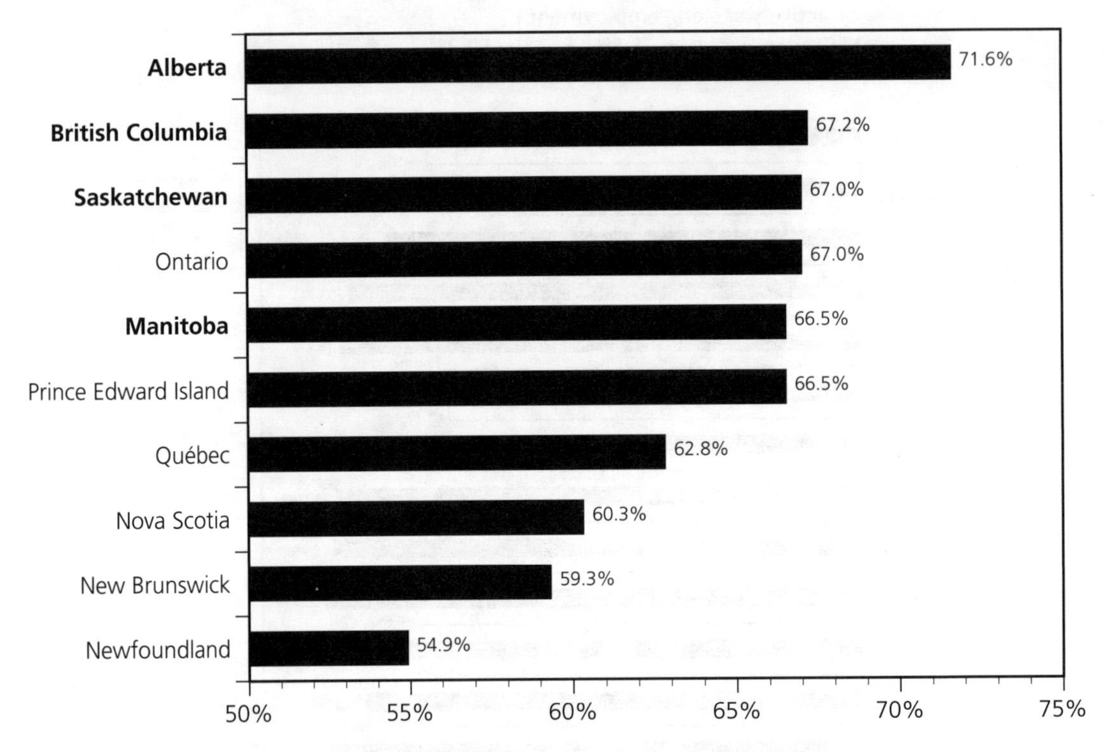

(Information not available for Yukon and Northwest Territories.)

Source: Statistics Canada

Figure 13.6 **Mexico**
Economically Active Population (EAP) by State, 1990
(Percent of the Total Population 15 and Older)

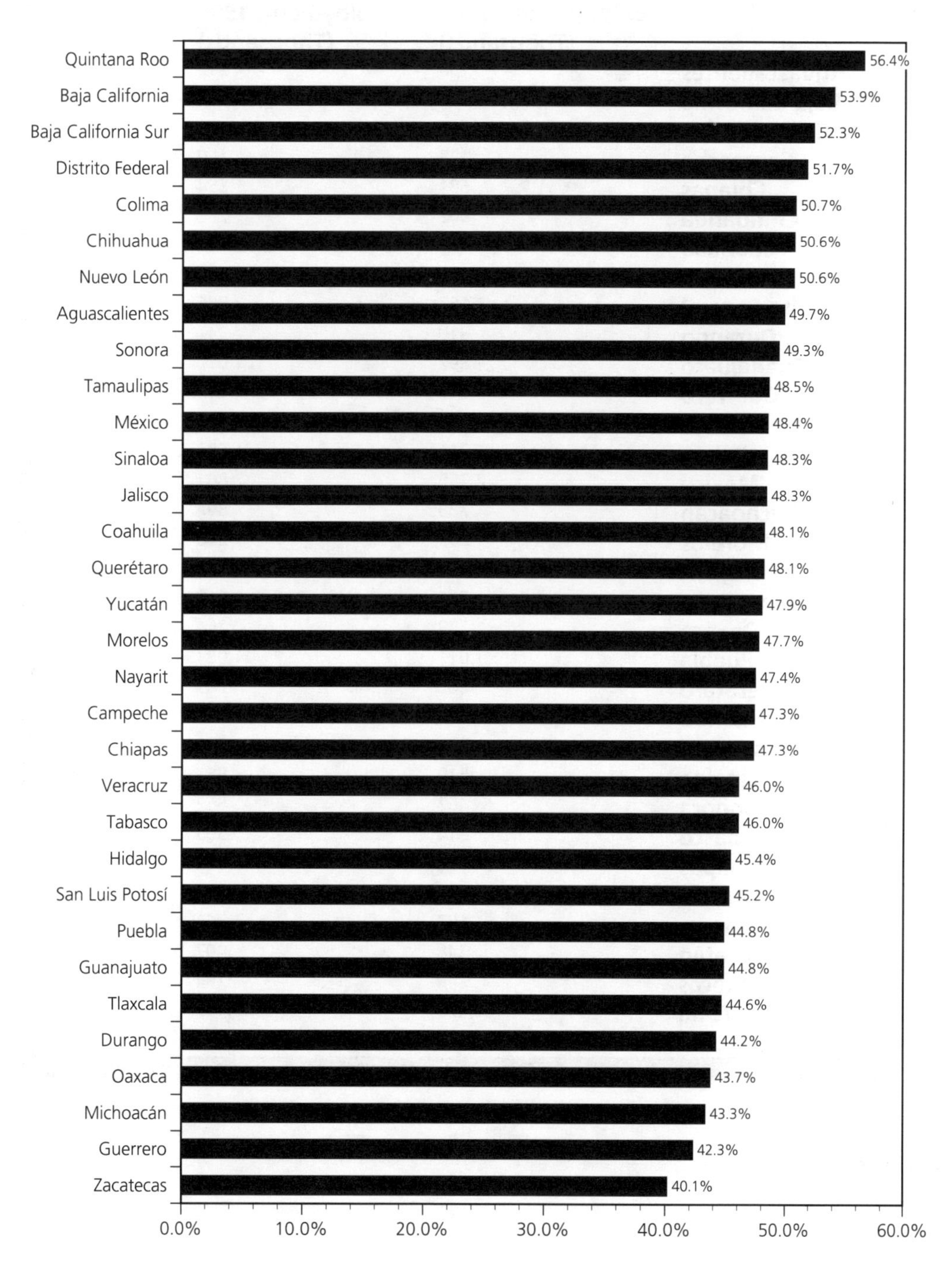

Source: INEGI, XI Censo General

Figure 13.7 **Mexico**
Employment by State, 1980 and 1990

	Employment, 1980 (Thousands)	Employment, 1990 (Thousands)	Percent Increase
Aguascalientes	150	212	41.3%
Baja California	381	565	48.2%
Baja California Sur	67	103	54.3%
Campeche	128	150	17.4%
Chiapas	681	854	25.4%
Chihuahua	602	773	28.5%
Coahuila	458	586	28.0%
Colima	105	133	27.2%
Distrito Federal	3,167	2,885	-8.9%
Durango	310	347	12.2%
Guanajuato	892	1,030	15.5%
Guerrero	617	612	-0.9%
Hidalgo	468	493	5.5%
Jalisco	1,311	1,553	18.4%
México	2,256	2,861	26.8%
Michoacán	788	892	13.1%
Morelos	283	348	23.3%
Nayarit	197	233	18.2%
Nuevo León	761	1,010	32.6%
Oaxaca	773	754	-2.4%
Puebla	1,011	1,084	7.2%
Querétaro	208	289	38.6%
Quintana Roo	75	163	116.6%
San Luis Potosí	484	529	9.3%
Sinaloa	519	661	27.3%
Sonora	452	562	24.3%
Tabasco	314	393	25.3%
Tamaulipas	578	685	18.3%
Tlaxcala	157	197	24.9%
Veracruz	1,697	1,742	2.7%
Yucatán	348	407	17.1%
Zacatecas	240	294	22.4%
Mexico Total	**20,478**	**23,403**	**14.0%**

Source: INEGI, XI Censo General

Figure 13.8 **Western United States Employment by State, 1980 and 1991**

	Employment, 1980 (Thousands)	Employment, 1991 (Thousands)	Percentage Increase
Alaska	169	243	43.8%
Arizona	1,014	1,498	47.7%
Arkansas	742	937	26.3%
California	9,849	12,497	26.9%
Colorado	1,251	1,542	23.3%
Hawaii	405	539	33.1%
Idaho	330	398	20.6%
Iowa	1,110	1,237	11.4%
Kansas	945	1,095	15.9%
Minnesota	1,770	2,136	20.7%
Missouri	1,970	2,295	16.5%
Montana	280	302	7.9%
Nebraska	628	736	17.2%
Nevada	400	633	58.3%
New Mexico	465	583	25.4%
North Dakota	245	271	10.6%
Oklahoma	1,138	1,202	5.6%
Oregon	1,045	1,251	19.7%
South Dakota	238	297	24.8%
Texas	5,851	7,167	22.5%
Utah	551	745	35.2%
Washington	1,608	2,171	35.0%
Wyoming	210	203	-3.3%
West	**32,214**	**39,978**	**24.0%**
East	**58,192**	**69,003**	**19.0%**
U.S. Total	**90,406**	**108,981**	**21.0%**

Source: U.S. Bureau of Labor Statistics

Figure 13.9 **Canada**
Employment by Province, 1982 and 1992

	Employment, 1982 (Thousands)	Employment, 1992 (Thousands)	Percent Increase
Alberta	1,127	1,240	10.0%
British Columbia	1,202	1,517	26.2%
Manitoba	454	484	6.6%
New Brunswick	242	289	19.4%
Newfoundland	173	188	8.7%
Northwest Territories	17	24	41.2%
Nova Scotia	313	361	15.3%
Ontario	4,063	4,714	16.0%
Prince Edward Island	45	53	17.8%
Québec	2,574	2,953	14.7%
Saskatchewan	426	440	3.3%
Yukon	12	15	25.0%
West	**3,238**	**3,720**	**15.0%**
East	**7,410**	**8,558**	**15.0%**
Canada Total	**10,648**	**12,278**	**15.0%**

Source: Statistics Canada

Sector Employment

Throughout western North America, the service sector is the primary source of employment. Wholesale and retail trade is second. Services led wholesale and retail trade as the number one employer in twelve of the twenty-three western U.S. states, and in every Mexican state and Canadian province.

Figure 13.10 **Western United States Employment by Industry, Nonform Establishments, by State, 1991**

(May not total 100% due to rounding.)

	Construction	Manufacturing	Transportation & Public Utilities	Wholesale & Retail	Finance, Insurance, & Real Estate	Services	Government
Alaska	4.5%	7.4%	9.1%	19.3%	4.5%	21.4%	29.2%
Arizona	5.2%	11.8%	5.4%	24.9%	6.2%	27.5%	18.1%
Arkansas	3.8%	25.0%	6.0%	22.0%	4.1%	21.1%	17.5%
California	4.4%	16.2%	5.0%	23.3%	6.6%	27.6%	16.6%
Colorado	4.3%	12.1%	6.4%	24.3%	6.3%	27.2%	18.3%
Hawaii	6.3%	3.7%	8.0%	25.2%	7.1%	29.3%	20.2%
Idaho	5.0%	15.8%	5.0%	25.4%	5.3%	21.6%	21.4%
Iowa	3.7%	18.8%	4.4%	25.2%	5.8%	24.0%	17.8%
Kansas	3.8%	16.8%	5.9%	24.6%	5.3%	22.6%	20.0%
Minnesota	3.6%	18.5%	5.1%	24.3%	6.0%	26.2%	16.1%
Missouri	3.8%	18.1%	6.6%	23.9%	5.9%	25.2%	16.2%
Montana	3.6%	7.3%	6.6%	26.8%	4.6%	25.5%	23.5%
Nebraska	3.8%	13.5%	6.5%	25.4%	6.7%	24.5%	19.8%
Nevada	6.5%	4.1%	5.2%	20.4%	4.6%	44.4%	12.8%
New Mexico	4.8%	7.2%	5.0%	23.7%	4.5%	26.1%	25.9%
North Dakota	3.7%	6.6%	6.3%	26.6%	4.8%	26.2%	24.4%
Oklahoma	3.2%	14.1%	5.7%	23.6%	5.1%	22.6%	22.1%
Oregon	4.2%	16.9%	5.2%	25.2%	6.6%	23.7%	18.1%
South Dakota	4.0%	11.8%	4.7%	26.6%	5.7%	25.3%	21.2%
Texas	4.8%	13.8%	6.1%	24.2%	6.0%	24.8%	17.8%
Utah	4.3%	14.2%	5.6%	23.9%	4.8%	25.4%	20.7%
Washington	5.4%	16.2%	5.2%	24.2%	5.4%	24.6%	18.9%
Wyoming	5.9%	4.9%	7.4%	22.7%	3.4%	19.7%	27.1%

Source: U.S. Bureau of Labor Statistics

Figure 13.11 **Canada**
Employment by Industry, by State, 1990

	Agriculture*	Manufacturing	Construction	Transportation, Communication, & Other Utilities	Wholesale & Retail Trade	Finance, Insurance, & Real Estate	Services	Government
Alberta	13.0%	7.4%	6.6%	7.6%	17.3%	5.2%	36.4%	6.5%
British Columbia	4.7%	10.9%	7.3%	8.4%	18.9%	6.9%	37.2%	5.8%
Manitoba	10.1%	10.5%	4.3%	8.9%	17.4%	5.4%	36.0%	7.4%
New Brunswick	6.6%	12.5%	5.5%	8.7%	19.7%	4.5%	34.6%	8.3%
Newfoundland	8.0%	8.5%	5.3%	9.0%	19.1%	3.7%	36.7%	9.0%
Nova Scotia	6.6%	11.4%	5.0%	8.0%	19.1%	5.5%	35.2%	9.1%
Ontario	3.3%	17.8%	5.4%	7.0%	17.0%	7.1%	35.8%	6.7%
Prince Edward Island	7.5%	7.5%	0.0%	7.5%	13.2%	0.0%	35.8%	9.4%
Québec	3.4%	17.5%	5.0%	7.5%	17.8%	5.8%	36.1%	7.0%
Saskatchewan	20.0%	5.9%	4.3%	7.0%	16.4%	5.5%	34.5%	6.8%

*Includes fishing, trapping, logging, forestry, and mining.

(May not total 100% due to rounding.)
(Information not available for Yukon and Northwest Territories.)

Source: Statistics Canada

Figure 13.12 **Mexico**
Employment by Industry, by State, 1990

	Agriculture	Manufacturing	Utilities, Transport, & Communication	Construction	Commerce	Financial Services	Services	Public Administration and Defense
Aguascalientes	15.5%	24.7%	5.8%	8.6%	14.2%	1.3%	22.9%	5.1%
Baja California	10.5%	23.2%	5.1%	7.5%	16.4%	1.9%	27.6%	3.8%
Baja California Sur	19.7%	8.7%	6.1%	7.7%	14.6%	1.6%	29.3%	8.8%
Campeche	36.8%	9.2%	4.1%	7.0%	11.0%	0.9%	21.4%	5.6%
Chiapas	58.7%	5.9%	2.9%	4.5%	7.6%	0.5%	13.9%	2.9%
Chihuahua	18.4%	26.4%	4.6%	7.6%	13.0%	1.4%	21.7%	3.1%
Coahuila	14.9%	25.6%	5.5%	8.2%	13.3%	1.5%	24.8%	3.2%
Colima	26.0%	9.9%	5.7%	8.3%	12.0%	1.3%	27.1%	6.6%
Distrito Federal	1.4%	21.3%	7.4%	4.3%	17.2%	3.9%	33.4%	7.3%
Durango	30.4%	17.0%	4.5%	7.0%	11.7%	1.0%	21.6%	4.1%
Guanajuato	24.2%	25.0%	4.0%	8.2%	13.4%	1.1%	18.6%	2.2%
Guerrero	36.8%	9.2%	4.5%	6.7%	10.3%	0.7%	23.9%	3.8%
Hidalgo	39.0%	15.4%	4.2%	7.3%	9.9%	0.5%	17.1%	2.9%
Jalisco	15.4%	24.0%	4.8%	8.0%	15.6%	1.8%	24.2%	2.8%
México	9.2%	28.4%	6.6%	7.1%	15.5%	1.5%	23.2%	4.9%
Michoacán	34.2%	15.2%	3.7%	7.3%	12.6%	1.0%	18.2%	2.4%
Morelos	20.7%	16.2%	4.7%	10.6%	13.3%	1.0%	27.4%	3.7%
Nayarit	38.5%	10.0%	4.0%	6.8%	10.6%	0.8%	21.3%	3.8%
Nuevo León	6.9%	29.8%	5.5%	8.9%	14.6%	2.2%	25.8%	3.1%
Oaxaca	54.1%	10.1%	3.1%	4.8%	7.5%	0.4%	14.6%	3.2%
Puebla	37.5%	17.8%	4.0%	6.1%	11.6%	0.9%	16.9%	2.3%
Querétaro	18.6%	25.4%	4.0%	10.7%	11.8%	1.4%	21.9%	3.2%
Quintana Roo	19.8%	6.3%	6.6%	8.3%	13.7%	1.5%	31.9%	5.9%
San Luis Potosí	32.5%	17.3%	4.4%	6.9%	11.6%	1.0%	20.1%	2.8%
Sinaloa	37.1%	10.6%	4.5%	5.7%	12.1%	1.4%	21.3%	3.5%
Sonora	24.2%	16.1%	5.4%	7.2%	14.2%	1.9%	23.9%	4.4%
Tabasco	41.0%	8.4%	4.0%	6.0%	10.0%	0.8%	20.1%	5.2%
Tamaulipas	19.2%	19.0%	6.3%	7.7%	13.9%	1.4%	25.2%	4.1%
Tlaxcala	28.7%	25.5%	4.4%	7.9%	9.7%	0.5%	18.2%	3.1%
Veracruz	42.5%	11.5%	4.9%	5.7%	11.0%	0.7%	18.4%	2.8%
Yucatán	27.3%	15.5%	4.4%	8.1%	13.8%	1.5%	24.3%	3.3%
Zacatecas	42.3%	8.8%	2.7%	9.8%	10.1%	0.8%	18.8%	3.6%

(Table does not include 3.43% "not specified" category.)

Source: INEGI, Anuario Estadistico

Manufacturing employment was the third largest source of jobs in 1991. Regionally, manufacturing employment was proportionately the greatest in the Mexican border state of Nuevo León, and lowest in Hawaii.

MANUFACTURING

California and Texas accounted for 51 percent of the value of manufactured goods shipped in 1990 from states in the western United States. Ontario accounted for more than half the value of all of Canada's manufacturing, less than 20 percent coming from western Canada, where British Columbia and Alberta dominate manufacturing activity. Mexico City and environs accounted for 36 percent of the value of Mexico's manufactures.

Figure 13.13 **Western United States Value of Manufactures Shipments by State, 1990**

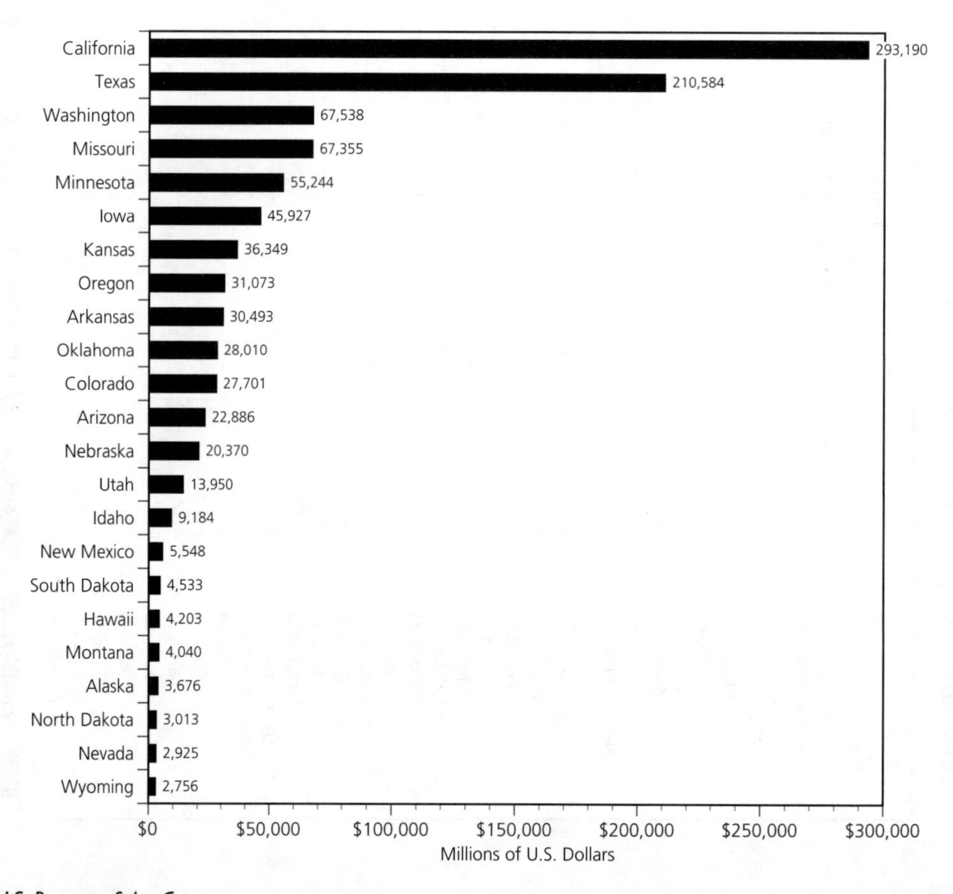

Source: U.S. Bureau of the Census

Figure 13.14 **Canada**
Value of Manufactures Shipments by Province, 1990

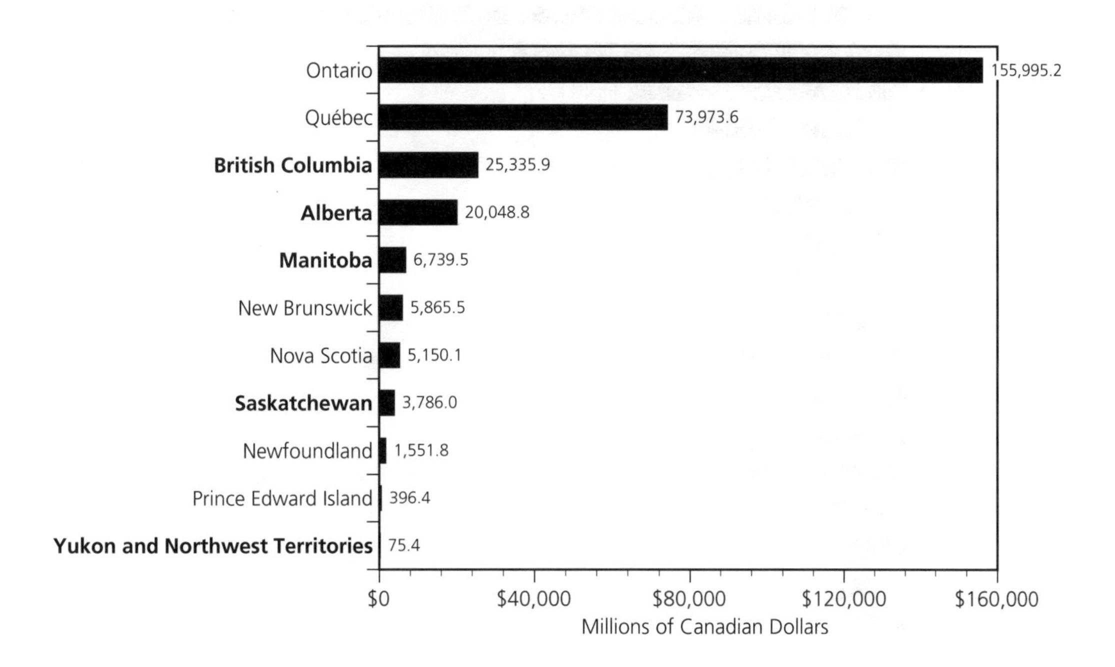

Source: Statistics Canada, Canada Yearbook

Figure 13.15 **Mexico**
Manufactures Production by State, 1988

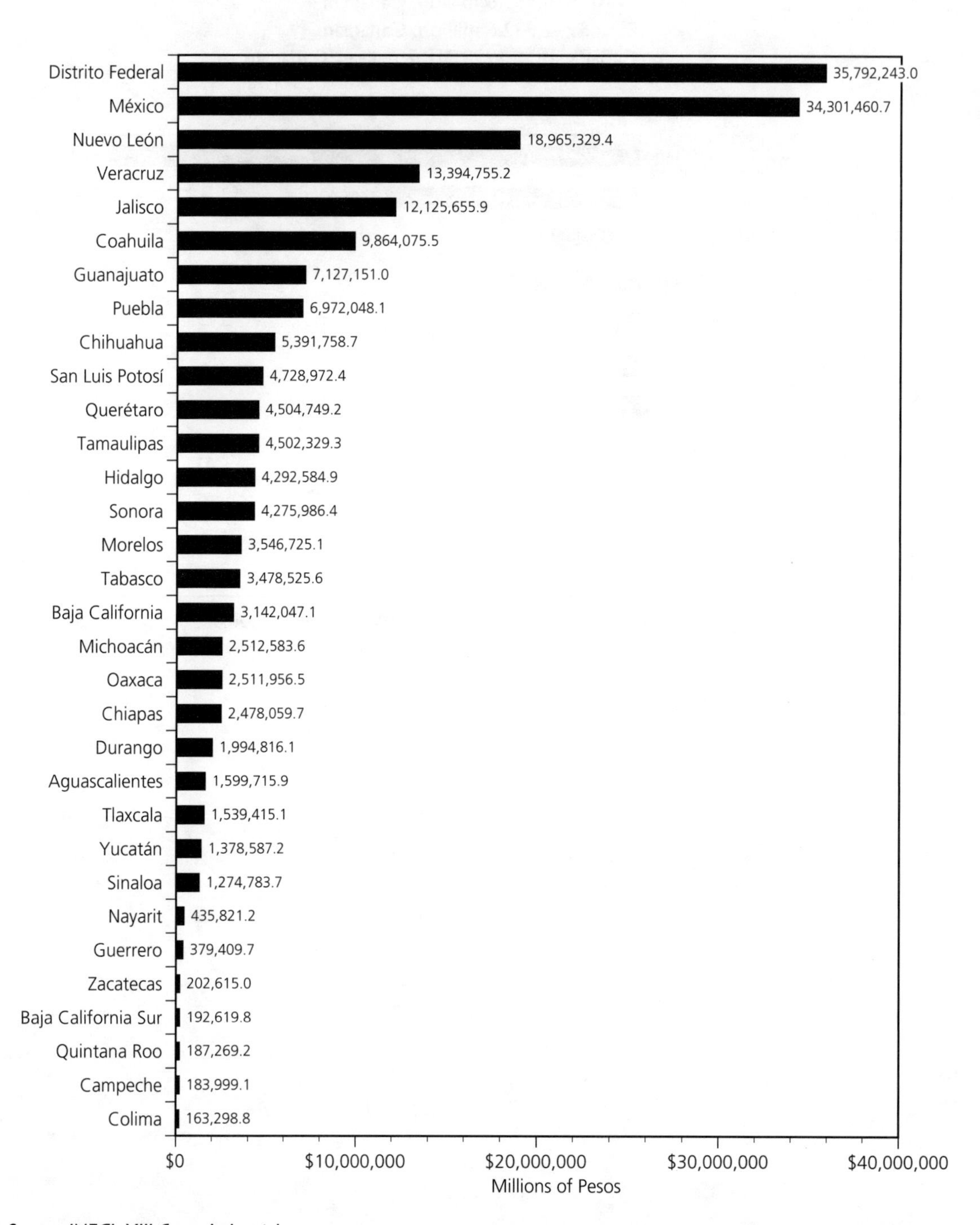

Source: INEGI, XIII Censo Industrial

Business Startups and Failures

Ontario has more business incorporations and business failures than any other Canadian province or western U.S. state. The ratio of business failures to new incorporations is much higher in Canada than it is in the western United States.

Figure 13.16 **Canada**
New Incorporations by Province, 1990

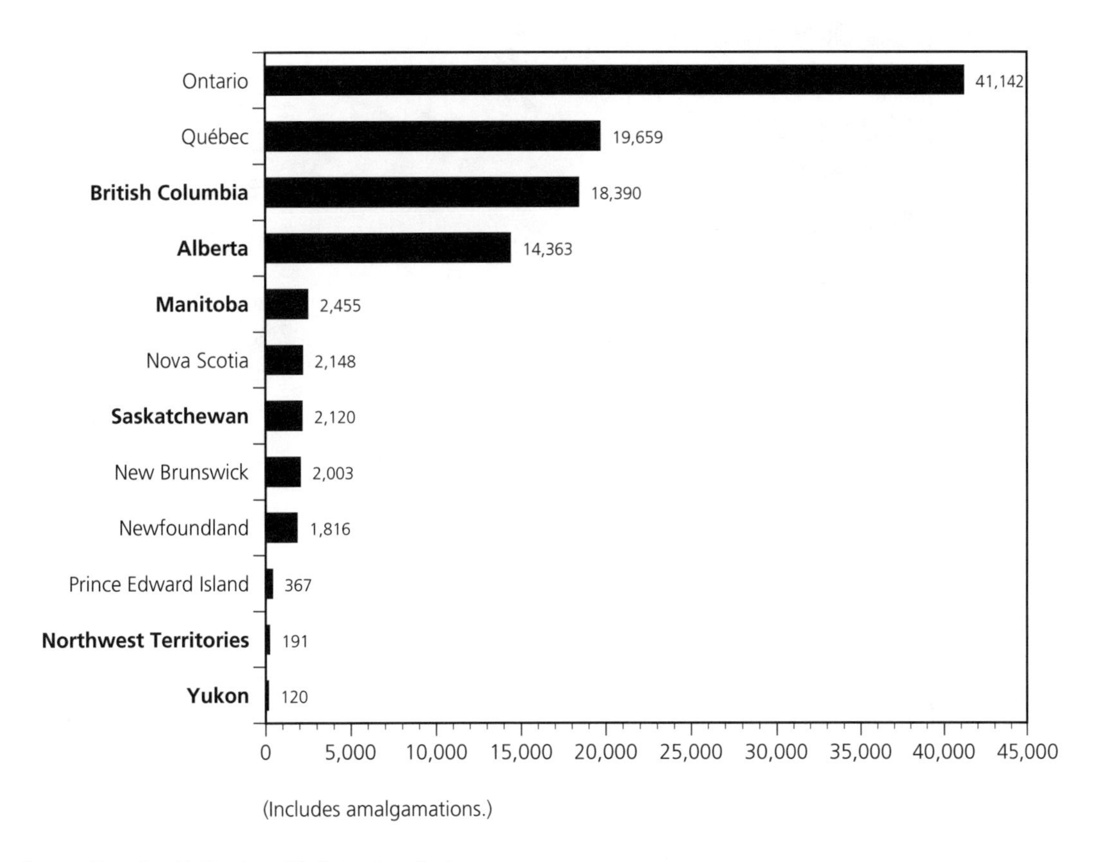

Source: Canadian Federation of Independent Business

Figure 13.17 **Western United States Incorporations by State, 1991**

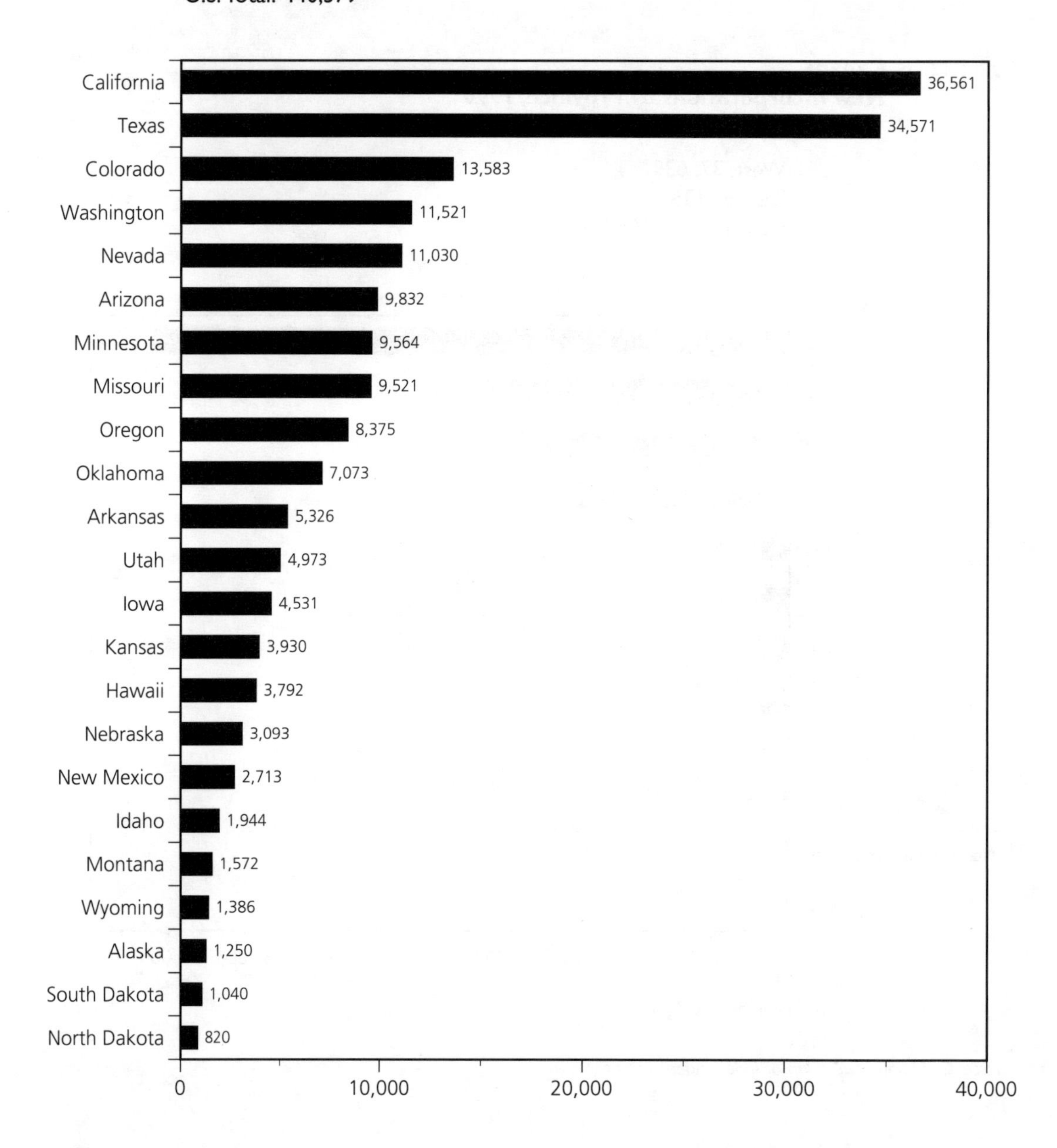

Source: U.S. Bureau of Economic Analysis

Figure 13.18 **Canada**
Bankruptcies by Province, 1991

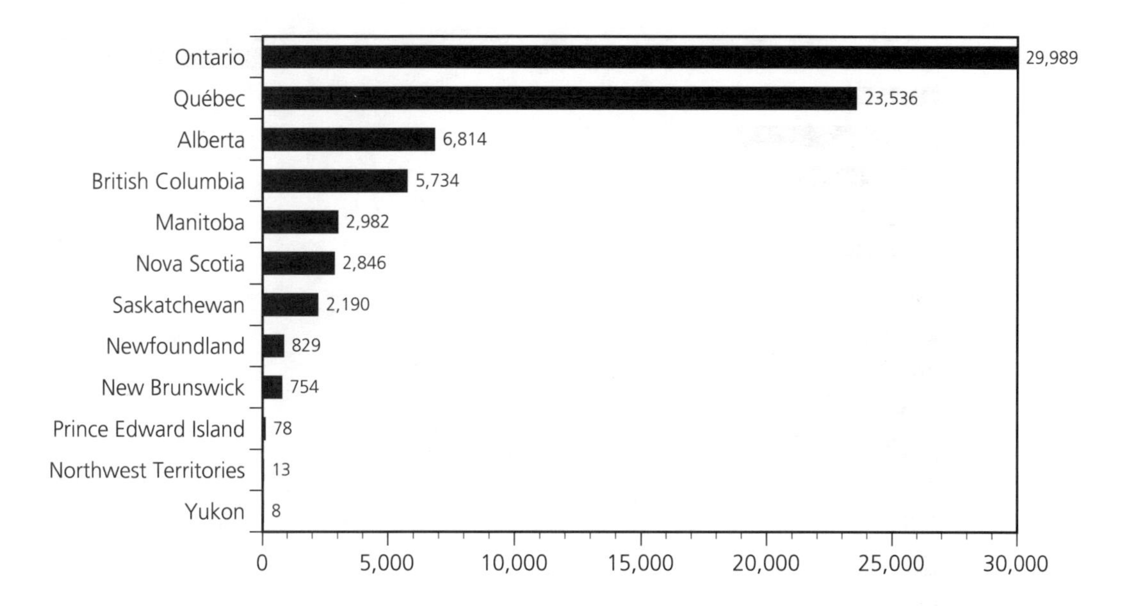

Source: Statistics Canada, Provincial Economic Accounts

Figure 13.19 **Western United States Business Failures by State, 1990**

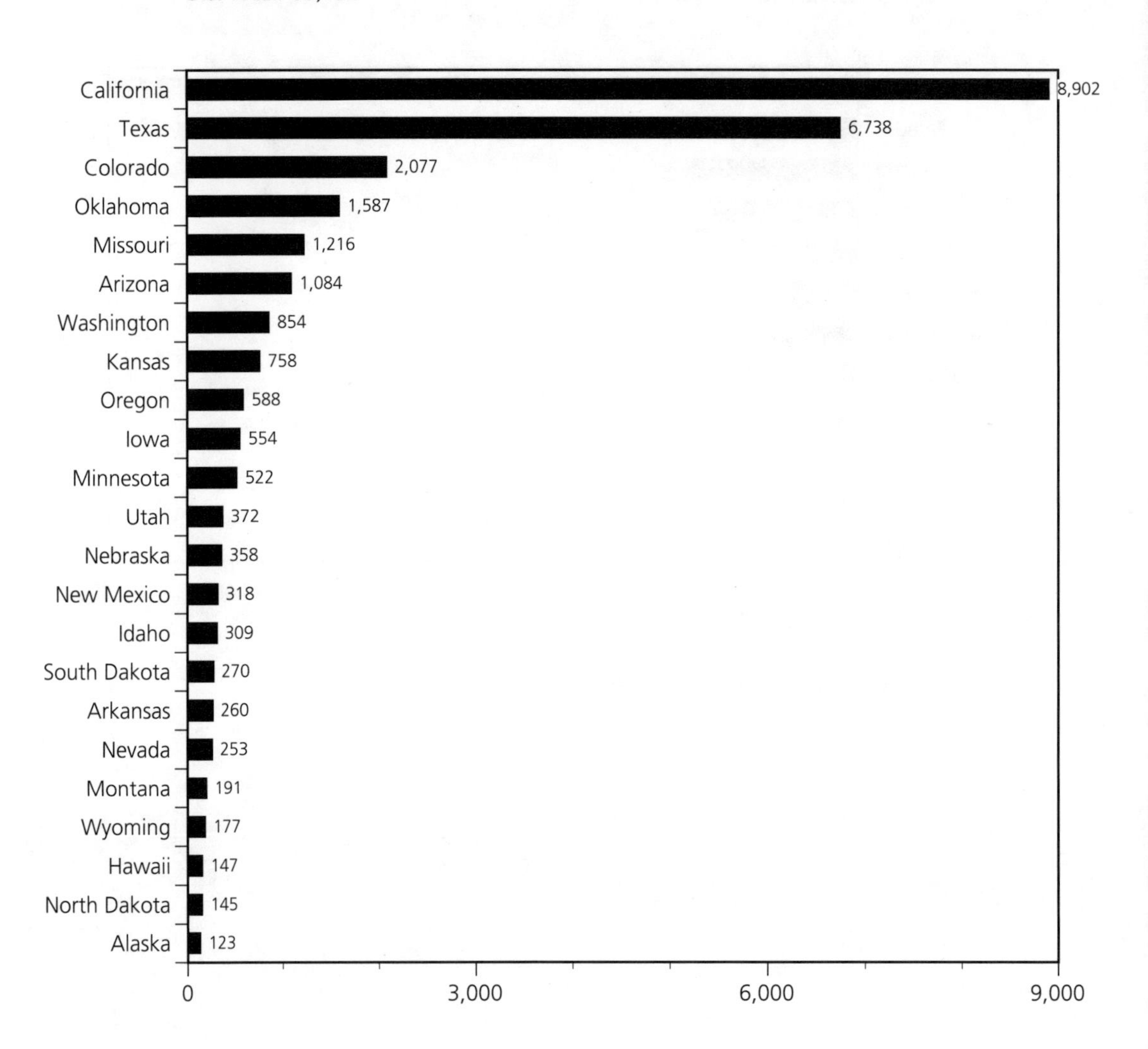

Source: U.S. Bureau of the Census

Corporate Headquarters

The western United States is home to 31 of the largest 100 industrial firms in the United States, 147 of the top 500 firms, and 181 of the top 1,000 firms, according to *Business Week* in 1993. Of these, 117 were headquartered in California.

Figure 13.20 Western United States
Twenty-Five Largest *Fortune* 500 Industrial Corporations, 1992

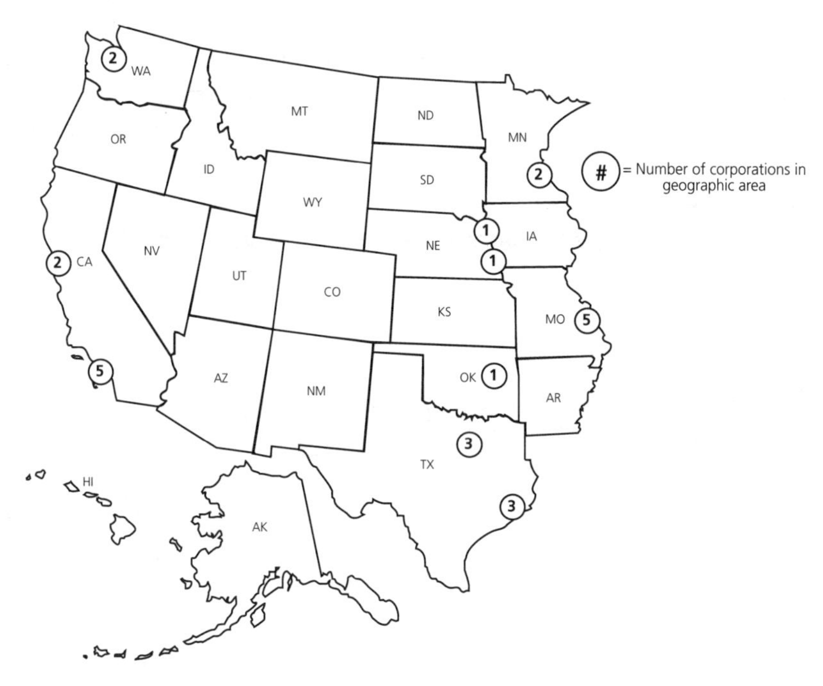

Source: Fortune, April 19, 1993

	Headquarters	National Rank
Exxon	Irving, TX	2
Chevron	San Francisco, CA	9
Boeing	Seattle, WA	12
Shell Oil	Houston, TX	17
Conagra	Omaha, NE	18
Atlantic Richfield	Los Angeles, CA	22
McDonnell-Douglas	St. Louis, MO	23
Hewlett-Packard	Palo Alto, CA	24
3M	St. Paul, MN	28
Tenneco	Houston, TX	30
Anheuser-Busch	St. Louis, MO	41
IBP	Dakota City, NE	42
Rockwell International	Seal Beach, CA	43

	Headquarters	National Rank
Lockheed	Calabasas, CA	45
Coastal	Houston, TX	46
Weyerhaeuser	Tacoma, WA	51
Citgo Petroleum	Tulsa, OK	53
Unocal	Los Angeles, CA	55
Occidental Petroleum	Los Angeles, CA	56
Monsanto	St. Louis, MO	60
General Mills	Minneapolis, MN	68
Ralston Purina	St. Louis, MO	69
Emerson Electric	St. Louis, MO	70
Texas Instruments	Dallas, TX	71
Kimberly-Clark	Dallas, TX	75

Although the western United States has a small proportion of total U.S. corporate headquarters, it has a much higher proportion of the high-technology businesses on which prosperity in the 1990s chiefly will depend. Of the twenty-two computer and peripheral manufacturers in the *Business Week* 1000 for 1993, fourteen are located in the West. Of thirty-one computer software firms in the *Business Week* 1000, eighteen—including the two largest—are in the West. Of fourteen semiconductor manufacturers in the *Business Week* 1000, eleven are in the West. The most important of these is Microsoft, headquartered in the Seattle suburb of Redmond, Washington. In 1980 Microsoft didn't exist. In 1993, Microsoft had a net worth comparable to that of the Ford Motor Company.

"Silicon Valley" in and around San Jose and the Los Angeles area has the largest concentration of computer and computer-related firms in the western United States, but mini-Silicon Valleys have formed along Utah's Wasatch Front Range from Provo to Salt Lake City, along Colorado's Front Range from Ft. Collins to Colorado Springs, and in Austin, Texas.

Figure 13.21 **Western United States High-Technology Corporate Headquarters by State, 1992**

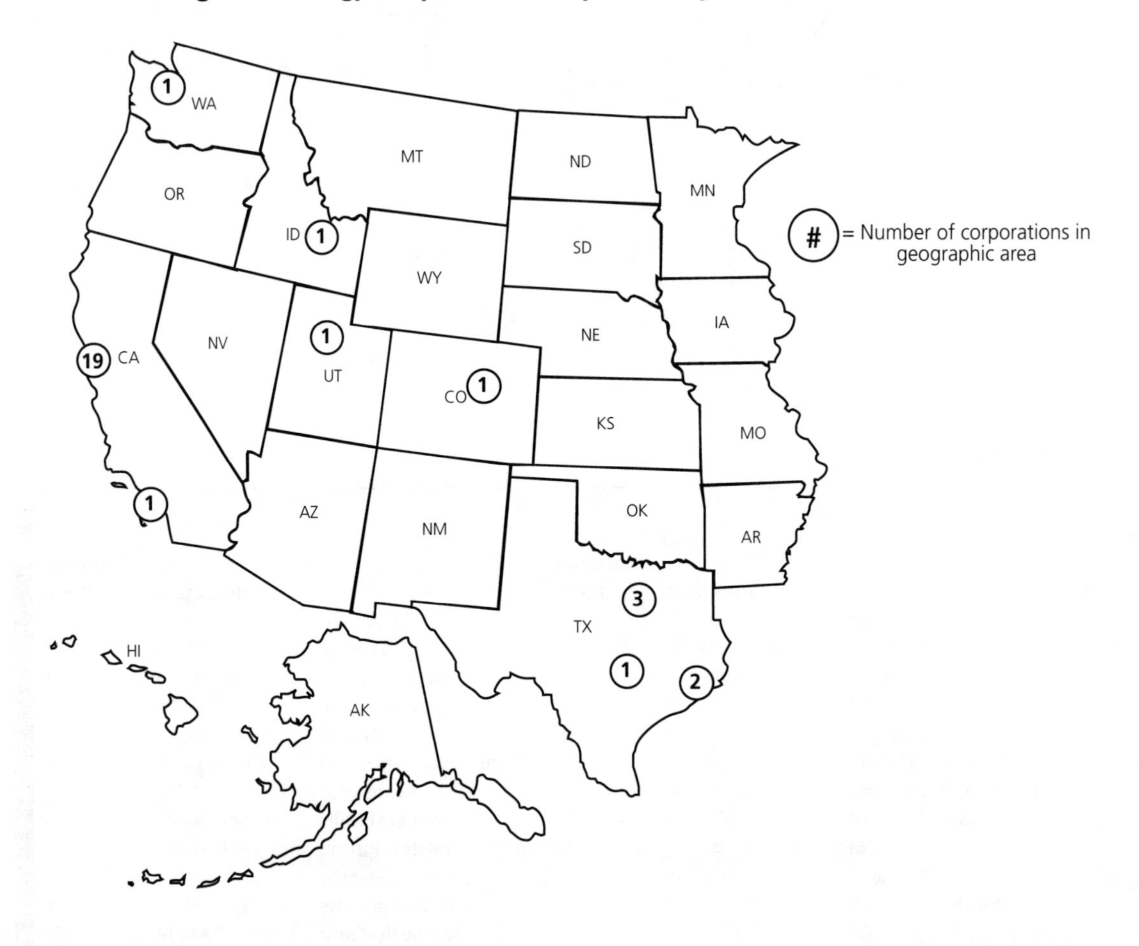

Source: Business Week 1000, 1993

Figure 13.21a **Western United States**
Ten Largest Computer and Peripheral Companies, 1992

	Headquarters	National Sector Rank
Hewlett-Packard	Palo Alto, CA	2
Apple	Cupertino, CA	4
Compaq	Houston, TX	7
Silicon Graphics	Mountain View, CA	9
Tandy	Fort Worth, TX	10
Tandem	Cupertino, CA	11
Dell Computer	Austin, TX	12
Seagate Technology	Scotts Valley, CA	13
Storage Technology	Louisville, CO	14
Amdahl	Sunnyvale, CA	16

Figure 13.21b **Western United States**
Ten Largest Semiconductor Companies, 1992

	Headquarters	National Sector Rank
Intel	Santa Clara, CA	2
Texas Instruments	Dallas, TX	4
Advanced Micro Devices	Sunnyvale, CA	7
National Semicondutor	Santa Clara, CA	9
Micron Technology	Boise, ID	10
Read-Rite	Milpitas, CA	11
Linear Technology	Milpitas, CA	12
Xilinx	San Jose, CA	13
Adeptec	Milpitas, CA	14
Solectron	Milpitas, CA	16

Source: Business Week 1000, 1993

Figure 13.21c **Western United States**
Ten Largest Software Companies, 1992

	Headquarters	National Sector Rank
Microsoft	Redmond, WA	1
Novell	Provo, UT	2
Electronic Data Systems	Dallas, TX	4
Cisco Systems	Menlo Park, CA	5
Oracle Systems	Redwood Shores, CA	6
Synoptics Communications	Santa Clara, CA	10
BMC Software	Sugar Land, TX	13
Sybase	Emeryville, CA	15
Computer Sciences	El Segundo, CA	16
Electronic Arts	San Mateo, CA	17

Source: Business Week 1000, 1993

Figure 13.22 **Canada**
Twenty-Five Largest Corporations, 1992

Source: Canadian Business

	Headquarters
BCE, Inc.	Montréal, Québec
General Motors of Canada, Ltd.	Oshawa, Ontario
Ford Motor Co. of Canada, Ltd.	Oakville, Ontario
George Weston, Ltd.	Toronto, Ontario
Imasco, Ltd.	Montréal, Québec
Alcan Aluminum, Ltd.	Montréal, Québec
Chrysler Canada, Ltd.	Windsor, Ontario
Loblaw Companies, Ltd.	Toronto, Ontario
Imperial Oil, Ltd.	Toronto, Ontario
Canadian Pacific, Ltd.	Montréal, Québec
Noranda, Inc.	Toronto, Ontario
Northern Telecom, Ltd.	Mississauga, Ontario
Brascan, Ltd.	Toronto, Ontario

	Headquarters
Bell Canada	Montréal, Québec
Seagram Co., Ltd.	Montréal, Québec
Thomson Corp.	Toronto, Ontario
Ontario Hydro	Toronto, Ontario
Hydro-Québec	Montréal, Québec
IBM Canada, Ltd.	Markham, Ontario
Univa, Inc.	Montréal, Québec
Power Corp. of Canada	Montréal, Québec
Hudson's Bay Co.	Toronto, Ontario
Oshawa Group, Ltd.	Etobicoke, Ontario
Petro-Canada	Calgary, Alberta
Shell Canada, Ltd.	Calgary, Alberta

A look at the corporate structure of Canada and Mexico shows how far along the economic integration of North America really is. Some of the largest private-sector firms in both Canada and Mexico are subsidiaries of U.S. corporations.

Very few major Canadian corporations are headquartered in the West.

Many of the largest corporations in Mexico are government owned (though privatization is rapidly changing that), or are subsidiaries of foreign—chiefly U.S.—firms.

Figure 13.23 **Mexico**
Twenty-Five Largest Corporations, 1991

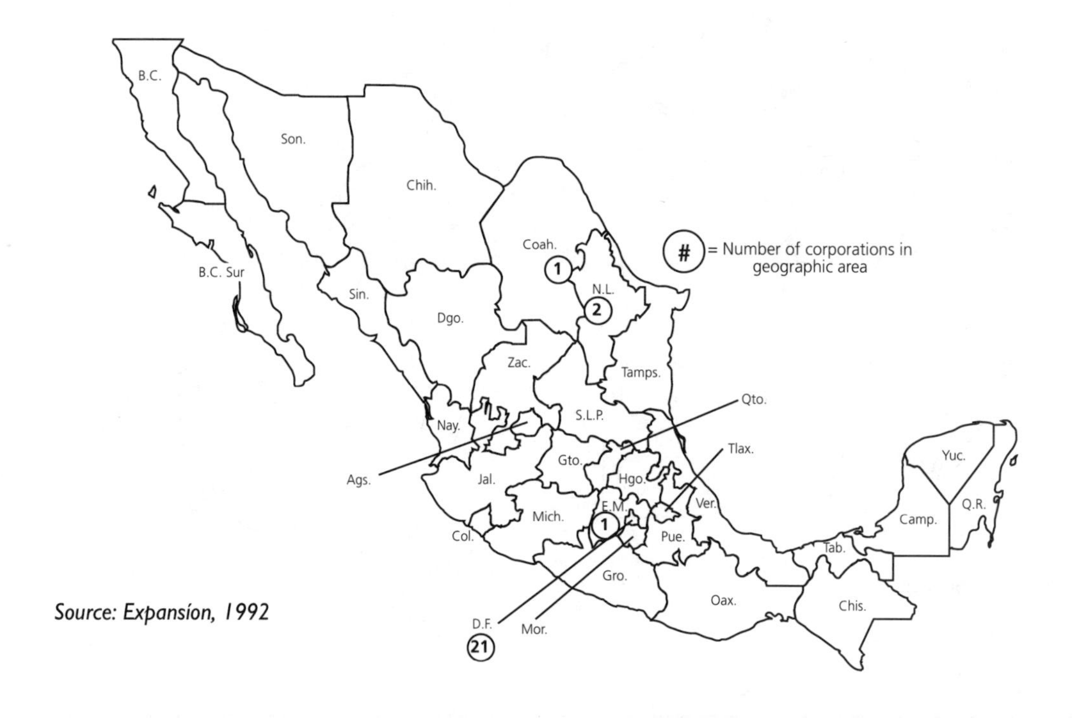

Source: Expansíon, 1992

	Headquarters
Petroleos Mexicanos	Distrito Federal
Telefonos de Mexico	Distrito Federal
General Motors de Mexico	Distrito Federal
Ford Motor Co.	Distrito Federal
Chrysler de Mexico	Distrito Federal
Anchor Glass Containers Corp.	Distrito Federal
IBM de Mexico	Distrito Federal
CIA Nestle	Distrito Federal
CIA Mexicana de Aviacion	Distrito Federal
Altos Hornos de Mexico	Monclova, Coahuila
Celanese Mexicana	Distrito Federal
Hylsa	Toluca, México
Kimberly-Clark de Mexico	Distrito Federal
El Puerto de Liverpool	Distrito Federal
Aerovias de Mexico	Distrito Federal
Empresas la Moderna	Monterrey, Nuevo León
Spicer	Distrito Federal
Seguros la Provincial	Distrito Federal
American Express Co.	Distrito Federal
Tolmex	Distrito Federal
Fomento Economico	Monterrey, Nuevo León
NADRO	Distrito Federal
Tiendas de Descuento	Distrito Federal
Cigarros la Tabacalera	Distrito Federal
Ingenieros Civiles Asociados	Distrito Federal

Chapter Fourteen
Government Finance

All three North American nations are federal republics, but there are considerable differences in how they are structured and governed.

The United States

The Constitution of the United States was approved and signed on September 17, 1787 and sent to the states for ratification. The United States has operated as a constitutional democracy longer than any other country in the world.

The federal government is divided into three equal branches: the legislative, executive, and judicial. All legislative powers are vested in the Congress of the United States, which consists of a Senate and a House of Representatives. Each state, independent of its size, has two senators, elected for a six-year term, with a third of the one hundred members elected every two years. The House of Representatives is elected every other year for a two-year term. Its membership is apportioned among the states based on each state's population. Currently there are 465 House members, with California having the most representatives (52).

For a bill to become law, it must be passed by both Houses of Congress and signed by the president, or if the president vetoes it, be passed by a two-thirds overide vote of each House. All bills for raising revenue must originate in the House of Representatives.

The executive power is vested in the President of the United States, who is elected every four years. The president and vice president are elected by direct vote. However, under the electoral system, the leading candidates in each state receive votes equal to the number of senators and representatives from that state (for example, California would have 54 electoral votes, Wyoming 3). If no candidate has a majority vote in the electoral college, then the House selects the president from the leading candidates and the Senate selects the vice president. The 22nd amendment to the Constitution, which was adopted in February 1951, limits the president to two terms in office.

The president has many responsibilities, including acting as Commander-in-Chief of the Army and Navy; conducting foreign affairs; making treaties, nominating ambassadors, judges, and other officers of the United States; and generally supervising the operation of the federal government.

The governments of the fifty states are organized in a similar manner, with bicameral legislatures, except for Nebraska, which has a unicameral legislature. The executive officer is the governor. The United States has traditionally been a two-party system, although third and fourth parties have appeared from time to time.

Since World War II, the power of the Federal government has increased significantly and the power of the states has decreased.

Canada

Canada's federal government takes its form from the system used in British North America in 1867, when the British colonies of Nova Scotia, New Brunswick, and "Canada" (mostly present-day Ontario and Québec) were united into one "dominion."

In April, 1982, all formal subordination to the British crown was terminated, but the Queen of England is still the titular head of Canada's executive branch. Canada, like Britain, is a parliamentary system, and real political power is vested in the prime minister and the cabinet. Canada's Parliament consists of a 104-member Senate, whose members are appointed on a provincial basis, and a 295-member House of Commons. The Canadian Senate is much more important than Britain's House of Lords, but much less so than the U.S. Senate. The prime minister is a member of the House of Commons, and, as the leader of the party with the most seats, is asked to form the government. Both bodies may originate legislation, but only the House of Commons may introduce tax or spending bills.

In both law and practice, the ten Canadian provinces have more authority vis-à-vis the federal government than do the fifty U.S. states. Each Canadian province has a prime minister and a unicameral legislature.

Canada has three major national political parties: the Progressive Conservatives (who governed from 1984 to 1993), the Liberals (currently in power), and the New Democrats. The 1993 election saw the emergence of two strong regional parties, the Bloc Québecois in Québec and the Reform Party in the West. In the same election, official party status was lost by the Progressive Conservatives and the New Democratic Party in the House of Commons. However, the New Democrats remain the most left-leaning of the three parties, with support strongest in the West.

Mexico

The government of the United Mexican States (Estados Unidos Mexicanos) takes its shape from the Constitution of 1917, which supports socialism, capitalism, democracy, authoritarianism, and a host of provisions for specific social reforms, based mostly upon cultural traditions, some of which date from pre-Hispanic times.

As in the United States, the federal government is divided into three branches: executive, legislative, and judicial. But unlike the United States, the branches are not equally balanced. The president of Mexico has much more power relative to his government than the president of the United States has relative to his. The legislative branch may pass as many laws as it wishes, but only the president can put a law into effect. There is no constitutional way for a Mexican president to be forced to sign a law and order its publication. Legislation sponsored by the executive branch takes precedence over other business, and the Constitution gives the president wide authority to issue basic rules that give effect to the more general provisions of a statute. Such regulations have the same legal force as the statute itself, as the Mexican Constitution is based on the Napoleonic Code, a legal code that relies on statutory law. United States law, on the other hand, is based upon English common law, or case law.

The Mexican federal legislature consists of a sixty-four-member Senate (two for each of the thirty-one Mexican states and two from the Distrito Federal) and a five-hundred-member Chamber of Deputies, three hundred elected by relative majority or plurality and two hundred elected by proportional representation (as in many European nations). The latter are selected by the number of votes cast for each competing political party rather than for the individual. No single party can hold more than 315 seats (63 percent).

The Chamber of Deputies is relatively more powerful in relation to the Mexican Senate than is the U.S. House of Representatives in relation to the U.S. Senate. As in Canada and the United States, all spending and tax measures must originate in the Chamber of Deputies.

Beginning in 1994, the Mexican Senate will increase the number of its members to 128, or four per state. Three will be elected directly (by plurality) and the fourth will represent the political party that finished second in that state's elections.

The Mexican president and members of the Mexican Senate serve six-year terms. Members of the Chamber of Deputies serve a three-year term. The president cannot ever be re-elected, while members of the Senate or the Chamber of Deputies may skip one or more terms and then run for office again.

The government of each of the thirty-one Mexican states is headed by a governor, who is elected to a single six-year term, and a Chamber of Deputies, which ranges in size according to the state's population. The Regent of the Distrito Federal, the functional equivalent of a governor, is appointed by the Mexican president, as are members of his or her cabinet. Beginning in 1997, however, these appointments will have to be ratified by a qualified two-thirds vote of the local congress of the Distrito Federal, the Assembly of Representatives. This body consists of sixty-six members, forty-four of whom are elected by plurality and twenty-two through proportional representation.

Governors have proportionately the same power relative to state governments in Mexico as the president does to the federal government. However, the Mexican states have far less power vis-à-vis the federal government than do U.S. states or Canadian provinces, chiefly because the Mexican federal government controls the vast majority of tax revenues. Local governments have even less power. Governors must approve municipal budgets.

Mexico has been practically a one-party state since the 1910 revolution. The Institutional Revolutionary Party (PRI) has won every election for president, nearly every election for governor and senator, and most seats in the federal Chamber of Deputies since 1929, when its predecessors, the PNR and the PRM, were created. The fact that he is also the strongest member of the PRI boosts the *de facto* power of the Mexican president beyond the generous provisions of the Mexican Constitution. The principal opposition parties are the conservative National Action Party (PAN) and the center left Democratic Revolutionary Party (PRD). PAN is strongest in northern Mexico, particularly in the states of Baja California del Norte and Chihuahua, where the governors are PAN members, and in the central state of Guanajuato. The PRD is strongest in Michoacán, the State of México, several southern states, and the Distrito Federal. There are six other smaller parties.

Federal Government Spending and Debt

Government spending and public debt have been rising in the United States and Canada, and falling in Mexico. Public spending in Mexico is also included in the investment component of Gross Domestic Product because of the number of government-owned enterprises. Because government is not separated from private investment in the national product accounts, both lines have been included. Nor does the relatively low government spending include debt service.

Federal spending as a portion of total public spending is highest in Mexico, and lowest in Canada.

Figure 14.1 **United States**
Government Expenditures (Federal, State, and Local) as Percent of Gross Domestic Product, 1980–1992*

(*Tabulated from fourth-quarter results, 1989 through 1991, and third-quarter results, 1992.)

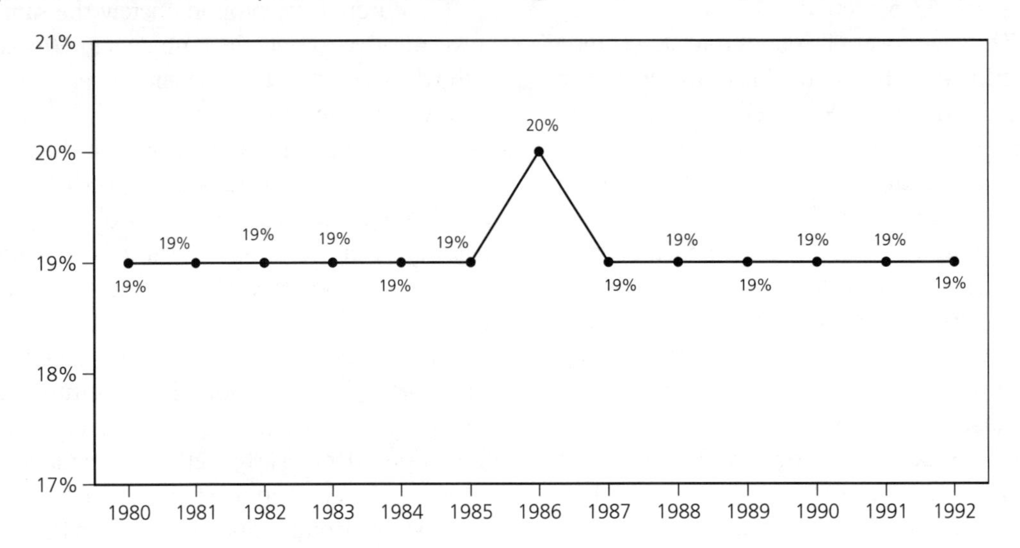

Source: Economic Report of the President, 1993
(GDP does not include transfer payments.)

Transfer payments (monetary payments, such as social security, made to individuals are not counted in Gross Domestic Product. However, if added to other government expenditures that are included in Gross Domestic Product, the combined total is more than one-third of the current U.S. Gross Domestic Product.

Figure 14.2 **Canada**
Government Expenditures as Percent of Gross Domestic Product, 1980–1992

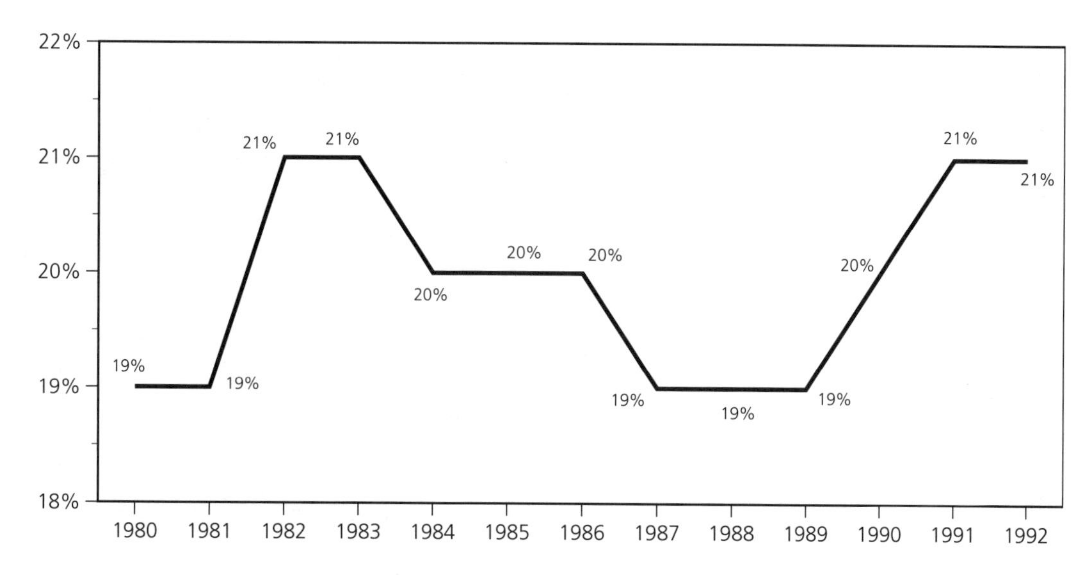

Source: International Monetary Fund, National Product Accounts

Figure 14.3 **Mexico**
Government Expenditures as Percent of Gross Domestic Product, 1980–1991

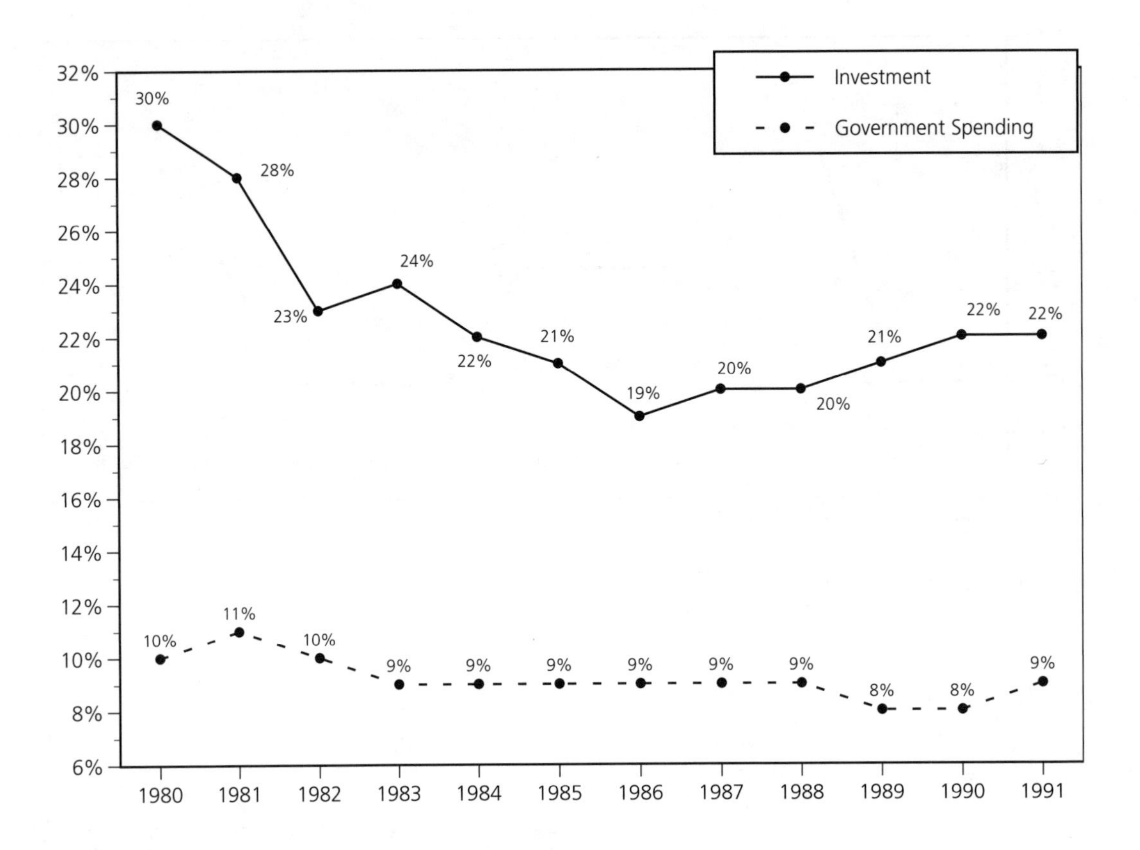

(* Investment represents a substantial portion of government spending in Mexico.)

Source: International Monetary Fund, National Product Accounts

Figure 14.4a **United States**
Government Expenditures and Receipts, 1980–1991

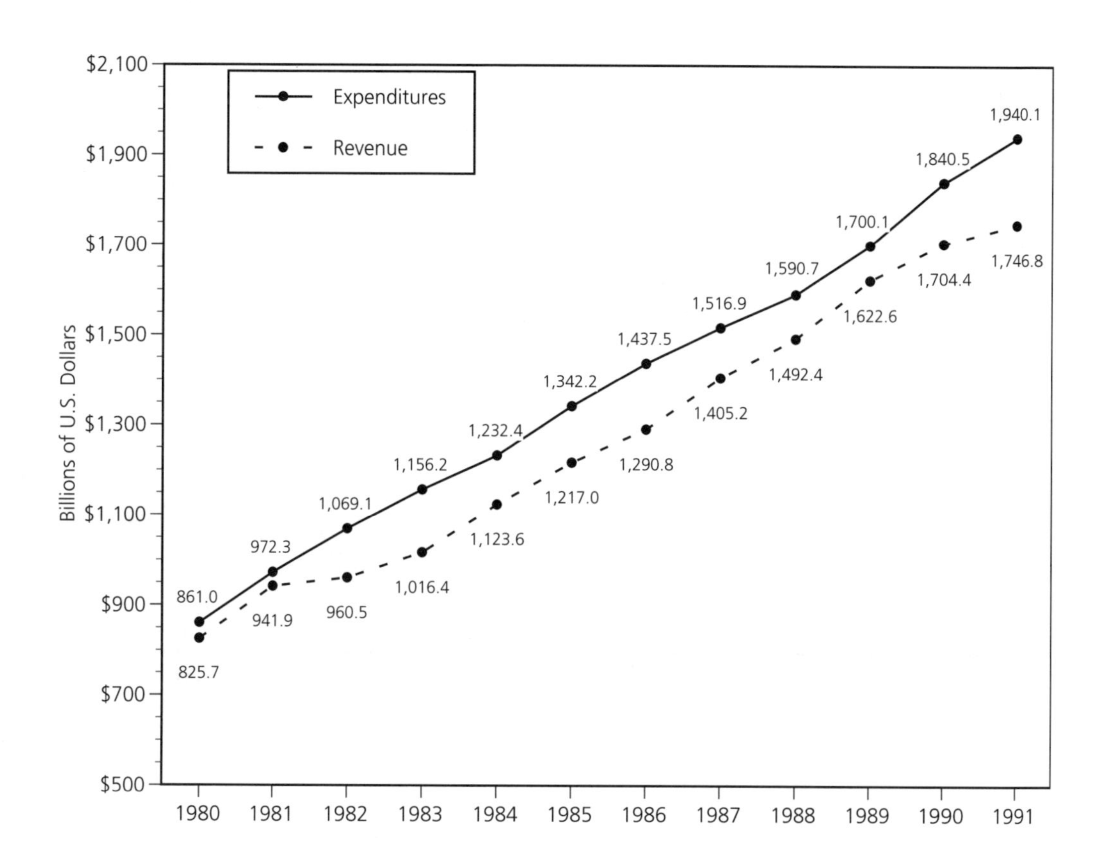

Source: Economic Report of the President

Figure 14.4b **United States**
Government Expenditures as a Percent of Revenue, 1980–1993

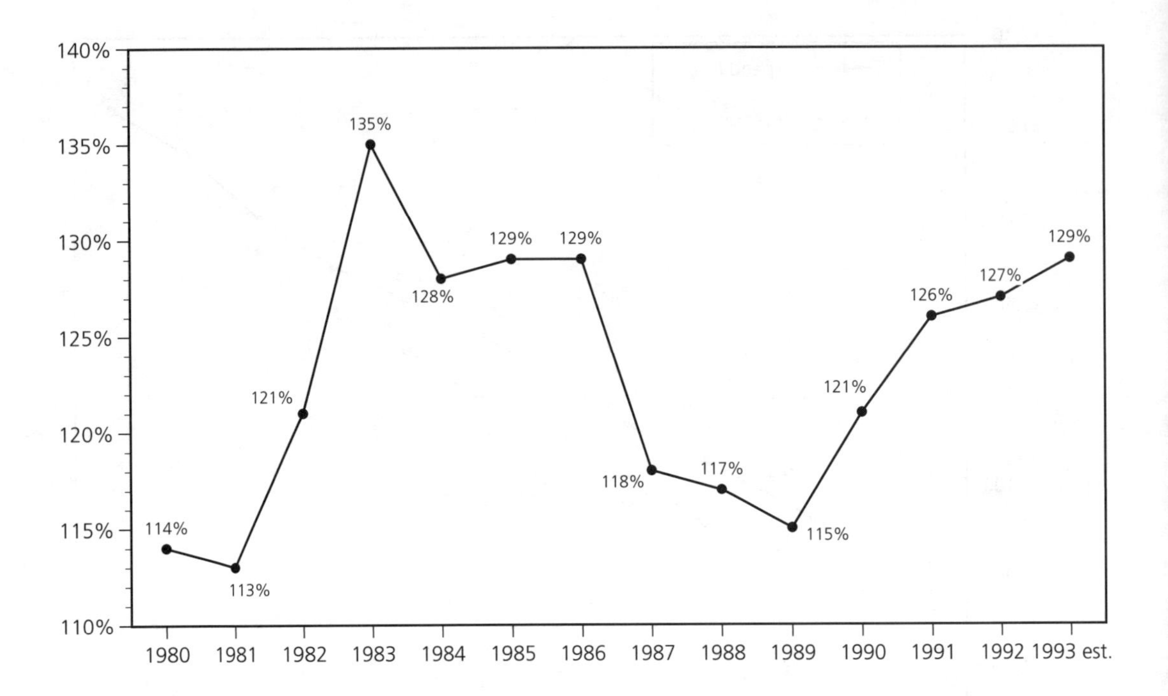

Source: Economic Report of the President

Figure 14.5a **Canada**
Government Expenditures and Revenue, 1980–1989

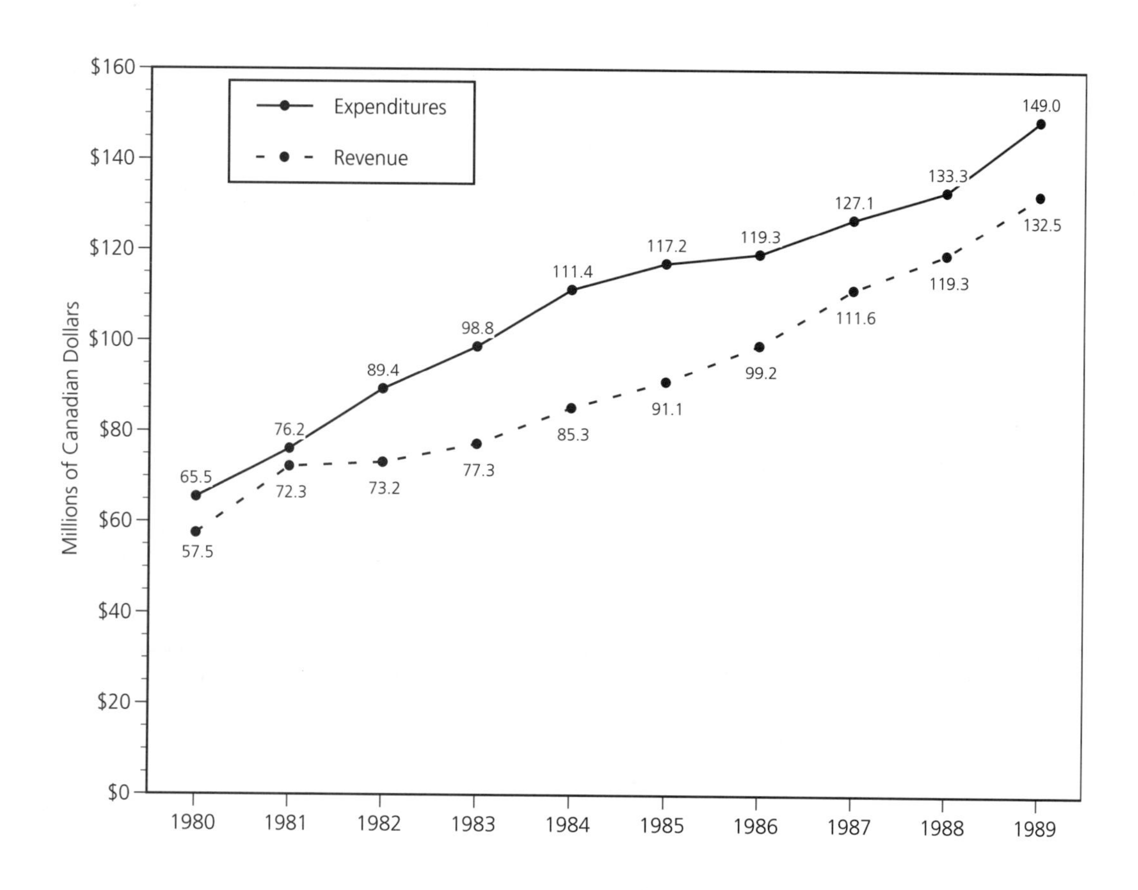

Source: International Monetary Fund, National Product Accounts

Figure 14.5b **Canada**
Government Expenditures as a Percent of Revenue, 1980–1989

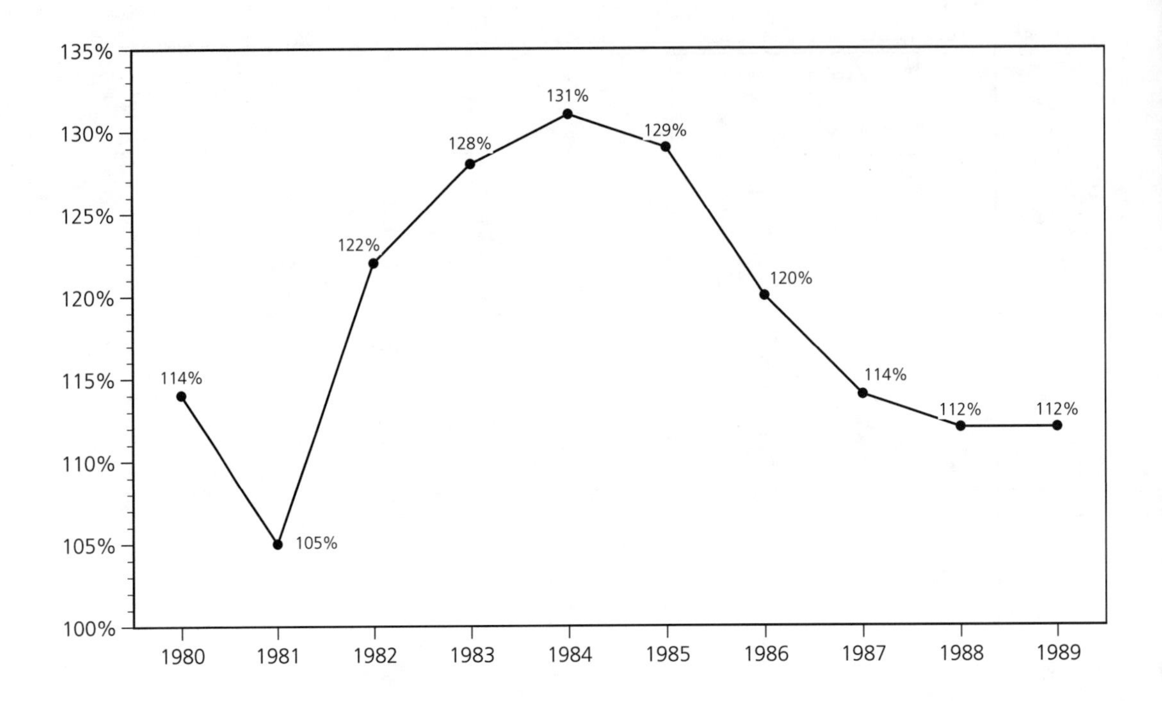

Source: International Monetary Fund, National Product Accounts

Figure 14.6a **Mexico**
Government Expenditures and Revenue, 1980–1990 preliminary

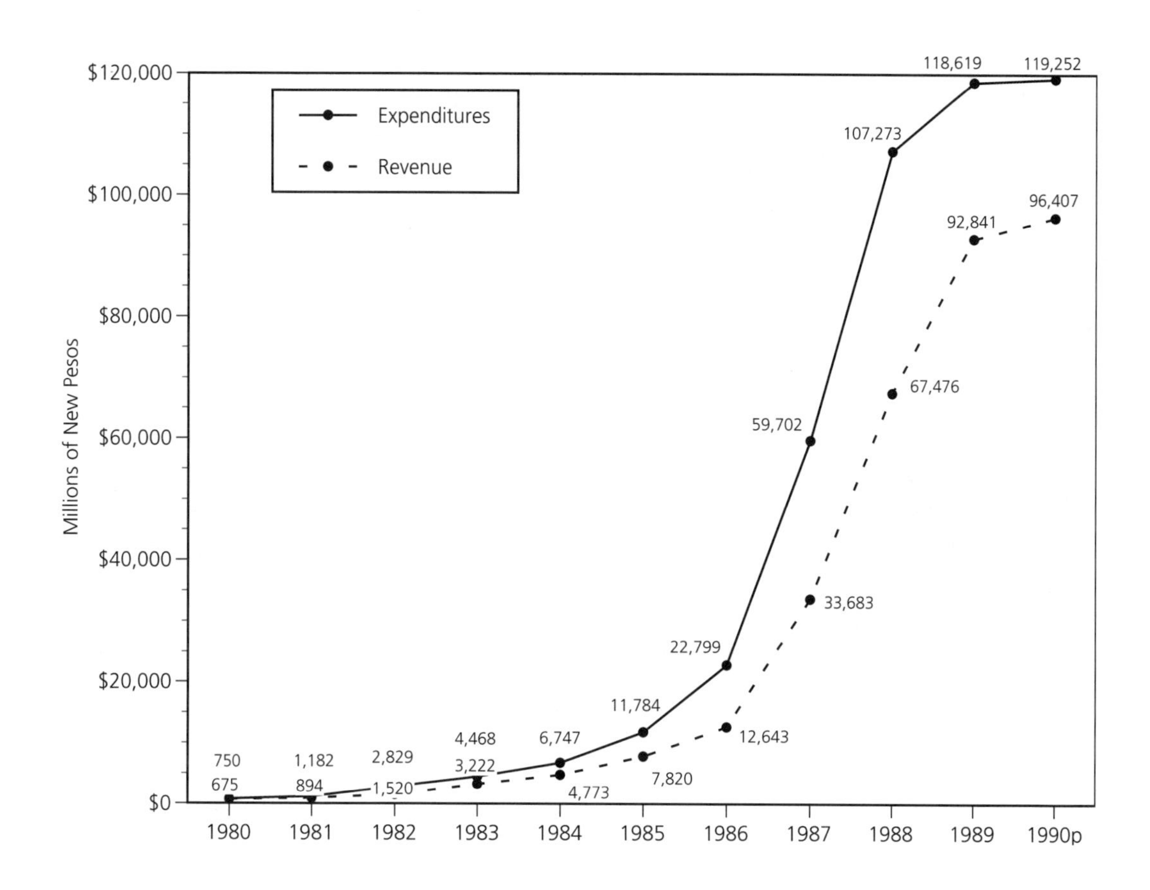

Source: International Monetary Fund, National Product Accounts

Figure 14.6b **Mexico**
Government Expenditures as a Percent of Revenue, 1980–1990 preliminary

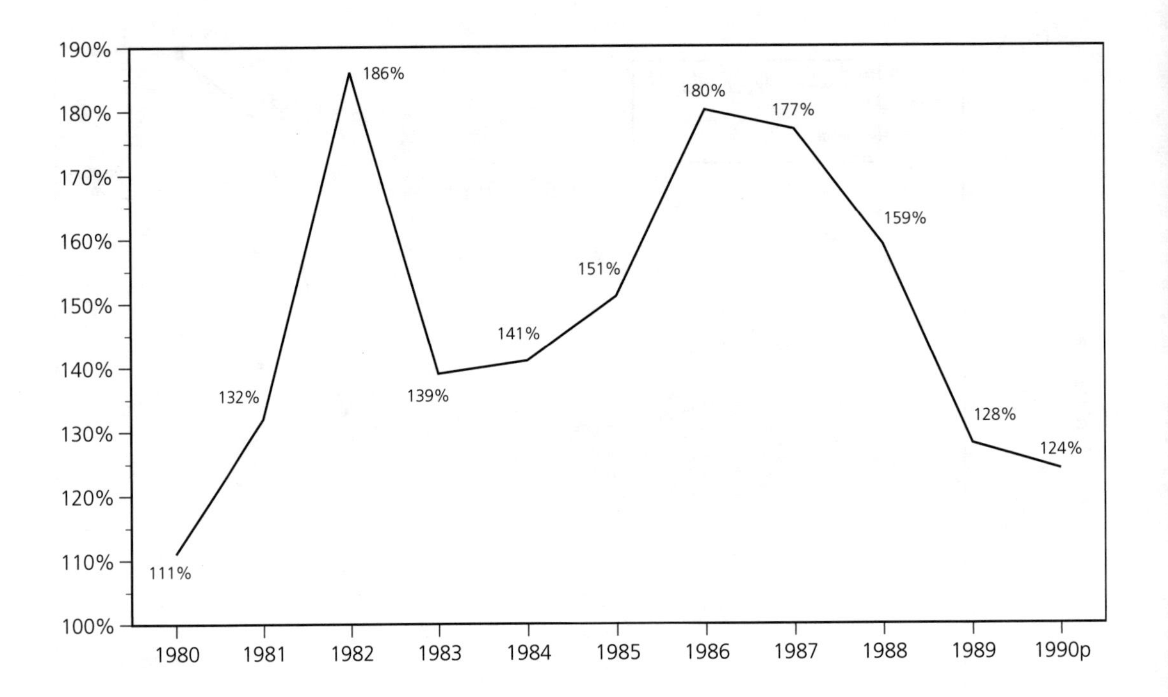

Source: International Monetary Fund, National Product Accounts

Income tax, both personal and corporate, is the primary source of federal revenues in the United States, Canada, and Mexico. Canada's relatively recent GST also contributes to federal revenues.

Figure 14.7 **United States Federal Receipts by Source, September 1992 (Billions of Dollars at Seasonally Adjusted Rate)**

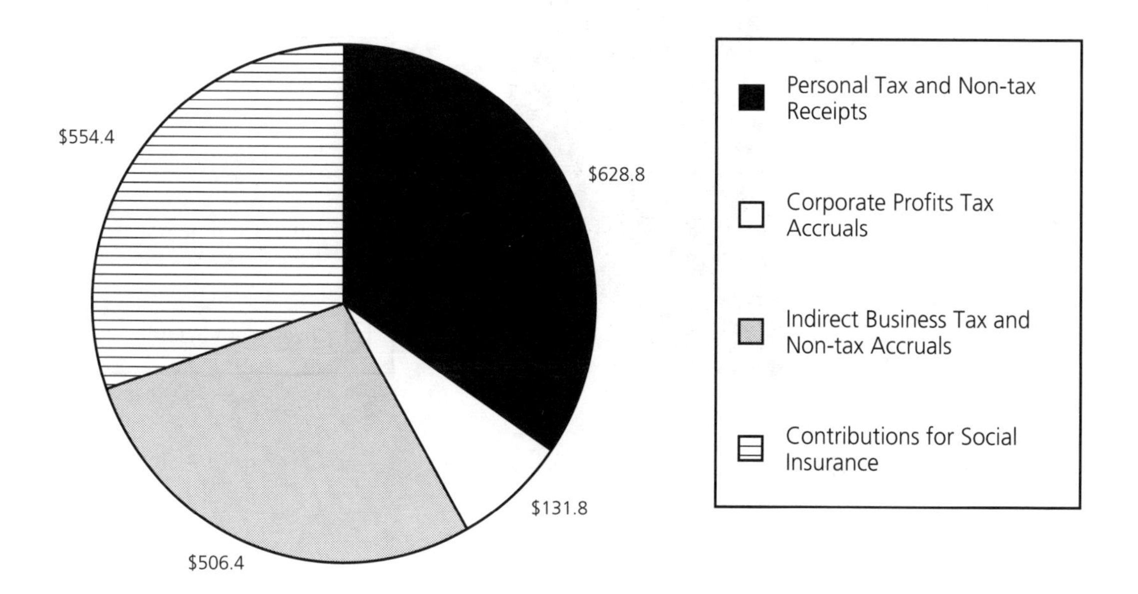

Source: Department of Commerce

Figure 14.8 **Canada**
Government Receipts by Source, September 1992
(Millions of Canadian Dollars)

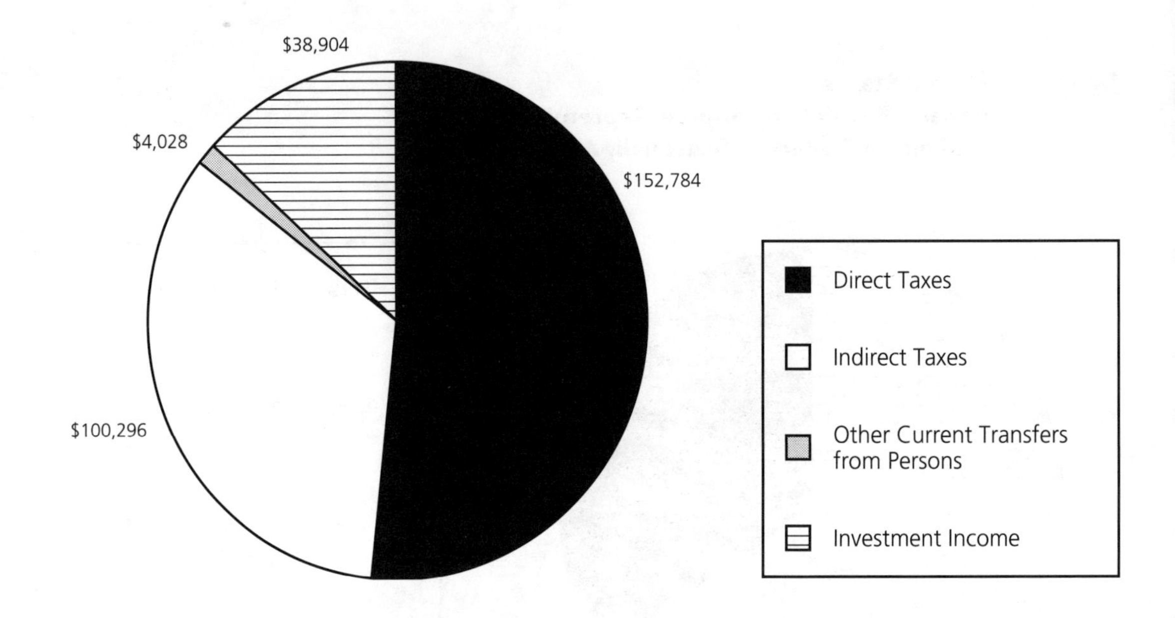

Source: Statistics Canada, Canadian Economic Observer

Figure 14.9 **Mexico**
Government Receipts by Source, December 1992 preliminary
(Millions of New Pesos)

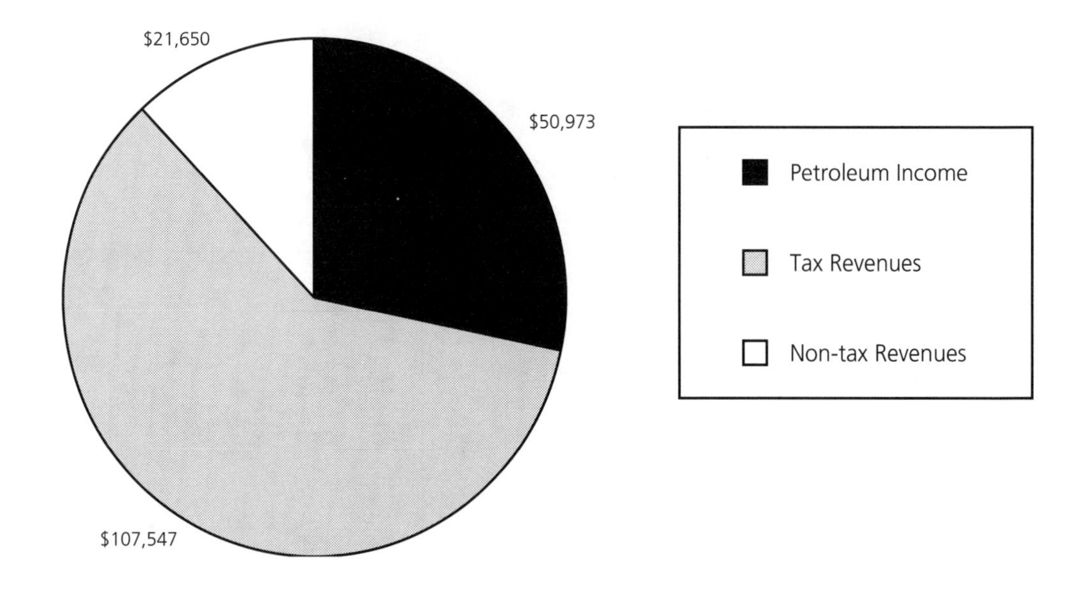

Source: Finance Secretariat, Planning Department

Figure 14.10 **United States**
Gross Domestic Product by Sector, 1991
(Billions of U.S. Dollars)

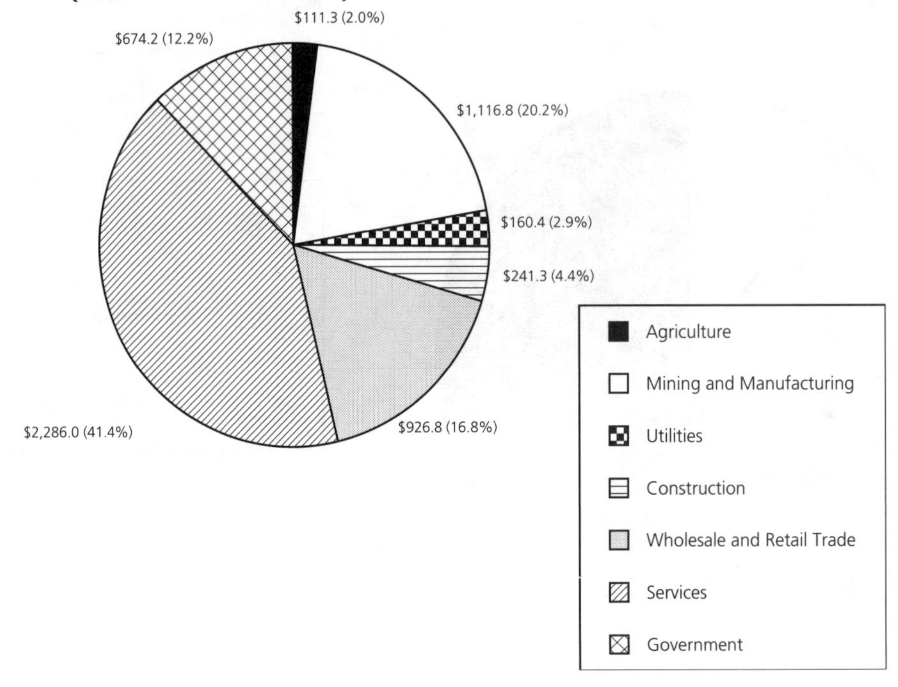

Source: Bureau of Economic Analysis, U.S. Department of Commerce

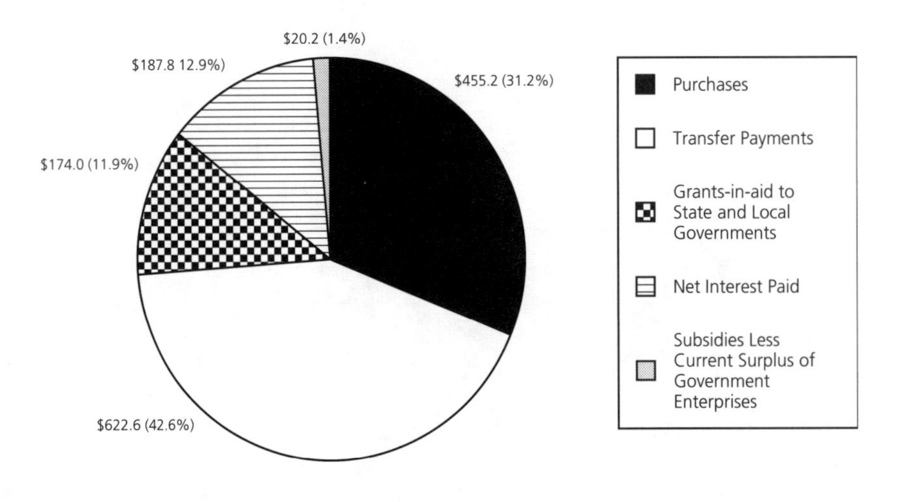

Source: U.S. Department of Commerce and Office of Management and Budget

Federal Government Outlays

Social services spending is the largest, and fastest-growing, component of the federal budgets of all three North American countries.

In the United States and Canada, federal expenditures for health care have been rising especially rapidly.

Figure 14.11 **Canada**
Gross Domestic Product by Industry at Factor Cost in 1986 Prices, September 1992
(Millions of Canadian Dollars)

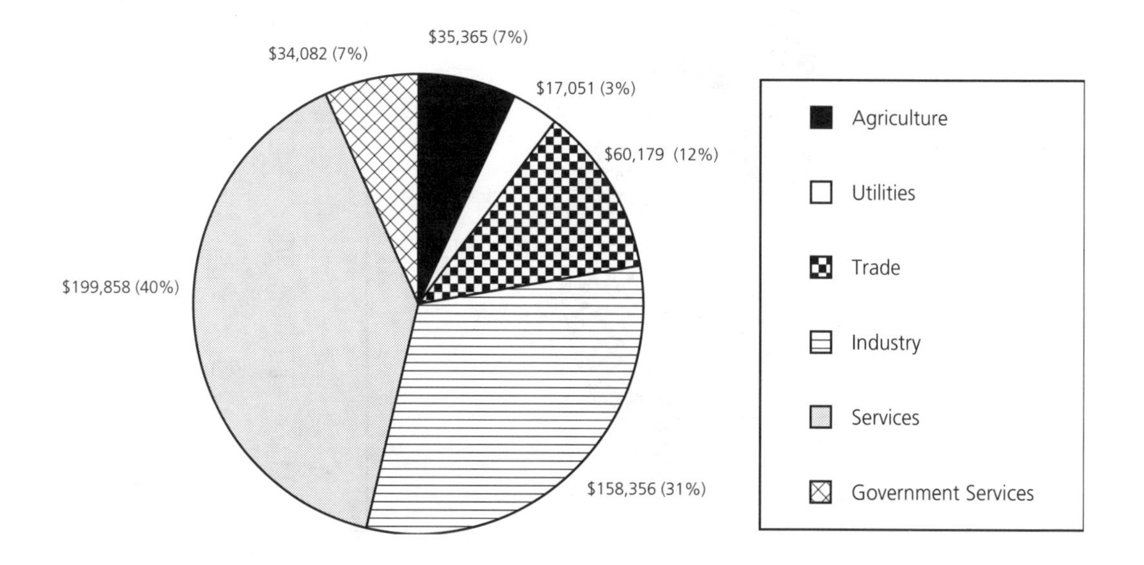

Source: Statistics Canada, Canadian Economic Observer

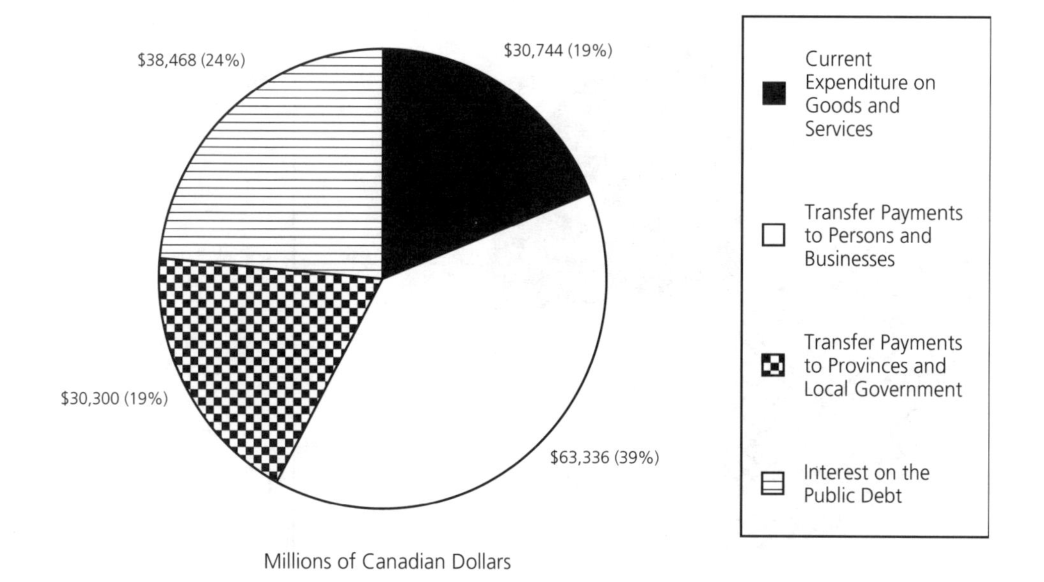

Millions of Canadian Dollars

Source: Statistics Canada

Figure 14.12 **Mexico**
Gross Domestic Product by Area, Third Quarter 1992
(Millions of 1980 Constant New Pesos)

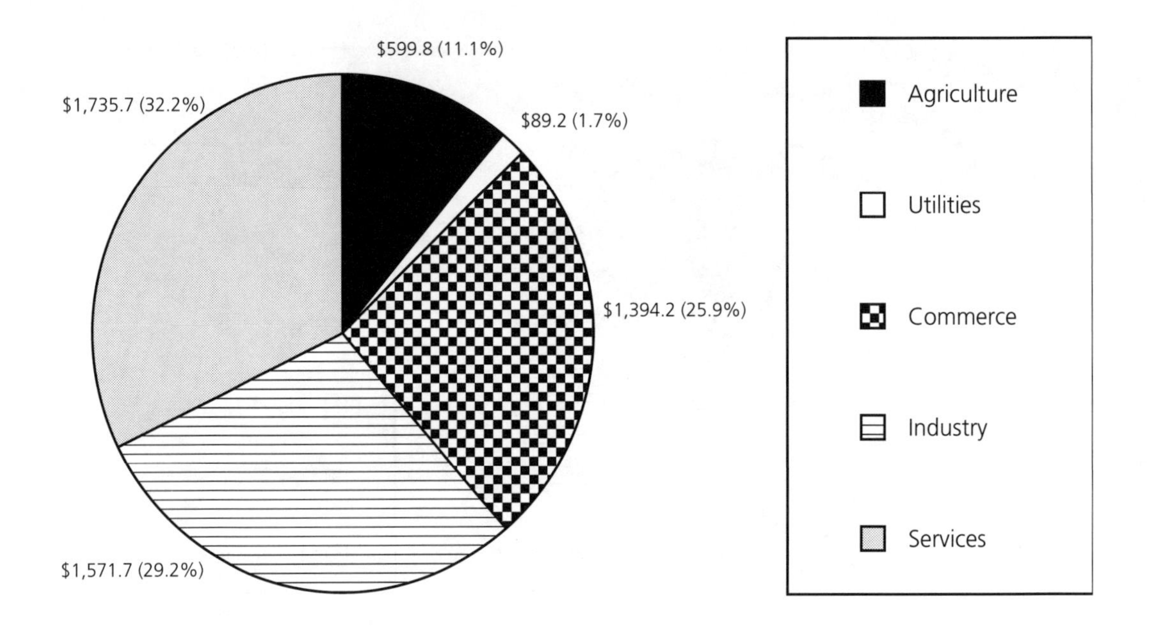

Source: INEGI

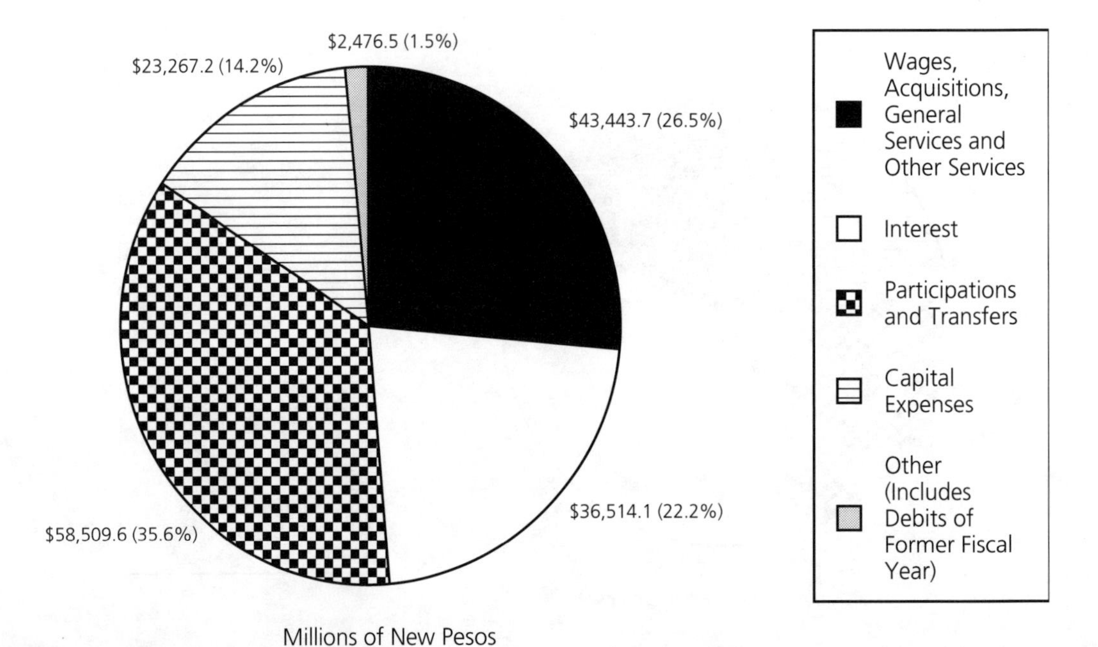

Source: Secretariat of Finance, Planning Department

Figure 14.13 **United States and Canada**
Health Expenditures as a Percent of Gross Domestic Product, 1980–1991

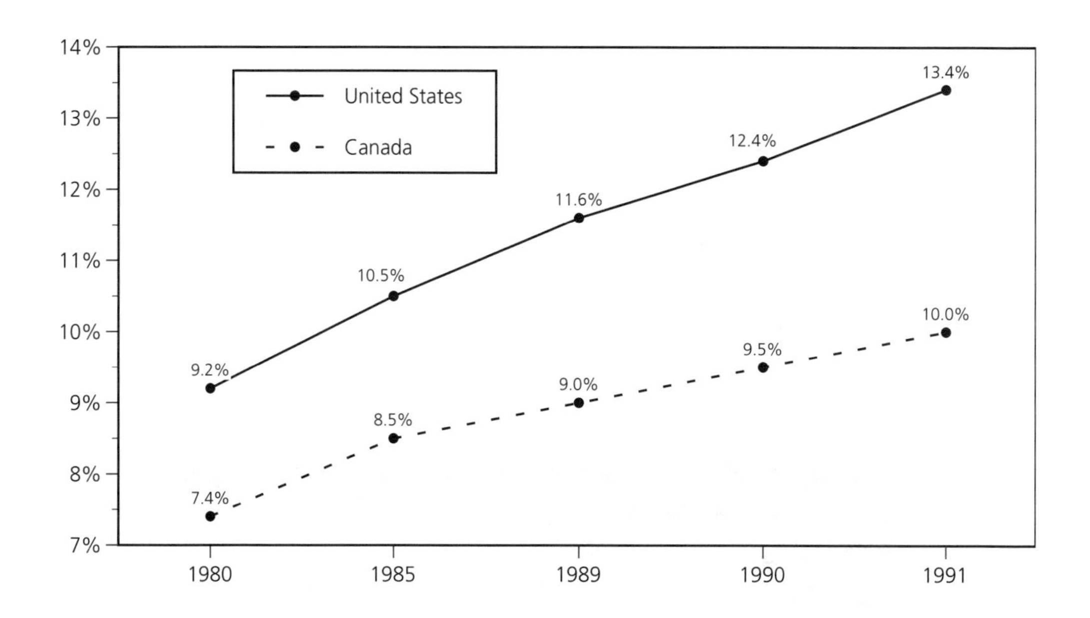

Source: OECD, Paris, France.

Figure 14.14a **Western United States State Revenues, Fiscal Year 1991**

West: **$252,803** million
East: **$407,386** million
U.S. Total: **$660,189** million

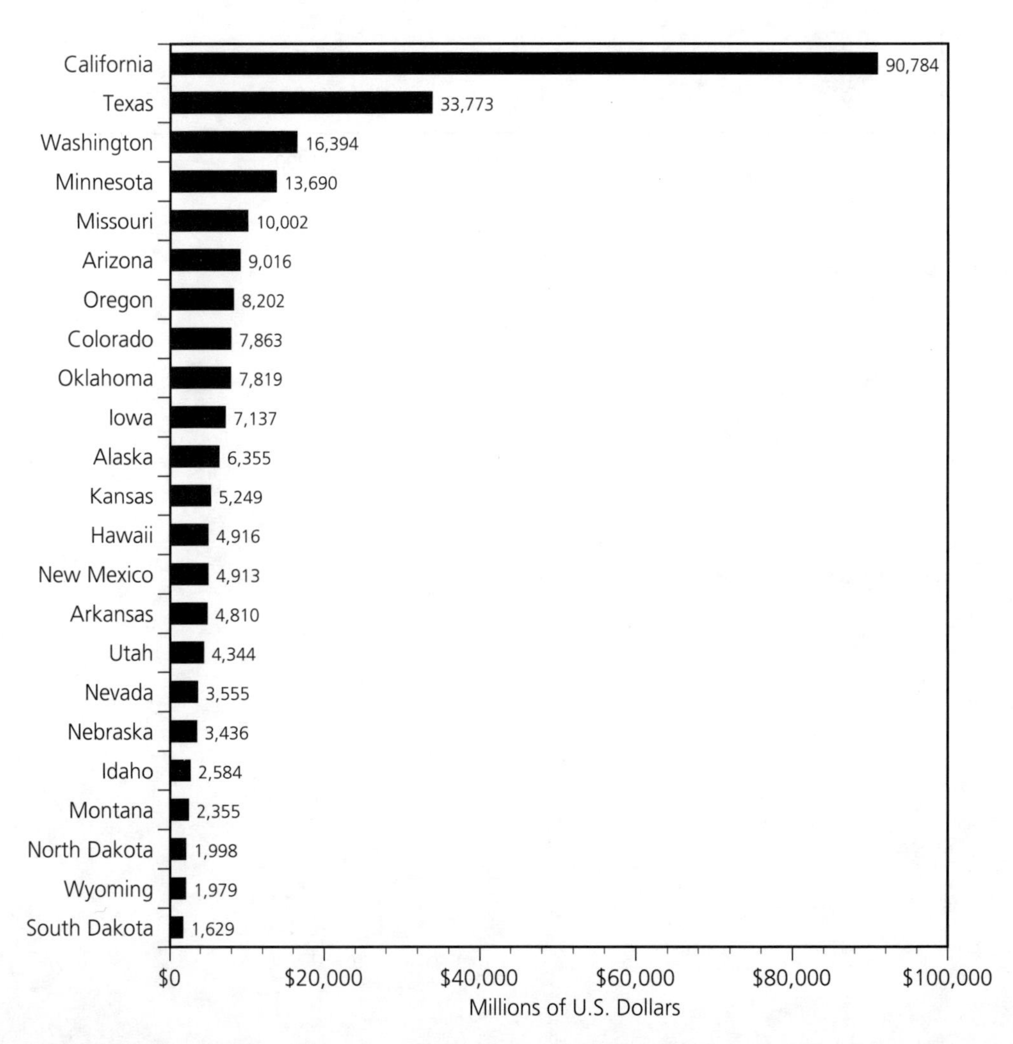

Source: U.S. Census Bureau

Figure 14.14b **Western United States State Expenditures, Fiscal Year 1991**

West: $233,460 million
East: $395,335 million
U.S. Total: $628,795 million

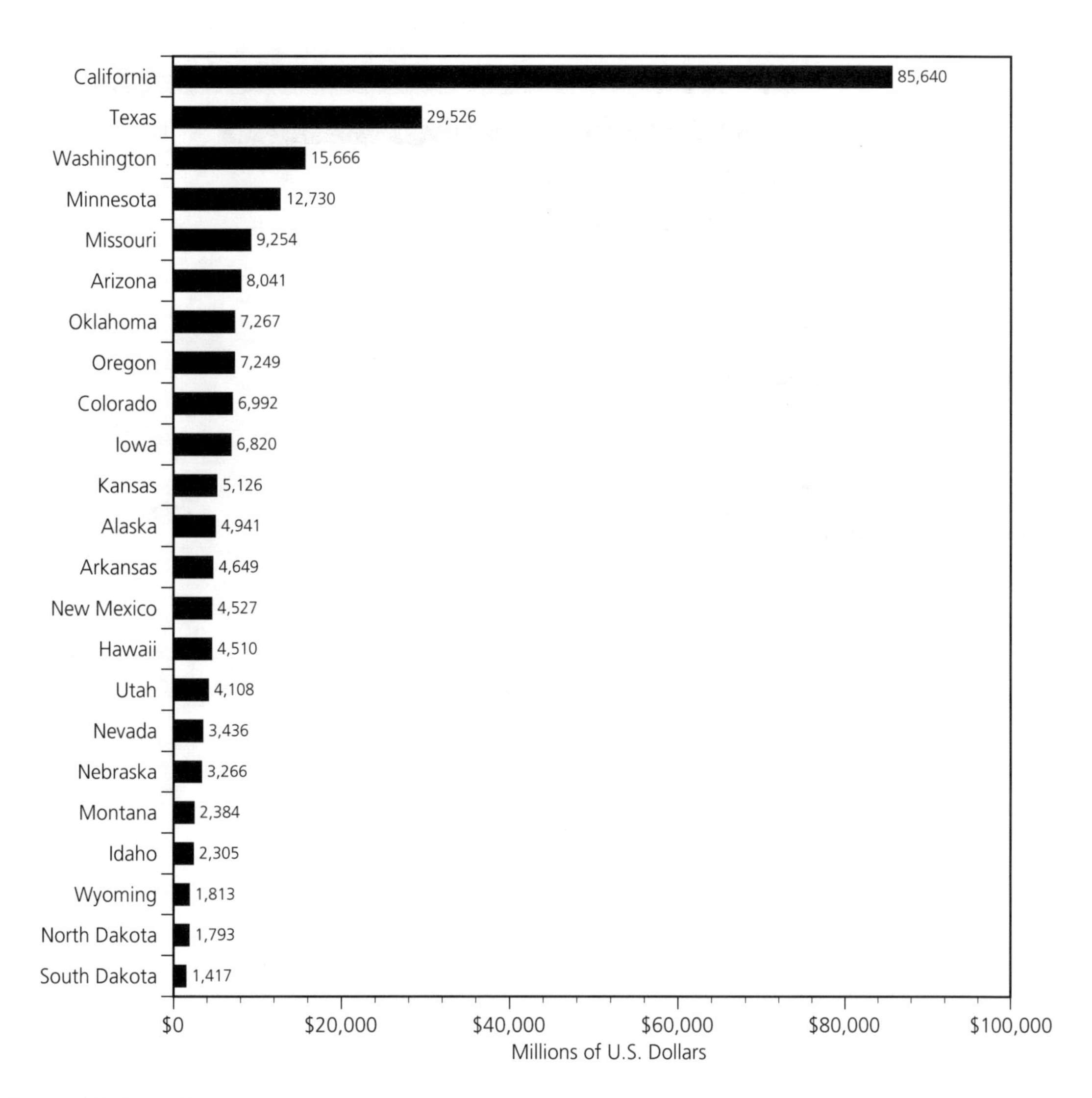

Source: U.S. Census Bureau

Figure 14.15a **Canada**
Provincial Revenues (Revised Estimate), 1991–1992

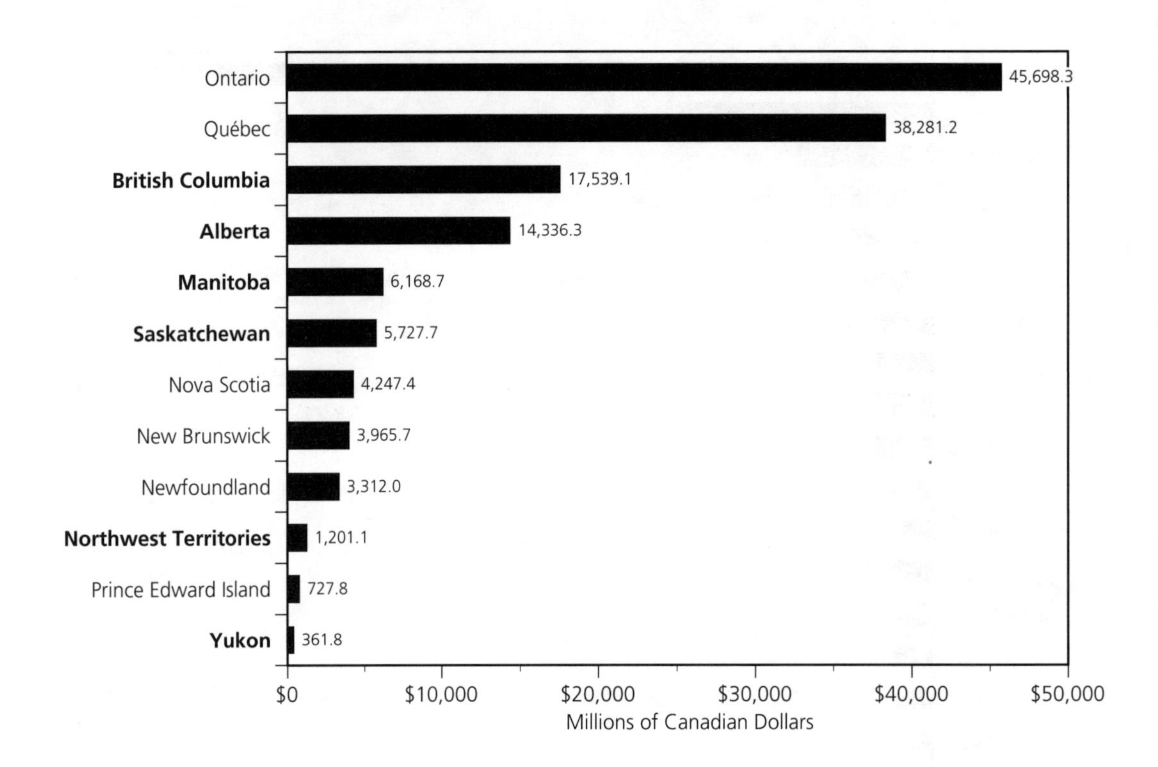

Source: Statistics Canada

Figure 14.15b **Canada**
Provincial Expenditures (Revised Estimate), 1991–1992

West: $49,328.9 million
East: $115,478.1 million
Canada Total: $164,807 million

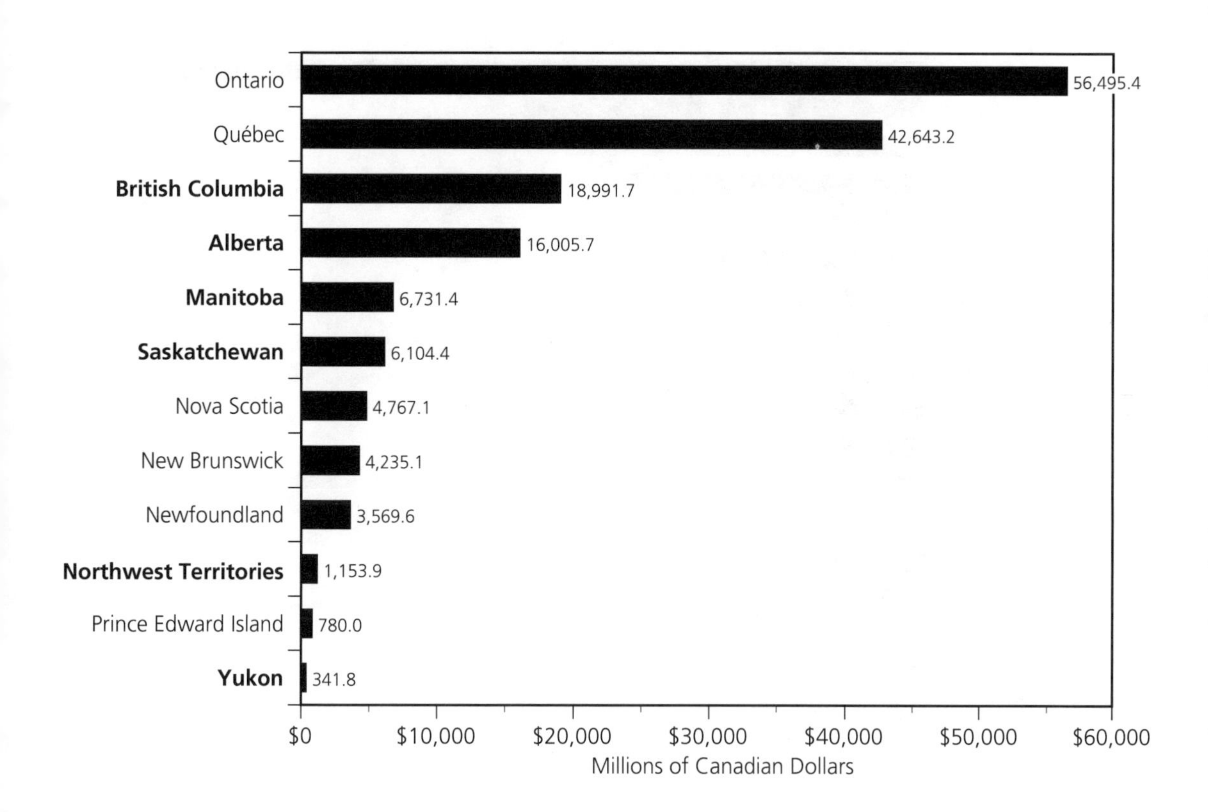

Source: Statistics Canada

Figure 14.16 **Mexico**
Revenue/Expenditures by State, 1989*

Mexico Total: $22,458,414 million

(*Revenue and expenditures at the state level are presented as a balance sheet, so they are equivalent.)

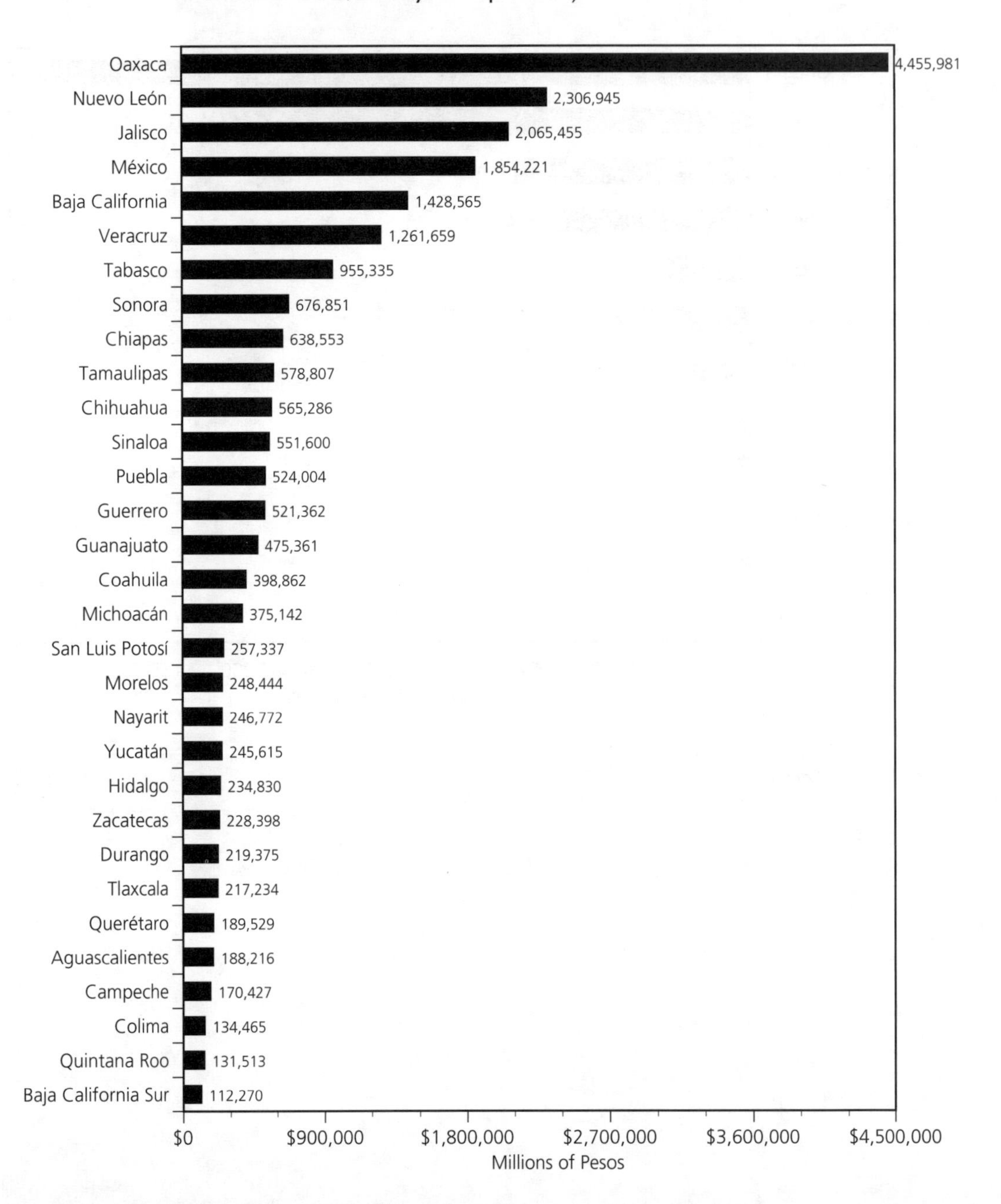

Source: INEGI, Finanzas Públicas Estatales y Municipales de México

Chapter Fifteen
Financial Institutions

COMMERCIAL BANKS—WESTERN UNITED STATES

Commercial bank asset growth in the western United States is generally correlated with population growth, although it also reflects the impact of the real-estate bust in states such as California, Texas, Oklahoma, and Colorado.

Figure 15.1 **Western United States Assets of FDIC-Insured Commercial Banks and Trust Companies, December 31, 1992**

West: 27.4%
East: 61%
U.S. Average: 49.4%

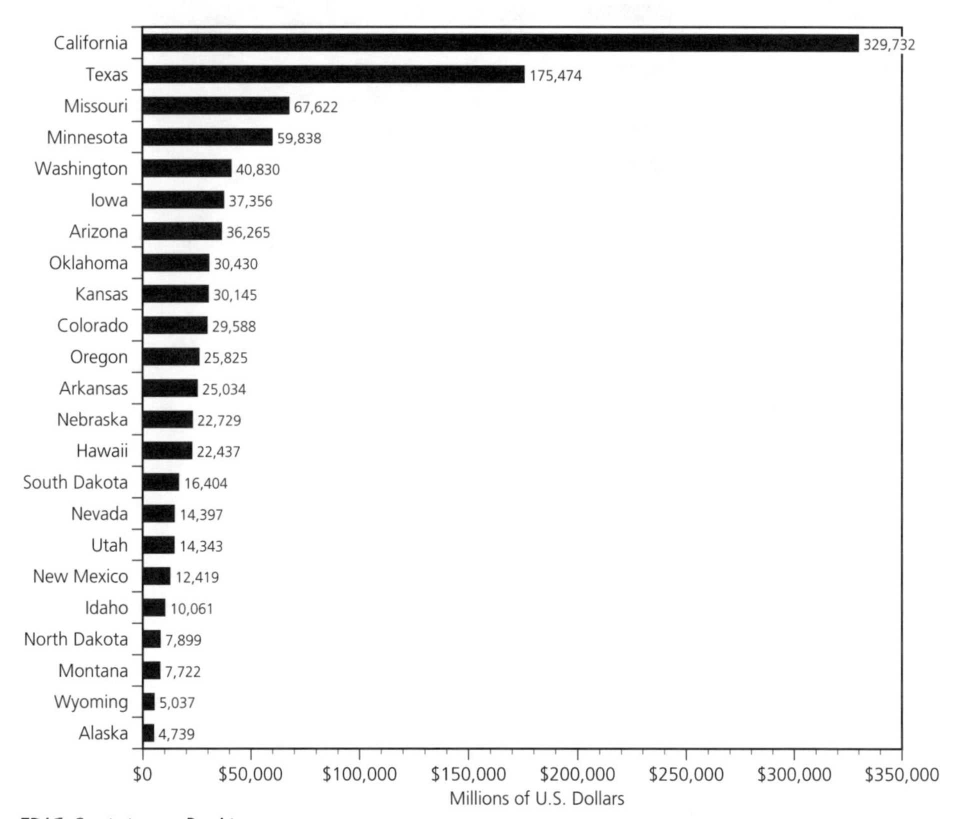

Source: FDIC, Statistics on Banking

Figure 15.2 **Western United States**
Asset Growth of Insured Commercial Banks and Trust Companies, 1983–1992

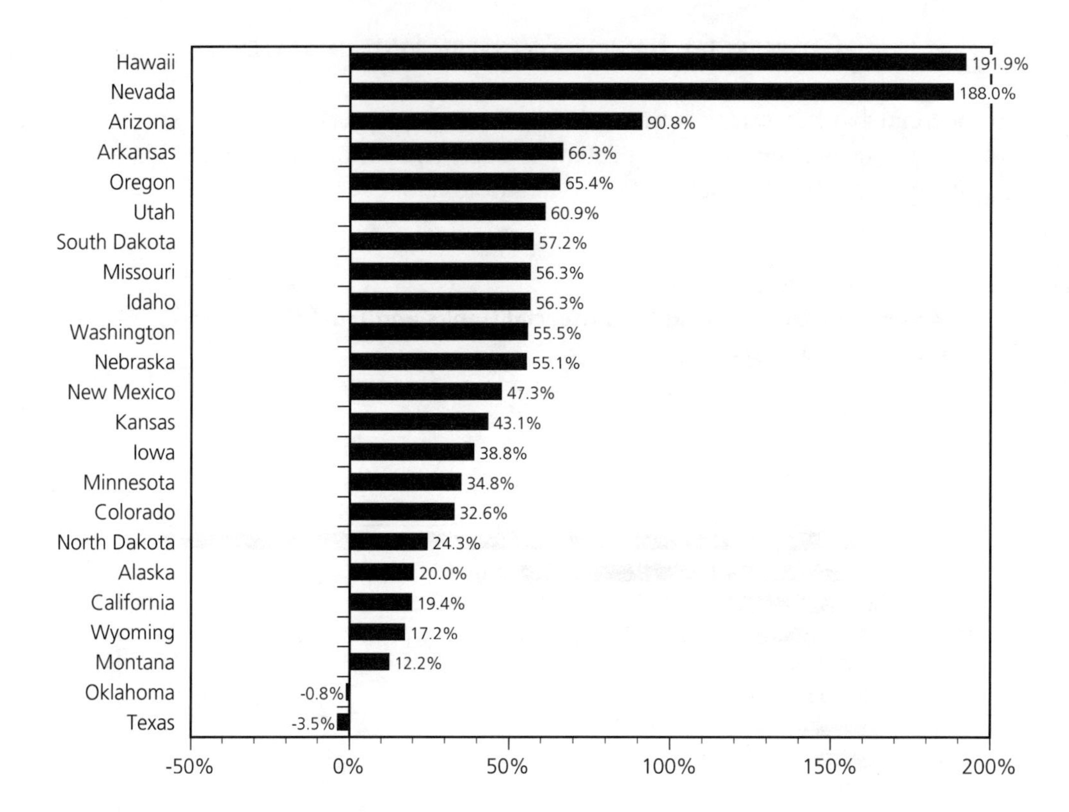

Source: FDIC, Statistics on Banking

Figure 15.3 **Western United States Total Loans and Leases, FDIC-Insured Commercial Banks and Trust Companies, December 31, 1992**

West: $601,462 million
East: $1,376,080 million
U.S. Total: $1,977,542 million

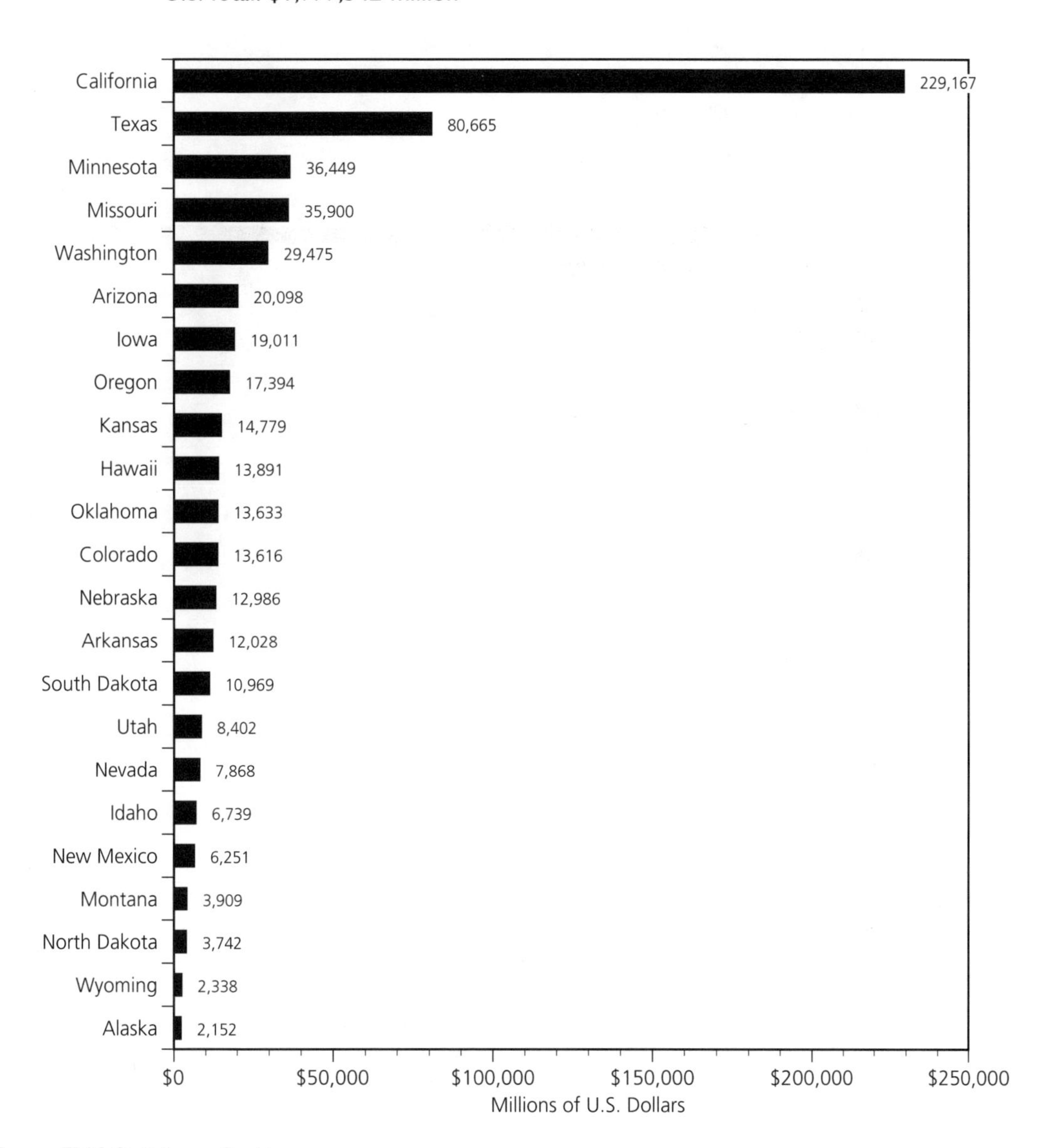

Source: FDIC, Statistics on Banking

For the decade ending in 1992, commercial bank deposits grew most in Nevada and Hawaii. They shrunk in Oklahoma and Texas.

Figure 15.4 **Western United States**
Deposits at FDIC-Insured Commercial Banks and Trust Companies, 1992

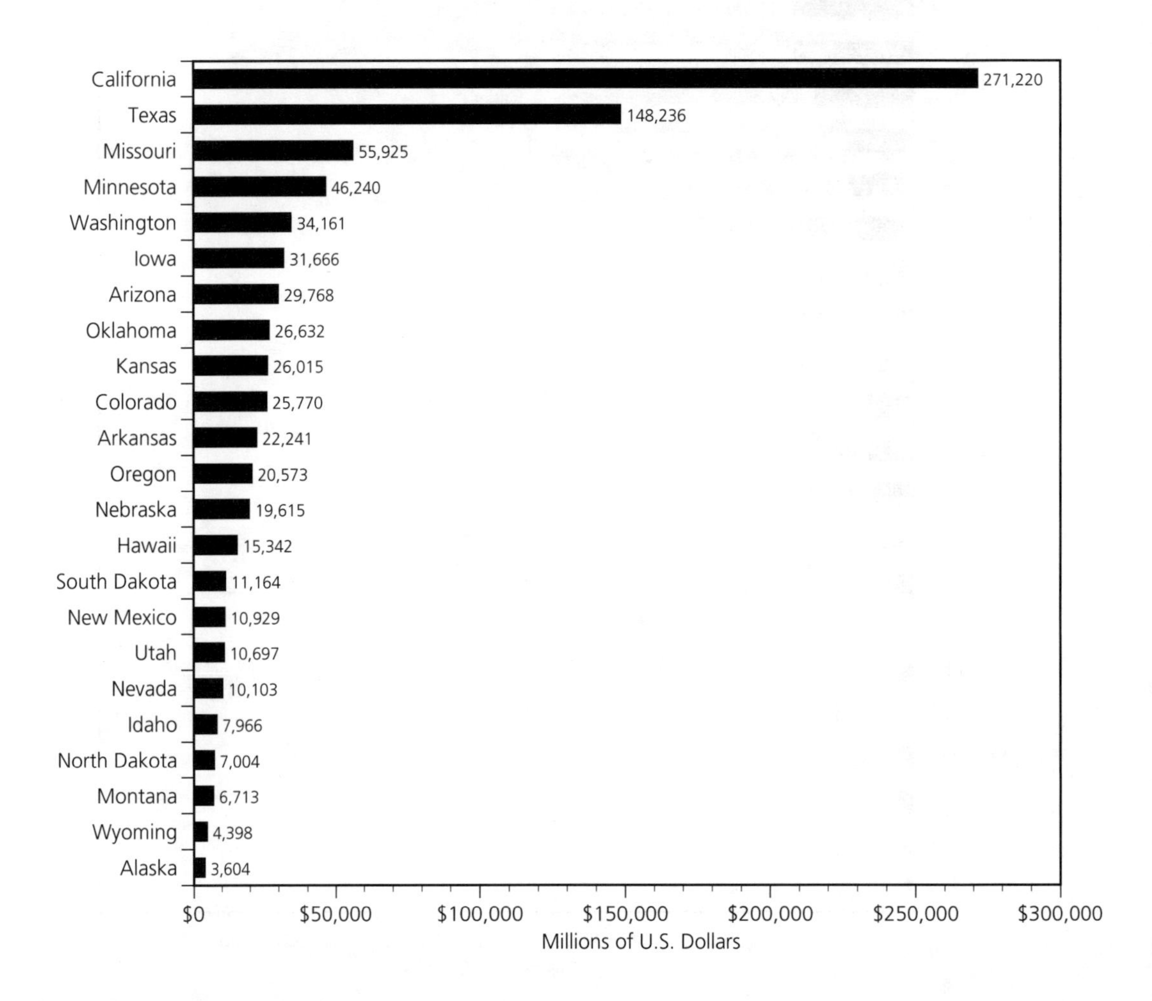

Source: FDIC, Statistics on Banking

Figure 15.5 **Western United States Deposit Growth of Insured Commercial Banks and Trust Companies, 1983–1992**

West: 27.5%
East: 56.6%
U.S. Average: 46.1%

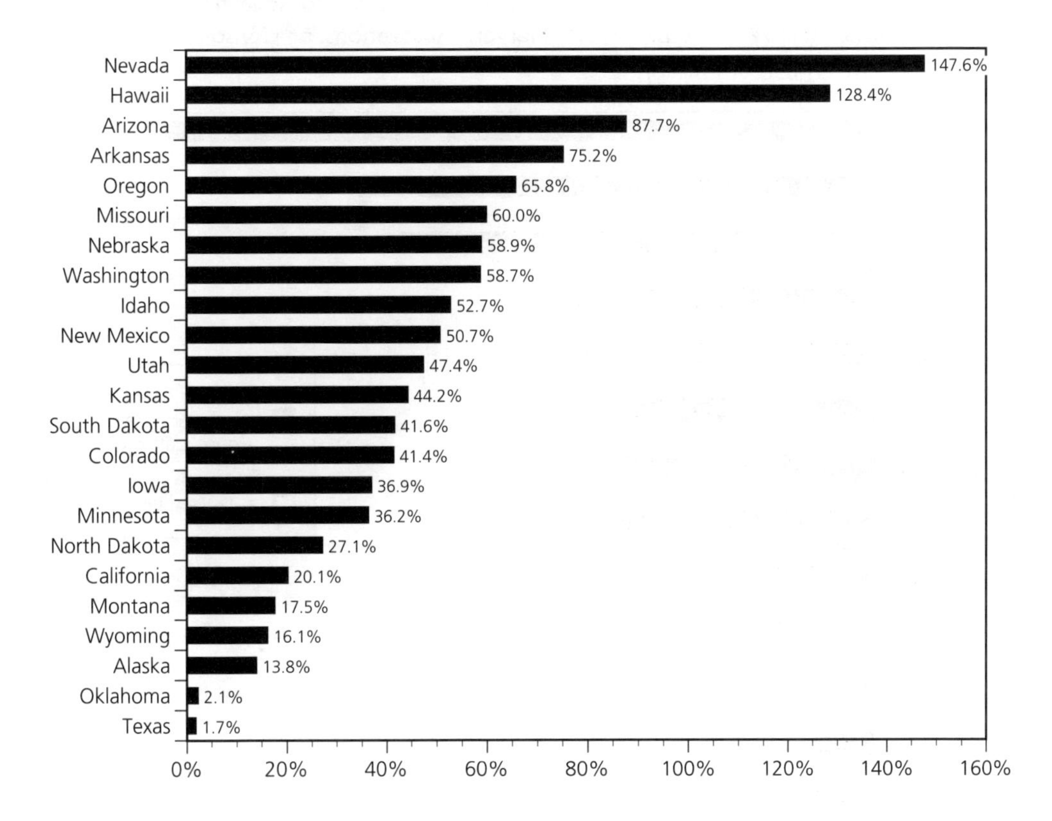

Source: FDIC, Statistics on Banking

Figure 15.6 **Western United States**
FDIC-Insured Commercial Banks and Trust Companies
Ceasing Operation, 1992

(Alaska, New Mexico, and Wyoming had no banks ceasing operation.)

West: 57 Failed Banks; 219 Mergers, Absorptions, or Consolidations
East: 40 Failed Banks; 210 Mergers, Absorptions, or Consolidations
U.S. Total: 97 Failed Banks; 429 Mergers, Absorptions, or Consolidations

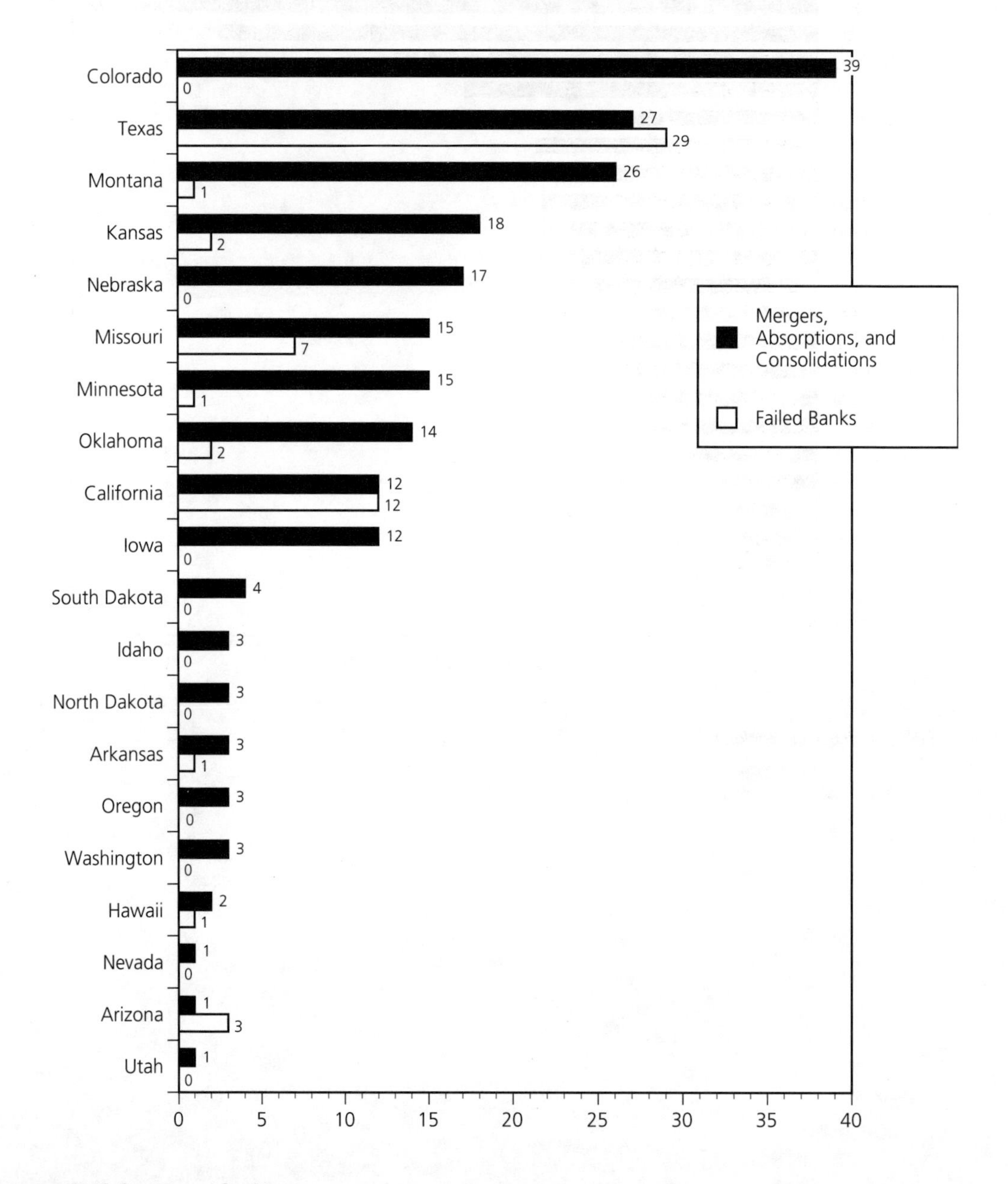

Source: FDIC, Statistics on Banking

Chartered Banks—Canada

Commercial bank assets in the province of Ontario in 1992 were nearly twice those of all of western provinces combined. But the value of savings and demand deposits was more evenly spread among the provinces.

Eastern Canada dominated demand for business and personal loans, but agricultural loan demand was greatest in western Canada.

Figure 15.7 **Canada**
Chartered Bank Total Assets by Province, Fourth Quarter 1992

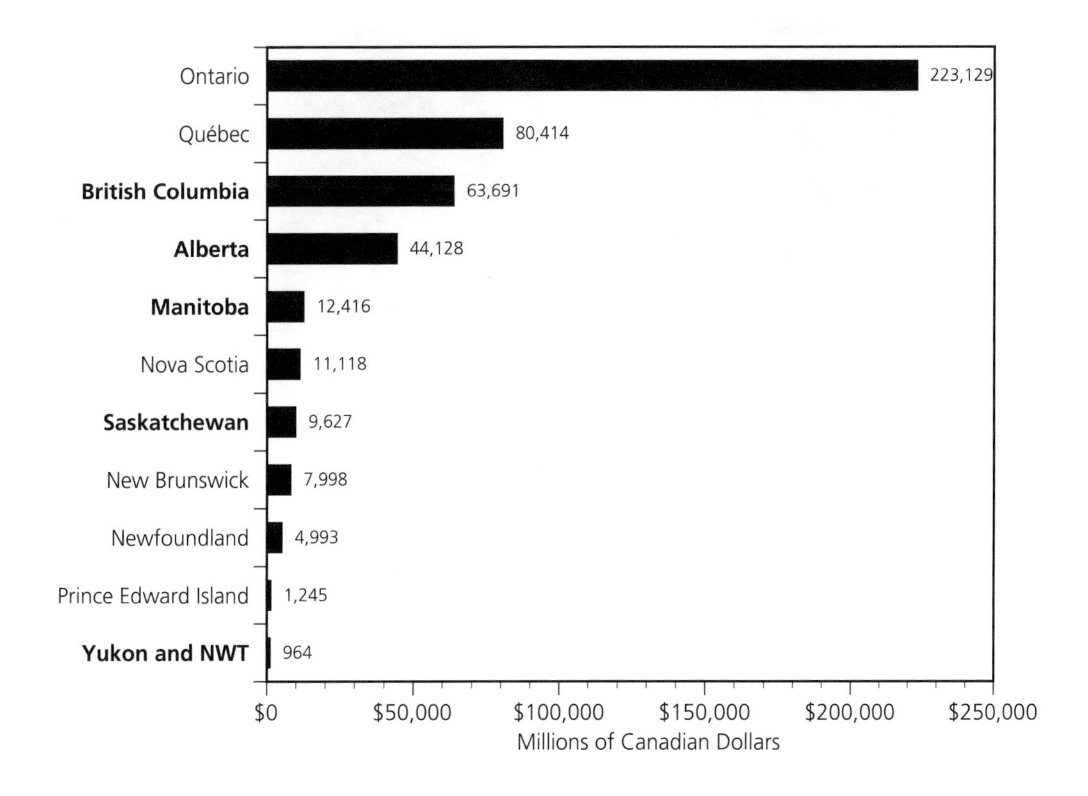

Source: Bank of Canada Review

Figure 15.8 **Canada**
Chartered Bank Personal Savings Deposits by Province, Fourth Quarter 1992

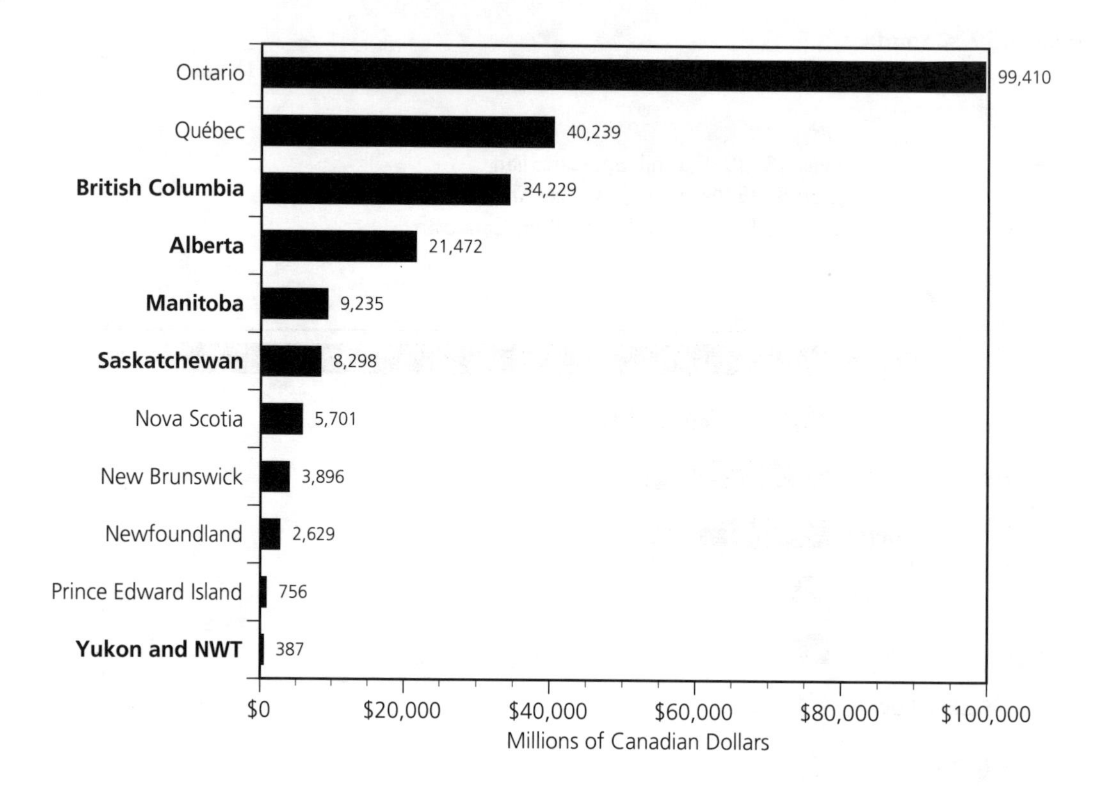

Source: Bank of Canada Review

Figure 15.9 **Canada**
Chartered Bank Gross Demand Deposits, Fourth Quarter 1992

West: $7,064 million, Canadian
East: $14,346 million, Canadian
Canada Total: $21,410 million, Canadian

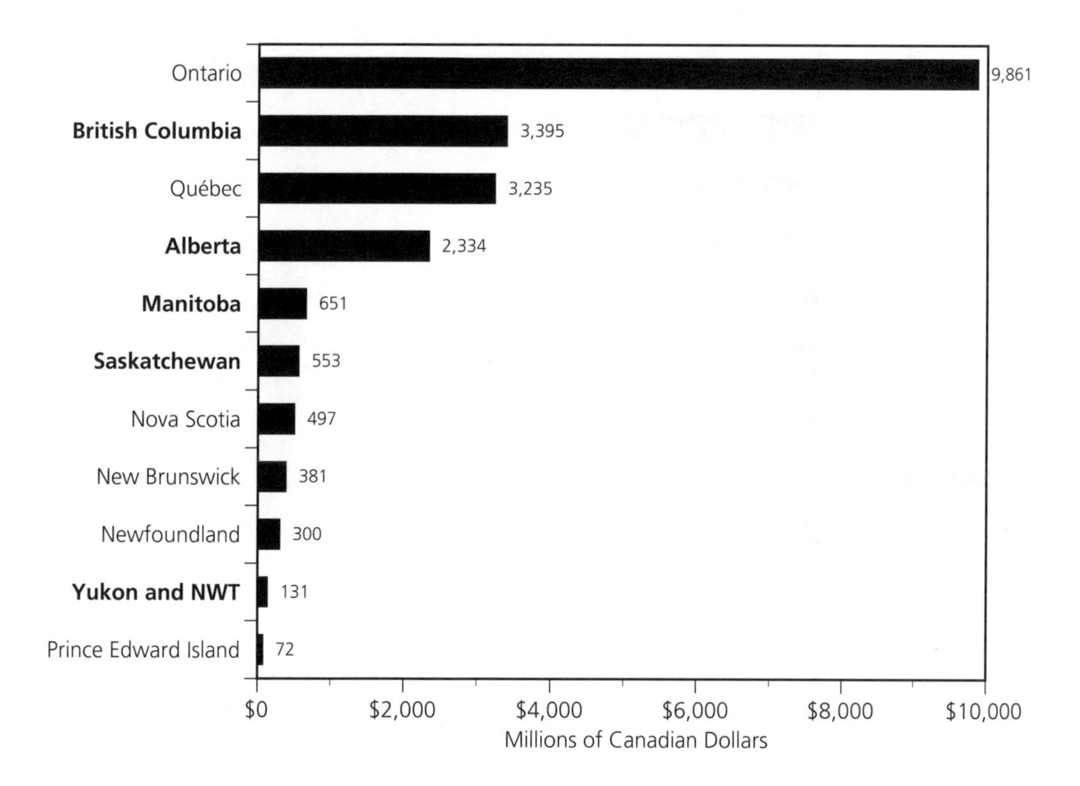

Source: Bank of Canada Review

Figure 15.10 **Canada**
Personal Loans by Chartered Banks, by Province, Fourth Quarter 1992

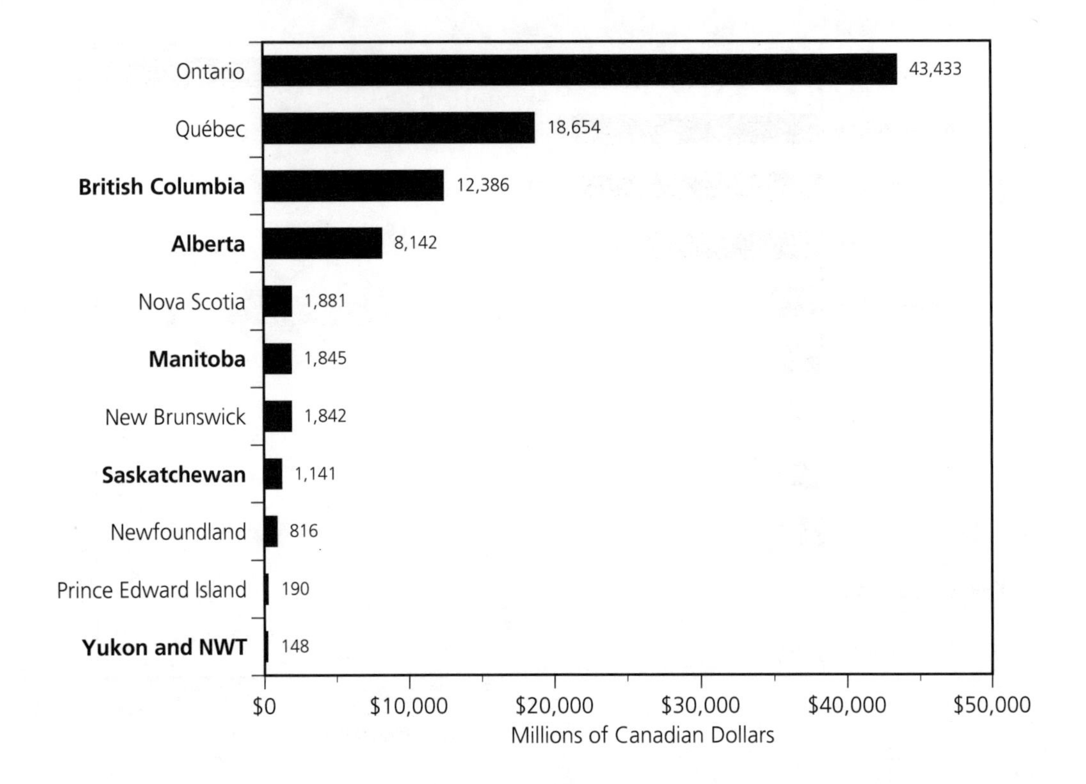

Source: Bank of Canada Review

Figure 15.11 **Canada**
Business Loans by Chartered Banks, by Province, Fourth Quarter 1992

West: $23,662 million, Canadian
East: $66,816 million, Canadian
Canada Total: $90,478 million, Canadian

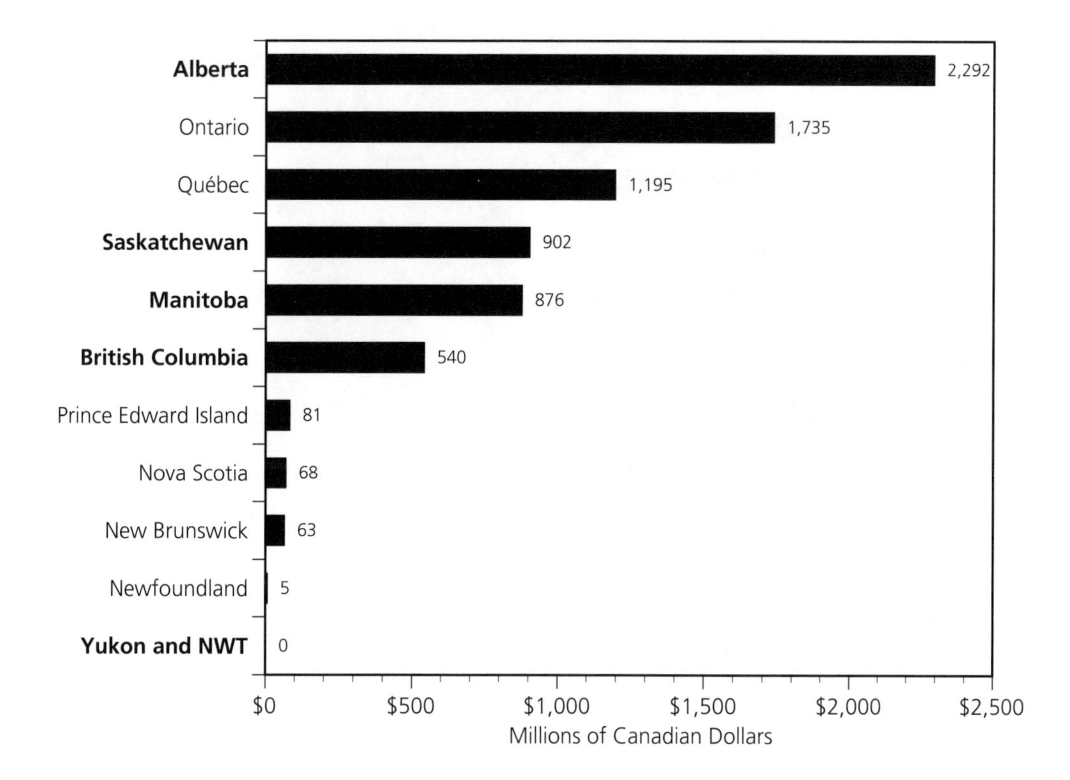

Source: Bank of Canada Review

Figure 15.12 **Canada**
Agricultural Loans by Chartered Banks, by Province, Fourth Quarter 1992

West: $4,610 million, Canadian
East: $3,147 million, Canadian
Canada Total: $7,757 million, Canadian

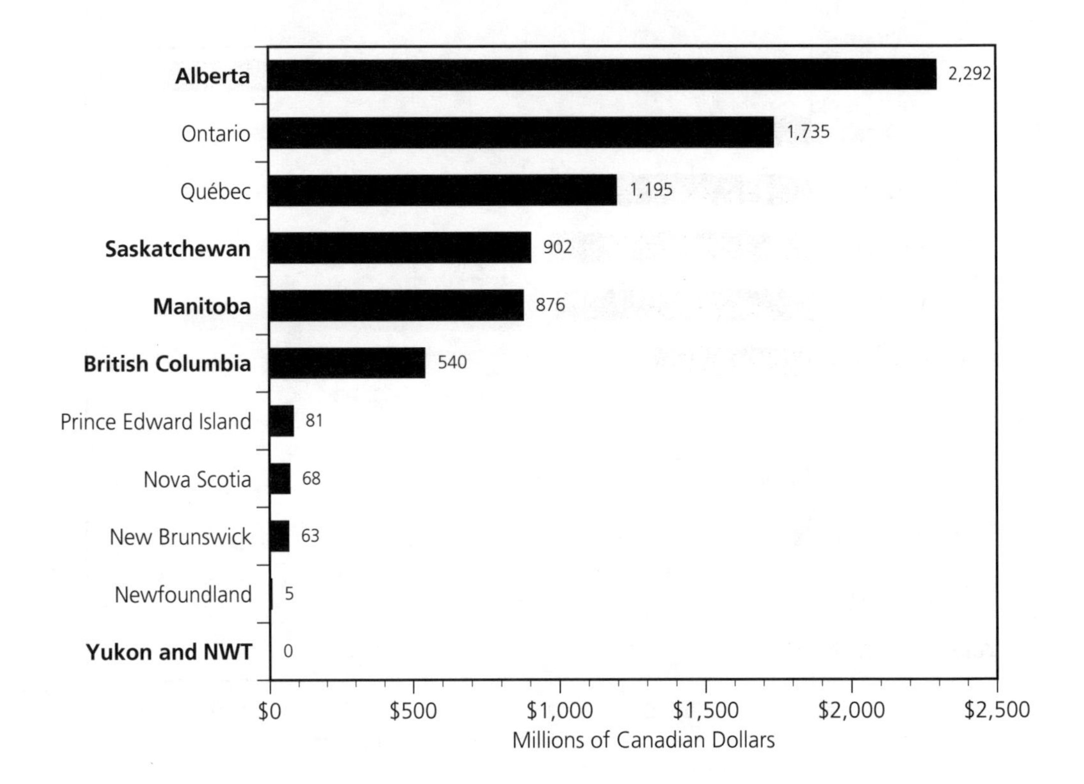

Source: Bank of Canada Review

Commercial Banks—Mexico

Mexico had eighteen commercial banks in 1993 (the number of banks is expected to increase). Two—the Banco Nacional de Mexico and Bancomer, both headquartered in Mexico City—dominate banking in Mexico.

Deposit growth has been greatest at two medium-sized Mexican banks, Banco Mexicano and Banpaís.

The proportion of nonperforming loans is much greater at Mexican banks than at U.S. or Canadian banks.

Figure 15.13 **Mexico**
Total Assets of Principal Banks, March 1993

Total: $532,774

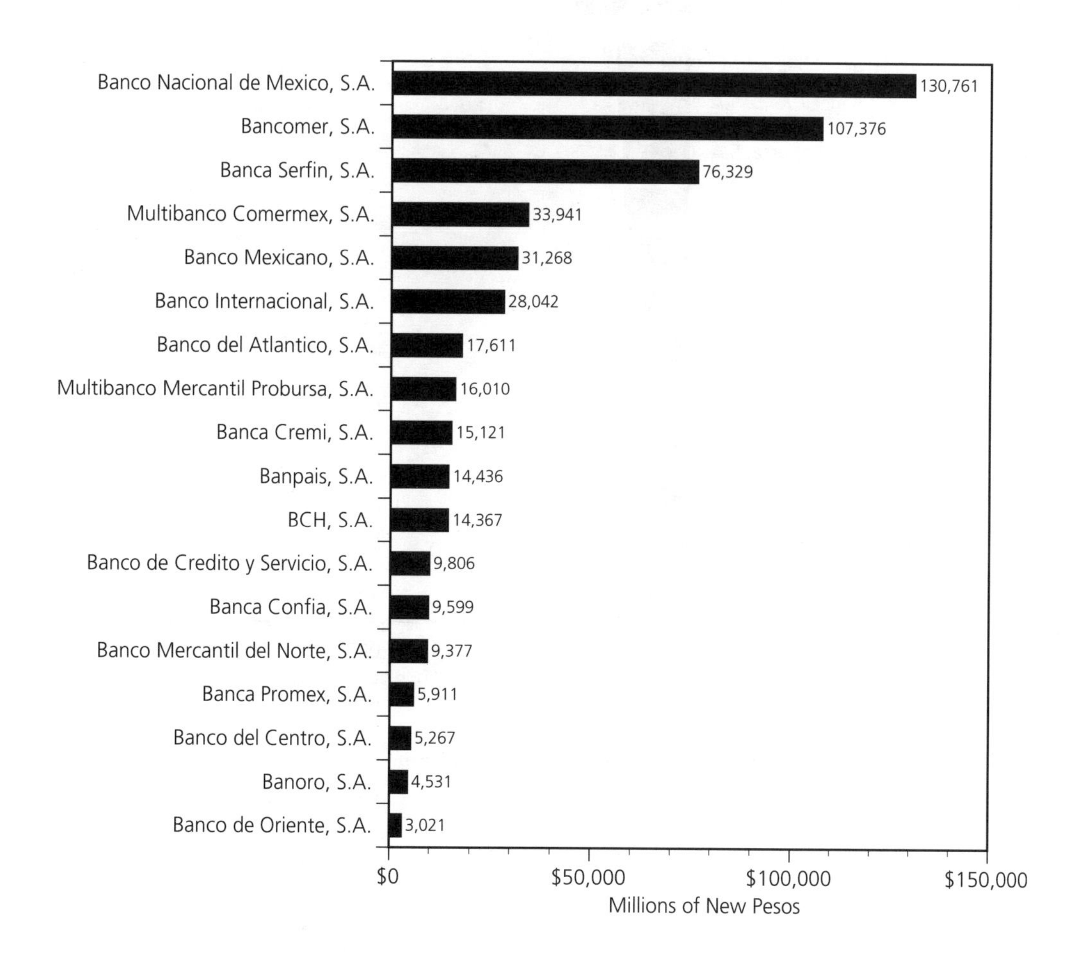

Source: National Banking Commission, Buletin Estadistivo de Banca Multiple

Figure 15.14 **Mexico**
Total Deposits of Principal Banks, March 1993

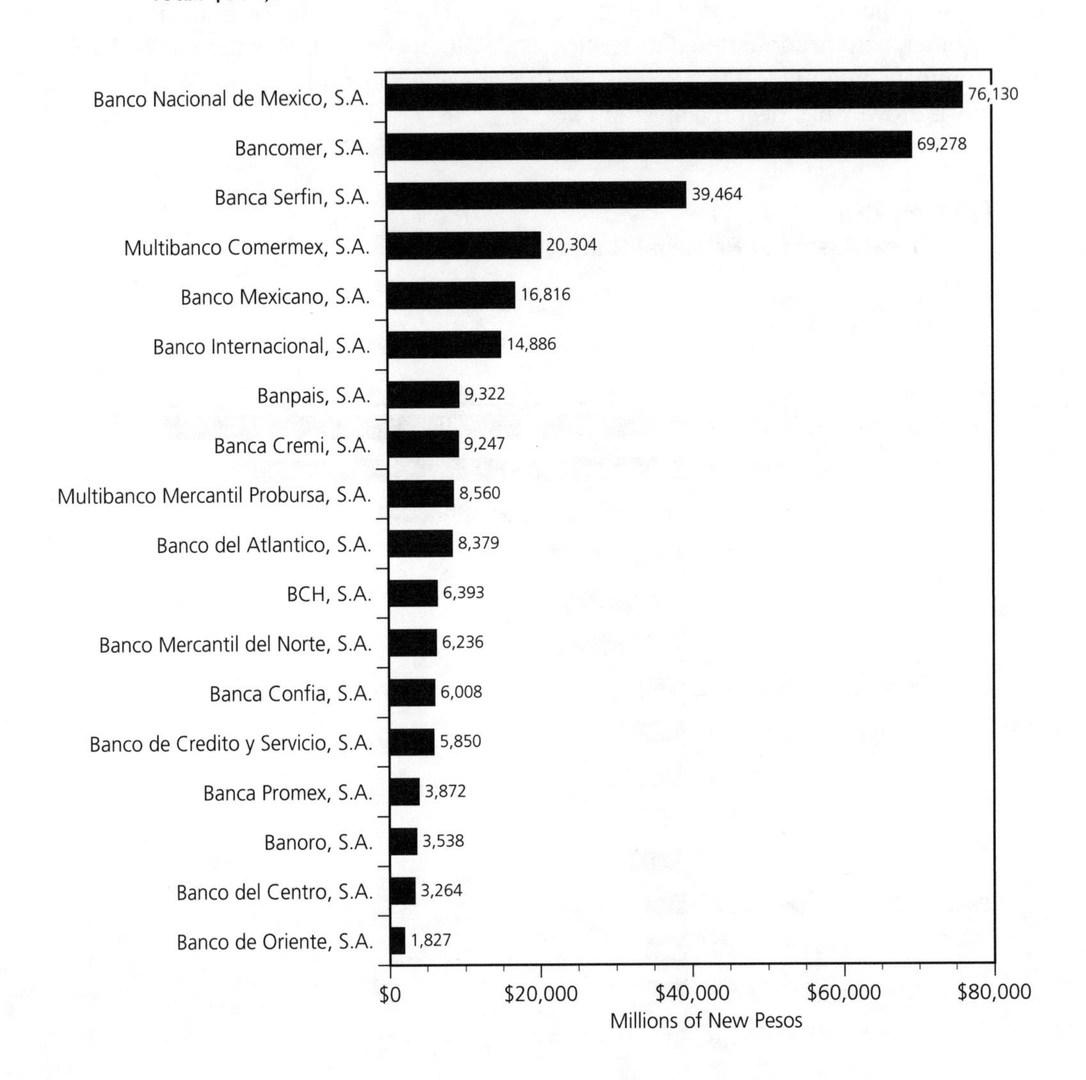

Source: National Banking Commission, Buletin Estadistivo de Banca Multiple

Figure 15.15 **Mexico**
Total Loans of Principal Banks, March 1993

Total: $362,472

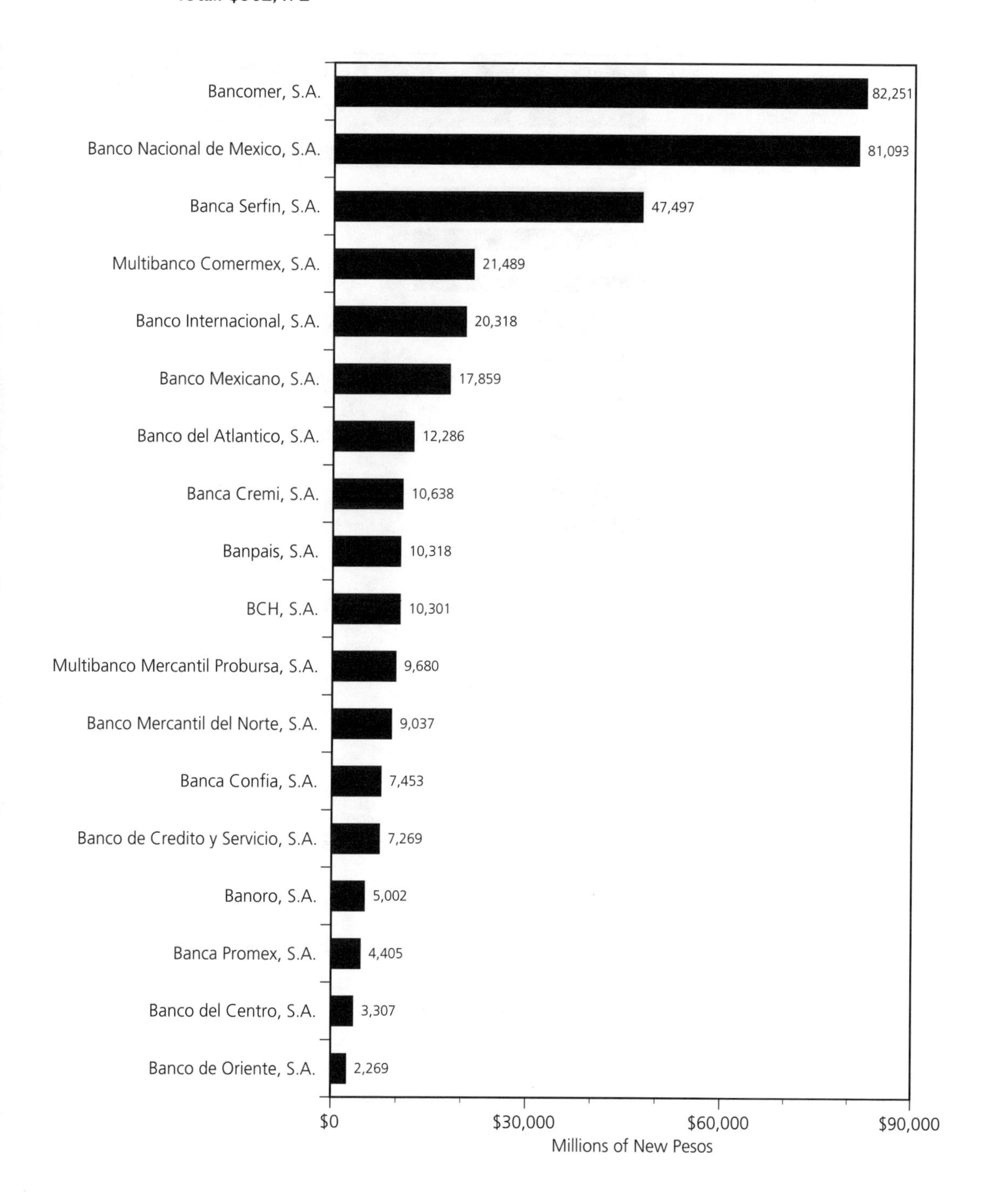

Source: National Banking Commission, Buletin Estadistivo de Banca Multiple

Figure 15.16 **Mexico**
Branches of Principal Banks, March 1993

Total: 4,462

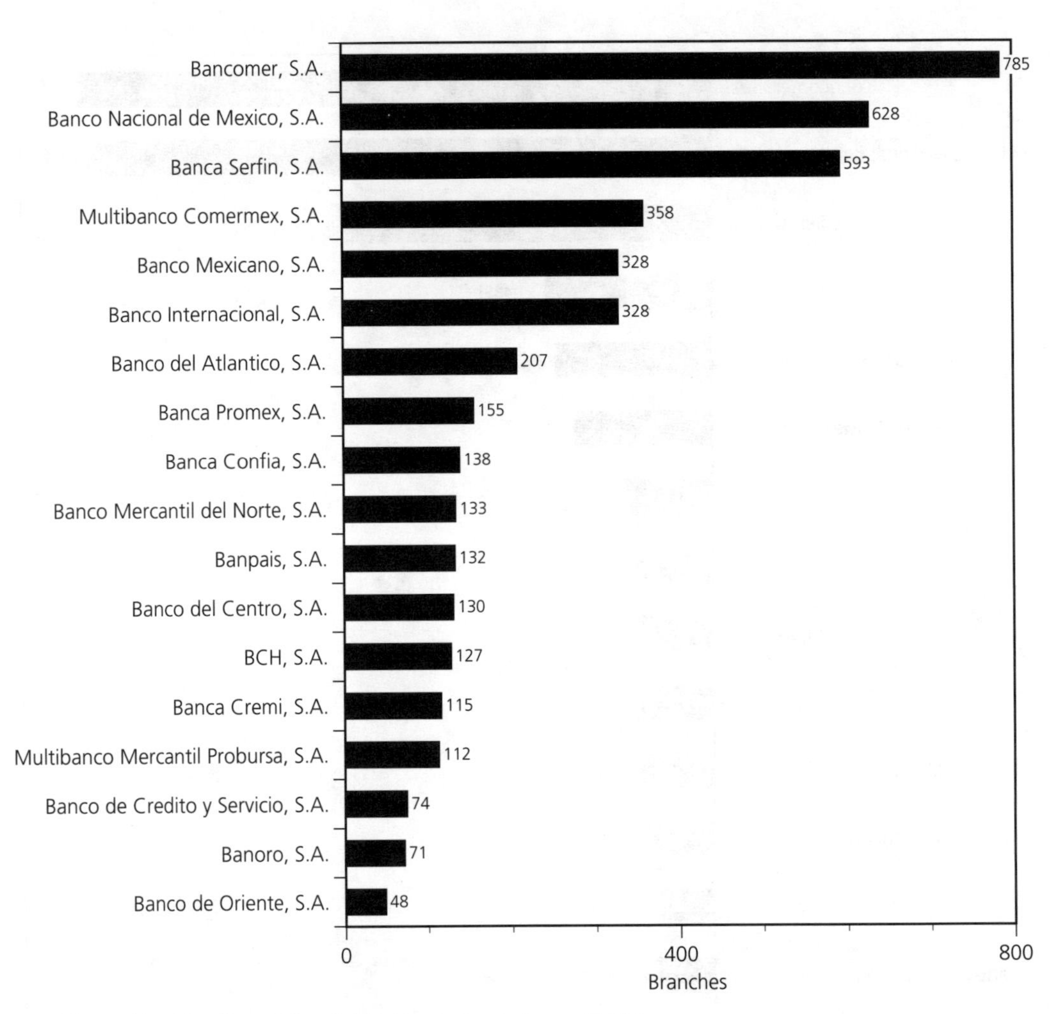

Source: National Banking Commission, Buletin Estadistivo de Banca Multiple

Figure 15.17 **Mexico**
Growth of Assets and Deposits at Principal Banks, March 1992–March 1993

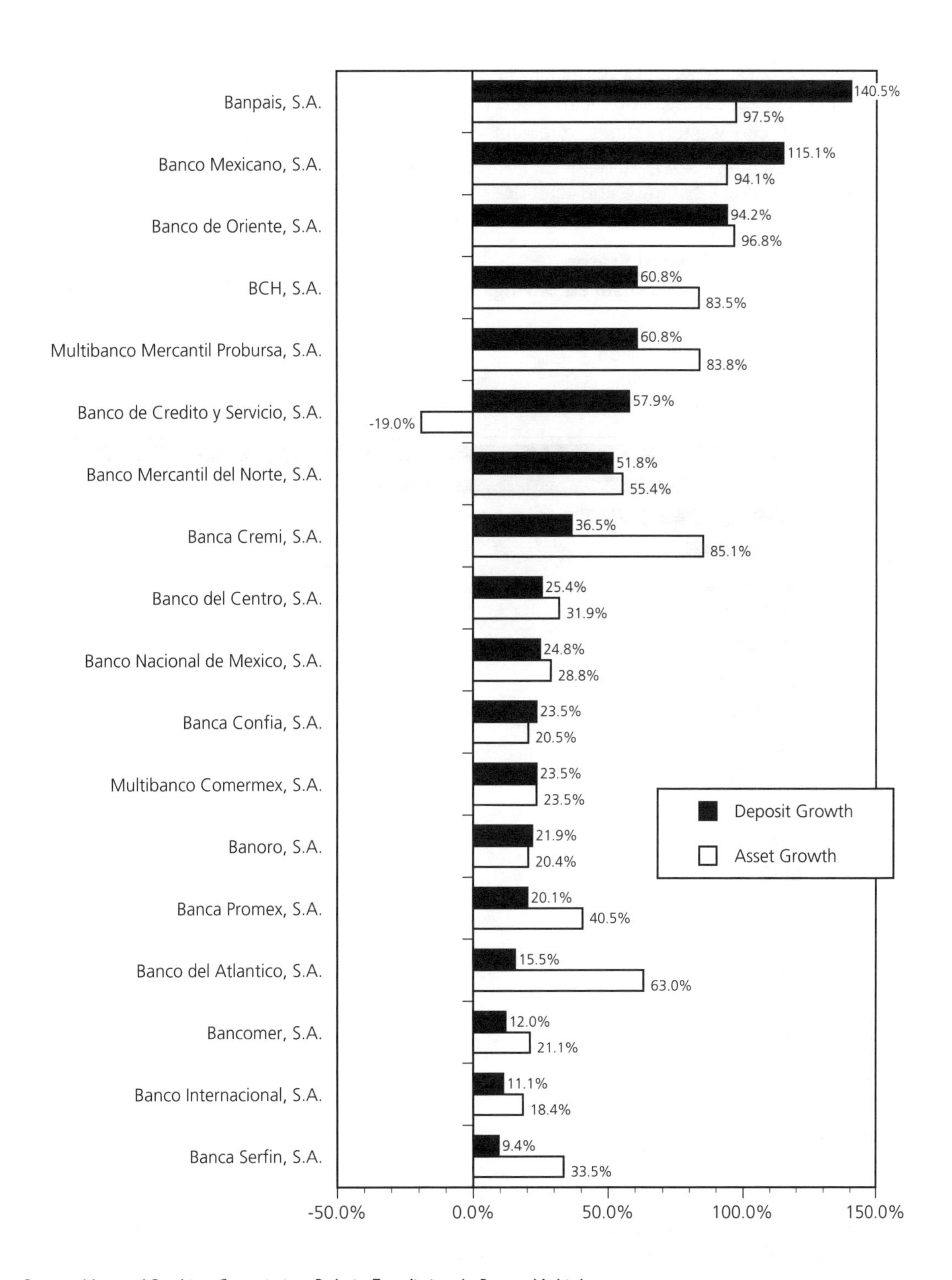

Source: National Banking Commission, Buletin Estadistivo de Banca Multiple

Savings and Loans

California dominates the declining, but still substantial, savings-and-loan industry in the western United States. Between December 1991 and December 1992, savings-and- loan assets declined in eighteen western states, deposits declined in all but four—Oregon, Idaho, Montana, and North Dakota. Savings- and-loan deposits declined by one-third in Hawaii, but rose an astonishing 72.4 percent in Oregon. Overall, savings-and-loan deposits were 43 percent the size of commercial bank deposits in the twenty-three western states. California accounted for 66 percent of all savings-and-loan deposits in the western United States. A smaller proportion of savings-and- loan loans were nonperforming in 1992 than were commercial bank loans.

Figure 15.18a **Western United States**
Assets of FDIC-Insured Savings Institutions, December 31, 1992* (Excluding California)

West: **$430,890** million
East: **$599,799** million
U.S. Total: **$1,030,689** million

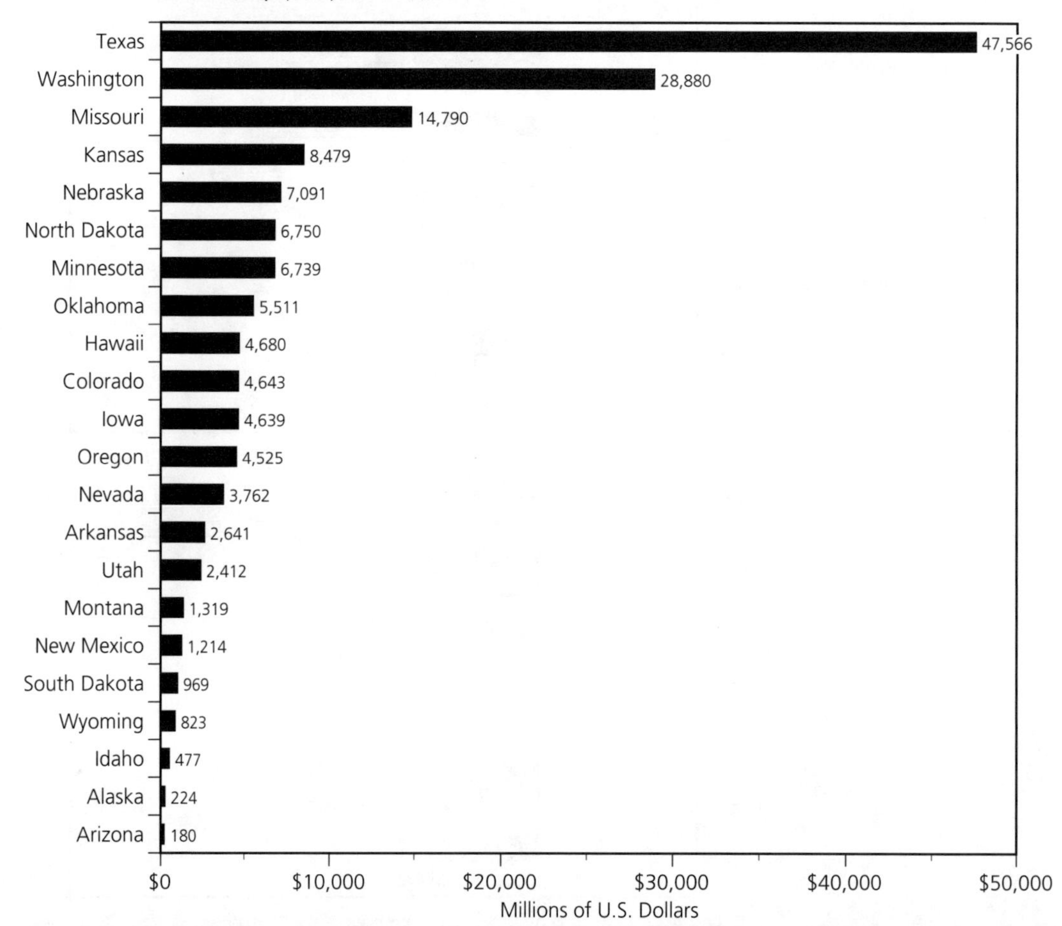

(* Includes SAIF-insured savings institutions regulated by the Office of Thrift Supervision; does not include institutions under Conservatorship of the Resolution Trust Corporation.)

Source: FDIC, Statistics on Banking

Figure 15.18b **Western United States**
Assets of FDIC-Insured Savings Institutions, December 31, 1992

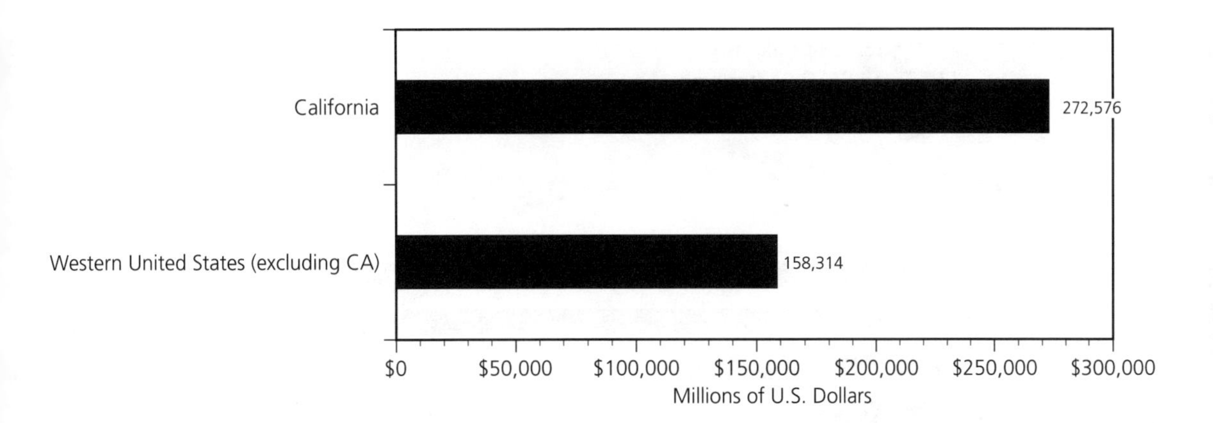

Source: FDIC, Statistics on Banking

Figure 15.19a **Western United States**
Total Loans and Leases, FDIC-Insured Savings Institutions, December 31, 1992* (Excluding California)

West: $288,599 million
East: $362,038 million
U.S. Total: $650,637 million

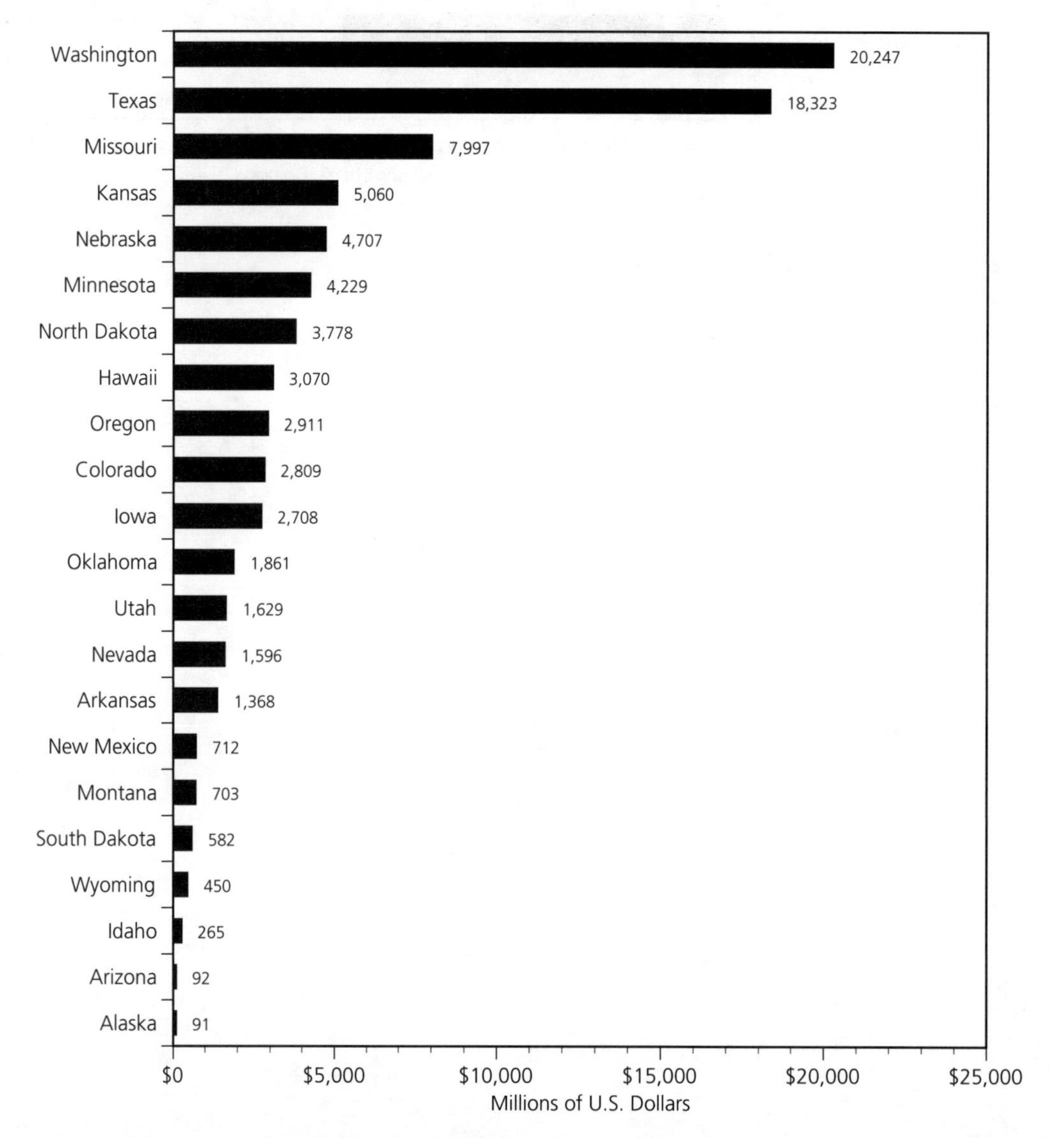

(* Includes SAIF-insured savings institutions regulated by the Office of Thrift Supervision; does not include institutions under Conservatorship of the Resolution Trust Corporation.)

Source: FDIC, Statistics on Banking

Figure 15.19b **Western United States**
Total Loans and Leases, FDIC-Insured Savings Institutions, December 31, 1992

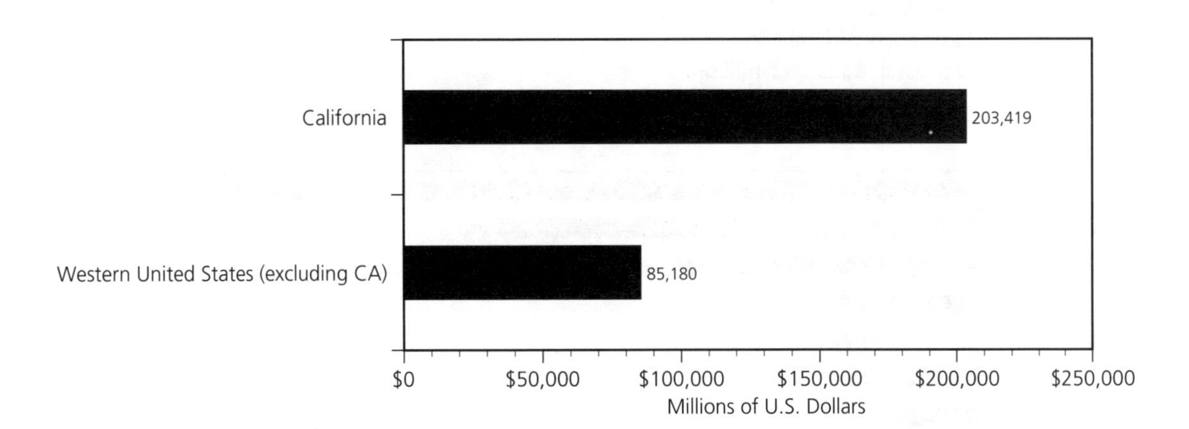

Source: FDIC, Statistics on Banking

Figure 15.20a **Western United States**
Deposits at FDIC-Insured Savings Institutions, December 31, 1992*
(Excluding California)

West: **$323,732** million
East: **$501,361** million
U.S. Total: **$825,093** million

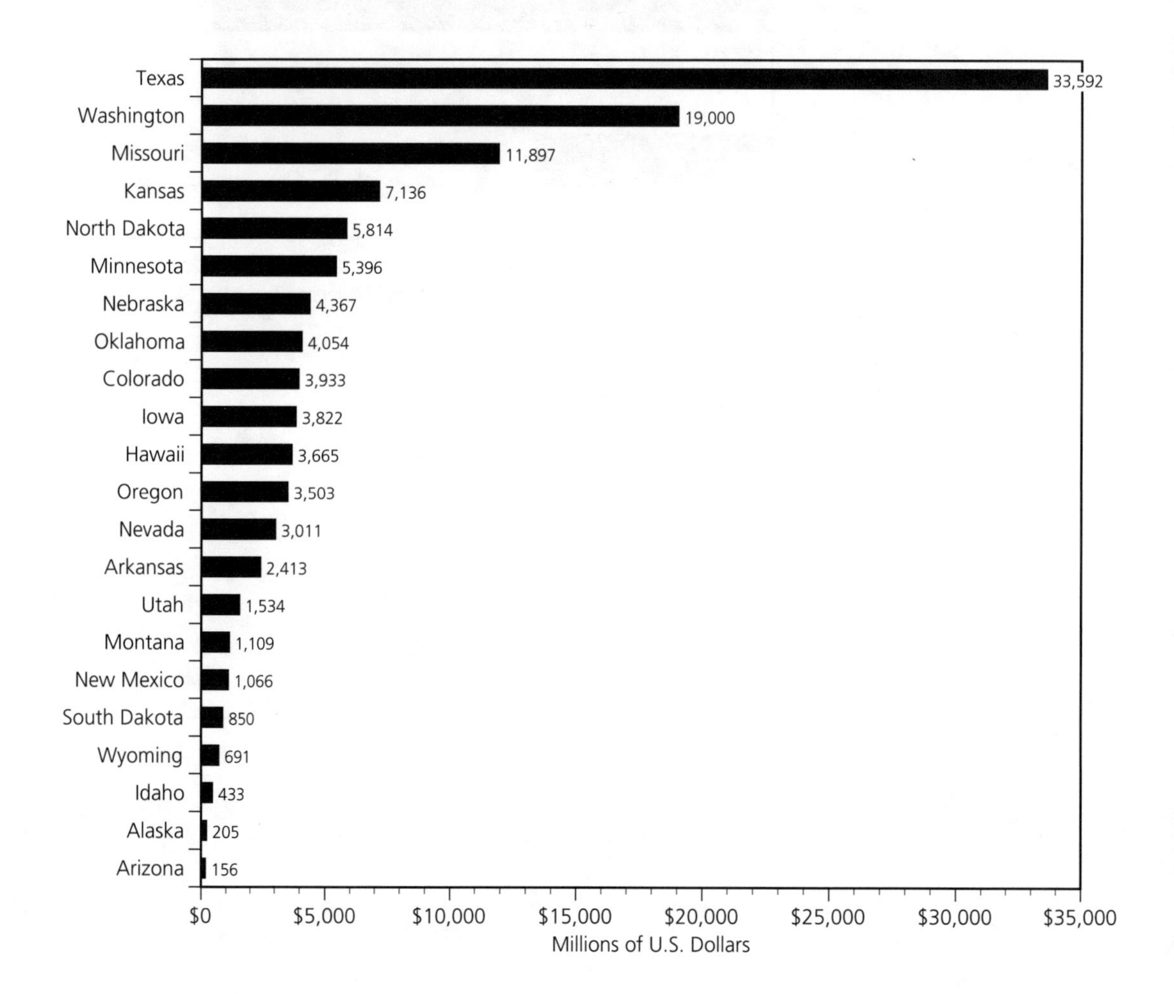

(* Includes SAIF-insured savings institutions regulated by the Office of Thrift Supervision; does not include institutions under Conservatorship of the Resolution Trust Corporation.)

Source: FDIC, Statistics on Banking

Figure 15.20b **Western United States**
Deposits at FDIC-Insured Savings Institutions, December 31, 1992

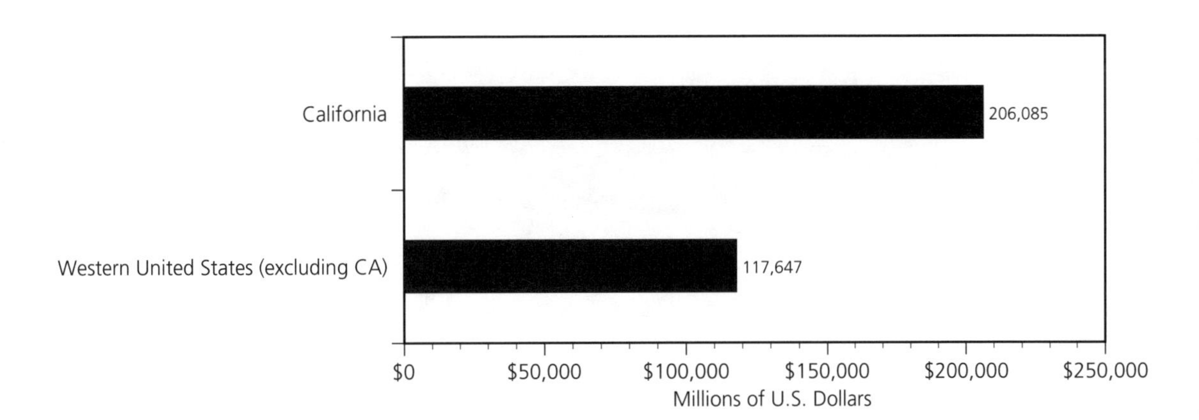

Source: FDIC, Statistics on Banking

Chapter Sixteen
International Trade

Trade in North America is large and growing. Continental trade approached $270 billion in 1992. This includes total two-way trade of $189 billion between the United States and Canada, the world's largest trading relationship; $75 billion between the United States and Mexico, America's third-largest trading partner (Japan is second), with whom trade is growing very rapidly—increasing by more than 300 percent since 1986; and $2.5 billion between Canada and Mexico in 1992, a small but rapidly growing trade relationship.

Concentrated Regional Trade in North America

Trade is concentrated by origin and destination. Though U.S. trade with Canada and Mexico affects every U.S. state, the bulk of North American trade is accounted for by the border states. Measured by the origin and destination of trade, *seven states dominate U.S.-Canadian trade*—Michigan, New York, California, Illinois, Texas, Ohio, and Washington. *Five states dominate U.S.-Mexican trade*—Texas, California, Arizona, Michigan, and Illinois. Measured on a per capita basis, however, many smaller states and interior states have a large stake in the continuing expansion of North American trade. Moreover, the interior states are experiencing the most rapid rates of trade growth.

Trade growth is concentrated in the West. Nearly all trade growth has been in western North America. The greatest increases are in U.S. trade with Mexico, nearly all of which passes through one of the western border states, especially Texas. Nearly all the increases in U.S.-Canadian trade have been from the western provinces—especially Alberta and British Columbia. Most of the growing trade between Canada and Mexico originates in Ontario.

Trade is concentrated by business sector. Manufactured or processed goods dominate U.S. exports to Canada. Energy and other raw materials dominate Canadian exports to the United States. U.S. exports to Mexico and Mexican exports to the United States are both dominated by machinery and transportation equipment—a boomerang trade pattern that reflects the importance of joint production agreements between U.S. companies and the *maquiladoras*—and the fact that the United States provides most of the supplies to the maquiladoras. Maquiladoras are essentially assembly plants for (primarily) U.S. manufacturers. Parts are shipped from U.S. factories into Mexico, where they are assembled into finished products and shipped back into the United States.

Trade bottlenecks are concentrated at the borders. There are 113 border crossings between the United States and Canada—53 in the West. There are 37 border crossings between the United States and Mexico, all of which are in the West. Most of the major trade bottlenecks are located at these border crossings—not in the interior national transportation systems.

U.S.-Canadian Trade

The U.S. and Canada have the largest bilateral trading relationship in the world. In 1992, merchandise trade between the two North American neighbors totaled $189 billion, with U.S. imports from Canada exceeding exports by 9 percent. U.S.- Canadian merchandise trade is larger than total U.S. trade with France, Italy, Britain, Germany, and South Korea combined. U.S.-Canadian trade exceeds U.S. trade with Japan by nearly 30 percent.

Western states account for about $40 billion of that total trade, and western U.S.-Canadian trade is growing faster than trade with Canada as a whole. Western U.S. states' imports from Canada increased 18 percent between 1988 and 1992. Imports from Canada by states east of the Mississippi increased by less than 3 percent during this period. Western states' exports to Canada increased by 3 percent from 1988 to 1992, while eastern states' exports to Canada declined by 3 percent during this period. Exports from the western Canadian provinces increased by 14 percent between 1988 and 1992, while exports from the eastern provinces to the United States increased by only 4 percent during this period.

Most of the trade gains are by the western U.S. states and western Canadian provinces. Of thirty-two U.S. states with import gains of more than 5 percent between 1988 and 1992, seventeen are in the West. Of eight states with import increases of 40 percent or more, half are in the West. The value of exports from the western provinces and territories of Canada increased by $2.6 billion between 1988 and 1992—an increase of 14 percent. This was more than triple the percentage increase of exports from eastern Canada.

The Northern border has a unique feature—the domination of trade and traffic by a region called Cascadia—and the example set by the economic and political integration that is developing in that region. Cascadia is the trade zone surrounding the transportation corridor that includes one increasingly integrated market comprised of two states and one province (Washington, Oregon, and British Columbia) and four major cities (Vancouver, Seattle, Tacoma, and Portland).

U.S.-Mexican Trade

Since 1986, when Mexico lowered its tariffs, eliminated most import permits and many other non-tariff barriers, and joined the General Agreement on Tariffs and Trade (GATT), U.S.-Mexican trade has been growing faster than the economy of either country. Two-way trade grew from $30 billion in 1986 to more than $75 billion in 1992, and U.S. exports to Mexico grew from $12.3 billion in 1986 to more than $40 billion in 1992. Since 1986, trade has been growing at a rate of more than 16 percent a year compounded.

U.S.-Mexican trade stimulates jobs and economic growth in nearly every state on both sides of the border. But most trade flows between the Mexican and U.S. border states and between the border region and the industrial Midwest and Northeast. A unique element of U.S.-Mexican trade, the maquiladora trade, is a major reason for this concentration.

Maquiladoras, also known as twin plants, are factories located in Mexico. Maquiladoras manufacture products that are made primarily with U.S. components for export back into the U.S. market. Mexico originates less than 3 percent of the supplies used for maquiladoras.

Maquiladoras account for nearly half of the total trade between the United States and Mexico. Because 90 percent of the maquiladora manufacturing employment is located in the Mexican border states, maquiladora trade movements are concentrated in the border region. The growth of maquiladora factories in Mexico has helped diversify the Mexican economy—moving manufacturing and population away from the overcrowded Mexico City area. But maquiladoras contribute to traffic congestion in the border regions.

The principal origin of U.S. exports to Mexico is Texas, with more than $17 billion in 1992. California is second, followed by Arizona, Michigan, and Illinois. The principal destinations for U.S. exports to Mexico are the Distrito Federal (Mexico City), the state of México, the state of Jalisco, and the border states of Baja California del Norte, Chihuahua, Nuevo León, and Tamaulipas.

The principal points of origin of Mexican exports to the United States is the Distrito Federal (Mexico City's metropolitan area) and the border states of Baja California del Norte, Chihuahua, and Tamaulipas, and the state of Jalisco. The principal destinations for Mexican exports to the United States are Texas, California, and the industrial Midwest and Northeast.

Trading Partners

All three nations of North America are working hard to increase the volume and diversity of their exports, but their imports are also continuing to grow. The benefits of additional goods and services, increased employment, and the leveling effect on prices of internationally traded goods have served as strong economic stimuli in all three countries. With NAFTA in place, trade among the three nations is likely to outpace growth of global trade and will continue to generate jobs and new wealth in all the countries.

Figure 16.1 **United States**
Top Ten Trading Partners, 1992
(Figures include domestic and foreign merchandise, free alongside ship [f.a.s.], as well as general imports and customs.)

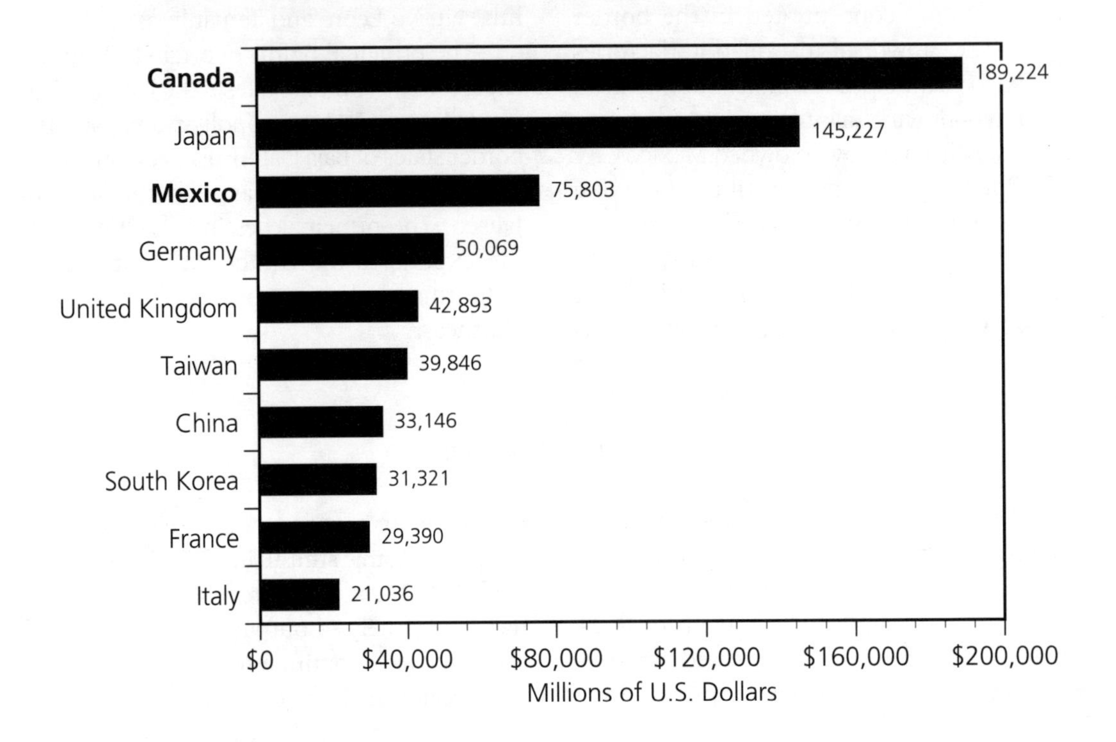

Source: U.S. International Trade Commission

Figure 16.2 **Canada**
Top Ten Trading Partners, 1991

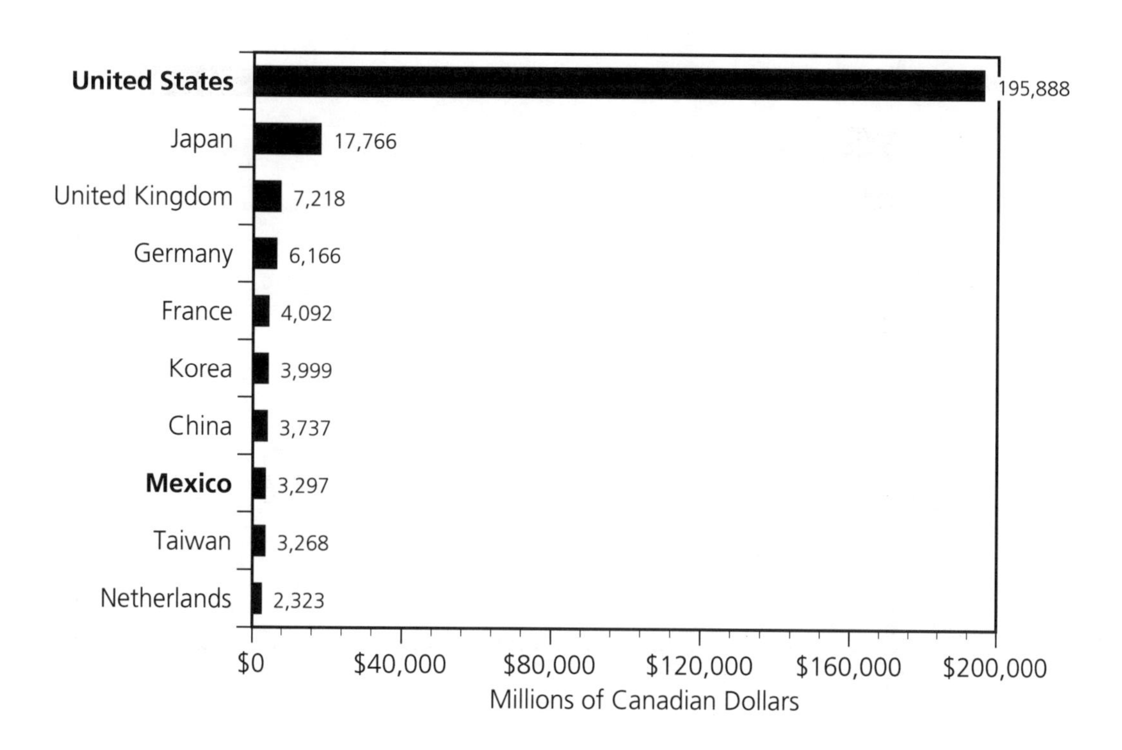

Source: Europa Yearbook, 1993

Figure 16.3 **Mexico**
Top Ten Trading Partners, 1990

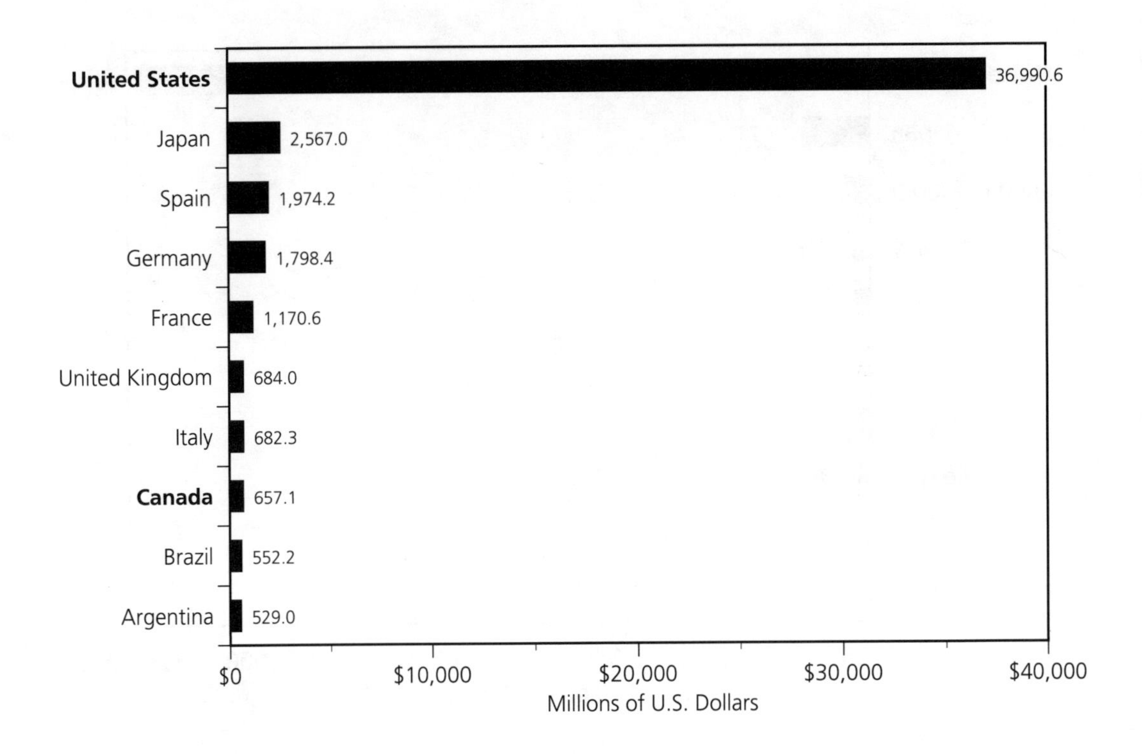

Source: Europa Yearbook, 1993

The principal trading partners for the North American nations are each other, and although the United States trades more with Canada, U.S. trade with Mexico is growing faster. Mexico may soon surpass Japan as the second-largest trading partner of the United States.

Canada is America's number one trading partner, but the states in the western United States trade more with Mexico. This is chiefly because of Texas, which does twice as much business with Mexico as the rest of the region combined.

Trade between Canada and Mexico is small, but increasing.

Figure 16.4 **United States: Trade with Canada, 1986–1992 (Figures include general imports, customs, and domestic and foreign merchandise, f.a.s.)**

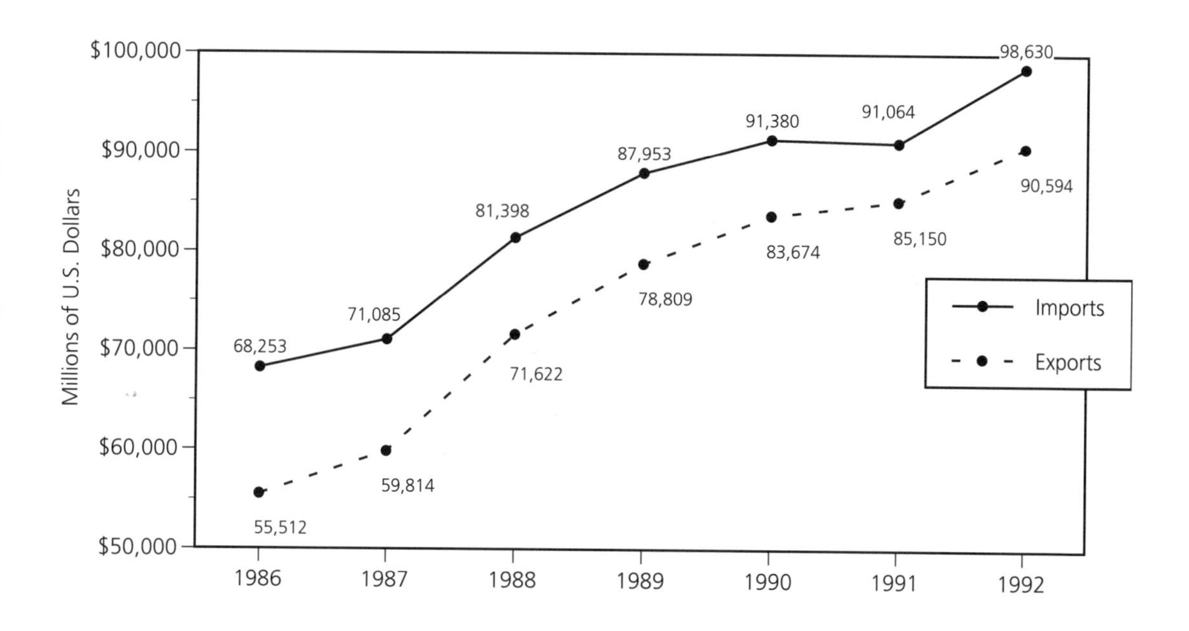

Source: U.S. International Trade Commission

Figure 16.5 **United States: Trade with Mexico, 1986–1992 (Figures include general imports, customs, and domestic and foreign merchandise, f.a.s.)**

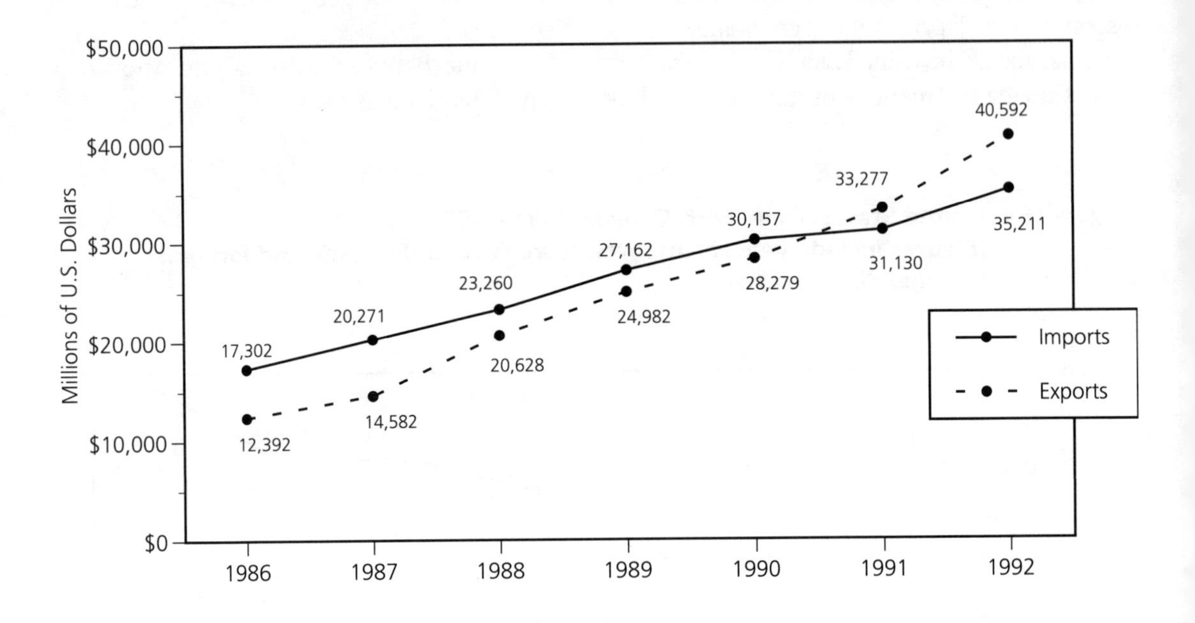

Source: U.S. International Trade Commission

Figure 16.6 **Western United States Exports to Mexico by State, 1990–1991**

West: $23,691,568,000
East: $9,489,721,000
U.S. Total (Includes values of merchandise shipped from Puerto Rico and the Virgin Islands.): $33,275,780,000

	Exports, 1991 (Thousands of U.S. Dollars)	National Rank	Percent of Total State Exports, 1990
Texas	$15,485,379	1	32.1%
California	$5,526,877	2	8.0%
Arizona	$990,787	5	17.8%
Washington	$290,573	15	0.3%
Missouri	$288,245	16	7.2%
Kansas	$258,266	19	7.3%
Minnesota	$216,964	22	2.6%
Iowa	$108,261	27	3.2%
Arkansas	$95,929	29	4.3%
Colorado	$90,148	31	4.2%
Oklahoma	$80,354	32	2.8%
Nebraska	$64,401	33	3.9%
Oregon	$55,401	34	0.8%
Utah	$39,340	36	2.2%
Idaho	$32,925	38	3.1%
New Mexico	$18,219	41	6.4%
Nevada	$11,304	44	6.8%
Montana	$9,716	45	3.4%
Hawaii	$6,535	46	0.0%
Wyoming	$6,224	47	4.3%
South Dakota	$6,105	48	2.3%
Alaska	$6,045	49	0.1%
North Dakota	$3,570	51	10.2%

Source: U.S. Department of Commerce

Balance of Trade

The United States is the world's largest exporter. But the United States is an even larger importer. Consequently, the United States has been running trade deficits since the 1970s.

Figure 16.7 **United States: Merchandise Trade Balance, 1980–1992**

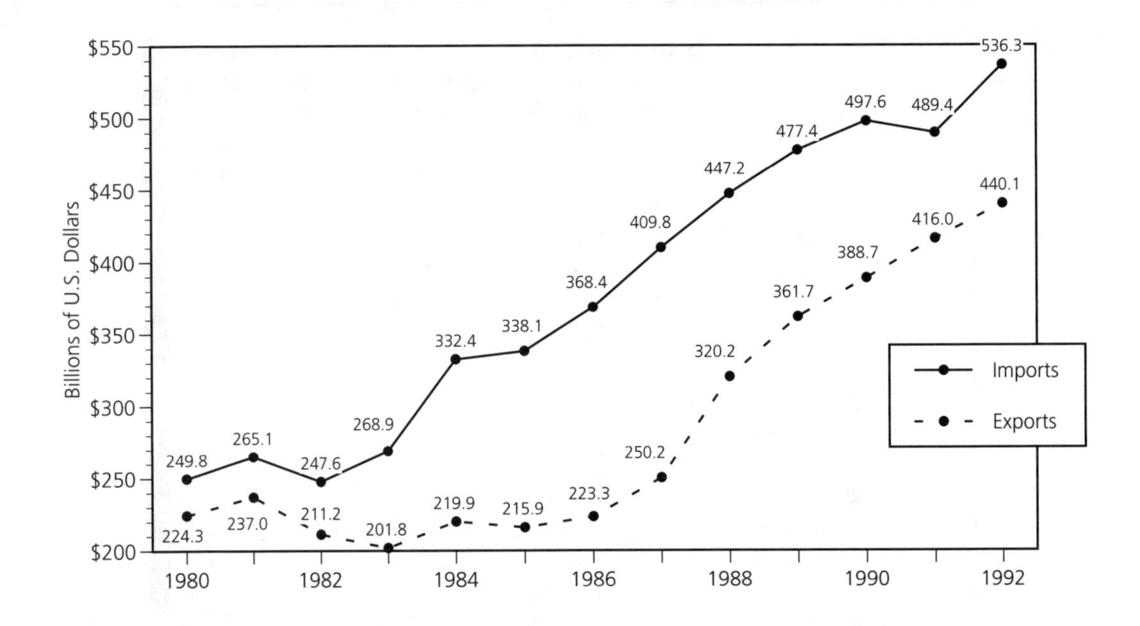

Source: U.S. Department of Commerce, Bureau of Economic Analysis

Figure 16.8 **Canada: Merchandise Trade Balance, 1980–1992**

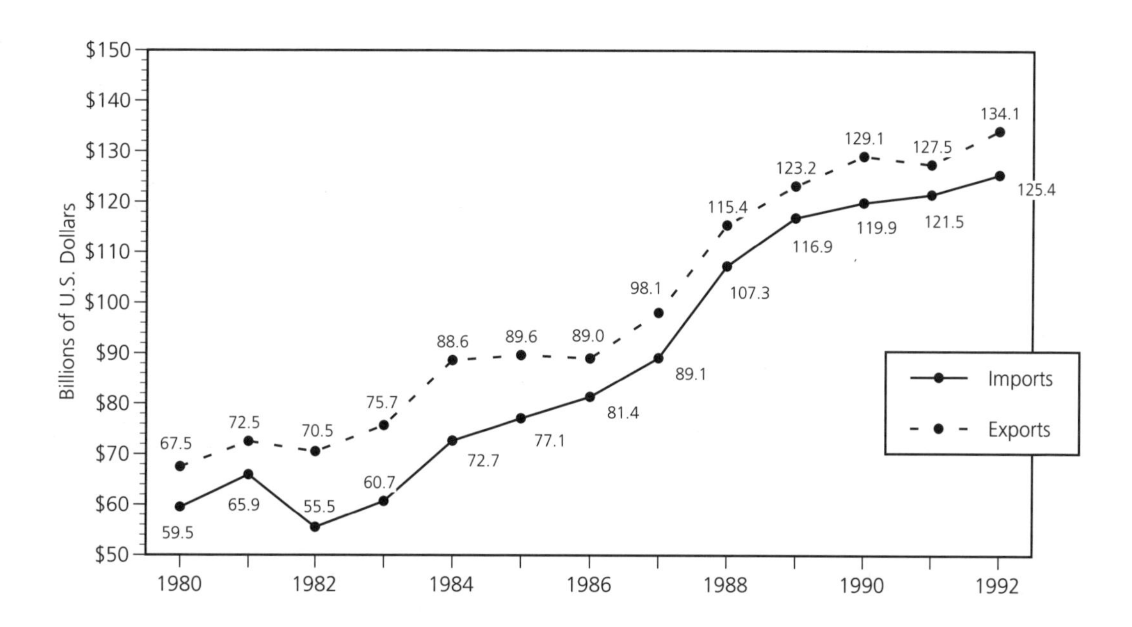

Source: International Monetary Fund

Figure 16.9 **Mexico: Merchandise Trade Balance, 1980–1991**

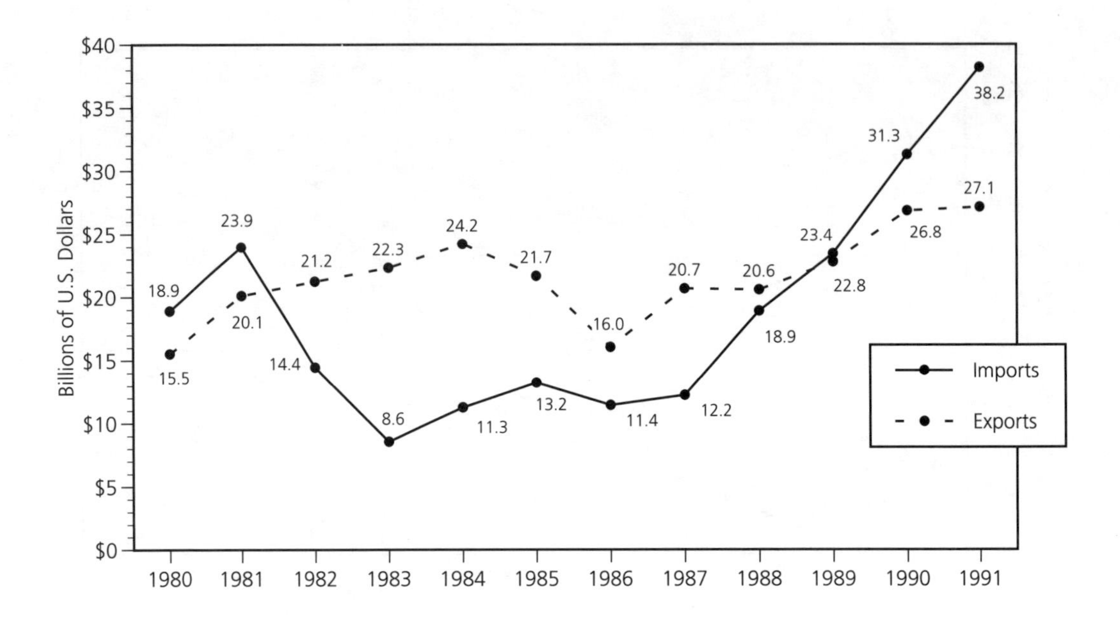

Source: International Monetary Fund

Mexican imports have tripled since 1986. Exports between 1986 and 1991 have increased 69 percent. But as living standards increase, so does demand for imports. Consequently, Mexico has run a modest trade deficit since 1989.

Mexico is one of the few major trading partners with which the United States has a positive balance of trade—U.S. exports to Mexico exceed imports from Mexico.

Principal Exports/Imports

The principal U.S. exports are machinery and high-technology goods, chiefly aerospace. Principal U.S. imports are petroleum and motor vehicles.

Canada's principal exports in 1991 were machinery and equipment, automotive products, industrial goods, and forest products, chiefly wood products. Canada's imports largely paralleled her exports. Machinery and equipment were the largest category of imports, followed by automotive products, industrial goods, and consumer goods.

Figure 16.10 **United States: Top Five Exports, 1991 (N.E.S.; free alongside ship [f.a.s.] transaction basis.)**

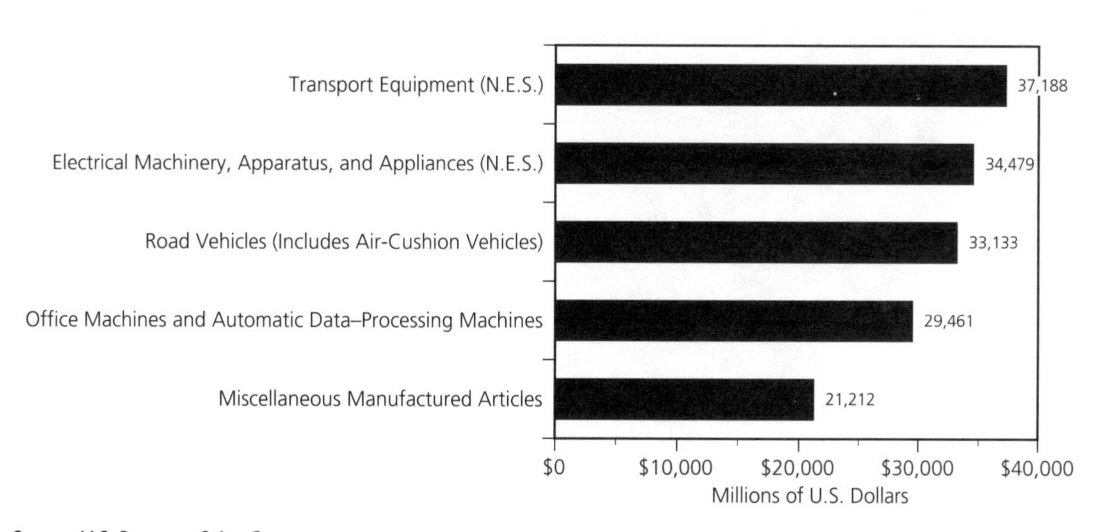

Source: U.S. Bureau of the Census

Figure 16.11 **United States: Top Five Imports, 1990 (N.E.S. customs value basis.)**

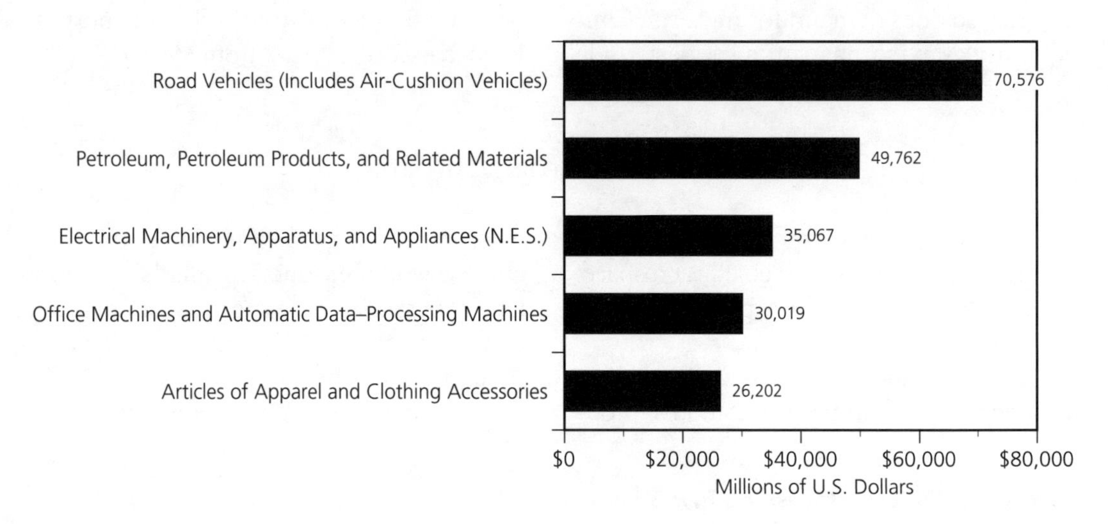

Source: U.S. Bureau of the Census

Figure 16.12 **Canada: Top Five Exports, 1991**
Balance of payment basis.

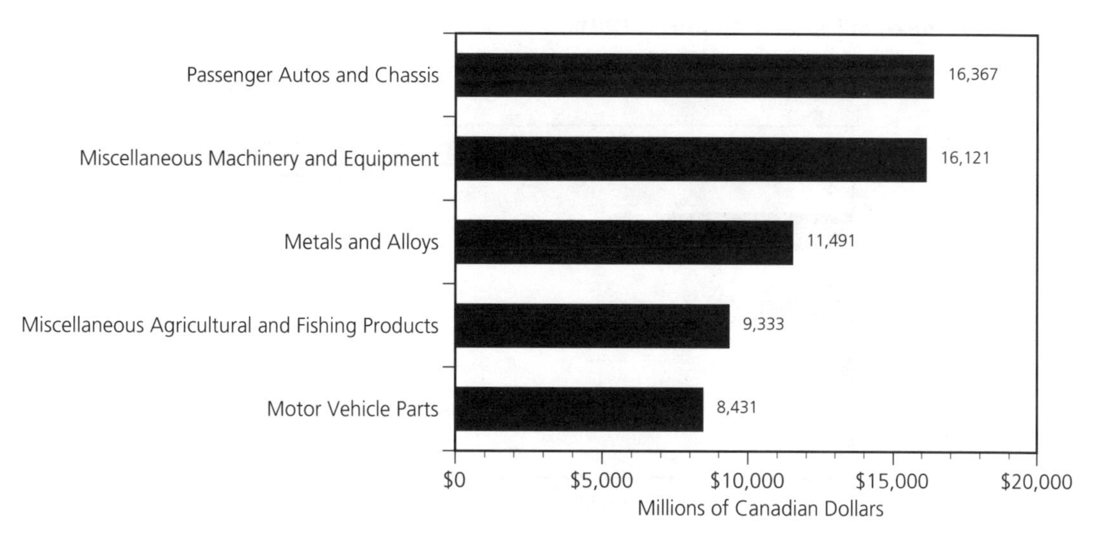

Source: Statistics Canada, Canada Yearbook

Figure 16.13 **Canada: Top Five Imports, 1991**
Balance of payment basis.

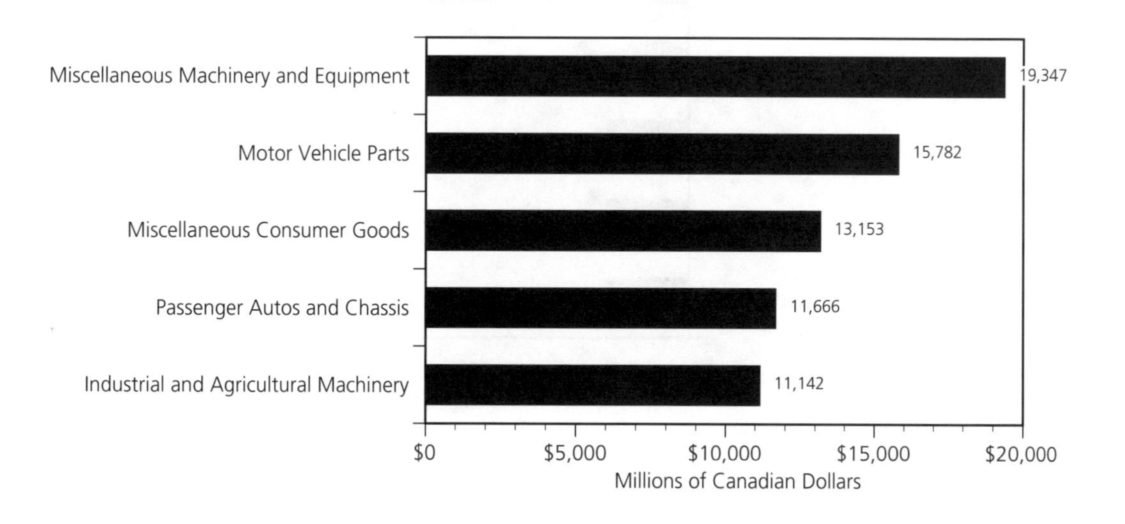

Source: Statistics Canada, Canada Yearbook

Mexico's chief exports are oil, automobiles and automobile parts, and fresh vegetables. Assembly materials for automobiles are the leading import.

Figure 16.14 **Mexico: Top Five Exports, 1991**

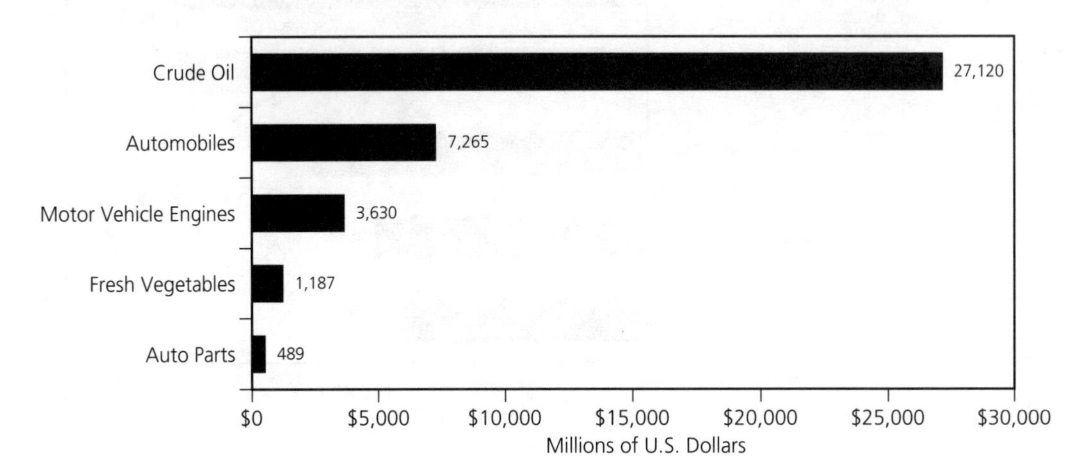

Source: INEGI

Figure 16.15 **Mexico: Top Five Imports, 1991**

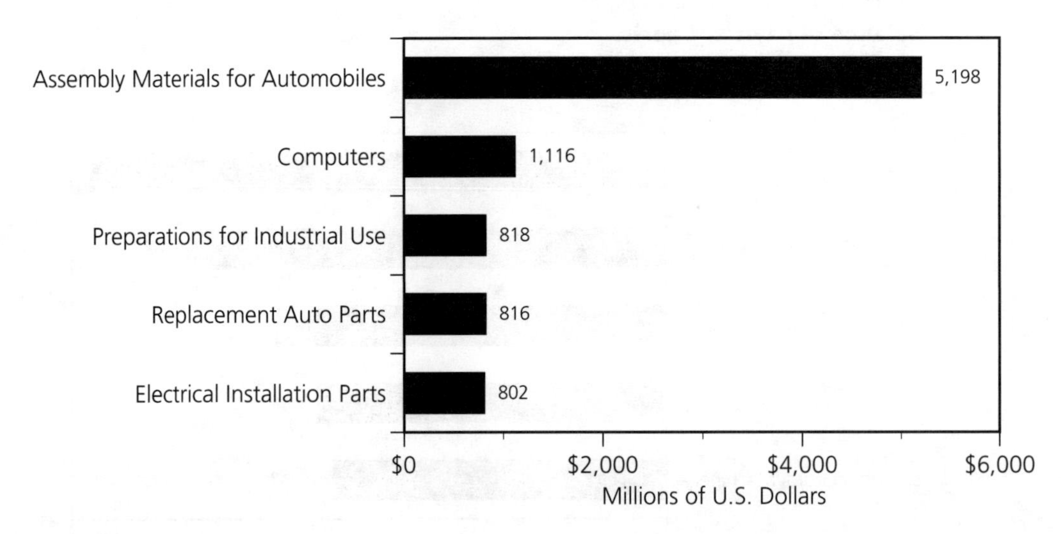

Source: INEGI

Trade Flows

The principal "river of trade" in North America flows from southern Ontario through Michigan and the U.S. industrial heartland to the border crossing at Laredo, Texas, to Monterrey and down to the region around Mexico City, where most of Mexico's population and industry are located.

Border Gateways

There are fifty-three border crossings from Canada into the western United States. The busiest for commercial traffic is the Pacific Highway-Blaine crossing between Seattle, Washington, and Vancouver, British Columbia.

There are thirty-seven border crossings on the U.S.-Mexican border. The busiest for auto traffic is San Ysidro in California. But more than half of the value of all commercial goods shipped from the United States to Mexico, and from Mexico to the United States, pass through the ports of Houston and Laredo in south Texas.

Figure 16.16 **Western United States Merchandise Exports and Imports by Coastal Area and Customs District, 1992**

West:	$207.5 Billion in Exports
	$208.1 Billion in Imports
East:	$240 Billion in Exports
	$317 Billion in Imports
U.S. Totals:	$447.5 Billion in Exports
	$525.1 Billion in Imports

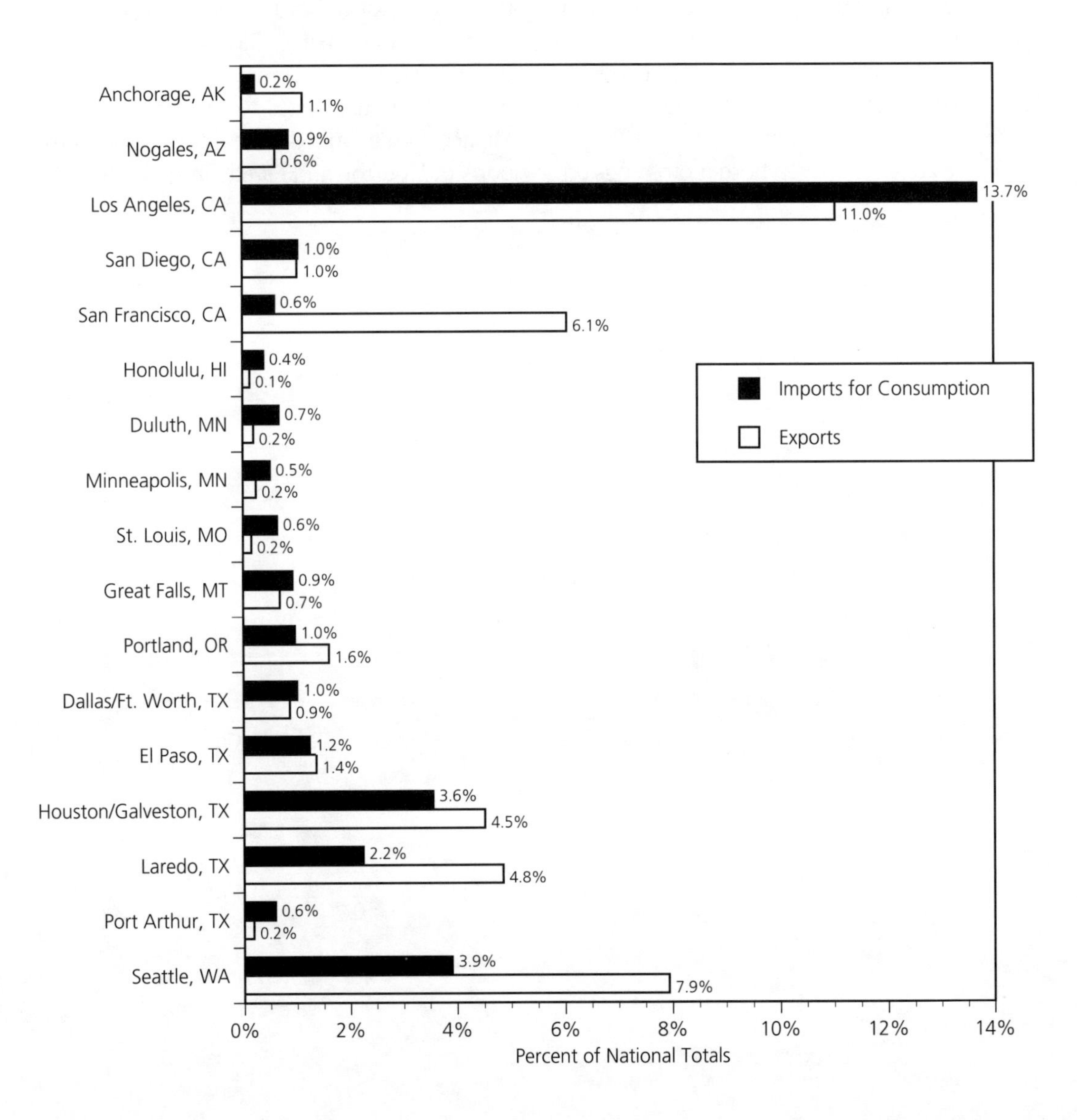

Source: U.S. Bureau of the Census

Maquiladoras

Maquiladoras are manufacturing plants located in Mexico. More than 1,500 maquiladoras manufacture goods made up of foreign components, the vast majority of which are produced in the United States. Mexico provides less than 3% of the supplies for the maquiladoras. Most of the products of these factories are exported, primarily to the United States.

The maquiladora program emerged in the mid-1960s as a Mexican economic development program. Originally an agreement between two Mexican Cabinet officers to relax strict foreign investment, customs, and immigration laws in 1966, this initial policy was so successful that in 1971 it was formalized as the Border Industrialization Program.

Although an increasing number of maquiladoras are being established in the Mexican heartland, 80 percent of the plants and 90 percent of maquiladora employment are located in the Mexican border states. The concentration is greatest near El Paso/Ciudad Juarez.

Maquiladoras produce a wide variety of products, but production is dominated by four sectors: electrical and electronic goods, automotive parts, textiles and apparel, and furniture.

Figure 16.17 **Mexico**
Maquiladora Industry, Value-Added, October 1989 and October 1990

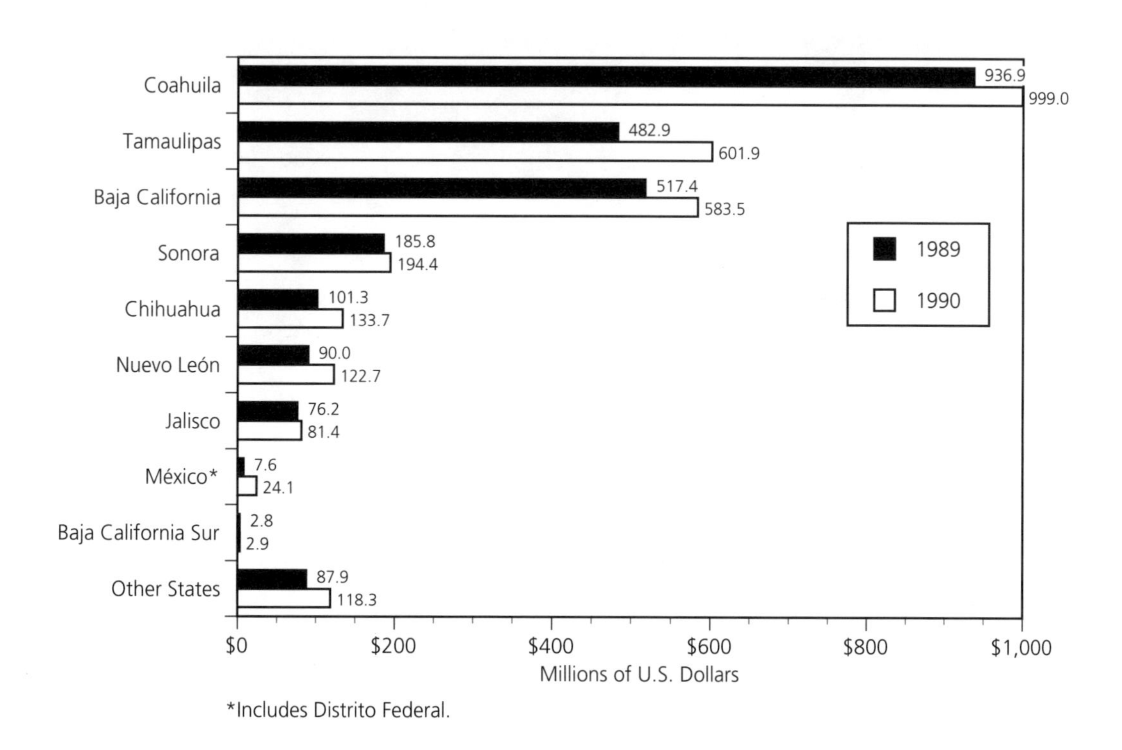

Source: Secretariat of Trade and Industrial Development (SECOFI)

Between 1985 and 1990, total maquiladora value-added production grew $3.6 billion, an average annual growth rate of more than 20 percent. The number of maquiladoras increased by nearly 20 percent in that period, and employment grew at an average annual rate of 17.3 percent. Exports from maquiladoras rose from 6 percent of total Mexican merchandise exports to 14 percent.

Figure 16.18 **Mexico**
Maquiladora Industry, Number of Plants, October 1990

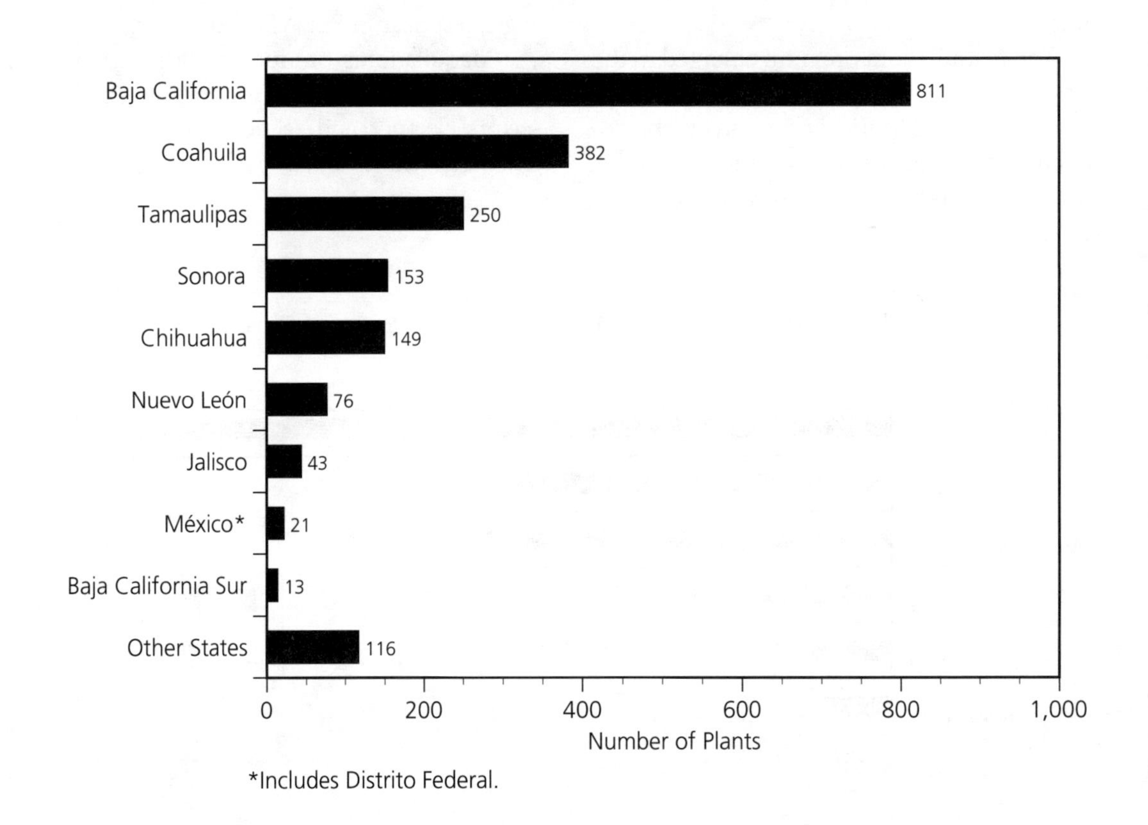

Source: Secretariat of Trade and Industrial Development (SECOFI)

Figure 16.19 **Mexico**
Maquiladora Industry, Workers Employed by State, October 1990

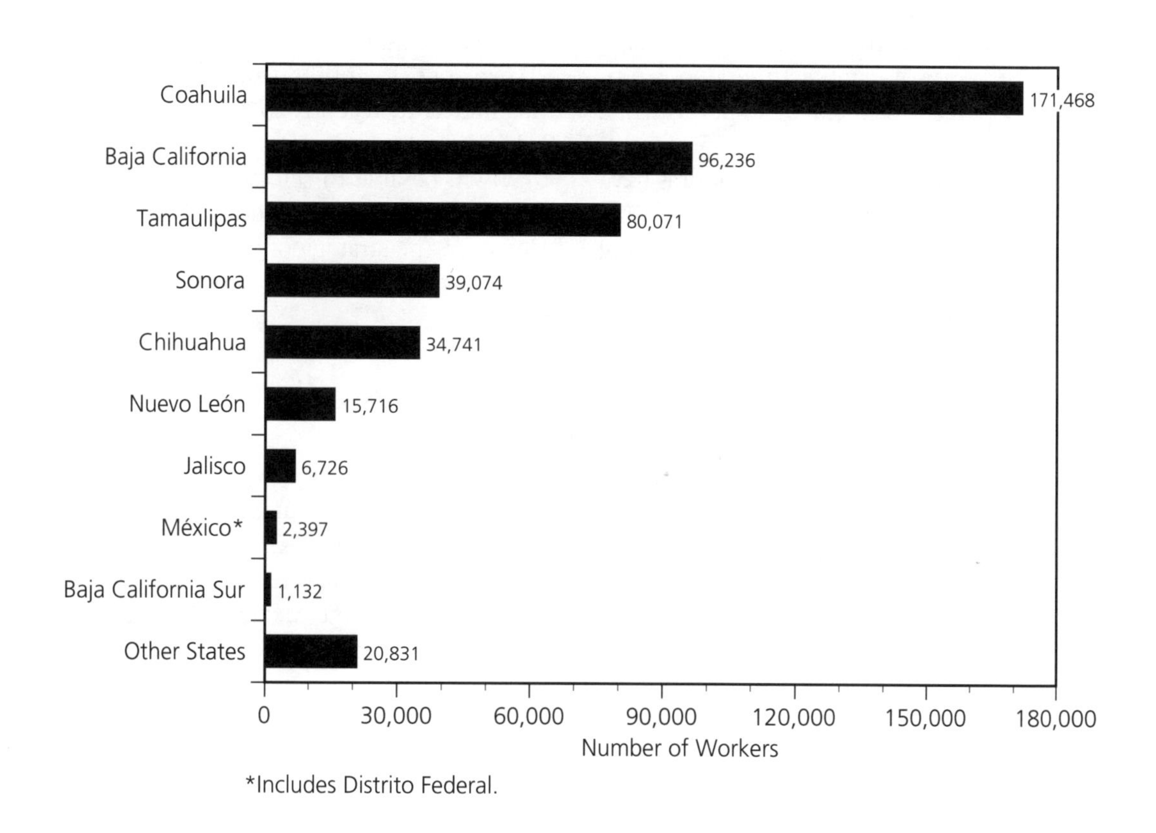

*Includes Distrito Federal.

Source: Secretariat of Trade and Industrial Development (SECOFI)

The maquiladoras are a major destination for U.S. exports to Mexico. Between 1976 and 1986, crude oil exports from Mexico represented almost 70 percent of Mexico's total exports. Since then, those trends have reversed. In 1994, exports of manufactured goods represent 75 percent of Mexico's exports, while crude oil exports reflect less than 25 percent.

Also, imports from the maquiladoras have grown from 20 percent to 52 percent of total U.S. imports from Mexico, reflecting growing reliance by U.S. manufacturers on production-sharing arrangements with Mexico.

Begun strictly as production for export, maquiladoras are becoming integrated into the Mexican economy. Prior to 1983, by law, all maquiladora production had to be exported. In 1983, maquiladoras were permitted to sell 20 percent of their production in Mexico. This was increased recently to 33 percent. Under NAFTA, maquiladoras will disappear since they will be able to sell their final goods all over North America, including Mexico itself. However, the strategy of using joint production arrangements in Mexico is likely to increase.

Tourism

Tourism is one of the world's fastest growing industries, and it's big business for North America. Travel by citizens of the United States, Canada, and Mexico has increased steadily throughout the past decade. U.S. tourists are visiting Canada less but Mexico more; however, revenues from U.S. tourists to Canada are up 40 percent from ten years ago. Receipts from U.S. tourists to Mexico have more than doubled in the same period.

Figure 16.20 **Canada**
Exchange Rate with U.S. Dollar, 1980–October 1992

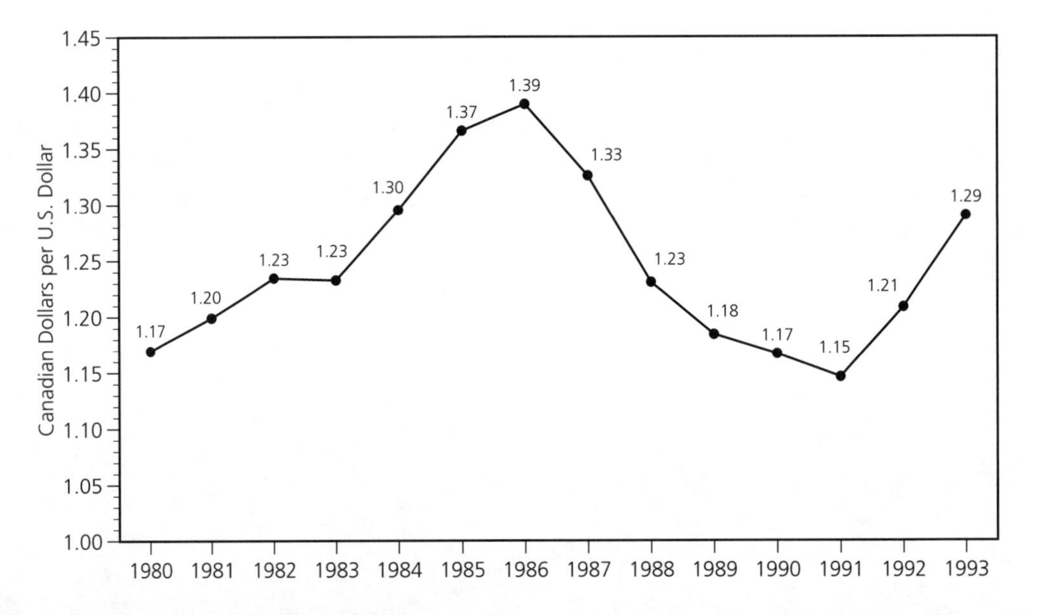

Source: U.S. Bureau of Labor Statistics and Federal Reserve

Figure 16.21 **Mexico**
Exchange Rate (Market) with U.S. Dollar, March 1990–December 1992

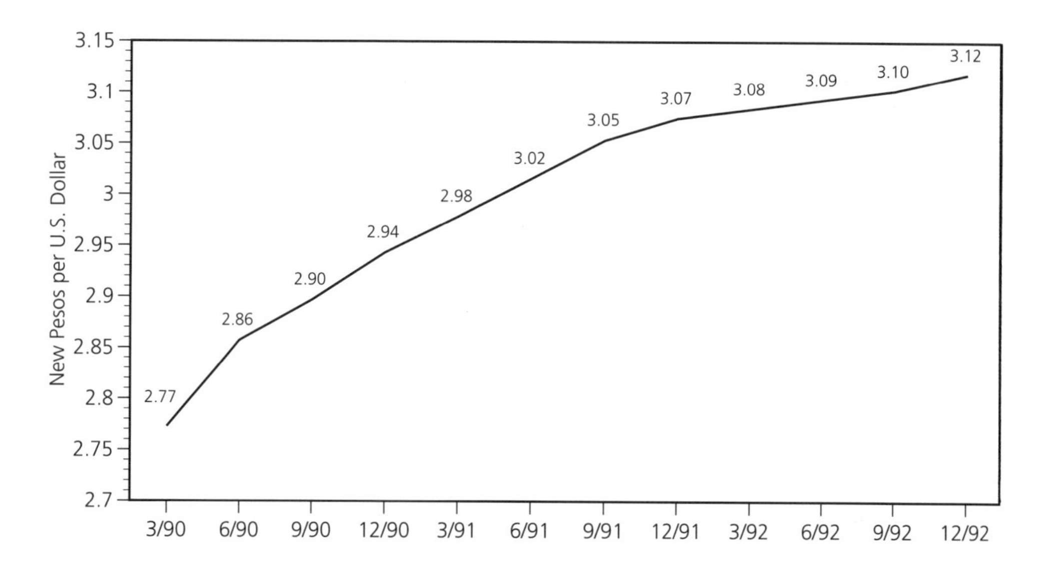

Source: Banco de Mexico

Figure 16.22 **United States**
Travel Industry Total Business Receipts, 1980–1990

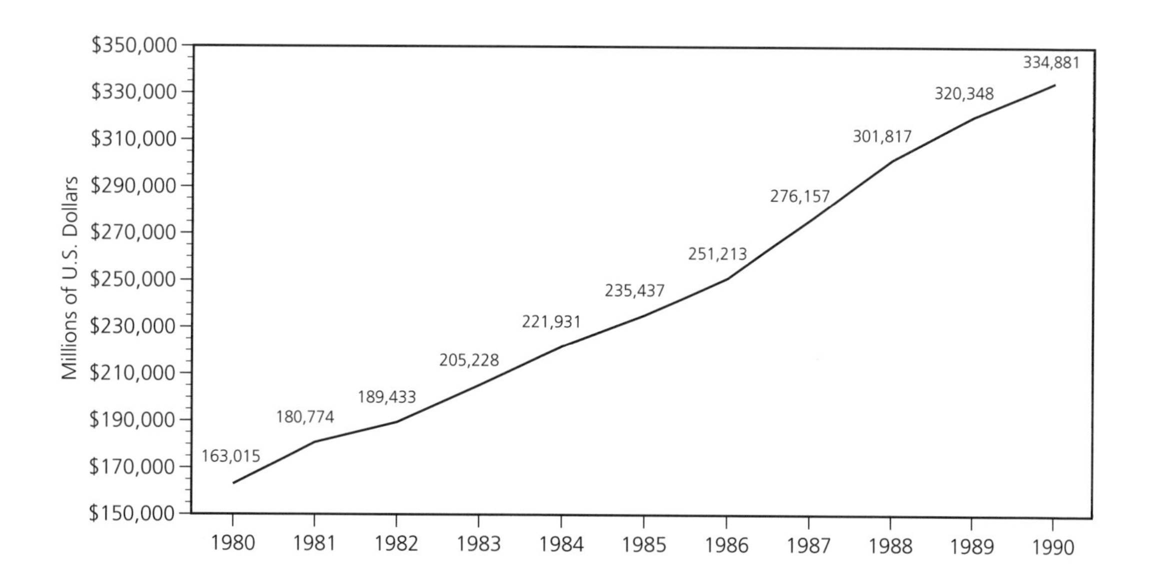

Source: U.S. Travel Data Center, Washington, D.C.

Receipts from tourism in the United States have more than tripled since 1985; income from Canadian tourists has grown even faster, and income from Mexican tourists has nearly tripled.

Figure 16.23a Canadian and Mexican Travel to the United States Number of Travelers, 1985–1992

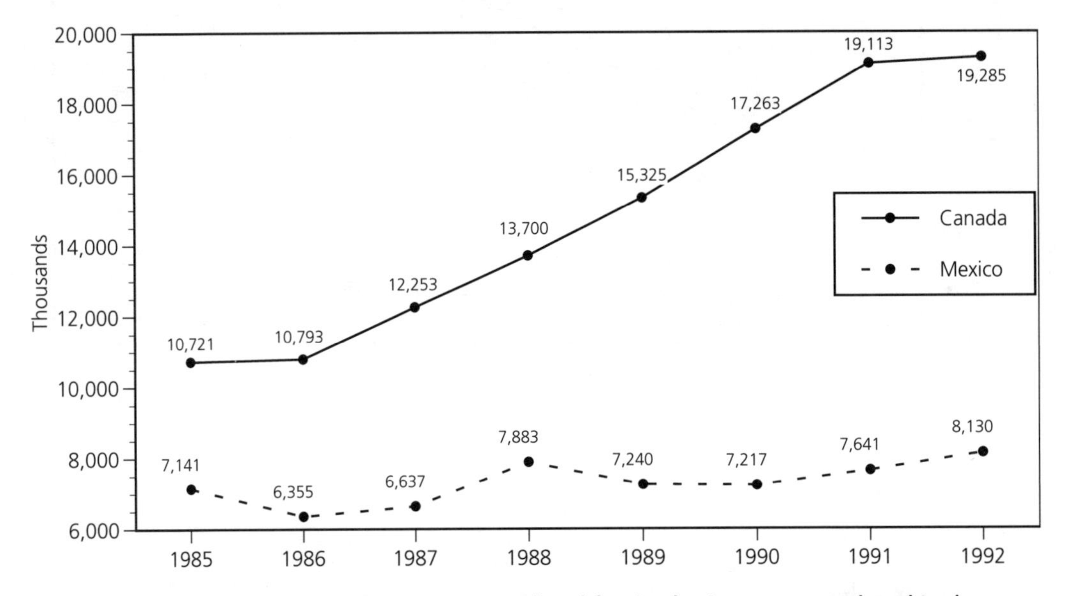

(Excludes foreign government personnel and foreign businessmen employed in the United States. Includes travelers for business and pleasure, foreigners in transit through the United States, and students.)

Source: U.S. Immigration and Naturalization Service

Figure 16.23b **Canadian and Mexican Travel to the United States Travel Receipts, 1985–1992**

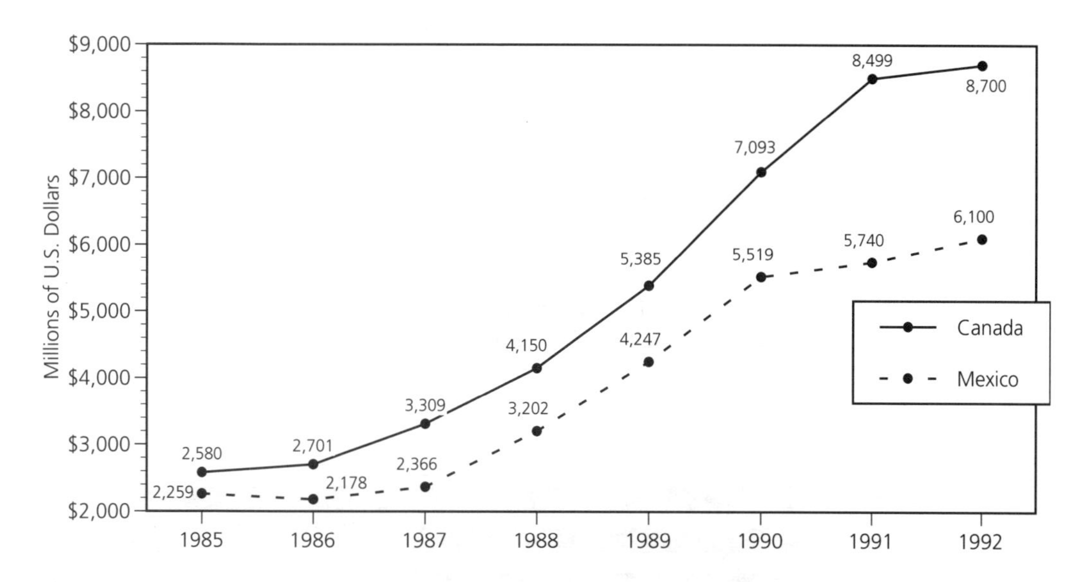

Source: U.S. Travel and Tourism Administration

Figure 16.24 **Canadian and Mexican Tourists Admitted into the United States, 1985–1991**

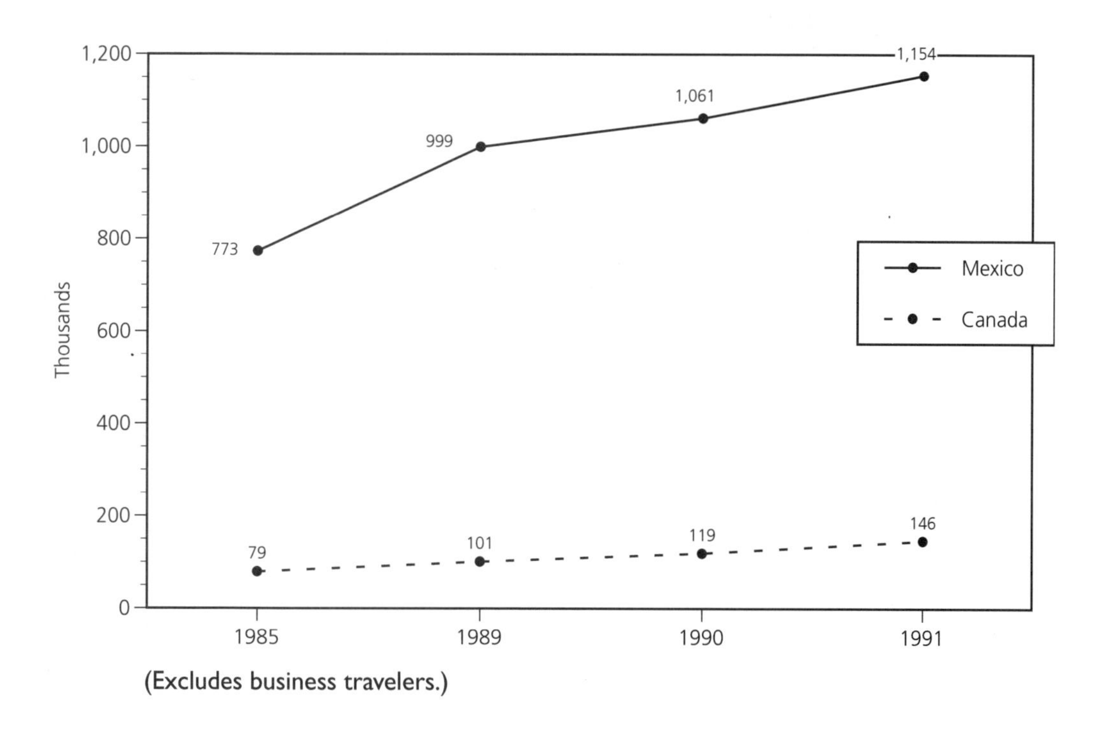

(Excludes business travelers.)

Source: U.S. Travel and Tourism Administration

The effects of tourism are felt in numerous industries, from food, lodging, travel, and entertainment to the retail trade in clothing, jewelry, shoes, sports equipment, and souvenirs.

Canada's dollars from tourism are reflected in the huge growth of revenue in the food and beverage industries as well as the accommodation service industries.

Figure 16.25 **Canada**
Revenues of Accommodation Service Industry by Province, 1990

West: **$3,863** million, Canadian
East: **$4,759.8** million, Canadian
Canada Total: **$8,622.8** million, Canadian

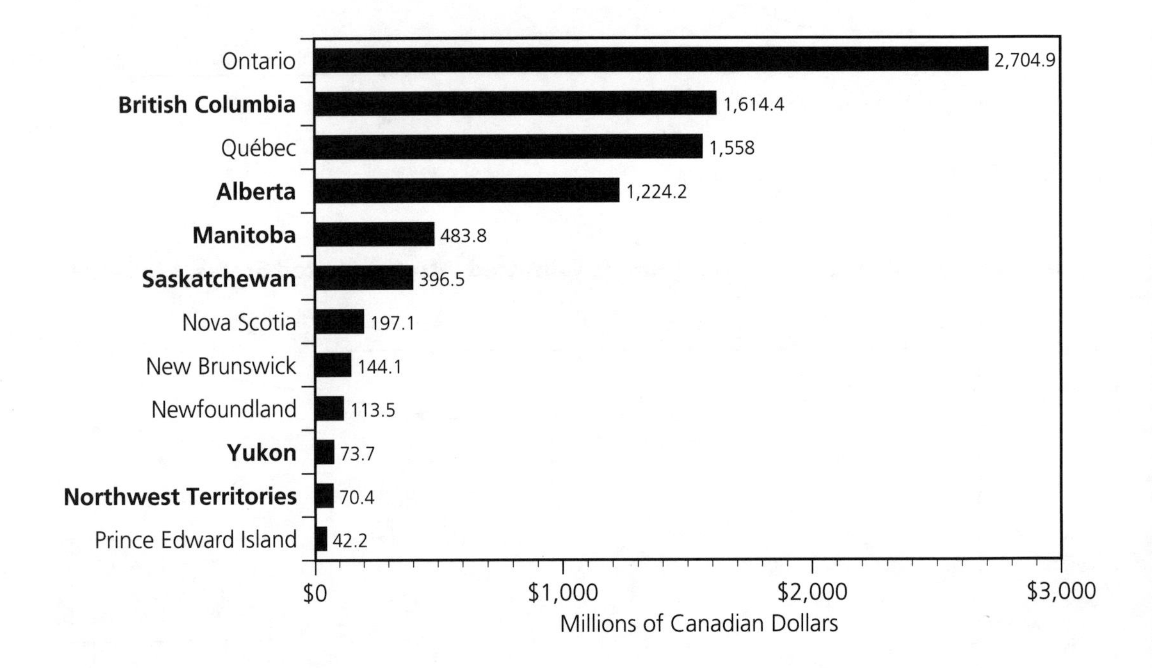

,Source: Statistics Canada, Canada Yearbook

Figure 16.26 **Expenditures in Canada by Travelers from United States and by Travelers from Other Countries, 1989–1991**

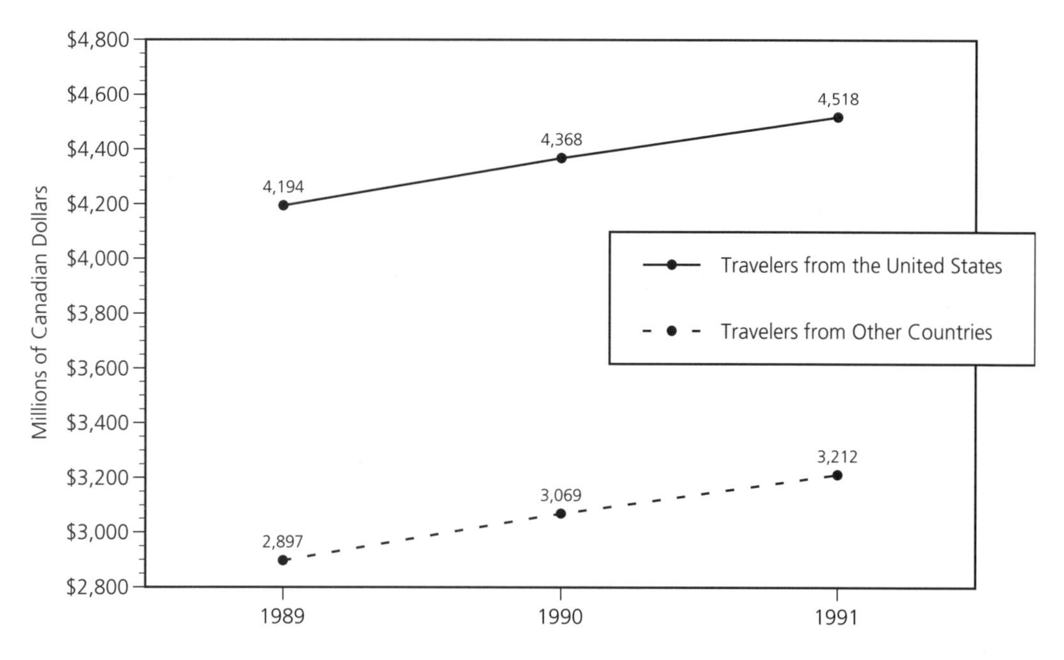

Source: Europa Yearbook

Figure 16.27 **Mexico**
Tourism Industry Revenues and Expenditures, March 1990–September 1992

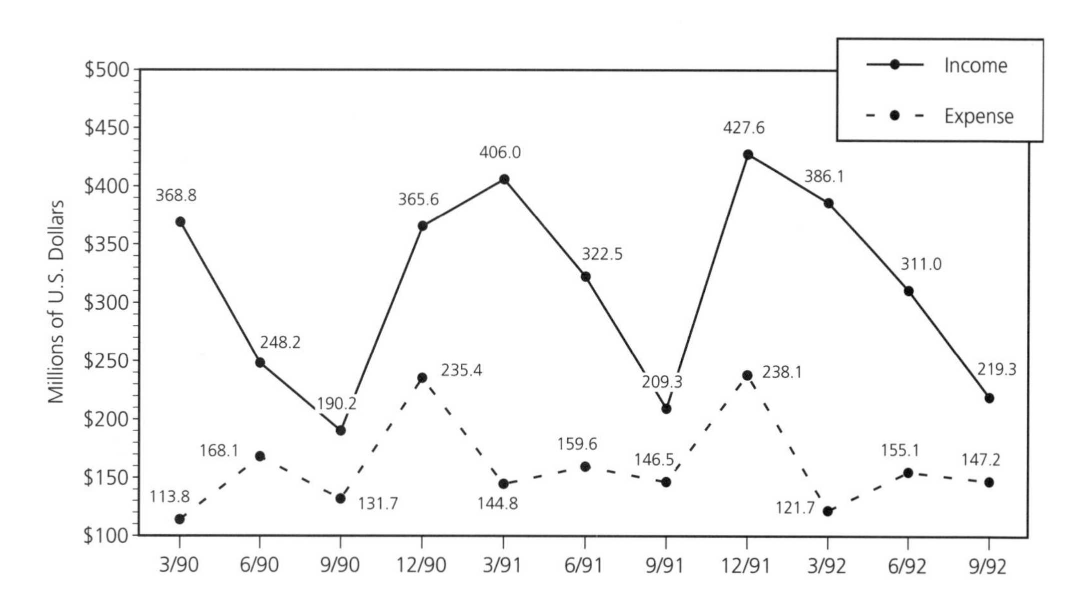

Source: Banco de Mexico

Figure 16.28 **Mexico**
Tourist Inflow and Outflow, March 1990–September 1992

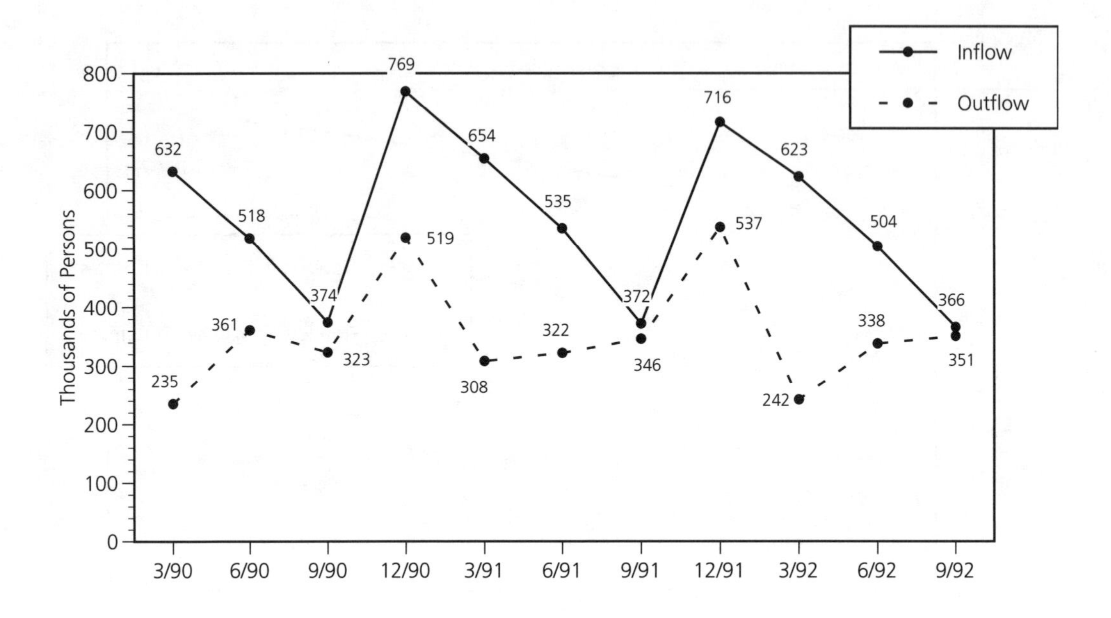

Source: Banco de Mexico

Figure 16.29 **Mexico**
Number of Hotel Rooms Rented and Number of Visitors to Principal Tourist Destinations, 1990 preliminary

	Number of Hotel Rooms	Thousands of Domestic Visitors	Thousands of Foreign Visitors
Tijuana, B.C.	4,733	1,080.9	315.2
La Paz, B.C.Sur	1,412	179.0	45.9
Loreto, B.C.Sur	455	12.0	26.1
Los Cabos, B.C.Sur	2,531	27.7	228.0
Manzanillo, Col.	2,987	288.0	50.2
Cd. Juárez, Chih.	2,794	793.1	72.4
Distrito Federal	18,138	1,630.3	793.8
Guanajuato, Gto.	1,717	296.3	27.3
Acapulco, Gro.	17,001	1,049.2	417.1
Ixtapa–Zihuatanejo, Gro.	4,169	191.5	109.3
Guadalajara, Jal.	13,092	2,113.8	157.3
Puerto Vallarta, Jal.	8,646	380.8	307.1
Morelia, Mich.	2,678	629.8	17.3
Monterrey, N.L.	3,988	733.6	97.0
Bahías de Huatulco, Oax.	1,310	77.3	41.7
Oaxaca, Oax.	2,672	342.8	110.5
Cancún, Q.R.	17,470	395.2	1,180.5
Cozumel, Q.R.	2,875	48.0	170.3
Mazatlán, Sin.	7,935	632.2	243.9
Reynosa, Tamps.	1,491	255.0	7.2
Veracruz, Ver.	4,269	809.7	19.8
Mérida, Yuc.	3,188	305.7	164.0
Zacatecas, Zac.	1,031	248.0	12.1

Source: Secretaria de Turismo, FONATUR

Appendix One
The North American Free Trade Agreement (NAFTA)

International trade is growing at a much faster rate than are world economies as a whole. That's why trade is a major engine of economic growth and job creation. According to the International Monetary Fund (IMF), the value of world exports plus imports rose from $6 billion in 1948 to more than $6 trillion in 1990—an increase of 1,000 percent. Half of that increase has been in the last fifteen years.

International commerce is growing even faster than trade. Trade is the largest, but only one of four components of international commerce. The others are overseas economic activities by multinational corporations, financial flows, and travel and tourism.

As it grows, the composition of international trade is changing. Trade in services is the most rapidly growing sector of international trade. Trade in services—for example, transportation, telecommunications, construction, and financial services—now accounts for one-fourth of world trade. Trade in services increased more than 60 percent between 1985 and 1990. Travel and tourism, which now generates annual revenues of $3.4 trillion, is the world's largest industry. In 1991, travel and tourism became the number one source of foreign exchange earnings for the U.S.

Figure A.1 **World Exports of Commercial Services (Includes Transport, Travel, Tourism, Banking, and Insurance)**

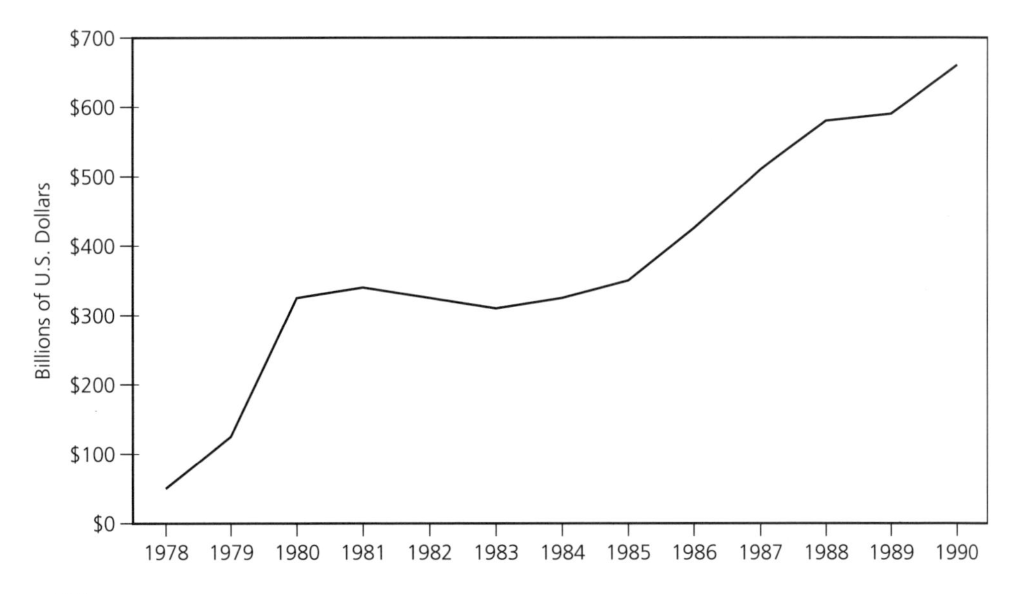

Source: GATT

While the United States is a substantial net importer of goods, the United States is a net exporter of services, and U.S. citizens earn more money on U.S. investments abroad than foreigners earn on their investments in the United States.

Figure A.2 **United States Balance of Trade, 1992**

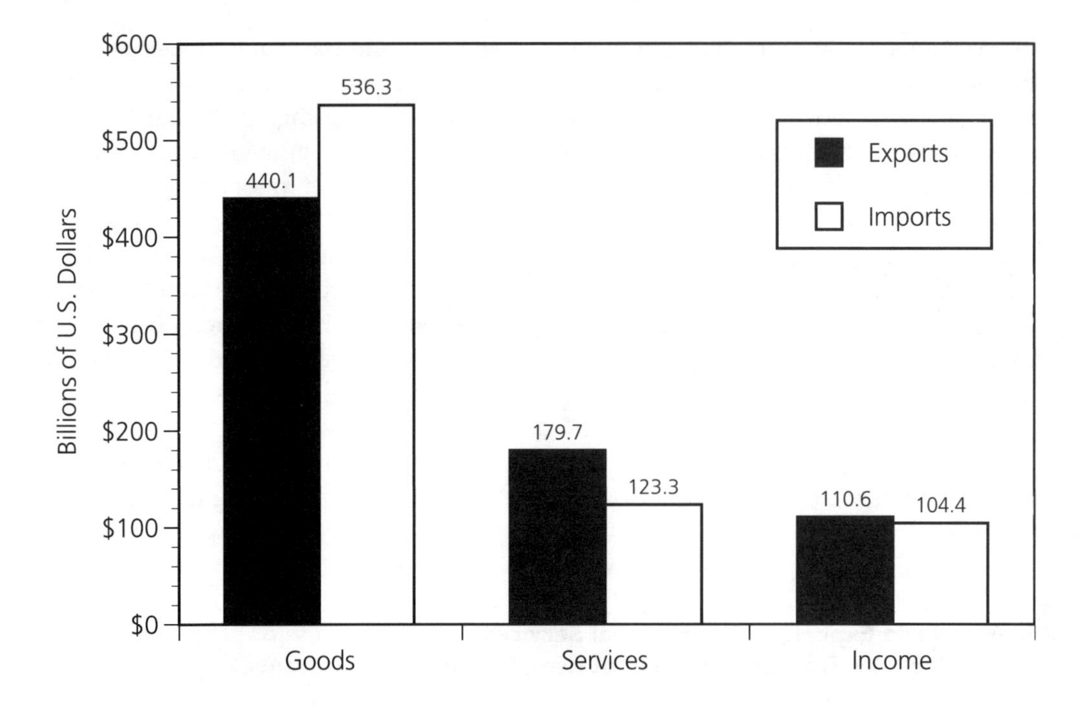

Source: U.S. Department of Commerce

The direction of trade flows has changed. In the immediate post–World War II era, when current international trade and finance systems were established, the overwhelming flow of international trade was across the Atlantic—to and from the United States and Western Europe, and within Western Europe.

This is no longer true. The major growth of world trade is in the Asia-Pacific region. U.S. trade across the Pacific exceeded trade across the Atlantic in 1983. By 1991, trade across the Pacific was more than double the trade across the Atlantic.

Information Revolution Impacts on Trade and Commerce

Throughout history, competitive advantage among cities and states has been shaped by technology and infrastructure. As Massachusetts Institute of Technology historian Warner Schilling has pointed out, the glory of Athens rested on silver mines; the might of Sparta on a process for making steel; the Romans through roads; and the Assyrians overran Babylon and Egypt with the chariot. Europe's colonization of much of the world depended on clocks, the compass, gunpowder, and improvements in the design of sailing ships.

Today, the technologies of advantage are information and telecomputing—the combination of telephones, computers, and software. Present market economies run chiefly on information. The new computer and telecommunications technologies have greatly expanded the amount of information available, and the speed at which it can be disseminated. Bits and bytes transmitted by satellite have bound together disparate locations into a single marketplace that can be monitored and accessed by managers, investors, and consumers located anywhere. These technologies also permit managers and service-providers to be more responsive to customers located anywhere.

These new information technologies have fueled an explosive growth in world financial markets. The volume of shares traded on the New York Stock Exchange (NYSE) has increased from 524.8 million in 1950 to 2.94 billion in 1970, and to 51.4 billion in 1992.

Canada has stock exchanges in Toronto, Montréal, Vancouver, Winnipeg, and Calgary. The largest is in Toronto. Trading volume is significantly less than trading volume on the NYSE, but it is rising proportionately. The volume of shares traded on the Toronto exchange has increased from 522.91 million shares in 1970 to 7.33 billion shares in 1992. In 1992, 1.7 billion shares were traded on the Montréal exchange, 3.89 billion shares were traded on the Vancouver exchange, 868 million shares on the Alberta (Calgary) exchange, and 15,112 shares on the Winnipeg exchange.

Figure A.3 **United States**
Average Daily Volume of Shares Traded on the New York Stock Exchange (NYSE) and the NASDAQ, 1980, 1984–1992

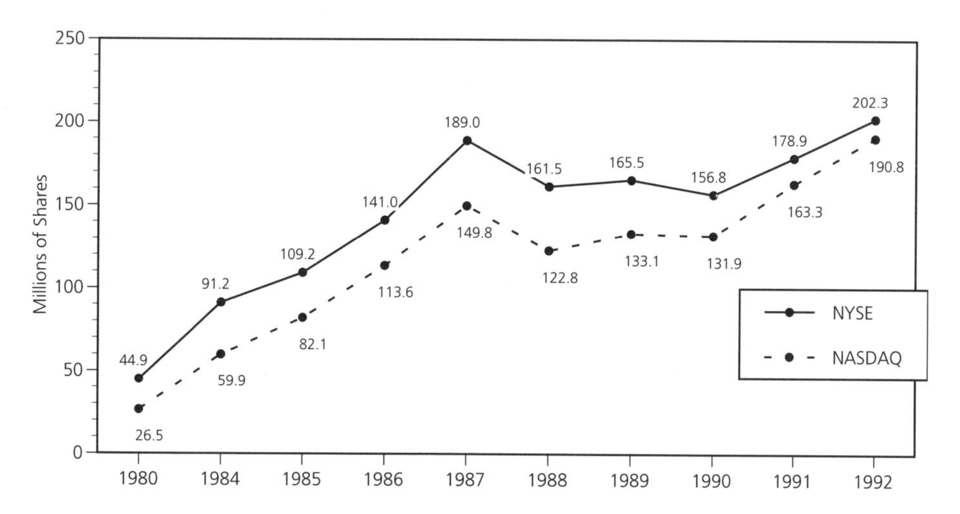

Source: New York Stock Exchange, Fact Book and National Association of Securities Dealers, Fact Book

Figure A.4 **Canada**
Combined Volumes of the Montréal and Toronto Stock Exchanges, December 1990–October 1992

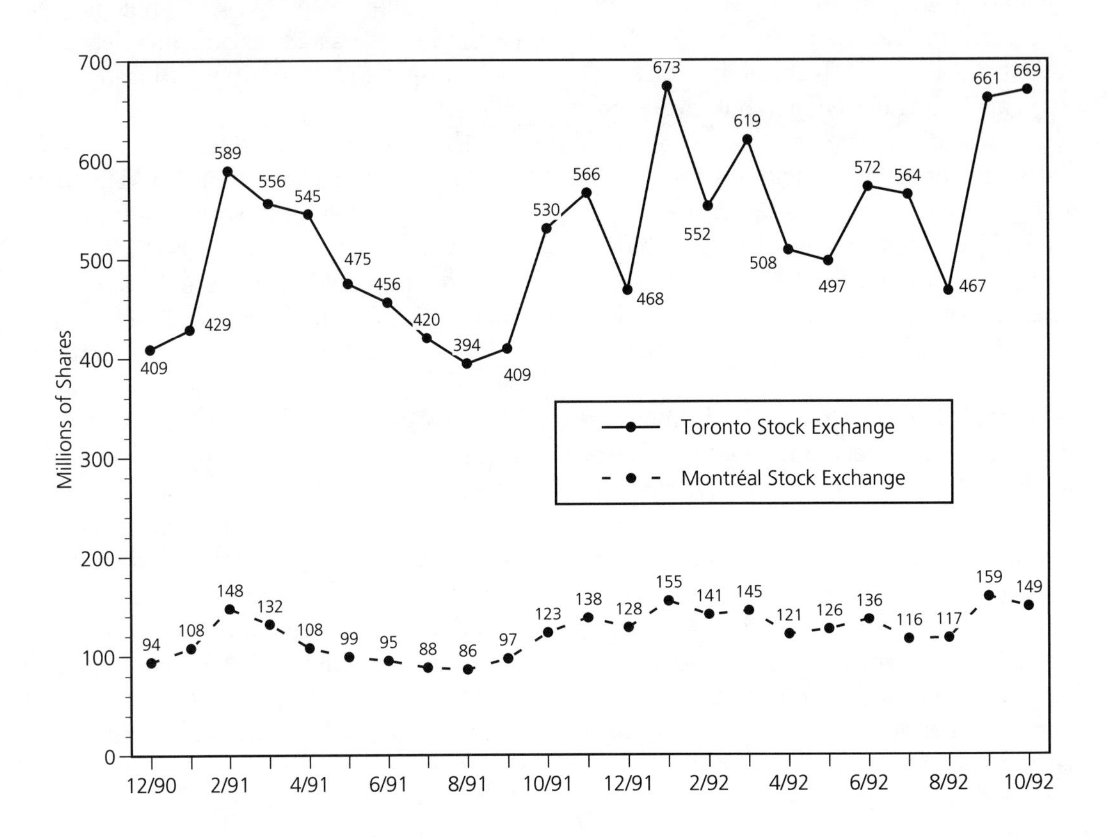

Source: Statistics Canada, Canadian Economic Observer

Figure A.5 **Mexico**
Stock Market Index, February 1991–February 1993
(1978=100; Base 781.62, October 1978)

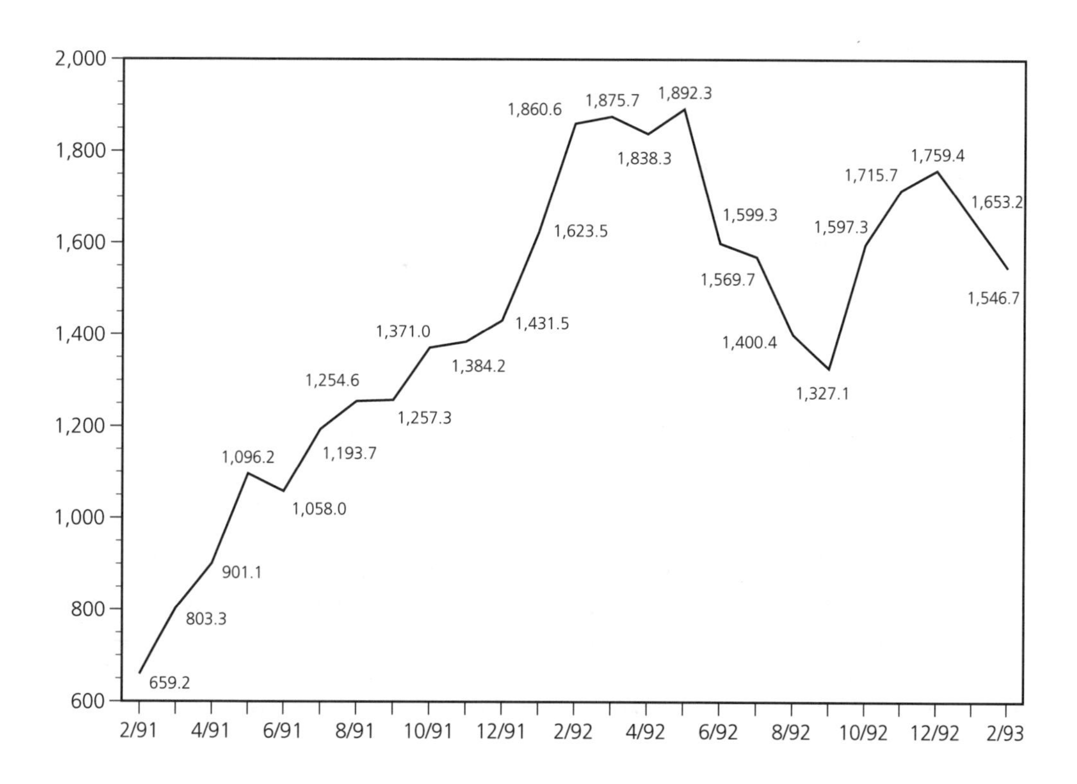

Source: Bolsa Mexicana de Valores, February 1993

CAD/CAM engineering and other computer-driven innovations are transforming manufacturing around the world. The whole concept of mass production is under siege as the new technologies make smaller and smaller production runs profitable. Niche markets are proliferating as custom production tailored to niche markets becomes affordable.

Remarkable improvements in bulk transportation systems and in transportation management—especially the rise of intermodalism—make a wide dispersion of functions within the same industrial concern possible, practical, and affordable. The speed and low cost of modern bulk transportation systems, coupled with "just-in-time" inventory management practices made possible by telecomputing, allows assembly plants to be located thousands of miles from primary production facilities, taking advantage of favorable labor markets, taxes, or other elements of the local business climate.

New Players Move to Center Stage

In the new world economy, both national and global political entities are increasing in numbers but declining in relative influence. Even though the G-7 nations—the United States, Canada, Japan, and the European Union (EU) countries of Britain, France, Germany, and Italy—are considered the major economic powers, a recent IMF study shows a different picture. Both Brazil and Mexico rank above Canada, and China and India have more economic muscle than four of the G-7 nations.

At the same time as countries are being reshuffled in economic importance, new players are emerging. State and provincial governments are playing a more assertive role in international trade. For example, forty-three U.S. state governments now have trade representation in Tokyo. Regional and sub-regional economic alliances, both within and across national borders, are expanding their economic and trade development activities. An ancient form of political organization—the city-state—is again rising in prominence. Great metropolitan regions often overshadow states and provinces, and, in some instances, even the countries in which they are located. Cities are beginning to play a larger role in international relations, independent of national and provincial governments.

Figure A.6 **World's Largest Cities, by Country in the Year 2000**

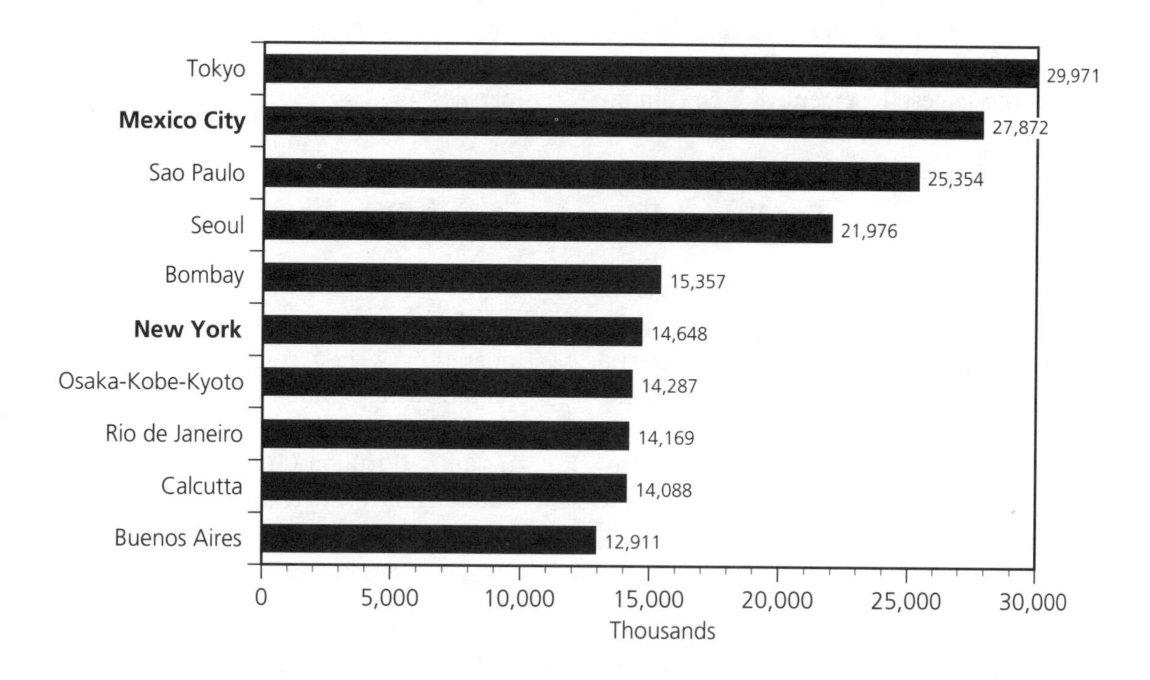

Source: U.S. Bureau of the Census

Figure A.7 **Top Markets by Purchasing Power, ca. 1992**

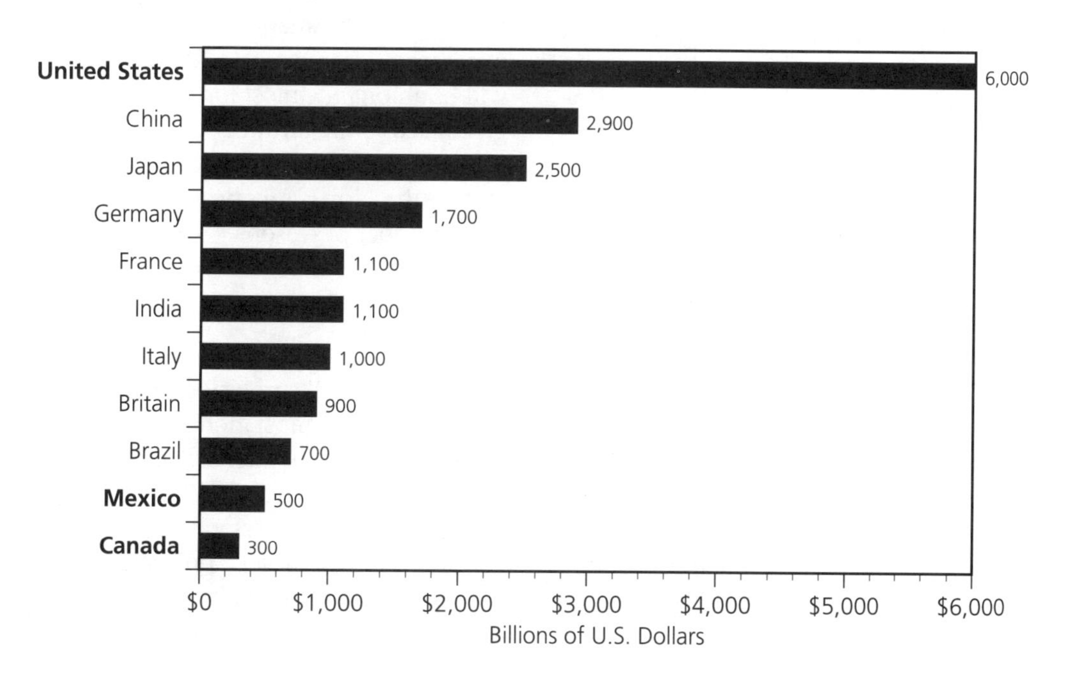

Source: IMF

The number and types of commercial enterprises which engage in international trade are expanding dramatically. In 1965, 80 percent of all U.S. international trade was generated by the largest 1 percent of U.S. companies—Boeing, General Electric, General Motors, and the like. By 1990, more than 50 percent of U.S. exports by dollar value were made by companies with fewer than five hundred employees.

The Rise of Regional Trading Blocs

The international organizations formed in the aftermath of World War II are also faltering. An important example is the General Agreement on Tariffs and Trade (GATT), the world's principal political arrangement for managing and expanding international economic cooperation and the world's free-trade markets. GATT has held eight "rounds" of multilateral trade negotiations. The most recent is the Uruguay round (1986 to the present). In December 1993, the Uruguay round reached a conclusion that is more of a whimper than a bang.

Concurrent with, and contributing to, the GATT negotiations and other international economic institutions and arrangements has been the rise of regional open-trading groups. The most important of these open-trade regions are the following:

- The European Union (EU), formerly the European Economic Community (EEC), currently consisting of twelve nations with a population of 325 million and a combined gross product of $5 trillion.

- The North American Free Trade Agreement (NAFTA) comprised of Canada, the United States, and Mexico, with a combined population of 370 million and a combined gross product of almost $7 trillion.

- The Asian Pacific Economic Cooperation (APEC), consisting of twelve nations with a combined population of 800 million and a combined gross product of $10 trillion ($4 trillion without counting the United States and Canada).

NAFTA has joined the United States, Canada, and Mexico in the world's largest, youngest, richest, and most endowed in natural resources open-trade group of any of the world's major trade regions. NAFTA is the instrument that established an open-trade region for North America. Once the region has come together under NAFTA, North America will be among the most formidable players in global economic forums.

NAFTA is important to North America, and to the North American West especially, for a number of reasons.

- NAFTA's provisions will reduce or eliminate most non-tariff barriers (taxes, fees, regulations, and other government regulations). As a result, "international" business will come to look more like "domestic" business, which will help level the playing field—not only among trading partners but also between large and small business enterprises. So new players—small enterprises in particular—can enter the scene, spurring job creation.

• NAFTA is expected to create jobs and increase economic growth in the United States, Canada, and Mexico—most dramatically in Mexico. That's good for the United States because as living standards rise, imports rise, and 70 cents of every dollar Mexico spends on imports is spent on U.S. products. Moreover, and perhaps most important, NAFTA gives North American industry crucial competitive advantages over the trade blocs emerging in Europe and Asia. These advantages will also provide effective bargaining tools for North American political leaders to use in negotiations with the EEC and other trade blocs.

• Expanding trade and new job creation raises living standards in all three countries. History shows rising living standards invariably increase pressures for democratization and political reform, which in turn increase the political capacity, the will, and the material resources available to achieve environmental cleanup.

What some in Canada and the United States see as a potential liability of NAFTA—the loss of labor-intensive jobs to Mexico—is in fact an advantage. U.S. and Canadian industries with very expensive labor-intensive processes have been losing ground to industrial conglomerates in Europe and Japan that are shifting their labor-intensive activities to Eastern Europe and to Southeast Asia, respectively. About 35 percent of Japanese-manufactured goods are produced through production-sharing agreements with other Asian countries. Only about 4 percent of U.S. products are produced by production-sharing ventures with Canadian or Mexican firms.

The economic evidence has shown that labor-intensive, low-wage jobs in the United States and Canada will continue to migrate, if not to Mexico, then to southeast Asia. Because NAFTA has relaxed or removed Mexican laws and regulations that caused U.S. and other foreign firms to establish subsidiaries in Southeast Asia, NAFTA should actually slow job migration. Jobs that otherwise might have migrated will stay in the United States or Canada because the company need no longer be located in Mexico to sell its products or services there. And if jobs do migrate, it is better that they go to Mexico than to other low-wage countries, because Mexico buys a much higher proportion of its imports from the United States and Canada than do the nations of Southeast Asia, where the principal beneficiary is Japan. Finally, there is evidence that as Mexican trade restrictions relax, multinational corporations that moved some manufacturing operations from the United States to Southeast Asia are beginning to move those operations out of Asia and into Mexico.

Much of the competitive advantage of NAFTA lies in the greater efficiency of transportation networks in Canada and the United States. In North America, we move goods faster and cheaper than our chief economic competitors in Europe and Japan, so North American competitive advantage is increasing in this area. According to a recent study by the EOP Group, the rate of improvement in the efficiency of automotive transportation in the United States is more than double the rate for Germany and Japan.

This is a critically important advantage as the trend toward joint production between high- and low-wage areas within the same regional trading areas becomes more pronounced. The EEC's low-wage areas in Eastern Europe and the Soviet Union are closer geographically than is Mexico to the manufacturing heartlands of the United States and Canada. But because our intermodal continental transportation system is more efficient, we can ship goods faster and more cheaply. The distances between Japan and its low-wage coproducers in Southeast Asia are greater, and involve transport over water as well as land.

If it came to a test of strength, Western Europe would have more to lose by being cut off from all North American markets than it could conceivably gain from restricting U.S. imports. NAFTA is a powerful means to insure that other emerging trading groups facilitate world trade rather than restrict it.

Elements of NAFTA

NAFTA and side-agreements on labor and the environment address six important issues:

Market access—tariffs and nontariff barriers such as import licenses or quotas, rules of origin, government procurement, agriculture, autos, and other industries.

Trade rules—including subsidies, health and safety standards, environmental protection, adjustment assistance for displaced workers, and other safeguards.

Services—including banking, insurance, real estate, telecommunications, land transportation, and other services.

Investment rules—the primary effect being in Mexico, which historically has restricted foreign investment.

Intellectual property—including patents, copyrights, and trademarks.

Dispute settlement—establishing a process for resolving disputes under the agreement, as a substitute for judicial appeal.

Most of the attention has been given to market access, where NAFTA would phase out 99 percent of all tariffs over ten years and do away with remaining tariffs on politically sensitive products over fifteen years. Mexican tariffs currently average about 10 percent, down from 29 percent in 1986. United States tariffs on Mexican imports average 4 percent.

Many especially difficult or highly technical issues such as land transportation safety, truck size, and weight limits will be left for "harmonization" negotiations that will take place after adoption of the agreement.

Since Mexican tariffs on U.S. and Canadian goods are approximately two-and-one-half times U.S. tariffs on Mexican goods, initially NAFTA will open far more Mexican markets to U.S. goods than it will threaten U.S. producers with Mexican imports. In a detailed analysis of thirty-nine sectors of the economies of the United States, Canada, and Mexico, the International Trade Commission, an agency of the U.S. government, concurred that NAFTA will spur economic and job growth in all three countries. Growth will be proportionately larger in Mexico, because it is a much smaller economy than the United States. Major provisions of the more than 2,000-page treaty include:

Motor vehicles and parts. Mexican tariffs on cars and light trucks have been cut in half. Within five years, duties on 75 percent of U.S. parts exports to Mexico will be eliminated. Mexican rules requiring a balance between imports of autos and auto parts will be phased out over ten years.

Auto rule of origin. To qualify for tariff cuts, automobiles and light trucks will have to derive at least 62.5 percent of their value from parts manufactured in North America. This provision is designed to prevent auto manufacturers in Japan and Europe from funneling their cars through Mexico to evade U.S. tariffs.

Telecommunications. Mexican restrictions on U.S. investment and sales in the Mexican market for telecommunications and services will be eliminated.

Textiles and apparel. Barriers to $250 million (more than 20 percent) of U.S. exports to Mexico have been eliminated. Restrictions on the remaining $700 million will be eliminated gradually over the next six years.

Financial services. U.S. banks and securities firms will be allowed to establish wholly owned subsidiaries in Mexico. All existing restrictions on financial services will be phased out by January 1, 2000. With regard to insurance, U.S. companies with existing joint ventures will be permitted to obtain 100 percent ownership by 1996. New entrants can obtain a majority stake in Mexican firms by 1998. All restrictions on equity and market share will be eliminated by the year 2000.

Investment. Mexican domestic content rules stating how much of the value of a product must be attributed to local parts and labor will be eliminated, permitting additional use of U.S. parts. U.S. companies operating in Mexico will receive the same treatment as Mexican-owned firms.

Agriculture. Mexican import licenses, which cover about 25 percent of U.S. agricultural exports, have been eliminated at once. All Mexican tariffs on farm products will be phased out over fifteen years.

Land transportation. U.S. trucking companies will be allowed to carry international cargo to the Mexican states contiguous to the United States by 1995 and will have cross-border access to all of Mexico by 2000.

Intellectual property rights. U.S. authors, movie makers, record producers, software designers, and other producers of high-tech entertainment and consumer goods will have greater protection for their patents, copyrights, and trademarks.

Environment. The United States may continue to block imports that do not meet U.S. standards. States and cities are allowed to enact even tougher safety and environmental standards. The United States, Mexico, and Canada have agreed not to lower health, safety, or environmental standards to attract investment.

Because the pact is inclusive rather than exclusive (other nations are eligible to join), NAFTA could become a means to get the world moving once again in the direction of free trade. While the presumption is that any additional members of NAFTA would be from Latin America, there are no geographic restrictions on which nations may join.

Appendix Two
Border Coalitions

Speeding the economic integration of North America has been the formation of cross-border, sub-continental regional organizations. These organizations foster economic growth in regions of common economic interests across national borders and develop strategies to overcome problems caused by increasing cross-border trade.

The largest and most advanced of these regional organizations is the *Pacific Northwest Economic Region (PNWER)*, also known as *Cascadia*. The PNWER was created by state legislatures in Washington, Oregon, Idaho, Montana, and Alaska, and provincial parliaments in British Columbia and Alberta. The council is comprised of twenty-eight member delegates and is staffed by the Northwest Policy Center at the University of Washington. PNWER promotes regional development through cross-border cooperation. Recent activities have focused on regional tourism, value-added wood processing, environmental technology, work-force retraining, telecommunications linkages, and market development for recyclables. For more information about the Pacific Northwest Economic Region, contact David Harrison at the Northwest Policy Center, 327 Parrington Hall DC-14, Seattle, Washington 98195. Phone: (206) 543-7900; FAX: (206) 543-1096.

A parallel effort is the *Cascadia Transportation and Trade Task Force*, formed under the umbrella of the Seattle-based Discovery Institute. The Task Force is chaired by the mayor of Seattle and has representatives from fifteen local governments, fourteen state and provincial agencies, and four U.S. and Canadian federal agencies. A top priority of the Task Force is construction of a high-speed rail system from Vancouver, British Columbia, to Portland, Oregon. For more information about the Cascadia Transportation and Trade Task Force, contact Bruce Agnew at the Discovery Institute, 1201 Third Avenue, 40th Floor, Seattle, Washington 98101. Phone: (206)-287-3144; FAX: (206) 583-8500.

The *Red River Trade Corridor* is a cooperative economic development effort among Minnesota, North Dakota, and Manitoba. Founded in 1990, the Red River Trade Corridor serves as a reference point for businesses to access economic development information in any of the three jurisdictions in the region. It is governed by a board of directors drawn from business and community leaders in the region. For more information, contact Jerry Nagel, at the University of Minnesota, Crookston, 208 Selvig Hall, Crookston, Minnesota 56716. Phone: (218) 281-8459; FAX: (218) 281-8050).

The *Border Trade Alliance* is a grass-roots coalition of business and community leaders, chambers of commerce, trade associations, and federal, state, and local government officials from the U.S. states of California, Arizona, New Mexico, and Texas. Established in 1986, the Border Trade Alliance operates as a network to develop a consensus on what is required to foster healthy economic growth on both sides of the U.S.-Mexico border. For more information about the Border Trade Alliance, contact Bill Stephenson, Chairman, P.O. Box 53999, M/S 8612, Phoenix, Arizona 85072. FAX: (602) 250-3360.

The *Border Governors Conference* represents the three U.S. and six Mexican border states of Texas, New Mexico, Arizona, Baja California del Norte, Sonora, Chihuahua, Coahuila, Nuevo León, and Tamaulipas. The Conference meets every two years. For further information, contact Marshall Kuykendall, Texas Department of Commerce, P.O. Box 12728, Austin, Texas 78711. Phone: (512) 472-5059.

The *Border Mayors Conference* is a binational group of U.S.-Mexico border mayors, who meet regularly to discuss matters of mutual concern. For further information, contact Victor Garcia, Bogart International, 401 West A. St., Suite 2500, San Diego, California 92101. Phone: (619) 232-8563.

DIRECTORY

United States, Western States Government and Private Organizations for Business Development, Commerce, and Trade

Alaska

Alaska State Chamber of Commerce
217 2nd Street, #201
Juneau, AK 99801
(907) 586-2323

Arizona

Douglas Industrial Development Authority
2740 9th Street
Douglas, AZ 85607
(602) 364-7981

Nogales/Santa Cruz County
Chamber of Commerce
Kino Park
Nogales, AZ 85621
(602) 287-3685

Border Industrial Development
P.O. Box 1688
Nogales, AZ 85614
(602) 287-3685

Arizona State Chamber of Commerce
1221 E. Osborn Road, #100
Phoenix, AZ 85014
(602) 248-9172

Yuma Economic Development Corporation
P.O. Box 1750
Yuma, AZ 85364
(602) 783-0193

Arkansas

Arkansas State Chamber of Commerce
P.O. Box 3645
410 S. Cross
Little Rock, AR 72203
(501) 374-9225

California

Calexico Chamber of Commerce
P.O. Box 948
1100 Imperial Avenue
Calexico, CA 92231
(619) 357-1166; 357-1365

Mexicali Industrial Development Commission
P.O. Box 6343
Calexico, CA 92231
011-52-655-2-67-80; 2-57-30; 2-66-10

Small Business and International
Trade Center
900 Otay Lakes Road
Chula Vista, CA 92010
(619) 421-6700

Western Maquiladora Trade Association
P.O. Box 3927
Chula Vista, CA 92011-0255
(619) 435-5869

Imperial County Regional Economic
Development, Inc.
1411 State Street
El Centro, CA 92243
(619) 353-5050

California State Chamber of Commerce
P.O. Box 1736
1201 K Street, 12th Floor
Sacramento, CA 95812-1736
(916) 444-6670

San Diego Economic Development
Corporation
701 B Street, Suite 1850
San Diego, CA 92101
(619) 234-8484

City of San Diego
Binational Affairs Department
202 C Street, M.S. 8-A
San Diego, CA 92101
(619) 696-3653

Colorado

Colorado Association of Commerce and
Industry
1776 Lincoln Street, #1200
Denver, CO 80203-1029
(303) 831-7411

Hawaii

Hawaii Chamber of Commerce
735 Bishop Street
Honolulu, HI 96813
(808) 522-8800

Idaho

Idaho Department of Commerce
International Business Development
700 W. State Street
Boise, ID 83720
(208) 334-2470

Iowa

Iowa Department of Economic Development
200 E. Grand Avenue
Des Moines, IA 50309
(515) 281-3251

Kansas

Kansas Chamber of Commerce and Industry
500 Bank IV Tower
Topeka, KS 66603
(913) 357-6321

Minnesota

Minnesota State Chamber of Commerce
480 Cedar, #500
St. Paul, MN 55101
(612) 292-4650

Missouri

Missouri State Chamber of Commerce
428 E. Capitol Avenue
P.O. Box 149
Jefferson City, MO 65102
(314) 634-3511

Montana

Montana State Chamber of Commerce
P.O. Box 1730
2030 11th Avenue
Helena, MT 59624
(406) 442-2405

Nebraska

Nebraska Chamber of Commerce and Industry
1320 Lincoln Mall
P.O. Box 95128
Lincoln, NE 68509
(402) 474-4422

Nevada

Nevada State Chamber of Commerce
P.O. Box 3499
Reno, NV 89505
(702) 786-3030

New Mexico

Association of Commerce and Industry of
New Mexico
2309 Renard Place S.E., #402
Albuquerque, NM 87106-4259
(505) 842-0644

Greater Las Cruces Economic
Development Council
400 S. Main Street
Las Cruces, NM 88001
(505) 524-1745

Border Research Institute
P.O. Box 3001/3BRI
Las Cruces, NM 88003-0001
(505) 646-3524

New Mexico Trade Division
1100 Saint Francis Drive
Santa Fe, NM 87503
(505) 827-0307

North Dakota

Greater North Dakota Association
P.O. Box 2467
808 3rd Avenue S.
Fargo, ND 58108
(701) 237-9461

Oklahoma

Oklahoma State Chamber of Commerce
4020 N. Lincoln Boulevard
Oklahoma City, OK 73105
(405) 424-4003

Oregon

Oregon Economic Development Department
International Trade Division
One World Trade Center
121 S.W. Salmon, Suite 300
Portland, OR 97204
(800) 452-7813

South Dakota

Industrial and Commerce Association of
South Dakota
P.O. Box 190
Pierre, SD 57501
(605) 224-6161

Texas

Texas State Chamber of Commerce
900 Congress Avenue, #501
Austin, TX 78701
(512) 472-1594

Brownsville Chamber of Commerce
Economic Development Council
1600 E. Elizabeth Street
Brownsville, TX 78520
(800) 552-5352

Port of Brownsville
Directors of Development
P.O. Box 3070
Brownsville, TX 78523-3070
(512) 831-4592

Corpus Christi Area Economic
Development Corporation
P.O. Box 640
Corpus Christi, TX 78403
(512) 883-5571

Del Rio Chamber of Commerce
1915 Avenue F
Del Rio, TX 78840
(512) 775-3551

El Paso Industrial Development Corporation
Nine Civic Center Plaza
El Paso, TX 79901
(915) 532-0523

Laredo Chamber of Commerce
P.O. Box 790
Laredo, TX 78042-0790
(512) 722-9895

South Texas Development Council
P.O. Box 2187
Laredo, TX 78044-2187
(512) 722-3995

Council for South Texas Economic Progress
520 Pecan
McAllen, TX 78501
(512) 682-1201

McAllen Economic Development Corporation
One Park Place, Suite 100
McAllen, TX 78503
(512) 682-2876

San Antonio World Trade Center
P.O. Box 899
San Antonio, TX 78293
(512) 225-5888

Middle Rio Grande Development Council
209 N. Getty
Uvalde, TX 78801
(512) 278-2527; 278-4161

Utah

Utah State Chamber of Commerce
3540 South 4000 W., #430
Salt Lake City, UT 84120
(801) 969-8755

Washington

Association of Washington Business
P.O. Box 658
1414 S. Cherry
Olympia, WA 98507
(206) 943-1600

Wyoming

Wyoming International Trade Office
Herschler Building, 2nd Floor West
Cheyenne, WY 82002
(307) 777-6412

United States Government National Offices for Commerce, International Trade, and Statistics

National Offices:

Export-Import (ExIm) Bank
811 Vermont Avenue, N.W.
Washington, D.C. 20571
(202) 566-8990 (Main)
(202) 566-8957 (Canadian Loan Officer)
(202) 566-8998 (Mexican Loan Officer)

Interstate Commerce Commission (ICC)
12th Street and Constitution Avenue, N.W.
Washington, D.C. 20423
(202) 927-7119

Office of the U.S. Trade Representative
600 17th Street, N.W.
Washington, D.C. 20506
(202) 395-3230

Securities and Exchange Commission (SEC)
450 5th Street, N.W.
Washington, D.C. 20549
(202) 272-2000; 272-3100

U.S. Bureau of the Census
Federal Center
Suitland, MD 20233
(301) 763-4040

U.S. Department of Commerce
14th and Constitution Avenue, N.W.
Washington, D.C. 20230
(202) 482-2000 (Main)
(202) 482-4111 (Export Administration)
(202) 482-4904 (U.S. Travel and Tourism Administration)

(202) 482-0300 (International Trade Administration, Mexican Desk)
(202) 482-3103 (International Trade Administration, Canadian Desk)
(800) 872-8723 (Trade Information Center)

U.S. Department of Commerce
Trade Development Office
SA 16, Room 309
Washington, D.C. 20523-1602
(703) 875-4357

District Offices:

U.S. Department of Commerce
International Trade Administration (ITA)
U.S. and Foreign Commercial Service (USFCS)

Alaska
4201 Tudor Centre Drive, Suite 319
Anchorage, AK 99508
(907) 271-6237

Arizona
Phoenix Plaza
2901 N. Central Avenue, Suite 970
Phoenix, AZ 86012
(602) 640-2513

Arkansas
TCBY Tower Building
425 W. Capitol Avenue, Suite 700
Little Rock, AR 72201
(501) 324-5794

California
One World Trade Center, Suite 1670
Long Beach, CA 90631
(310) 980-4560

11000 Wilshire Boulevard, Room 9200
Los Angeles, CA 90024
(310) 575-7104
6363 Greenwich Drive, Suite 230
San Diego, CA 92122
(619) 557-5395

250 Montgomery Street, 14th Floor
San Francisco, CA 94104
(415) 705-2300

5201 Great American Parkway, #456
Santa Clara, CA 95054
(408) 291-7625

Colorado
1625 Broadway, Suite 680
Denver, CO 80202
(303) 844-6623
(also serves Wyoming)

Hawaii
P.O. Box 50026
300 Ala Moana Boulevard
Honolulu, HI 96850
(808) 541-1782

Idaho
700 W. State Street
Boise, ID 83720
(208) 334-3857

Iowa
Federal Building
210 Walnut Street, Room 817
Des Moines, IA 50309
(515) 284-4222

Kansas
151 N. Volutsia
Wichita, KS 67214
(316) 269-6160

Minnesota
Federal Building
110 S. 4th Street, Room 108
Minneapolis, MN 55401
(612) 348-1638
(also serves North Dakota)

Missouri
601 E. 12th Street, Room 635
Kansas City, MO 64106
(816) 426-3141

8182 Maryland Avenue, Suite 303
St. Louis, MO 63105
(314) 425-3302

Nebraska
11133 O Street
Omaha, NE 68137
(402) 221-3664
(also serves South Dakota)

Nevada
1755 E. Plumb Lane, Room 152
Reno, NV 89502
(702) 784-5203

New Mexico
c/o New Mexico Department of Economic Development
1100 St. Francis Drive
Santa Fe, NM 87503
(505) 827-0350

North Dakota
Federal Building
110 S. 4th Street, Room 108
Minneapolis, MN 55401
(612) 348-1638
(also serves Minnesota)

Oklahoma
6601 Broadway Extension, Room 200
Oklahoma City, OK 73116
(405) 231-5302

440 S. Houston Street
Tulsa, OK 74127
(918) 581-7650

Oregon
One World Trade Center
121 S.W. Salmon, Room 242
Portland, OR 97204
(503) 326-3001

South Dakota
11133 O Street
Omaha, NE 68137
(402) 221-3664
(also serves Nebraska)

Texas
P.O. Box 12728
410 E. 6th Street, Suite 414-A
Austin, TX 78711
(512) 482-5939

P.O. Box 58130
2050 N. Stemmons Freeway, Suite 170
Dallas, TX 75256
(214) 767-0542

Number One Allen Center
600 Dallas, Suite 1160
Houston, TX 77002
(713) 229-2578

Utah
324 S. State Street, Suite 105
Salt Lake City, UT 84111
(801) 524-5116

Washington
320 N. Johnson Street, Suite 350
Kennewick, WA 99336
(509) 735-2751

3131 Elliott Avenue, Suite 290
Seattle, WA 98121
(206) 553-5615

U.S. Customs Service Regional Offices:

U.S. Customs Service
1301 Constitution Avenue, N.W.
Washington, D.C. 20229
(202) 927-6724

Pacific Region
One World Trade Center, Suite 705
Long Beach, CA 90831-0700
(310) 980-3110
(serves California, Nevada, Idaho, Oregon, Washington, Alaska, and Hawaii)

North Central Region
55 E. Monroe Street
Chicago, IL 60603-5790
(312) 886-3377
(serves Montana, Wyoming, Utah, Colorado, Kansas, Nebraska, North Dakota, South Dakota, Minnesota, Iowa, and Missouri)

South Central Region
423 Canal Street
New Orleans, LA 70130
(504) 589-2976
(serves Arkansas)

Southwest Region
5850 San Felipe Street
Houston, TX 77057
(713) 953-6825
(serves Oklahoma, Texas, New Mexico, and Arizona)

U.S. Customs Service District Offices:

Alaska
605 W. 4th Avenue
Anchorage, AK 99501
(907) 271-2675

Arizona
International and Terrace Streets
Nogales, AZ 85621
(602) 761-2010

California
880 Front Street, Room 5-S-9
San Diego, CA 92188
(619) 557-5360

P.O. Box 2450
555 Battery Street
San Francisco, CA 94126
(415) 705-4340

Terminal Island
300 S. Ferry Street
San Pedro, CA 90731
(301) 514-6001

Hawaii
P.O. Box 1641
335 Merchant Street
Honolulu, HI 96806
(808) 541-1725

Minnesota
515 W. 1st Street, #209
Duluth, MN 55802-1390
(218) 720-5201

110 S. 4th Street
Minneapolis, MN 55401
(612) 348-1690

Missouri
7911 Forsyth Boulevard, Suite 625
St. Louis, MO 63105
(314) 425-3134

Montana
P.O. Box 789
300 2nd Avenue S.
Great Falls, MT 59405
(406) 453-7631

North Dakota
Federal Building
P.O. Box 1610
Pembina, ND 58271
(701) 825-6201

Oregon
511 N.W. Broadway
Portland, OR 97209
(503) 326-2865

Texas
P.O. Box 619050
1215 Royal Lane
Dallas, TX 75261
(214) 574-2170

P.O. Box 9516
94 Viscount Street
El Paso, TX 79925
(915) 540-5800

Portway Plaza
1717 E. Loop, Suite 400
Houston, TX 77029
(713) 671-1000

Lincoln Juarez Bridge
P.O. Box 3130
Laredo, TX 78041-3130
(512) 726-2267

4550 75th Street
Port Arthur, TX 77642
(409) 724-0087

Washington
1000 2nd Avenue
Suite 2200
Seattle, WA 98104

U.S. Information Agency (USIA)
301 4th Street, S.W.
Washington, D.C. 20547
(202) 619-4700 (Main)
(202) 619-5864 (Mexican Desk)
(202) 619-6853 (Canadian Desk)

U.S. Small Business Administration Regional Offices:

U.S. Small Business Administration
409 3rd Street, S.W.
Washington, D.C. 20416
(800) 827-5722

(Arizona, California, Hawaii, and Nevada)
71 Stevenson Street, 20th Floor
San Francisco, CA 94105
(415) 744-6404

(Colorado, Montana, North Dakota, South Dakota, Utah, and Wyoming)
633 17th Street
North Tower, 7th Floor
Denver, CO 80202-3607
(303) 294-7186

(Iowa, Kansas, Missouri, and Nebraska)
911 Walnut Street
Kansas City, MO 64106
(816) 426-3608

(Arkansas, New Mexico, Oklahoma, and Texas)
8625 King George Drive
Building C
Dallas, TX 75235-3391
(214) 767-7633

(Alaska, Idaho, Oregon, and Washington)
2601 4th Avenue, Suite 440
Seattle, WA 98121-1273
(206) 553-5676

SBA District and Branch Offices:

Alaska
222 W. 8th Avenue, #67
Anchorage, AK 99513-7559
(907) 271-4022

Arizona
Central and One Thomas
2828 N. Central Avenue, Suite 300
Phoenix, AZ 85004-1025
(602) 640-2400

Arkansas
2120 Riverfront Drive, Suite 100
Little Rock, AR 72202
(501) 324-5277

California
2719 N. Air Fresno Drive, Suite 107
Fresno, CA 93727-1547
(209) 487-5791

330 N. Brand Boulevard, Suite 1200
Glendale, CA 91203-2304
(213) 894-2977

660 J Street, Suite 215
Sacramento, CA 95814-2413
(916) 551-1186

880 Front Street, Suite 4237
San Diego, CA 92101-8837
(619) 557-7252

211 Main Street, 4th Floor
San Francisco, CA 94105-1988
(415) 744-6801

901 W. Civic Center Drive, Suite 160
Santa Ana, CA 92703-2352
(714) 836-2494

Colorado
P.O. Box 660
721 19th Street, Suite 426
Denver, CO 80201-0660
(303) 844-3984

Hawaii
P.O. Box 50207
300 Ala Moana Boulevard, Room 2213
Honolulu, HI 96850-4981
(808) 541-2965

Idaho
1020 Main Street, Suite 290
Boise, ID 83702
(208) 334-1696

Iowa
373 Collins Road, N.E., Room 100
Cedar Rapids, IA 52402-3147
(319) 393-8630

210 Walnut Street, Room 749
Des Moines, IA 50309
(515) 284-4422

Kansas
100 E. English Street, Suite 510
Wichita, KS 67202
(316) 269-6273

Minnesota
100 N. 6th Street, Suite 610-C
Minneapolis, MN 55403
(612) 370-2324

Missouri
323 W. 8th Street, Suite 501
Kansas City, MO 64105
(816) 374-6708

815 Olive Street, Room 242
St. Louis, MO 63101
(314) 539-6600

620 S. Glenstone Street, Suite 110
Springfield, MO 65802-3200
(417) 864-7670

Montana
301 S. Park
Room 334, Drawer 10054
Helena, MT 59626-0054
(406) 449-5381

Nebraska
11145 Mill Valley Road
Omaha, NE 68154
(402) 221-4691

Nevada
Box 7527–Downtown Station
301 E. Stewart Street
Las Vegas, NV 89125-2527
(702) 388-6611

New Mexico
625 Silver Street, Suite 320
Albuquerque, NM 87102
(505) 766-1887

North Dakota
P.O. Box 3086
657 2nd Avenue N., Room 218
Fargo, ND 58108-3086
(701) 239-5131

Oklahoma
500 N.W. 5th Street, Suite 670
Oklahoma City, OK 73102
(405) 231-5237

Oregon
222 S.W. Columbia Street, Suite 500
Portland, OR 97201-6695
(503) 326-3329

South Dakota
110 S. Phillips Avenue, Suite 200
Sioux Falls, SD 57102
(605) 330-4231

Texas
606 N. Carancahua, Suite 1200
Corpus Christi, TX 78476
(512) 888-3301

4300 Amon Carter Boulevard, Suite 114
Dallas, TX 76155
(817) 355-1933

10737 Gateway W., Suite 320
El Paso, TX 79935
(915) 540-5676

222 E. Van Buren, Suite 500
Harlingen, TX 78550
(512) 427-8533

9301 S.W. Freeway, Suite 550
Houston, TX 77054
(713) 773-6518

1611 10th Street, Suite 200
Lubbock, TX 79401
(806) 743-7462

7400 Blanco Road, Suite 200
San Antonio, TX 78216
(512) 229-4503

Utah
Federal Building
125 S. State Street, Room 2237
Salt Lake City, UT 84138-1195
(801) 524-5804

Washington
915 2nd Avenue, Room 1792
Seattle, WA 98174-1088
(206) 220-6520

Farm Credit Building
W. 601 1st Avenue, 10th Floor E.
Spokane, WA 99204-0317
(509) 353-2820

Wyoming
Federal Building
100 E. B Street, Room 4001
Casper, WY 82601
(307) 261-5761

U.S. Department of State:

2201 C Street, N.W.
Washington, D.C. 20520
(202) 647-4000 (Main)
(202) 647-9894 (Mexican Affairs Office)
(202) 647-2170 (Canadian Affairs Office)

U.S. Embassy (Canada)
100 Wellington Street
Ottawa, Ontario
Canada K1P 5T1
(613) 238-5335

U.S. Embassy (Mexico)
Paseo de la Reforma 305
México, D.F.
011-52-5-211-0042

U.S. Consulate General
615 Macleod Trail, S.E., Suite 1080
Calgary, Alberta
Canada T2G 4T8
(403) 265-2116

Cogswell Tower
Scotia Square, Suite 910
Halifax, Nova Scotia
Canada B3J 3K1
(902) 429-2480

P.O. Box 65
Postal Station Desjardins
Montréal, Québec
Canada H6B 1G1
(514) 398-0673

2 Place Terrasse
C.P. 939
Québec City, Québec
Canada G1R 4T9
(418) 692-2096

480 University Avenue, Suite 602
Toronto, Ontario
Canada M5G 1V2
(416) 595-5414

1095 W. Pender Street
Vancouver, British Columbia
Canada V6E 4E9
(604) 688-4311

(There are no consulates in Mexico.)

U.S. Department of Treasury
15th and Pennsylvania Avenue, N.W.
Washington, D.C. 20220
(202) 622-2000 (Main Desk)
(202) 622-1269 (Mexican Desk)
(202) 622-0095 (Canadian Desk)

Canadian Government and Private Organizations

Canada Customs and Excise:

International Programs Division
Connaught Building, 8th Floor
Ottawa, Ontario
Canada K1A 0L5
(613) 984-7186

Tariff Programs Appraisal
360 Coventry Road
Ottawa, Ontario
Canada K1K 2C6
(613) 993-0534

Canadian Exporters Association
99 Bank Street, Suite 250
Ottawa, Ontario
Canada K1P 6B9
(613) 238-8888

Consumer and Corporate Affairs:

Consumer Products Branch
Place du Portage, Phase 1
50 Victoria Street, 16th Floor
Hull, Québec
Canada K1A 0C9
(819) 997-1591 (food products)
(819) 997-1177 (non-food products)

Corporations Branch
Place du Portage, Phase II, 4th Floor
Hull, Québec
Canada K1A 0C9
(819) 997-1142

Patents, Trademarks, Copyrights, and Industrial Design Office
Place du Portage
Hull, Québec
Canada K1A 0C9
(819) 997-1936 (patent)
(819) 997-1725 (copyright)
(819) 997-1420 (trademark)
(819) 997-1725 (industrial design)

Department of External Affairs
Export and Import Permits Bureau
Lester B. Pearson Building
4C-125 Sussex Drive
Ottawa, Ontario
Canada K1A 0G2
(613) 992-1362 (textiles, apparel)
(613) 995-7762 (agricultural products)
(613) 996-5623 (other commodities)

Embassies:

Embassy of Canada (United States)
801 Pennsylvania Avenue, N.W.
Washington, D.C. 20001
(202) 682-1740

Embassy of Canada (Mexico)
Apartado Postal 105-05
11580 Mexico, D.F.
Mexico
011-52-5-724-7900

Canadian Consulates:

(Arizona, Colorado, Hawaii, Utah, Wyoming, Nevada, and California)
300 S. Grand Avenue, 10th Floor
California Plaza
Los Angeles, CA 90071
(213) 687-7432

(Missouri and the quad-city region of Iowa)
310 S. Michigan Avenue, #1200
Chicago, IL 60604-4296
(312) 427-1031

(Iowa, Nebraska, Minnesota, North Dakota, South Dakota, and Montana)
701 4th Avenue S.
Minneapolis, MN 55415-1899
(612) 333-4641

(Texas, Arkansas, Kansas, New Mexico, and Oklahoma)
760 N. Paul Street
St. Paul Place, #1700
Dallas, TX 75201-9990
(214) 922-9812

(Alaska, Idaho, Oregon, and Washington)
412 Plaza 600
Sixth and Stewart
Seattle, WA 98101-1286
(206) 443-1777

(There are no consulates in Mexico.)

Employment and Finance:

Employment and Immigration
140 Promenade du Portage
Ottawa, Ontario
Canada K1A J9
(613) 682-7100

Federal Business Development Bank
800 Place Victoria
Postal Box 335
Montreal, Québec
Canada H4Z 1L4
(514) 283-8904

Financial Post Information Services
Survey of Markets
777 Bay Street
Toronto, Ontario
Canada M5W 1A7
(416) 595-8287

Investment Canada
P.O. Box 2800
Postal Station D
Ottawa, Ontario
Canada K1P 6A5
(613) 998-0645

Revenue Canada and Excise GST Office
1730 St. Laurent Boulevard
P.O. Box 8257
Ottawa, Ontario
Canada K1G 3H7
(613) 990-8584

Standards Council of Canada:

Standards Information Service
350 Sparks Street
Suite 1200
Ottawa, Ontario
Canada K1P 6N7
(613) 238-3222

Statistics:

Advisory Services
North American Life Centre
1770 Market Street
Halifax, Nova Scotia
Canada B3J 3M3
(800) 565-7192
(902) 426-5331
(serves Newfoundland and Labrador, Nova Scotia, Prince Edward Island, and New Brunswick)

Advisory Services
E. Tower, Suite 412
Guy Favreau Complex
200 René Lévesque Boulevard W.
Montréal, Québec
Canada H2Z 1X4
(800) 361-2831
(514) 283-5725

Statistical Reference Centre (NCR)
Lobby, R.H. Coats Building
Holland Avenue
Tunney's Pasture
Ottawa, Ontario
Canada K1A 0T6
(613) 951-8116

Advisory Services
Arthur Meighen Building
25 St. Clair Avenue E., 10th Floor
Toronto, Ontario
Canada M4T 1M4
(800) 263-1136
(416) 973-6586

Advisory Services
MacDonald Building
344 Edmonton Street, Suite 300
Winnipeg, Manitoba
Canada R3B 3L9
(800) 563-7828
(204) 983-4020

Advisory Services
Avord Tower
2002 Victoria Avenue, 9th Floor
Regina, Saskatchewan
Canada S4P 0R7
(800) 563-7164
(306) 780-5405

Advisory Services
First Street Plaza, Room 401
138 4th Avenue S.E.
Calgary, Alberta
CanadaT2G 4Z6
(800) 563-7828
(403) 292-6717

Advisory Services
Park Square
10001 Bellamy Hill, 8th Floor
Edmonton, Alberta
Canada T5J 3B6
(800) 563-7828
(403) 495-3027
(serves northern Alberta and Northwest Territories)

Advisory Services
Sinclair Centre
757 W. Hastings Street, Suite 300
Vancouver, British Columbia
CanadaV6C 3C9
(800) 663-1551
(604) 666-3691
(serves British Columbia and the Yukon)

Tourism:

Industry, Science, and Technology Canada
235 Queen Street
Ottawa, Ontario
Canada K1A 0H5
(613) 954-3810

Provincial Ministries:

Alberta Consumer and Corporate Affairs
10025 Jasper Avenue, 22nd Floor
Edmonton, Alberta
Canada T5J 3Z5
(403) 422-3935

British Columbia Ministry of Development, Trade, and Tourism
1770 Pacific Boulevard S., 2nd Floor
Vancouver, British Columbia
Canada V6B 5E7
(604) 660-3908

Manitoba Department of Industry, Trade, and Tourism
155 Carlton Street, 4th Floor
Winnipeg, Manitoba
CanadaR3C 3H8
(204) 946-2468

New Brunswick Department of Economic Development and Tourism
P.O. Box 6000
Fredericton, New Brunswick
Canada E3B 6H1
(506) 729-5600

Department of Economic Development and Tourism (Northwest Teritories)
Yellowknife, Northwest Territories
Canada X1A 2L9
(403) 873-7229

Nova Scotia Department of Economic Development
P.O. Box 519
Halifax, Nova Scotia
Canada B3J 2R7
(902) 424-4211

Ontario Ministry of Consumer and Commercial Relations
Companies Branch
393 University Avenue
Toronto, Ontario
Canada M7A 2H6
(416) 596-3757

Prince Edward Island Enterprise
West Royalty Industrial Park
Charlottetown, Prince Edward Island
Canada C1E 1B0
(902) 368-6324

Quebec Ministry of Industry, Commerce, and Trade
700 Sherbrooke Street W., 8th Floor
Montréal, Québec
Canada H3A 1G1
(514) 982-3013

Saskatchewan Department of Economic Development
1919 Saskatchewan Drive
Regina, Saskatchewan
Canada S4P 3V7
(306) 787-2232

Department of Economic Development, Mines, and Small Business (Yukon Territory)
P.O. Box 2703
Whitehorse, Yukon
Canada Y1A 2C6
(403) 667-5466

Mexican Government

Chihuahua Industrial Promotion Organization
Don Quixote de la Mancha No. 1
Apartado Postal No. I-1
31109 Chihuahua
Chihuahua, Mexico
011-52-14-17-58-88

Ciudad Juárez Economic Development Corporation
Adolfo de la Huerta No. 742-3
32340 Ciudad Juárez
Chihuahua, Mexico
011-52-161-6-32-68; 6-56-33

Ciudad Juárez Maquiladora Association
Rio Nilo 4049-10, Esquina Lopez Mateos
32310 Ciudad Juárez
Chihuahua, Mexico
011-52-161-3-42-57; 3-42-58; 6-14-61

Embassy of Mexico (United States)
1911 Pennsylvania Avenue, N.W.
Washington, D.C. 20036
(202) 728-1600

Embassy of Mexico (Canada)
1800-130 Albert Street
Ottawa, Ontario
Canada K1P 5G4
(613) 233-8988

Mexican Consulates:

Arizona
553 Stone Avenue
Tucson, AZ 85701
(602) 882-5595

1990 W. Camelback, Suite 110
Phoenix, AZ 85015
(602) 242-7398

137 Terrace Avenue
Nogales, AZ
(602) 287-2521

California
331 W. 2nd Street
Calexico, CA 92231
(619) 357-3863

905 N. Fulton Street
Fresno, CA 93721
(209) 233-3065

2401 W. 6th Street
Los Angeles, CA 90012
(213) 351-6800

201 E. 4th Street
Oxnard, CA 93030
(805) 483-4684

P.O. Box 710
Sacramento, CA 95814
(916) 446-4696

588 W. 6th Street
San Bernadino, CA 92401
(714) 888-2500

610 A Street, 1st Floor
San Diego, CA 92101
(619) 231-0337

870 Market Street, Suite 528
San Francisco, CA 94102
(415) 392-6576

380 N. 1st Street, Suite 102
San Jose, CA 95112
(408) 294-3415

406 W. 4th Street
Santa Ana, CA 92701
(714) 835-3749

Colorado
707 Washington Street, Suite A
Denver, CO 80203
(303) 830-0523

Missouri
1015 Locust Street, Suite 922
St. Louis, MO 63101
(314) 436-3233

New Mexico
401 5th Street, N.W., Suite 1710
Albuquerque, NM 87102
(505) 247-2139

Texas
200 E. 6th Street, Suite 200
Austin, TX 78701
(512) 478-2866

P.O. Box 1711
Brownsville, TX 78520
(512) 542-2051

800 N. Shoreline
410 N. Tower
Corpus Christi, TX 78401
(512) 882-5694

1349 Empire Central, Suite 100
Dallas, TX 75247
(214) 630-2024

P.O. Box 1275
Del Rio, TX 78841
(512) 775-2352

P.O. Box 4230
Eagle Pass, TX 78852
(512) 773-9255

910 E. San Antonio Street
El Paso, TX 79901
(915) 533-3645

4200 Montrose Boulevard, Suite 120
Houston, TX 77006
(713) 524-2300

1612 Farragut Street
Laredo, TX 78040
(512) 723-6369

1418 Beech Street, Suites 102–104
McAllen, TX 78501
(512) 686-0243

511 W. Ohio Street, Suite 121
Midland, TX 79701
(915) 687-2334

127 Navarro Street
San Antonio, TX 78205
(512) 227-9145

Utah
182 S. 600 Street, Suite 202
Salt Lake City, UT 84102
(801) 521-8502

Washington
2132 3rd Avenue
Seattle, WA 98121
(206) 448-6819

Canada
Commerce Court West
P.O. Box 266
199 Bay Street, Suite 4440
Toronto, Ontario
Canada M5L 1E9

2000 rue Mansfield, Suite 1015
Montréal, Québec
Canada H3A 2Z7

1130 W. Pender Street, Suite 810
Vancouver, B.C.
Canada V6E 4A4

Instituto Nacional de Estadística, Geografía, y Información (INEGI):

Central Office
Av. Prolongación Héroe de Nacozari No. 2301 Sur
CP 202290 Ciudad Industrial
Aguascalientes, Mexico
011-52-4-918-0034

Mexican Government Tourism Offices:

10100 Santa Monica Boulevard, Suite 234
Los Angeles, CA 90067
(310) 203-8328

1911 Pennsylvania Avenue, N.W.
Washington, D.C. 20036
(202) 728-1750

2707 N. Loop
W 450
Houston, TX 77008
(713) 880-5153

Office of Free Trade Negotiation:

Avenida Alfonso Reyes, No. 30
9o. Piso
Colonia Hipódromo Condesa
06179 Mexico, D.F.
011-52-5-286-3493

Secretariat of Trade and Industrial Promotion (SECOFI):

Av. Alfonso Reyes, No. 30
10o. Piso
Colonia Hipodromo Condesa
06179 Mexico, D.F.
011-52-5-286-1823; 286-1483

1911 Pennsylvania Avenue, N.W.
Washington, D.C. 20036
(202) 728-1700

Federal Delegations and Sub-delegations:

Hospitabilidad 107 Centro
20000 Aguascalientes
Aguascalientes, Mexico
011-52-491-52024

Palacio Federal Centro Civico
Cuerpo A, Tercero. Piso
21000 Mexicali
Baja California Norte, Mexico
011-52-65-574273

Ignacio Comonfort 15
4o. Piso, Desp. 401 y 402
Zona del Río
22320 Tijuana
Baja California Norte, Mexico
011-52-66-340155

5 de Mayo y Madero
Col. Centro
23000 La Paz
Baja California Sur, Mexico
011-52-682-28056

Av. 16 de Septiembre s/n
Planta Baja, Palacio Fed.
24000 Campeche
Campeche, Mexico
011-52-981-63365

Perif. Luis Echeverría
460 Ote., Col. Républica
25280 Saltillo
Coahuila, Mexico
011-52-84-167212

Morelos 211 Nte. Centro, 2o. Piso
26000 Piedras Negras
Coahuila, Mexico
011-52-878-22642

Boulevard Independencia 2029
Ote. Col. Nva. San Isidro
27100 Torreón
Coahuila, Mexico
011-52-17-173783

Palacio Federal, Planta Baja
Av. Madero y Gral. Nuñez
28000 Colima
Colima, Mexico
011-52-331-23766

Palacio Federal, tercero Piso
Centro
29000 Tuxtla Gutiérrez
Chiapas, Mexico
011-52-961-26298

A. Rialfer s/n Blvd.
Díaz Ordaz
30740 Tapachula
Chiapas, Mexico
011-52-962-53189

A. Universidad 3705
Col. San Felipe
31170 Chihuahua, 2o. Piso
Chihuahua, Mexico
011-52-14-138047

Av. de la Raza 4519
Fracc. del Colegio
32340 Ciudad Juárez
Chihuahua, Mexico
011-52-16-167214

Av. Normal y A. de Circunv.
99 Edificio Paz Guadiana
34070 Durango
Durango, Mexico
011-52-181-20905

Morelos 326 Nte.
Edificio Durango
35000 Góez Palacio, 5o. Piso
Durango, Mexico
011-52-17-142565

Paseo de la Presa 30
36000 Guanajuato
Guanajuato, Mexico
011-52-473-23945

Constitucíon 103
Esq. Pino Suárez
37000 León, 1er. Piso
Guanajuato, Mexico
011-52-47-147197

5 de Mayo y Hidalgo 9
39000 Chilpancingo, 2o. Piso
Guerrero, Mexico
011-52-747-22077

Av. Costera Miguel Alemán
No. 54-1 Col. Azul
39670 Acapulco
Guerrero, Mexico
011-52-748-45596

Allende No. 603, Piso 2 y 3
Col. Centro
42000 Pachuca
Hidalgo, Mexico
011-52-771-5228

Av. Mariano Otero 3431
Nivel 1 y 2 Col. Verde Valle
44550 Guadalajara
Jalisco, Mexico
011-52-3-6210644

Paseo Tollocan 504
Col. Universidad
50130 Toluca
México, Mexico
011-52-72-195760

214 Col. Chapultepec Rte
58260 Morelia
Michoacán, Mexico
011-52-451-56601

Av. H. Colegio Militar s/n
Col. Buenavista
62130 Cuernavaca
Morelos, Mexico
011-52-73-170741

Allende 110 Oriente
Centro
63000 Tepic, 1er. Piso
Nayarit, Mexico
011-52-321-25082

Edificio Termex Av. Fundidora
Nivel 1 Local 88
64000 Monterrey
Nuevo León, Mexico
011-52-83-696480

1a. Privada de E. Zapata 108
Col. Reforma
68050 Oaxaca
Oaxaca, Mexico
011-52-951-55052

Av. Ferrocarril 54
Col. Morelos
70650 Oaxaca
Oaxaca, Mexico
011-52-971-44522

Calle 2 Sur 3910
Col. E. Carmen
Huexotitla
72530 Puebla
Puebla, Mexico
011-52-22-404509

Av. Héroes Esq. Lázaro
Denas, Piso 2, Edificio Plaza
Caracol
77000 Chetumal
Quintana Roo, Mexico
011-52-983-23056

Wenceslao de la Barquera 13
Col. Villas del Sur
76040 Querétaro
Querétaro, Mexico
011-52-42-120399

Zamarripa 1381
Col. Himno Nacional
78280 San Luis Potosí
San Luis Potosí, Mexico
011-52-48-150898

Calle Juárez 58 Pte.
Col. Centro
80000 Culiacán, Piso 2
Sinaloa, Mexico
011-52-67-139200

Periférico Pte. 310-A
Edificio Ocotillo
83200 Hermosillo
Sonora, Mexico
011-52-62-183176

Chihuahua 140 Nte.
Centro
85000 Ciudad Obregón, 2o. Piso
Sonora, Mexico
011-52-641-44044

Magdelena 12 Entre
Avenida Tecnol y Avenida Hermosillo
84000 Nogales
Sonora, Mexico
011-52-631-31455

Callejón Carranza y Calle
Tercera
83400 San Luis Río Colorado
Sonora, Mexico
011-52-653-40204

Av. Paseo Tabasco 1129
Col. Rivorosa
86050 Villahermosa
Tabasco, Mexico
011-52-931-59077

Matamoros 625
Entre Calle 11 y 12
Col. Centro
87000 Ciudad Victoria
Tamaulipas, Mexico
011-52-131-29133

Carr. Lauro Villar 600 y Honduras, Piso 2
Edificio Plaza Modelo
87420 Matamoros
Tamaulipas, Mexico
011-52-891-34122

Av. Guerrero 2902
Esq. Coahuila
Sector Central
88000 Nuevo Laredo
Tamaulipas, Mexico
011-52-871-40196

Av. Portes Gil. Esq. Colón
Col. Del Prado
88560 Reynosa
Tamaulipas, Mexico
011-52-892-25360

Av. Hidalgo 5004, Edificio Melik
tercero Piso Col. Sierra Morena
89210 Tampico
Tamaulipas, Mexico
011-52-12-132221

Avenida 20 de Noviembre 41-A. Altos
90000 Tlaxcala
Tlaxcala, Mexico
011-52-246-21065

Clavijero 1
Esq. Avila Camacho
91000 Jalapa, 1er. Piso
Veracruz, Mexico
011-52-281-72030

5 de Mayo y Ocampo
Palacio Federal Local 4
91700 Veracruz
Veracruz, Mexico
011-52-29-323207

Zaragoza 106, Centro
96400 Coatzacoalcos
Veracruz, Mexico
011-52-921-26400

8 Nte. 15, 2o. y 3o. Pisos
Col. Obrera
93260 Poza Rica
Veracruz, Mexico
011-52-782-21468

Av. Colón 501-C Edificio
Pza. Colón des. 301-C
Entre 60 y 62
97000 Mérida
Yucatán, Mexico
011-52-99-256822

Avenida González Ortega 126
Centro
98000 Zacatecas
Zacatecas, Mexico
011-52-492-21214

(The delegations assist in promotion of state economic affairs.)

Secretariat of Trade and Industrial Promotion Offices:

Office of Mexico
U.S. Department of Commerce
Room H-3026
Washington, D.C. 20230
(202) 482-0300

Office of Border Affairs
Boulevard Adolfo Lopez Mateos, No. 3025, 12 Piso
Colonia Héroes de Padierna
Mexico, D.F.
011-52-5-595-7917

Office of Export Promotion
Boulevard Adolfo Lopez Mateos, No. 3025, 3o. Piso
Colonia Héroes de Padierna
Mexico, D.F.
011-52-5-683-4344
011-52-5-683-5066

Office of Foreign Investment
Boulevard Manuel Avila Camacho, No. 1, 11 Piso
Co. Colmas de Chapultepec
11000 Mexico, D.F.
011-52-5-540-1331

Office of Foreign Trade Policy
Boulevard. Adolfo Lopez Mateos, No. 3025, 5o. Piso.
Colonia Héroes de Padierna
Mexico, D.F.
011-52-5-683-4755; 683-4659

Office of International Trade Negotiations
Boulevard Adolfo Lopez Mateos, No. 3025, 11 Piso
Colonia Héroes de Padierna
Mexico, D.F.
011-52-5-683-4035

State of Nuevo León–Proexport
Apartado Postal No. 3165
64000 Monterrey
Nuevo León, Mexico
011-52-83-45-73-53; 45-73-54; 45-73-55

Trade Commission of Mexico:

California
World Trade Center
350 S. Figueroa Street, Suite 296
Los Angeles, CA 90071
(213) 628-1220

Texas
2777 Stemmons Freeway, Suite 1622
Dallas, TX 75207
(214) 688-4096

1100 N.W. Loop 410, Suite 409
San Antonio, TX 78213
(512) 525-9748

Canada
1501 McGill College, Suite 1540
Montréal, Québec
Canada H3A 3M8

P.O. Box 32
66 Wellington Street W., Suite 2712
Toronto-Dominion Bank Tower
Toronto, Ontario
Canada M5K 1A1

Granville Street, 1365-200
Vancouver, B.C.
Canada V6C 1S4

Other Sources of Information:

Mexican Investment Board
Paseo de la Reforma 915
Lomas de Chapultepec
11000 Mexico, D.F
525-202-7804

Latin American Data Base
Latin American Institute
University of New Mexico
801 Yale NE
Albuquerque, NM 87131-1016
(505) 277-6839 or (800) 472-0888
Fax: 505-277-5989

Banco Nacional de México, S.A.
Aveida Juárez No. 90
Planta Baja, Mexico 06040
525-761-8588, x4006

Expansión, S.A.
Sinaloa 149
Mexico, D.F. 06700
525-208-7006, 525-207-2756 or 525-207-2066

United States *Fortune* 500 Industrial Companies Headquartered in the Western States

Adolph Coors
1819 Denver W. Drive
Golden, CO 80439

Advanced Micro Devices
1 AMD Place
Sunnyvale, CA 94088

Allergan
2525 DuPont Drive
Irvine, CA 92713

Alliant Techsystems
5901 Lincoln Drive
Edina, MN 55436

Amdahl
1250 E. Arques Avenue
Sunnyvale, CA 94088

Amgen
1840 Dehavilland
Thousand Oaks, CA 91320

Anheuser-Busch
1 Busch Place
St. Louis, MO 63118

Apple Computer
20525 Mariani Avenue
Cupertino, CA 95014

Applied Materials
3050 Bowers Avenue
Santa Clara, CA 95054

AST Research
16215 Alton Parkway
Irvine, CA 92713

Atlantic Richfield
515 S. Flower Street
Los Angeles, CA 90071

Avery Dennison
150 N. Orange Grove
Pasadena, CA 91103

Baker Hughes
3900 Essex Lane
Houston, TX 77027

Baroid
3000 N. Sam Houston E.
Houston, TX 77032

Beckman Instruments
2500 Harbor Boulevard
Fullerton, CA 92634

Bemis
222 S. 9th Street
Minneapolis, MN 55402

Berkshire Hathaway
1440 Kiewit Plaza
Omaha, NE 68103

Boeing
P.O. Box 3707
Seattle, WA 98124

Boise Cascade
1 Jefferson Square
Boise, ID 83728

Burlington Resources
999 3rd Avenue
Seattle, WA 98104

Chevron
225 Bush Street
San Francisco, CA 94104

Citgo Petroleum
P.O. Box 3758
Tulsa, OK 74102

Clark Oil and Refining
8182 Maryland Avenue
St. Louis, MO 63105

Clorox
1221 Broadway
Oakland, CA 94612

Coastal
9 Greenway Plaza
Houston, TX 77046

Compaq Computer
P.O. Box 692000
Houston, TX 77269

ConAgra
1 ConAgra Drive
Omaha, NE 68102

Conner Peripherals
3081 Zanker Road
San Jose, CA 95134

Cooper Industries
P.O. Box 4446
Houston, TX 77210

Cray Research
655 A Lone Oak Drive
Eagan, MN 55121

Cyprus Minerals
P.O. Box 3299
Englewood, CO 80155

Del Monte Foods
One Market Place
San Francisco, CA 94105

Dell Computer
9505 Arboretum Boulevard
Austin, TX 78759

Delux
1080 W. County Road F
Shoreview, MN 55126

Diamond Shamrock
P.O. Box 696000
San Antonio, TX 78269

Doskocil
P.O. Box 1570
Hutchinson, KS 67504

Dr. Pepper/Seven-Up
8144 Walnut Hill Lane
Dallas, TX 75231

Dresser Industries
1600 Pacific Building
Dallas, TX 75201

E-Systems
6250 L.B.J. Freeway
Dallas, TX 75240

Ecolab
Ecolab Center
St. Paul, MN 55102

Emerson Electric
8000 W. Florissant Avenue
St. Louis, MO 63136

Exxon
225 E. John Carpenter Freeway
Irving, TX 75062

Farmers Union (CENEX)
P.O. Box 64089
St. Paul, MN 55164

Farmland Industries
3315 N. Oak Trafficway
Kansas City, MO 64116

Fina
8350 N. Central Expressway
Dallas, TX 75206

Fleetwood Enterprises
3125 Myers Street
Riverside, CA 92513

Gateway 2000
610 Gateway Drive
North Sioux City, SD 57049

General Mills
1 General Mills Boulevard
Minneapolis, MN 55426

George A. Hormel
501 16th Avenue Northeast
Austin, MN 55912

H.B. Fuller
2400 Energy Park Drive
St. Paul, MN 55108

Hewlett-Packard
3000 Hanover Street
Palo Alto, CA 94304

Homestake Mining
650 California Street
San Francisco, CA 94108

Honeywell
2701 4th Avenue S.
Minneapolis, MN 55408

Hudson Foods
1225 Hudson Road
Rogers, AR 72756

IBP
IBP Avenue
Dakota City, NE 68731

Imperial Holly
8016 Highway 90-A
Sugar Land, TX 77478

Insilco
300 N. Marienfeld
Midland, TX 79701

Intel
3065 Bowers Avenue
Santa Clara, CA 95052

Interco
101 S. Hanley Road
St. Louis, MO 63105

International Multifoods
33 S. 6th Street
Minneapolis, MN 55402

Interstate Bakeries
12 E. Armour Boulevard
Kansas City, MO 64141

Jefferson Smurfit
8182 Maryland Avenue
St. Louis, MO 63105

Jostens
5501 Norman Center Drive
Minneapolis, MN 55437

Kellwood
600 Kellwood Parkway
Chesterfield, MO 63017

Kerr-McGee
123 Robert S. Kerr Avenue
Oklahoma City, OK 73102

Kimberly-Clark
P.O. Box 619100
Dallas, TX 75261

Land O' Lakes
4001 Lexington Avenue N.
Arden Hills, MN 55112

Leggett & Platt
1 Leggett Road
Carthage, MO 64836

Levi Strauss Associates
1155 Battery Street
San Francisco, CA 94111

Litton Industries
360 N. Crescent Drive
Beverly Hills, CA 90210

Lockheed
4500 Park Granada Boulevard
Calabasas, CA 91399

Longview Fibre
P.O. Box 639
Longview, WA 98632

Louisiana-Pacific
111 S.W. 5th Avenue
Portland, OR 97204

LSI Logic
1551 McCarthy Boulevard
Milpitas, CA 95035

Lyondell Petrochemical
1221 McKinney
Houston, TX 77010

Magma Copper
7400 N. Oracle Road
Tucson, AZ 85715

Magnetex
11150 Santa Monica
Los Angeles, CA 90025

Manville
P.O. Box 5108
Denver, CO 80217

Mapco
1800 S. Baltimore Avenue
Tulsa, OK 74119

Mary Kay Cosmetics
8787 Stemmons Freeway
Dallas, TX 75247

Mattel
333 Continental Boulevard
El Segundo, CA 90245

Maxtor
211 River Oaks Parkway
San Jose, CA 95134

Maxus Energy
717 N. Harwood Street
Dallas, TX 75201

Maxxam
P.O. Box 572887
Houston, TX 77257

Maytag
403 W. 4th Street N.
Newton, IA 50208

McDonnell Douglas
P.O. Box 516
St. Louis, MO 63166

Medtronic
7000 Central Avenue NE
Minneapolis, MN 55432

Mid-America Dairymen
3253 E. Chestnut Expressway
Springfield, MO 65802

Minnesota Mining
3M Center Building
St. Paul, MN 55144

Mitchell Energy and Development
2001 Timberloch Place
The Woodlands, TX 77387

Monsanto
800 N. Lindbergh Boulevard
St. Louis, MO 63167

Murphy Oil
200 Peach Street
El Dorado, AR 71730

National Cooperative Refinery Association
2000 S. Main Street
McPherson, KS 67460

National Semiconductor
2900 Semiconductor
Santa Clara, CA 95052

NCH
2727 Chemsearch Boulevard
Irving, TX 75062

Nerco
500 Northeast Multnomah
Portland, OR 97232

Newmont Mining
1700 Lincoln Street
Denver, CO 80203

NL Industries
3000 N. Sam Houston E.
Houston, TX 77032

Northrop
1840 Century Park E.
Los Angeles, CA 90067

Occidental Petroleum
10889 Wilshire Boulevard
Los Angeles, CA 90024

Oryx Energy
13155 Noel Road
Dallas, TX 75240

Paccar
P.O. Box 1518
Bellevue, WA 98009

Pennzoil
P.O. Box 2967
Houston, TX 77252

Pentair
1500 W. County Road B
St. Paul, MN 55113

Pet
400 S. 4th Street
St. Louis, MO 63102

Phelps Dodge
2600 N. Central Avenue
Phoenix, AZ 85004

Phillips Petroleum
Phillips Building
Bartlesville, OK 74004

Pilgrim's Pride
P.O. Box 93
Pittsburg, TX 75686

Potlatch
One Maritime Plaza
San Francisco, CA 94111

Quantum
500 McCarthy Boulevard
Milpitas, CA 95035

Ralston Purina
Checkerboard Square
St. Louis, MO 63164

Raychem
300 Constitution Drive
Menlo Park, CA 94025

Riceland Foods
P.O. Box 927
Stuttgart, AR 72160

Rockwell International
2201 Seal Beach Boulevard
Seal Beach, CA 90740

Rohr
P.O. Box 878
Chula Vista, CA 91912

Seagate Technology
920 Disc Drive
Scotts Valley, CA 95066

Shell Oil
4652 One Shell Plaza
Houston, TX 77002

Sigma-Aldrich
3050 Spruce Street
St. Louis, MO 63103

Silicon Graphics
2011 N. Shoreline
Mountain View, CA 94039

Stewart and Stevenson
P.O. Box 1637
Houston, TX 77251

Storage Technology
2270 S. 88th Street
Louisville, CO 80028

Sun Microsystems
2550 Garcia Avenue
Mountain View, CA 94043

Sun-Diamond Growers
5568 Gibraltar Drive
Pleasanton, CA 94588

Tandem Computers
19333 Vallco Parkway
Cupertino, CA 95014

Tektronix
26600 SW Parkway
Wilsonville, OR 97070

Teledyne
1901 Avenue of the Stars
Los Angeles, CA 90067

Temple-Inland
P.O. Drawer N.
Diboll, TX 75941

Tenneco
1010 Milam
Houston, TX 77001

Tesoro Petroleum
8700 Tesoro Drive
San Antonio, TX 78217

Texas Industries
7610 Stemmons Freeway
Dallas, TX 75247

Texas Instruments
13500 N. Central Expressway
Dallas, TX 75265

Thiokol
2475 Washington Boulevard
Ogden, UT 84401

Times Mirror
Times Mirror Square
Los Angeles, CA 90053

Toro
8111 Lyndale Avenue South
Minneapolis, MN 55420

Total Petroleum
P.O. Box 500
Denver, CO 80201

Tri-Valley Growers
1255 Battery Street
San Francisco, CA 94120

Trinity Industries
2525 Stemmons Freeway
Dallas, TX 75207

Tyson Foods
P.O. Box 2020
Springdale, AR 72762

Union Texas
P.O. Box 2120
Houston, TX 77252

Unocal 1201
W. 5th Street
Los Angeles, CA 90017

Valero Energy
530 McCullough Avenue
San Antonio, TX 78215

Valhi
5430 L.B.J. Freeway
Dallas, TX 75240

Valspar
1101 3rd Street S.
Minneapolis, MN 55415

Varian Associates
3050 Hanse Way
Palo Alto, CA 94304

Western Digital
8105 Irvine Center Drive
Irvine, CA 92718

Weyerhaeuser
33663 Weyerhaeuser Way South
Auburn, WA 98801

Willamette Industries
3800 1st Interstate Tower
Portland, OR 97201

United States *Fortune* 500 Service Companies Headquartered in the Western States

1st Interstate Bancorp
633 W. 5th Street
Los Angeles, CA 90071

20th Century Fox Film
10201 W. Pico Boulevard
Los Angeles, CA 90035

Airborne Freight
3101 Western Avenue
Seattle, WA 98121

Alaska Air Group
19300 Pacific Highway S.
Seattle, WA 98188

Albertson's
250 Park Center
Boise, ID 83706

Alexander and Baldwin
822 Bishop Street
Honolulu, HI 96813

America West Airlines
4000 E. Sky Harbor Boulevard
Phoenix, AZ 85034

American Freightways
2200 Forward Drive
Harrison, AR 72601

American General
2929 Allen Parkway
Houston, TX 77019

American Medical Holdings
8201 Preston Road
Dallas, TX 75225

American President
1111 Broadway
Oakland, CA 94607

American Stores
709 E. South Temple
Salt Lake City, UT 84102

AMR
4333 Amon Carter
Fort Worth, TX 76155

Arco Trans. Alaska
550 W. 7th Avenue
Anchorage, AK 99501

Argonaut Group
1800 Avenue of the Stars
Los Angeles, CA 90067

Arkansas' Best
1000 S. 21st Street
Fort Smith, AR 72901

Associated Milk Producers
6609 Blanco Road
San Antonio, TX 78279

Bancorp Hawaii
130 Merchant Street
Honolulu, HI 96813

Bank of California
400 California Street
San Francisco, CA 94104

Bankamerica Corporation
555 California Street
San Francisco, CA 94104

Bay View Capital Corporation
2121 S. El Camino Real
San Mateo, CA 94403

Bergen Brunswig
4000 Metropolitan Drive
Orange, CA 92668

Beverly Enterprises
1200 S. Waldron Road
Fort Smith, AR 72901

Boatmen's Bancshares
800 Market Street
St. Louis, MO 63101

BP Pipelines (Alaska)
900 E. Benson Boulevard
Anchorage, AK 99519

Browning-Ferris, Inc.
757 N. Eldridge
Houston, TX 77079

Burlington Northern
777 Main Street
Fort Worth, TX 76102

California Federal Bank
5700 Wilshire Boulevard
Los Angeles, CA 90036

Centex
3333 Lee Parkway
Dallas, TX 75219

Central and Southwest
1616 Woodall Rogers Freeway
Dallas, TX 75266

Ceridian
8100 34th Avenue S.
Minneapolis, MN 55425

Charles Schwab
101 Montgomery Street
San Francisco, CA 94104

Circle K
1601 N. 7th Street
Phoenix, AZ 85006

Citadel Holding Corporation
600 N. Brand Boulevard
Glendale, CA 91203

Coast Savings Financial
1000 Wilshire Boulevard
Los Angeles, CA 90017

Commerce Bancshares
1000 Walnut
Kansas City, MO 64016

Commercial Federal Corporation
2120 S. 72nd Street
Omaha, NE 68124

Computer Sciences
2100 E. Grand Avenue
El Segundo, CA 90245

Consolidated Freightways
3240 Hillview Avenue
Palo Alto, CA 94304

Continental Air Holdings
2929 Allen Parkway
Houston, TX 77019

Costco Wholesale
10809 120th Avenue Northeast
Kirkland, WA 98033

Dayton Hudson
777 Nicollet Mall
Minneapolis, MN 55402

Dial
Dial Tower
Phoenix, AZ 85077

Dillard Department Stores
1600 Cantrell Road
Little Rock, AR 72201

Dole Food
31355 Oak Crest
Westlake Village, CA 91361

Downey Savings and Loan Association
3501 Jamboree Road
Newport Beach, CA 92658

Electronic Data Systems
7171 Forest Lane
Dallas, TX 75230

Enron
1400 Smith Street
Houston, TX 77002

Enserch
300 S. St. Paul Street
Dallas, TX 75201

Exxon Pipeline
800 Bell Avenue
Houston, TX 77002

Farm and Home Financial Group
10100 N. Executive Hills
Kansas City, MO 64153

FHP International
9900 Talbert Avenue
Fountain Valley, CA 92708

First Bank System
First Bank Place
Minneapolic, MN 55480

First Hawaiian
P.O. Box 3200
Honolulu, HI 96847

First Security Corporation
79 S. Main Street
Salt Lake City, UT 84111

Firstfed Financial Corporation
401 Wilshire Boulevard
Santa Monica, CA 90401

Fleming
6301 Waterford
Oklahoma City, OK 73126

Fluor
3333 Michelson Drive
Irvine, CA 92730

Fourth Financial Corporation
100 N. Broadway
Wichita, KS 67202

Fred Meyer
3800 Southeast 22nd Avenue
Portland, OR 97202

Fremont General
2020 Santa Monica
Santa Monica, CA 90404

Gap
1 Harrison Street
San Francisco, CA 94105

General American Life
700 Market Street
St. Louis, MO 63101

Glenfed
700 N. Brand Boulevard
Glendale, CA 91203

Golden West Financial Corporation
1901 Harrison Street
Oakland, CA 94612

Great Western Financial Corporation
9200 Oakdale Avenue
Chatsworth, CA 91311

Greyhound Lines
15110 N. Dallas Parkway
Dallas, TX 75248

Gulf States Utilities
350 Pine Street
Beaumont, TX 77701

H & R Block
4410 Main Street
Kansas City, MO 64111

H.F. Ahmanson
4900 Rivergrade Road
Irwindale, CA 91706

HAL
531 Ohohia Street
Honolulu, HI 96819

Halliburton
500 N. Akard Street
Dallas, TX 75201

Harvest States Corporation
1667 N. Snelling Avenue
St. Paul, MN 55108

Hilton Hotels
9336 Civic Center Drive
Beverly Hills, CA 90210

Houston Industries
4400 Post Oak Parkway
Houston, TX 77027

IDS Life
80 S. 8th Street
Minneapolis, MN 55440

J.B. Hunt Transport
615 J.B. Hunt Drive
Lowell, AR 72745

J.C. Penney
6501 Legacy Drive
Plano, TX 75024

Kansas City South Industries
114 W. 11th Street
Kansas City, MO 64105

Magna Group
1401 S. Brentwood Boulevard
St. Louis, MO 63144

May Department Stores
611 Olive Street
St. Louis, MO 63101

McCaw Cellular Communications
5400 Carillon Point
Kirkland, WA 98033

McKesson
1 Post Street
San Francisco, CA 94104

Mercantile Bancorp.
Mercantile Towers
St. Louis, MO 63166

Merisel
200 Continental Boulevard
El Segundo, CA 90245

Mesa Airlines
2325 E. 30th Street
Farmington, NM 87401

Metropolitan Financial
6800 France Avenue S.
Minneapolis, MN 55435

Microsoft
1 Microsoft Way
Redmond, WA 98052

Minnesota Mutual Life
400 Robert Street N.
St. Paul, MN 55101

MNX
5310 St. Joseph Avenue
St. Joseph, MO 64505

Morrison Knudsen
720 Park Boulevard
Boise, ID 83729

Nash Finch
7600 France Avenue S.
Edina, MN 55435

National Intergroup
1220 Senlac Drive
Carrollton, TX 75006

National Medical Enterprises
2700 Colorado Avenue
Santa Monica, CA 90404

Nike
1 Bowerman Drive
Beaverton, OR 97005

Nordstrom
1501 5th Avenue
Seattle, WA 98101

Norwest Corporation
6th and Marquette
Minneapolis, MN 55479

NWA
5101 Northwest Drive
St. Paul, MN 55111

Pacific Enterprises
633 W. 5th Street
Los Angeles, CA 90017

Pacific Gas and Electric
77 Beale Street
San Francisco, CA 94177

Pacific Mutual Life
700 Newport Center
Newport Beach, CA 92660

Pacific Telesis Group
130 Kearney Street
San Francisco, CA 94108

Pacificare Health Systems
5995 Plaza Drive
Cypress, CA 90630

Pacificorp
700 Northeast Multnomah Street
Portland, OR 97232

Panhandle Eastern
5400 Westheimer Court
Houston, TX 77251

Pinnacle West Capital
400 E. Van Buren Street
Phoenix, AZ 85034

Pioneer Hi-Bred International
400 Locust Street
Des Moines, IA 50309

Price
4241 Jutland Drive
San Diego, CA 92117

Principal Mutual Life
711 High Street
Des Moines, IA 50392

Roosevelt Financial Group
900 Roosevelt Parkway
Chesterfield, MO 63017

Rykoff-Sexton
761 Terminal Street
Los Angeles, CA 90021

Safeco
Safeco Plaza
Seattle, WA 98185

Safeco Life Insurance
15411 Northeast 51st Street
Redmond, WA 98082

Safeway
4th and Jackson Streets
Oakland, CA 94660

Sanwa Bank California
601 S. Figueroa Street
Los Angeles, CA 90017

Scecorp
2244 Walnut Grove Avenue
Rosemead, CA 91770

SFFED Corporation
88 Kearney Street
San Francisco, CA 94108

Southern Pacific Trans.
1 Market Plaza
San Francisco, CA 94105

Southland
2711 N. Haskell Avenue
Dallas, TX 75204

Southwest Airlines
2702 Love Field Drive
Dallas, TX 75235

Southwestern Bell
175 E. Houston
San Antonio, TX 78205

Sprint
2330 Shawnee Mississippi Parkway
Westwood, MO 66205

St. Paul
385 Washington Street
St. Paul, MN 55102

Sumitomo Bank of California
320 California Street
San Francisco, CA 94104

Sun Life of America
11601 Wilshire Boulevard
Los Angeles, CA 90025

Supervalue
11840 Valley View Road
Eden Prairie, MN 55344

Sysco
1390 Enclave Parkway
Houston, TX 77077

Tandy
1800 One Tandy Center
Fort Worth, TX 76102

TCF Financial Corporation
801 Marquette Avenue
Minneapolis, MN 55402

TeleCommunications, Inc.
5619 DTC Parkway
Englewood, CO 80111

Texas Utilities
2001 Bryan Tower
Dallas, TX 75201

Transamerica
600 Montgomery Street
San Francisco, CA 94111

Transamerica Life and Annuity
1150 S. Olive Street
Los Angeles, CA 90015

Transamerica Occidental Life
1150 South Olive Street
Los Angeles, CA 90015

U.S. Bancorp
111 S.W. 5th Avenue
Portland, OR 97204

Unigroup
1 United Drive
Fenton, MO 63026

Union Bank
350 California Street
San Francisco, CA 94104

Union Electric
1901 Chouteau Avenue
St. Louis, MO 63166

Unionfed Financial
330 E. Lambert Road
Brea, CA 92621

United Healthcare
9900 Bren Road East
Minnetonka, MN 55343

United Missouri Bancshares
1010 Grand Avenue
Kansas City, MO 64106

Univar
6100 Carillon Point
Kirkland, WA 98033

US West
7800 E. Orchard Road
Englewood, CO 80111

USAA
USAA Building
San Antonio, TX 78288

Variable Annuity Life
2929 Allen Parkway
Houston, TX 77019

Vons
618 Michillinda Avenue
Arcadia, CA 91007

Wal-Mart Stores
702 S.W. 8th Street
Bentonville, AR 72716

Walt Disney
500 S. Buena Vista Street
Burbank, CA 91521

Washington Federal S and L Association
425 Pike Street
Seattle, WA 98101

Washington Mutual Savings
1201 3rd Avenue
Seattle, WA 98101

Wells Fargo and Company
420 Montgomery Street
San Francisco, CA 94163

Werner Enterprises
14507 Frontier Road
Omaha, NE 68137

West One Bancorp
101 S. Capitol Boulevard
Boise, ID 83702

Westcorp
23 Pasteur Road
Irvine, CA 92718

Western National Life
205 E. 10th Street
Amarillo, TX 79101

Western Resources
818 Kansas Avenue
Topeka, KS 66612

Williams
One Williams Center
Tulsa, OK 74172

Yellow Freight System
10777 Barkley
Overland Park, KS 66211

Top 25 Canadian Corporations

Alcan Aluminium, Limited
1188 Sherbrooke Street W., Suite 100
Montréal, P.Q.
Canada H3A 3G2

BCE, Incorporated
1000 de la Gauchetiere W., Suite 3700
Montréal, P.Q.
Canada H3B 4Y7

Bell Canada
200–1050 Beaver Hall Hill, 19th Floor
Montréal, P.Q.
Canada H2Z 1S4

Brascan, Limited
BCE Place
181 Bay Street, Suite 4400
Toronto, ON
Canada M5J 2T3

Canadian Pacific, Limited
P.O. Box 6042, Station A
Montréal, P.Q.
Canada H3C 3E4

Chrysler Canada, Limited
2450 Chrysler Centre
Windsor, ON
Canada N8W 3X7

Ford Motor Company of Canada, Limited
P.O. Box 2000
Oakville, ON
Canada L6J 5E4

George Weston, Limited
22 St. Clair Avenue E., Suite 1901
Toronto, ON
Canada M4T 2S7

General Motors of Canada, Limited
1908 Colonel Sam Drive
Oshawa, ON
Canada L1H 8P7

Hudson's Bay Company
401 Bay Street, 5th Floor
Toronto, ON
Canada M5H 2Y4

Hydro-Quebec
75 Rene-Levesque Boulevard W., 22nd Floor
Montréal, P.Q.
Canada H2Z 1S4

IBM Canada, Limited
3500 Steeles Avenue E.
Markham, ON
Canada L3R 2Z1

Imasco, Limited
600 de Maisonneuve W., Suite 1900
Montréal, P.Q.
Canada H3A 3K7

Imperial Oil, Limited
111 St. Clair Avenue W.
Toronto, ON
Canada M4V 1N5

Loblaw Companies, Limited
22 St. Clair Avenue E.
Toronto, ON
Canada M4T 2S7

Noranda, Incorporated
BCE Place
171 Bay Street, Suite 4100
Toronto, ON
Canada M5J 2T3

Northern Telecom, Limited
3 Robert Speck Parkway, Suite 1100
Mississauga, ON
Canada L4Z 3C8

Ontario Hydro
700 University Avenue, Suite 200
Toronto, ON
Canada M5G 1X6

Oshawa Group, Limited
302 The East Mall, Suite 200
Etobicoke, ON
Canada M9B 6B8

Petro-Canada
150 6th Avenue S.W.
Calgary, AB
Canada T2P 3E3

Power Corporation of Canada
751 Victoria Square
Montréal, P.Q.
Canada H2Y 2J3

The Seagram Company, Limited
1430 Peel Street
Montréal, P.Q.
Canada H3A 1S9

Shell Canada, Limited
400 4th Avenue S.W.
Calgary, AB
Canada T2P 0J4

The Thomson Corporation
T-D Bank Tower, Suite 2706
Toronto, ON
Canada M5K 1A1

Univa, Incorporated
1250 Rene-Levesque Boulevard W., Suite 4100
Montréal, P.Q.
Canada H3B 4X1

Top 16 Mexican Corporations

Aerovias de Mexico, S.A. de C.V.
Paseo de la Reforma 445
06500 Mexico
D.F., Mexico

Altos Hornos de Mexico, S.A. de C.V.
Prolongacíon Juárez s/n
77000 Monclova
Coahuila, Mexico

American Express Co., S.A. de C.V.
Hamburgo 75
D.F., Mexico

Celanese Mexicana, S.A.
Avenida Revolucíon, 1425
01040 Mexico
D.F., Mexico

Chrysler de Mexico, S.A.
Aptdo. Postal 53-951
Mexico
D.F., Mexico

Cigarros la Tabacalera Mexico, S.A. de C.V.
Manuel Salazar No. 132
40000 Mexico
D.F., Mexico

Compania Mexicana de Aviacion, S.A. de C.V
Balderas 36, 12. Piso
Av. Xola 535
03100 Mexico
D.F., Mexico

Compania Nestlé
Ejercito Nacional 453
11520 Mexico
D.F., Mexico

El Puerto de Liverpool, S.A. de C.V.
Avenida Mariano Escobedo No. 425
11570 Mexico
D.F., Mexico

Empresas la Moderna, S.A. de C.V.
Avenida. Francisco Madero, 2750 Pte.
Monterrey
Nuevo Léon, Mexico

Fomento Economico Mexicano, S.A. de C.V.
Cuauhtemoc 400 Sur
Monterrey
Nuevo León, Mexico

Ford Motor Company, S.A. de C.V.
Aptdo. Postal 39-bis
Mexico
D.F., Mexico

General Motors de Mexico, S.A. de C.V.
Aptdo. Postal 107-bix
Mexico
D.F., Mexico

Hylsa, S.A. de C.V.
Adolfo Lopez Mateos Esq. Co Cami
San Nicolas
6000 Toluca
México, Mexico

IBM de Mexico, S.A.
Mariano Escobedo 595
Col. Chapultepec
41560 Morales
D.F., Mexico

Kimberly-Clark de Mexico, S.A. de C.V.
Avenida 1, No. 9
Naucalpan de Juárez
D.F., Mexico

NADRO, S.A. de C.V.
Londres, 107-So. Piso
06600 Mexico
D.F., Mexico

Petroleos Mexicanos
Marina Nacional 329
30000 Mexico
D.F., Mexico

Seguros la Provincial, S.A.
Miguel Angel de Quevedo No. 915
04330 Mexico
D.F., Mexico

Spicer, S.A. de C.V.
Bosque de Ciruelos, 278
3o. Piso
11700 Mexico
D.F., Mexico

Teléfonos de Mexico, S.A. de C.V.
Parque Vía 190
10er. Piso
06599 Mexico
D.F., Mexico

Tolmex, S.A. de C.V.
Av. Constitucíon, 444 Pte.
64000 Monterrey
Nuevo León, Mexico

Chambers of Commerce

American Chamber of Commerce of
Mexico, A.C.
Lucerna 78, Colonia Juárez
06600 Mexico
D.F., Mexico
011-525-705-0995

Association of Chambers of Commerce of Latin
America (AACCLA)
1615 H Street, N.W.
Washington, D.C. 20062
(202) 463-5485

Canadian Chamber of Commerce
55 Metcalfe Street, Suite 160
Ottawa, Ontario
Canada K1P 6N4
(613) 238-4000

Chamber of Commerce of the United States
1615 H Street, N.W.
Washington, D.C. 20062
(202) 463-5460

National Chamber of Commerce of Mexico
(CANACO)
Reforma No. 42, 3o. Piso
06048 Mexico, D.F.
011-52-5-705-0549; 705-0424

U.S.-Mexico Chambers of Commerce:

4835 L.B.J. Freeway, Suite 750
Dallas, TX 75234
(214) 387-0404

515 S. Figueroa Street, Suite 1020
Los Angeles, CA 90071
(213) 623-7725

Manual Maria Contreras 133
Despachos 120 y 121
Delegacion Cuauhtemoc
06470 Mexico, D.F.
011-52-5-535-0613

Stock/Commodity Exchanges

The American Stock Exchange (AMEX)
86 Trinity Place
New York, NY 10006
(212) 306-1000

Bolsa Mexicana de Valores
(Mexican Stock Exchange)
Paseo de la Reforma 255
Col. Cuahtemoc
06500 Mexico, D.F.

Chicago Mercantile Exchange,
International Monetary Fund, and
Index and Option Market
30 S. Wacker Drive
Chicago, IL 60606
(312) 786-5600

New York Stock Exchange (NYSE)
11 Wall Street
New York, NY 10005
(800) 692-6973

Pacific Stock Exchange (PSE)
301 Pine Street
San Francisco, CA 94101
(415) 393-4000

Toronto Stock Exchange
2 First Canadian Place, Exchange Tower
Toronto, Ontario
Canada M5X 1J2
(416) 947-4700

Vancouver Stock Exchange
609 Granville Street
Vancouver, British Columbia
Canada V7Y 1H1
(614) 689-3334

The Winnipeg Commodity Exchange
500 Commodity Exchange Tower
360 Main Street
Winnipeg, Manitoba
Canada R3C 3Z4
(204) 949-0495

SELECTED BIBLIOGRAPHY

American Chamber of Commerce Researchers Association. *Cost of Living Index*. Alexandria, VA: American Chamber of Commerce, 1991.

Baer, M. Delal. "Nafta and U.S. National Security." Paper presented at the NAFTA Summit: Beyond Party Politics, Washington, DC: Center for Strategic and International Studies. June 28–29, 1993.

Banco de Mexico. "*Informe Anual.*" 1991.

Bank of Canada. *Bank of Canada Review*. Ottawa, Ontario: Bank of Canada, Secretary's Department, 1993.

Bosworth, Barry. "Economic Effects of the North American Free Trade Agreement." Paper presented at the NAFTA Summit: Beyond Party Politics, Washington, DC: Center for Strategic and International Studies. June 28–29, 1993.

Brown, Drustilla. "U.S. Labor and North American Integration." Paper presented at the NAFTA Summit: Beyond Party Politics, Washington, DC: Center for Strategic and International Studies. June 28–29, 1993.

Center for the New West. "A Continent Ascendant: North America in the 1990s." *Points West Symposium Papers*. June 1991.

Center for the New West. *Making Things Work: Transportation and Trade Expansion in Western North America*. Washington, DC: National Technical Informaion Service, 1993. (7 volumes).

Cornelius, Wayne A. and Ann L. Craig. "Politics in Mexico," in Gabriel A. Almond and G. Bingham Powell, Jr. *Comparative Politics Today*. New York: HarperCollins Publishers, 1992.

Dornbusch, Rudi. "NAFTA: Good Jobs at Good Wages." Paper presented at the NAFTA Summit: Beyond Party Politics, Washington, DC: Center for Strategic and International Studies. June 28–29, 1993.

Easterbrook, W. T. North American Patterns of Growth and Development: The Continental Context. Toronto: University of Toronto Press, 1990.

Economic Report of the President, 1993. Washington, DC: U.S. Government Printing Office, 1994.

Europa Publications, Limited. *The Europa World Year Book 1993*. London: Europa Publications, Limited, 1993.

Federal Deposit Insurance Corporation (FDIC). *Historical Statistics on Banking*. Washington, DC: FDIC Division of Research and Statistics, 1993.

———. *Statistics on Banking*. Washington, DC: FDIC Division of Research and Statistics, 1992.

Globerman, Steven and Michael Walker. *Assessing NAFTA: A Trinational Analysis*. Vancouver: The Fraser Institute, 1992.

Guide to Foreign Trade Statistics. Economics and Statistics Administration, Bureau of Census. USDOC, December 1992.

Hoffman, Mark S., ed. *World Almanac and Book of Facts*. New York: Pharos Books, 1993.

Hufbauer, Gary Clyde and Jeffrey J. Schott. *North American Free Trade: Issues and Recommendations*. Washington, DC: Institute for International Economics, 1992.

Instituto Nacional de Estadística, Geografía e Informatica (INEGI). *Anuario Estadístico de los Estados Unidos Mexicanos*. Aguascalientes, Mexico: INEGI, 1992.

———. *Estados Unidos Mexicanos, Resumen General, XI Censo General de Poblacion y Vivienda, 1990*. Aguascalientes, Mexico: INEGI, 1992.

———. *Finanzas Públicas Estatales y Municipales de México, 1989*. Aguascalientes, Mexico: INEGI, 1992.

———. *Mexican Bulletin of Statistical Information*, No. 7, January–March 1993. Aguascalientes, Mexico: INEGI, 1993.

———. *VII Censo Agropecuario, 1991*. Aguascalientes, Mexico: INEGI, 1992.

———. *XIII Censo Industrial*. Aguascalientes, Mexico: INEGI, 1992.

International Trade Commission. (1993a) "Potential Impact on the U.S. Economy and Selected Industries of the North American Free Trade Agreement." USITC Publication 2596.

Joint Economic Committee Congress of the United States. "Free Trade and the United States–Mexico Borderlands." A Regional Report by Baylor University. July 1, 1991.

Lowenthal, Abraham and Katrina Burgess. *The California-Mexico Connection*. Stanford: Stanford University Press, 1993.

McCray, John P. and Peggy. "Trade Transportation in the Lower Rio Grande Valley 1990–2020." San Antonio: McCray Reserach, 1992.

Pick, James et al. *Atlas of Mexico*. Boulder, CO: Westview Press, 1988.

Presidencia de la República, Dirección General de Comunicación Social. *Mexican Agenda*. Federal District, Mexico: Dirección de Publicaciones, 1992.

Southam Information and Technology Group (SITG). *Corpus Almanac and Canadian Sourcebook*. Don Mills, Ontario: SITG, 1993.

Statistical Office of the United Nations. *Energy Statistics*. New York: United Nations, 1991.

Statistics Canada. *Canada Yearbook, 1990–1994*. Ottawa, ON: Statistics Canada, 1990–1994.

Statistics Canada. *Canadian Economic Observer*. Catalog 11-010. Ottawa, ON: Statistics Canada, 1992.

Treasury Board Secretariat, Administrative Policy Branch, Bureau of Real Property Management. *The Directory of Federal Real Property*. Ottawa, ON: Minister of Supply and Services Canada, June 1992.

INDEX

ABOUT THE AUTHORS

Philip M. Burgess is President of the Center for the New West. He is a popular public speaker and writes a weekly column on national affairs that is distributed nationally by the Scripps Howard News Service.

Michael Kelly is Senior Fellow for national security policy at the Center for the New West.

Center for the New West, Inc., is an independent, nonprofit, nonpartisan policy research institute headquartered in Denver. The work of the Center focuses on trade, technology, and economic development. The Center has offices in Albuquerque, Los Angeles, Phoenix, and a national affairs office in Washington, D.C.